Symbol	Meaning
$a - b$	The subtraction of b from a
$a \div b$, $\frac{a}{b}$ a/b	The quotient of a and b
	Egyptian numerals
	Babylonian numerals
I, II, III, IV, X, C, M	Roman numerals
	Mayan numerals
a^n	Product using a as a factor n times
415_{six}	four, one, five base six
$a \mid b$	a divides b
g.c.d.	Greatest common divisor
l.c.m.	Least common multiple
^{-}n	The additive inverse of n
$\{\ldots, {}^{-}3, {}^{-}2, {}^{-}1, 0, 1, 2, 3, \ldots\}$	The set of integers
$\lvert a \rvert$	The absolute value of a
$\simeq$	Is equivalent to; is congruent to
1.643	A decimal point
$\approx$	Is approximately equal to
%	Per cent
$2.3\overline{14}$	2.3141414...
$\sqrt{n}$	The square root of n
$\overleftrightarrow{AB}$	Line through points A and B
$\overline{AB}$	Line segment AB
$\overrightarrow{AB}$	Ray from A through B
$\angle ABC$	Angle ABC
$\triangle ABC$	Triangle with vertices A, B, and C
$m(\overline{AB})$	Measure of line segment $\overline{AB}$
$40° \, 14' \, 30''$	40 degrees, 14 minutes, 30 seconds
$\widehat{ABC}$	Arc ABC
π	*Pi* which is approximately 3.1416
$\sim$	Is similar to
$P(A)$	The probability of A
$P(\overline{A})$	The probability of not A
4!	4-factorial; $4! = 4 \cdot 3 \cdot 2 \cdot 1$
${}_nC_r$	The combination of n things taken r at a time
$P(A \cap B)$	The probability of A and B
$P(A \cup B)$	The probability of A or B
$P(A \mid B)$	The conditional probability of A knowing that B
$\bar{x}$	The arithmetic mean of the x's
$s_x{}^2$	The variance of the x's
s_x	The standard deviation of the x's
$a \equiv b \pmod{m}$	a is congruent to b modulo m
$f(a)$	Function evaluated at a
$a + bi$	Complex number where $i = \sqrt{^{-}1}$

Third Edition

Modern Mathematics: An Elementary Approach

Ruric E. Wheeler

Samford University

Brooks/Cole Publishing Company
Monterey, California
A Division of Wadsworth Publishing Company, Inc.

Dedicated to Joyce, Eddy, & Paul

ISBN: 0-8185-0070-0

L.C. Catalog Card No: 72-86159

Printed in the United States of America

1 2 3 4 5 6 7 8 9 10—77 76 75 74 73

Preface

This book has been written with two different audiences in mind. The text is designed for both prospective elementary school teachers and liberal arts students who desire an appreciation and understanding of the basic structure of mathematics. This book should provide the mathematical concepts needed to teach any of the modern arithmetic courses and should prepare prospective teachers for inevitable future changes in mathematics curricula. The liberal arts student is shown that mathematics is interesting, exciting, and aesthetically pleasing through illustrations of some of the procedures mathematicians use to create mathematics.

The third edition makes improvements wherever needed and includes additional topics of contemporary mathematics. The most significant changes from the second edition are:

1. Complete revision of the exercise sets, listing problems in order of difficulty.
2. Inclusion of one or two challenging problems (marked with an asterisk) at the end of each exercise set.
3. The use of truth tables to validate properties of sets.
4. A new section in Chapter 2 on functions and relations.
5. The use of equivalence classes of ordered pairs of whole numbers as a second approach to the development of the system of integers.
6. Introduction and use of concepts such as absolute value, isomorphism, and factor trees.
7. A shortening of the chapters on integers and rational numbers, eliminating some duplication of properties.
8. A new section on approximate numbers and scientific notation.
9. A rearrangement of chapters so that geometry appears immediately after number systems.

10. A complete revision of the chapter on metric geometry assuming an informal approach.

11. A new section on geometric transformations.

12. A new section on counting schemes, including combinations.

13. A new introduction to the chapter on statistics emphasizing the use of statistics in everyday affairs.

14. Inclusion of new material on analytic geometry.

The latest recommendations of CUPM on the mathematical training of prospective elementary school teachers have been carefully studied and every effort has been made to include in this book more of the material recommended by CUPM than can be found in any other textbook today.

The users of the second edition have said that the language and style throughout the text appeal to students with diverse mathematical backgrounds. The book's simplicity of language, intuitive discussion, and numerous examples tend to facilitate study and understanding.

Since one objective of the book is to encourage the student to appreciate the many years of work necessary for the evolution of our number system, historical ideas have been incorporated throughout rather than relegated to a separate chapter.

The problems in the exercise sets vary from easy to very difficult. The more difficult problems are indicated by asterisks and may be omitted at the discretion of the instructor.

An up-to-date list of readings at the end of each chapter suggests reference sources by pages dealing with the subject matter under study. A complete bibliography is given at the end of the book.

The material in the book is arranged to offer a great deal of flexibility. For those who want to emphasize mathematical developments, the first eight chapters present an interesting development of number systems from counting numbers through real numbers. Yet the book can be successfully taught without proving a single theorem, since all mathematical development is supported by intuitive discussion and illustrated by examples.

Again, flexibility is evidenced by the fact that Chapters 9, 10, 11, 12, 13, and 14 are self-contained and can be taught in any order. The six course outlines presented at the end of this preface also illustrate the flexibility of the text.

Finally, sections that are not essential for the study of the remainder of the book are indicated by asterisks preceding the section number. In addition, each exercise set contains a large number of problems varying from easy to difficult.

An instructor's manual containing readiness tests for each chapter is available. These tests may be reproduced and used along with the multiple-choice and true-false questions for each chapter. In addition, the instructor's manual contains answers to all problems not answered in the back of this book.

Three programmed self-study manuals for students are or will soon be available. *A Programmed Study of Number Systems* covers the first seven chapters of this text. *Special Topics in Mathematics for Elementary School Teachers* will cover the chapters on geometry, probability, and statistics. The third self-study manual will discuss the structure of the real number system and algebra.

I am grateful for suggestions from many who studied and taught the first and second editions of this book. I am particularly indebted to the following people, who offered valuable suggestions on both revisions: Bill Bompart, Augusta College; Alan Clark, Ricks College; John C. Holland, David Lipscomb College; John K. Moulton, Gorham State College of the University of Maine; Margaret W. Perisho, Mankato State University; Harry E. Wickes, Brigham Young University; Dick Wood, Seattle Pacific College; and Letitia Yeager, Samford University.

To those who wrote letters or responded to a questionnaire and gave valuable suggestions for improving the second edition, I am grateful: Donald P. Alevine, Western Illinois University; Leon E. Arnold, Community College of Delaware County; August Arndt, Central Michigan University; Kathryn Ainsworth, University of Louisville; C. P. Barton, Stephen F. Austin State College; Helen Bass, Southern Connecticut State College; James Bierden, Rhode Island College; Jean A. Blake, Alabama A & M University; Melvin R. Breakiron, Pennsylvania State University; Raymond F. Bryant, Florissant Valley Community College; Verne Byers, University of Maine at Farmington; Robert Carlton, Georgetown College; Robert L. Clam, Central Michigan University; R. A. Close, Pan American University; Richard K. Coburn, Church College of Hawaii; James L. Courser, Farmington State College; Jane Craneros, Oral Roberts University; J. H. Croy, Mankato State College; Arthur Daniel, Malcomb County Community College; Vilas E. Deanne, Grace College; Kenneth E. Deen, Louisiana College; Donald R. Devine, Western Illinois University; Frances Dieshek, Del Mar College; Roy A. Dobgans, Georgetown College; Phyllis A. Dunlery, Texas Tech University; Carlton Lee Evans, Adams State College; Donald L. Evans, Polk Community College; Barney L. Erickson, Central Washington State College; Ray W. Fleischmann, Southwestern State College; James F. Fleming,

Hilbert College; Robert C. Frascatore, State University of New York at Buffalo; J. William Friel, University of Dayton; Z. T. Gallion, Mississippi State College for Women; William M. Gentry, Virginia Polytechnic Institute; Armen Gnepp, Burlington Community College; Evelyn B. Granville, California State University, Los Angeles; Russell W. Grover, New Mexico State University of San Juan; Ray Haertel, Central Oregon College; Rosemary R. Hamner, Northeast Louisiana University; B. J. Harmon, Monroe County College; F. L. Harmon, Northeast Louisiana University; Allen E. Hansen, Riverside City College; Viggo Hansen, California State University, Northridge; Hank Harmeling, North Shore College; Margaret Hartwig, Auburn University; John G. Harvey, University of Wisconsin; Lavoy J. Hatchett, Oral Roberts University; Virginia S. Hawn, Lenoir Rhyne; Rev. Oliver Herbert, St. Francis College; Margaret Hockensmith, St. Francis College; Patricia A. Hoover, Duquesne University; Jack A. Howard, Le Tourneau College; Mary Hudson, Samford University; Lottchen L. Hunter, Kansas State Teacher's College; Denis M. Hyams, University of Alabama in Huntsville; Leslie P. Jordan, Grossmont College; James Kasum, University of Wisconsin at Milwaukee; Dorothy Kennedy, S.U.N.Y. at Buffalo; Sarah Kennedy, Texas Tech University; J. Maurice Kingston, University of Washington; William Kirshner, Florida Atlantic University; Sister Agnes Rose Kokke, Dominican College; Philip J. Lafer, Washington State University; R. E. Lee, University of West Florida; Gilbert A. Lewis, University of Wisconsin at Milwaukee; C. Michael Lohn, Virginia Commonwealth at Richmond; Jean Luding, Bob Jones University; Bill Maneer, Wayne State University; James D. Mann, Morehead State University; John J. Matejcie, Florissant Valley College; Robert T. Meyer, California State University, Los Angeles; Merle Mitchell, University of New Mexico; J. Richard Morris, Virginia Commonwealth University; J. Albert Mosley, Keene State College; Julia H. Murphy, Auburn University; Linda Musco, Middlesex Community College; Henry Nameling, Jr., North Shore Community College; Donald G. Ohl, Bucknell University; R. L. Persky, Christopher Newport College of the College of William and Mary; Ann C. Peters, Keene State College; Arvine Phelps, Dalton Junior College; Robert Phillips, Lynchburg College; Sister Presentia, Hilbert College; Marvin L. Proctor, Pratt Community College; Amalee B. Ritchie, Lenoir Rhyne; Jack M. Robertson, Washington State University; Theron Rockhill, S.U.N.Y. at Brockport; Al Roy, Bristol Community College; Jayne Ryoti, Parkland College; Helen E. Salzberg, Rhode Island College; Neil

W. Seidl, Dominican College; Sister M. Leontius Schulte, College of St. Teresa; Harold F. Simmons, California State Polytechnic; Norma M. Simmons, Hinds Junior College; Mrs. Alex F. Smith, Middlesex Community College; Frank E. Smith, Nebraska Wesleyan University; Gaston Smith, William Carey College; Lehi T. Smith, Arizona State University; James L. Southam, Stanislaus State College; June R. Sparks, Dalton Junior College; Donald G. Spencer, Northeast Louisiana University; C. Michael Stein, Virginia Commonwealth Institute; Vance D. Stine, California State University, Los Angeles; Jerry D. Taylor, Campbell College; Virginia M. Tripp, Genesee Community College; Dewey Turner, Ricks College; Dean R. Wagner, Lock Haven State College; August W. Waltmann, Wartburg College; E. F. Ward, Tennessee Technological University; Amy G. West, Berkshire Community College; James C. Westrope, State University of New York at Buffalo; Mark F. Wiever, West Chester State College; Gilbert H. Wilson, Central Wyoming College; and Melvin C. Withnell, Southern Colorado State College.

Special thanks for constructive criticism of the first edition are due to Professors H. Bell, West Chester State College; S. P. Brown, Springfield College; W. D. Clark, Stephen F. Austin State College; Robert Colling, Kearney State College; H. Foisy, State University College at Potsdam; Roberta Gasparonis, Newark State College; B. Gee, Brigham Young University; Elsie S. Giegerich, Newark State College; William M. Gulas, Fairmont State College; Carlon A. Krantz, Newark State College; J. Kilroy, State University College of Potsdam; Robert O. Kimball, University of New Hampshire; Charles R. Leake, Wagner College; Charles Miller, American River College; C. W. McClure, Slippery Rock State College; Henry M. Mailloux, Bridgewater State College; Roy D. Mazzagatti, Miami-Dade Junior College; Billy R. Nail, Morehead State University; Roger Osborn, University of Texas; Charles G. Pickens, Kearney State College; Doyle Robertson, Glendale Community College; Ethel A. Robinson, California State University, Fresno; H. V. Sellers, University of Chattanooga; W. H. Spragens, University of Louisville; Kenneth Stephens, Le Tourneau College; Martha Stobbe, Bethel College; and Nelson C. Wood, State University College of Fredonia.

Particular gratitude goes to Robert J. Wisner, New Mexico State University, who read the complete manuscript of the third edition and contributed many valuable suggestions, and to James Smart, California State University, San Jose, for his reviews of the chapters on geometry.

I would also like to express appreciation to Kathy Hinkle, Linda Stewart, and Glenda Thorpe for their assistance in compiling this revision. Without their help the revision would have been delayed a year. Finally, I would like to thank Martha Bowman, Joyce Hughey, and Pam Collins for patience and understanding while typing the manuscript.

It is my hope that this book will be teachable and exciting.

Ruric E. Wheeler

Course Outlines

One-semester courses	*Students with average mathematical background (or less)*	*Better-prepared students*
Courses of liberal arts majors	Sections 1–5 of Chapter 1 Chapter 2 Sections 1–5 of Chapter 3 Sections 1, 2, 5, 6, 7 of Chapter 4 Chapter 5 Sections 1–5 of Chapter 7 Sections 1–4 of Chapter 8 Sections 1–3 of Chapter 11 Chapter 12	Chapters 1, 2, 3 Sections 5, 6, 7 of Chapter 4 Chapter 5 Chapter 6 Sections 1–5 of Chapter 7 Chapters 11, 12
Courses for elementary education majors	Sections 1–3 of Chapter 1 Chapters 2, 3, 4, 5 Sections 1–4 of Chapter 6 Chapter 7 Sections 1–5 of Chapter 9	Sections 1–3 of Chapter 1 Chapters 2, 3, 4, 5 Sections 1–4 of Chapter 6 Chapters 7, 9 Sections 1–5 of Chapter 10
Combination courses for liberal arts majors and elementary education majors	Sections 1–5 of Chapter 1 Chapters 2, 3, 4, 5 Sections 1–4 of Chapter 6 Chapter 7	Chapters 1, 2, 3, 4, 5 Sections 1–4 of Chapter 6 Chapters 7, 9
Two-semester courses		
Courses for liberal arts majors	Sections 1–5 of Chapter 1 Chapters 2, 3, 4, 5 Sections 1–4 of Chapter 6 Chapters 7, 8, 11, 12, 13, 14	Chapters 1, 2, 3, 4, 5, 6, 7, 8, 11, 12, 13, 14
Courses for elementary education majors	Sections 1–3 of Chapter 1 Chapters 2, 3, 4, 5 Sections 1–4 of Chapter 6 Chapters 7, 8, 9, 10 Sections 1–3 of Chapter 11 Chapter 12 Sections 1–4 of Chapter 13 Sections 1–5 of Chapter 14	Chapters 1, 2, 3, 4, 5 Sections 1–4 of Chapter 6 Chapters 7, 8, 9, 10, 11, 12, 13 Sections 1–5 of Chapter 14
Combination courses for liberal arts majors and elementary education majors	Sections 1–5 of Chapter 1 Chapters 2, 3, 4, 5 Sections 1–4 of Chapter 6 Chapters 7, 8, 9, 10 Sections 1–3 of Chapter 11 Chapters 12, 13 Sections 1–3 of Chapter 14	Chapters 1, 2, 3, 4, 5, 6, 7, 8, 9, 10, 11, 12, 13

These outlines are given in great detail in the instructor's manual for this book.

Contents

4

Numeration Systems
121

5

The System of Integers
169

6

Elementary Number Theory
203

7

Fractions and Rational Numbers
241

8

Decimals and Real Numbers
295

9

Informal Nonmetric Geometry
333

10

Informal Metric Geometry
367

11

Concepts of Probability
429

12

Introduction to Statistics
461

13

Mathematical Systems
513

14

Algebra and Geometry
541

1

Logic and an Introduction to Mathematical Reasoning

1

Mathematics Is Many Things to Many People

Approach your college friends with this question: "What is mathematics?" Ask your faculty adviser, or one of your teachers. Ask a neighbor; ask your parents. Ask *yourself* the question "What is mathematics?" Compare the answers. Are you surprised at the varied responses?

Some will say that mathematics is an operation with numbers used in answering the questions "How many?" and "How much?" In this role, mathematics is a *tool*, a collection of skills which may be used for calculation and problem solving. Thus, mathematics as a tool becomes indispensable to our modern world of business transactions, industrial production, and scientific research. In a similar manner, mathematics assumes importance for you. As a competent citizen, you must be able to cope with taxes, interest, budgets, grocery bills, and so on, all of which depend on mathematics. Yet mathematics is more than this.

To some of your friends, mathematics may be a *science*. This assumption is again correct. In the sense of its precision, in its rigor of development, in its search for truth, mathematics is the ultimate science, a science of logical reasoning. A study of mathematics involves a study of methods for drawing conclusions from assumed premises. From some of the greatest minds of every age have come contributions to the systematized mathematical knowledge that is in use today.

Some people compare mathematics to a *game*. In any game, the players must know certain rules and regulations in order to play. If at any time a rule is broken or ignored, the game ceases to be fair. The same is true in mathematics. One begins with given *rules* or *laws* (sometimes called *postulates*) and plays the game of logical reasoning. If at any time a law is broken, the mathematical reasoning is no longer valid.

The reply of a research mathematician to the question "What is mathematics?" may be "Mathematics is an *art*." Certainly there is aesthetic satisfaction in the development of new mathematical theories, in the construction of new systems, new concepts, new ideas. It is difficult to describe creative activity in mathematics. However, new mathematical concepts are being developed daily; indeed, man has developed more mathematics since 1900 than in all previous time. These facts indicate that mathematics is something man has created rather than something he has discovered.

Does anyone classify mathematics as a *language*? For most people, whether they realize it or not, mathematics is most definitely a language for expressing their ideas. In order to compare the weight of two football players, the average grades of students in a class, and innumerable other subjects, one must employ mathematics as a language. Just as the artist expresses certain feelings and thoughts through the media of painting and sculpturing, one uses mathematics to express ideas of quantity and order. The language of mathematics is needed to converse fluently, to express relationships, to make comparisons, to quote statistics, to reach conclusions.

Now are you thoroughly confused? You have read that mathematics is a tool, an art, a game, a science, and a language. Thus, mathematics means different things to different people. However, the important concept is that mathematics enters everyone's life in some way.

Consequently, this text attempts to present a clear explanation of some modern mathematics by demonstrating how mathematical systems are formed; and through this approach to the study of mathematics, you will be able to retain and use for a longer period of time the mathematics that you learn.

2

Inductive Reasoning

Discovering patterns, recognizing the ordered forms of a display, and discovering the regularity of a sequence are important in the study of mathematics. What skills do we need to develop in order to acquire the ability to accomplish these objectives? Scanning selectively can be helpful in determining a rule, particularly if one has some possible rules in mind as he scans.

Example: Detect the regularity of the numbers in the square. (*Hint:* Look at the sum of each row, column, and diagonal.)

1	15	14	4
12	6	7	9
8	10	11	5
13	3	2	16

Did you note that the sum of each row, column, and diagonal is 34?

Detecting the regularity of a pattern or abstracting what is common to a number of examples is an example of *inductive reasoning.* Reasoning, or drawing conclusions, can be classified into two categories—*inductive reasoning* and *deductive reasoning.* When, as above, a person makes observations and on the basis of these observations arrives at a conclusion, he is reasoning *inductively.* A small child feels the heat coming from a stove and after a few observations concludes that the stove is hot. Arriving at a conclusion on the basis of repeated scientific experimentation is sometimes called empirical inference or inductive inference, but it still involves inductive reasoning. A statistician collects and organizes his observations and then uses the information obtained to reach conclusions. Similarly, a chemist performs the same experiment many times under the same initial conditions. When he obtains the same result each time, he is convinced that this result will be obtained each time he performs the experiment. He is reasoning inductively: *arriving at a general conclusion from particular observations.*

This type of reasoning is not new in mathematics. Prior to the classical Greek civilization, mathematics was largely inductive in nature. Most mathematical formulas were "rule of thumb" procedures that were found to be approximately correct. Many were obtained by trial and error processes and later proved incorrect. For instance, at one time the circumference of a circle was thought to be three times the diameter. An Egyptian formula used the square of 8/9 of a circle's diameter as the area of a circle. We know now that both conclusions were wrong, and the formulas we use for these measurements have been proved to be true by methods of thought not involving physical measurements.

One makes repeated use of inductive reasoning in everyday affairs. You have heard the statement "Experience is the best teacher." For example, on a cloudy day you carry your umbrella because you have noticed a relationship between clouds and rain. However, inductive reasoning is not always as simple as this example indicates. To be good at inductive reasoning, one must train himself to notice the important elements in a situation and to understand these elements. A doctor often uses inductive reasoning when he examines a patient and makes a diagnosis. The doctor solves the problem, but the untrained person will fail because he does not recognize or understand the important facts of his observation.

It is unfortunate that many people believe that mathematics uses deductive reasoning exclusively. This is due in part to the fact that most *published* mathematics involves deductive reasoning. However, for every page of published mathematics, dozens of pages of unsuccessful attempts have been discarded. Inductive reasoning is essential to the creation of new mathematics. The development of new mathematics usually starts with conjectures. A *conjecture* is a statement which is thought, usually with good reason, to be true but which has yet to be proved true. Months or years often elapse before some conjectures are proved or disproved. Some conjectures remain unproved even after hundreds of years.

A very famous unsolved problem is the Goldbach conjecture for prime numbers. In 1742 Goldbach stated his conjecture that every even number greater than 4 is the sum of two odd primes. (A counting number other than 1 is said to be a *prime* if it is divisible only by itself and 1.) For example, 8 is the sum of 3 and 5, 10 is the sum of 3 and 7, 12 is the sum of 5 and 7, and so on. Goldbach never found a proof for this conjecture, and, despite efforts of outstanding mathematicians, neither has anyone else.

Consider the conjecture that the formula $N^2 - N + 41$ will give only prime numbers if one substitutes counting numbers for N. If you substitute 1 for N, you will get 41, which is a prime; if you substitute 2, you get 43; if you substitute 3, you get 47. Continue this process by substituting 4, 5, 6 to obtain 53, 61, and 71, respectively. Now 41, 43, 47, 53, 61, and 71 are all primes. Does it follow that if any counting number is substituted for N a prime is produced? Is the conjecture correct? (Try substituting 41 for N.)

It is hoped that this discussion will emphasize the fact that it is good to make conjectures—to reason inductively—but conclusions drawn from this type of reasoning may not be correct. In mathematics, one must prove conjectures by processes called *deductive reasoning*, which will be discussed later on in this chapter.

Exercise Set 1-1

1. Explain in a short paragraph
   (a) why mathematics might be considered as an invention and not as just a discovery.
   (b) the inductive method of reasoning.
2. Give three examples of inductive reasoning.
3. Use the mathematics library to find grounds for calling mathematics
   (a) a tool. (b) an art. (c) a language.
4. Discover a pattern and find the next three members of each set.
   (a) 2, 7, 12, 17, ...
   (b) 1, 4, 7, 10, ...
   (c) $A, B, C, B, C, D, C, D, E, D, \ldots$
   (d) 3, 5, 9, 15, 23, ...
   (e) 1, 3, 1, 8, 1, 13, 1, ...
   (f) 2, 4, 8, 16, ...
   (g) △, □, ⬠, ...
   (h) 2, 8, 26, 80, 242, ...
5. If Tom's father is Dick's son, how are Tom and Dick related?
6. "Brothers and sisters have I none, but this man's father is my father's son." Find the relationship between the two men.
7. Arrange 1, 2, 3, 4, 5, 6, 7, 8, 9 in the squares so that the sums of all the rows, columns, and diagonals are equal.

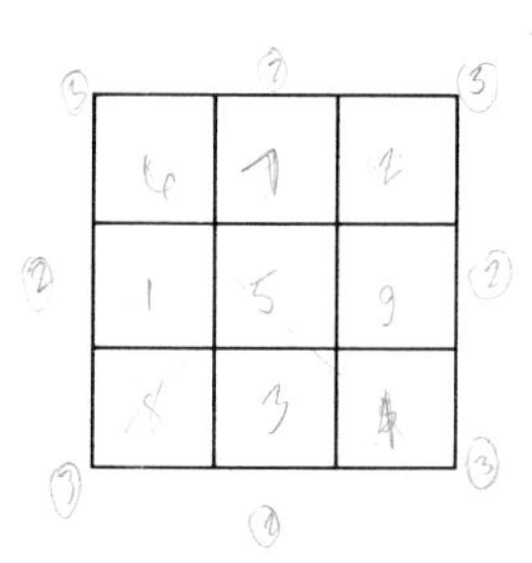

8. Can you answer the following questions?
   (a) A clock strikes three in 2 seconds. How long does it take to strike nine?

(b) A rectangular house is so built that every wall has a window opening on the south. A bear is seen from one of the windows. What color is the bear?

(c) Is it legal for a man to marry his widow's sister?

(d) How many cubic feet of dirt are in a hole 4 feet long, 2 feet wide, and 6 feet deep?

(e) Two United States coins total 30¢, yet one of the coins is not a nickel. Can you explain this?

*9. (a) Three students—John, James, and Edward—agree to be a part of an experiment in reasoning. First of all, the students are shown five ribbons, three of which are blue and two are red. The students are told that a ribbon will be placed on each of their backs and each is to try to determine the color of his ribbon. John can see the ribbons on James and Edward; and James can see the ribbon on Edward; but Edward cannot see the other ribbons. When asked the color of the ribbon on his back, John says he does not know. Likewise, James cannot reason the answer. However, when Edward is finally asked the question, he is able to give a correct answer. Why?

(b) Of five weights, four are the same and one is somewhat heavier than the other four. How can we determine the heavier weight if we are allowed to put all or some of them on a balance only twice?

(c) Three couples were out on a hike. It became necessary to cross the river in a small boat which had a maximum capacity of two. The boys were extremely jealous, so no girl could be left with a boy unless her date was present. How did they manage to cross the river?

(d) A man with a five gallon and a three gallon bucket went to the well to get exactly four gallons of water. How was he able to get the four gallons of water by using only two buckets?

(e) Each of four students named Aea, Bea, Sea, and Dea comes from a different place and each is the owner of a dog. One student, who lives at Whereitsat, has a dog named Spot. The student Aea, from Brooklyn, and another student each have a poodle, one poodle being white and the other being black. The white poodle lives in Atlanta near the home of the mutt, who is owned by Sea. A bird dog is named Joe, the same name as the white poodle. One poodle is named Fifi. Dea lives far from the other students, for she lives out West in Here. Name each student, his hometown, and the breed and name of his dog. (*Hint:* Making

7

Logic and an Introduction to Mathematical Reasoning

a chart and filling in the facts given here is a good method for solving this problem.)

(f) A beetle is at the bottom of a bottle 6 inches deep. Each day he climbs up 1 inch and each night slides back $\frac{1}{2}$ inch. How long will it take the beetle to climb out of the jar?

3

Propositions

In elementary mathematics, as well as in ordinary communication, one encounters assertions similar to the following:

(a) Paul ordered hamburgers, and Ed selected hot dogs.
(b) The pollution count will go above 300 or the high pressure cell will leave the locality.
(c) It is not true that George failed statistics.
(d) If the birth rate continues to increase, then greater emphasis must be placed on planned parenthood.

These examples illustrate sentences involving the words "and," "or," "not," and "if ... then." These and other connectives play an important role in one of the basic problems of elementary logic—determining whether complicated statements are true or false when you know the truth or falsity of their parts. This type of logic in everyday language is called "common sense" whether it is demonstrated by a politician, a sociologist, or an educator.

We introduce the concepts of logic through a discussion of terms that are the foundations upon which logic is built.

Definition 1-1: A *proposition* is an assertion that can be assigned a truth value; that is, it can be meaningfully classified either as true or as false.

Since some sentences are neither true nor false, not all sentences are propositions. Consider the following examples:

(a) Keep your eyes on the road. (A command)
(b) What time is it? (A question)
(c) Oh, what a gorgeous sunset! (An exclamation)

If a proposition is true, its truth value is denoted by T. If a proposition is false, its truth value is denoted by F. Every proposition has either truth value T or truth value F. Notice that the following sentences assert something and have truth values in the sense that they can be classified as true or false.

(a) President Lincoln was born in Texas. (F)
(b) $2 + 3 = 5$. (T)
(c) 8 is less than 7. (F)

"He is president of the Owl's Club" and "$x + y = 9$" are propositions only if we know to whom *He* refers and what numbers x and y represent. As long as these facts are unknown, the sentences cannot be definitely classified either as true or as false.

Having introduced the idea of propositions, it is easy to show how new propositions can be formed from existing propositions.

Definition 1-2: The *negation* of proposition p is the proposition "It is not true that p" (denoted by $\sim p$).

Thus, the negation of a proposition is the expression "It is false that," followed by the proposition itself. The negation of "New York City is the capital of the United States" is "It is not true that New York City is the capital of the United States." The use of the grammatical form "It is not true that p" sometimes results in an awkward sentence structure. For many propositions, the negation may be more simply expressed by negating the predicate. For example, the negation of the preceding proposition could be written as "New York City is not the capital of the United States." The negation of a proposition is, of course, false when the proposition is true and true when the proposition is false.

Table 1-1 summarizes the two possibilities for p and $\sim p$. Such a table is called a *truth table.* The table emphasizes the fact that p and $\sim p$ cannot be simultaneously true or simultaneously false.

p	$\sim p$
T	F
F	T

Table 1-1

Sometimes it is difficult to state the negations of propositions involving the word "all" or the word "some." In mathematics the word "some" always means *at least one and perhaps all.* On the other hand, "all" means *every one.* The negation of the proposition "All men have black hair" can be written "It is not true that all men have black hair," or "Not all men have black hair," or "Some men do not have black hair." This is not the same as the statement "All men do not have black hair" because this statement implies that *every* man does not have black hair. Consider the following propositions and their negations.

Propositions	*Negations*
Some women have red hair.	No woman has red hair.
All bananas are yellow.	Some bananas are not yellow.
No professor is baldheaded.	Some professors are baldheaded.
Some students do not work hard.	All students work hard.

Consider the proposition "Paul ordered hamburgers, and Ed selected hot dogs." We accept this proposition as true provided Paul did order hamburgers and Ed did select hot dogs. Otherwise we consider the statement as false. There are three ways that the proposition can be false: (1) Paul did not order hamburgers, and Ed did select hot dogs; (2) Paul did order hamburgers, and Ed did not select hot dogs; and (3) Paul did not order hamburgers, and Ed did not select hot dogs. Given any statements p and q connected by "and," the proposition "p and q" is true only when both p and q are true; otherwise, "p and q" is false.

Propositions utilizing "and," such as "Sugar is sweet, and the earth is flat," may be denoted by $S \wedge E$ where S denotes "Sugar is sweet" and E represents "The earth is flat." From the preceding discussion, $S \wedge E$ is considered to be true when both S is true and E is true. If either S or E is false (or if both are false), then $S \wedge E$ is false. In this case, $S \wedge E$ is false because the earth is not flat.

Therefore, since p is either true or false and since q is either true or false, there are four possibilities for the combined truth values of p and q. Table 1-2 illustrates these possibilities. In the third column of the table, we note the truth value of $p \wedge q$ for each of the four possibilities. When p is true and q is true, then $p \wedge q$ is true. When p is true and q is false, $p \wedge q$ is false. When p is false and q is true, $p \wedge q$ is false. Finally, when p is false and q is false, $p \wedge q$ is false.

p	q	$p \wedge q$	*Examples*
T	T	T	$2+3=5$, and 1 is less than 2.
T	F	F	$2+3=5$, and 4 is less than 3.
F	T	F	December has 30 days, and Christmas is on December 25.
F	F	F	December has 32 days, and January has 30 days.

Table 1-2

If you had a true-false question in political science such as "New York is located in the United States, or Moscow is located in Russia," would your answer be true or false? Propositions using the connective "or" are false only when both parts are false. The proposition "p or q" is true if p is true, if q is true, or if both p and q are true.

Propositions utilizing "or," such as "The earth is flat, or Moscow is located in Russia," may be denoted by $E \vee M$ where E represents "The earth is flat" and M represents "Moscow is located in Russia." (Note that the symbol representing "and" is $\wedge$ while the symbol representing "or" is $\vee$.) Table 1-3 illustrates the truth values for the proposition $p \vee q$ for all possible combinations of truth values of p and q.

p	q	$p \vee q$	*Examples*
T	T	T	Sugar is sweet, or carrots are yellow.
T	F	T	Sugar is sweet, or the earth is flat.
F	T	T	$2+3=7$, or 2 is less than 7.
F	F	F	$2+3=7$, or 2 is greater than 5.

Table 1-3

Sometimes, instead of making an outright assertion or statement, one may wish to restrict the statement. For example, instead of saying "I will carry my umbrella today," one may wish to say "If it rains, I will carry my umbrella today." This statement contains a condition concerning the carrying of an umbrella. "If the sun shines, we will go on a picnic," "If I get a bonus, then I will take the family on a vacation," and "If a number greater than 2 is prime, then it is odd" are additional examples of implications.

Definition 1-3: An *implication* is a proposition of the form "If p then q" where p and q are propositions. It is denoted by $p \rightarrow q$.

An example of an implication is "If it snows, then the streets will be slick." Here the p is "It snows" and the q, "The streets will be slick." Such an implication can be worded in many different ways, some of which are listed for comparison.

(a) If it snows, then the streets will be slick.
(b) If it snows, the streets will be slick.
(c) It will snow implies that the streets will be slick.
(d) The streets will be slick if it snows.
(e) The streets will be slick is implied by the fact that it will snow.

Using the letters p and q, the same implications may be written as follows:

(a) If p, then q (b) If p, q (c) p implies q
(d) q if p (e) q is implied by p

In the implication $p \rightarrow q$, p is sometimes called the *hypothesis* and q the *conclusion*. These terms are important since one can translate many expressions into the mathematical form $p \rightarrow q$ or "if ... then." Consider the following examples:

(a) I will take the family on a vacation when I get my bonus.
If I get my bonus, I will take the family on a vacation.
(b) Unless I study, I will fail this course.
If I do not study, I will fail this course.
(c) All illegal acts are immoral.
If an act is illegal, then it is immoral.
(d) I go to the movies only when I have a date.
If I go to the movies, then I have a date.
(e) John plays baseball only in the summer.
If John plays baseball, then it is summer.

In order to decide when to classify an implication as true and when to call it false, we will think of $p \rightarrow q$ as a promise. I promise my wife that "If I get a bonus, we will take a vacation." The promise will be broken (i.e., implication false) when and only when I get a bonus and do not take a vacation (i.e., p is true and q is false). If I do not get a bonus, I can either take or not take a vacation and the implication is true because the promise was made relative to what would happen if the bonus were granted.

This example illustrates the fact that a *false hypothesis leads to a true implication regardless of the truth or falsity of the conclusion. An implication is false only when the hypothesis is true and the conclusion is false.* These facts are illustrated in Table 1-4.

p	q	$p \to q$	*Examples*
T	T	T	If there is water pollution, the fish will be affected.
T	F	F	If $2+3=5$, then $2+4=7$.
F	T	T	If the earth is flat, then the moon is smaller than the earth.
F	F	T	If the earth is flat, then San Francisco is located on the edge of the earth.

Table 1-4

Closely related to an implication are its *converse*, its *inverse*, and its *contrapositive*. First we will define the converse.

Definition 1-4: The *converse* of the implication "p implies q" is the implication "q implies p" (i.e., the converse of $p \to q$ is $q \to p$).

"If two angles are equal, then they are right angles" is the converse of "If two angles are right angles, then they are equal." This example shows that the truth of an implication in no way ensures the truth of its converse. In fact, assuming that the truth of an implication ensures the truth of its converse is one of the common fallacies in thinking. Similarly, the truth of an implication certainly does not require that its converse be false. The converse of the implication "If a triangle has two equal sides, it has two equal angles" is "If a triangle has two equal angles, it has two equal sides." You will remember from your high school geometry that both statements are true.

When both an implication and its converse are true, this fact is often expressed by the phrase "If and only if" or "Necessary and sufficient." Symbolically, this is written $p \leftrightarrow q$. For $p \leftrightarrow q$ to be true, p and q must be simultaneously true or simultaneously false, since $p \leftrightarrow q$ is true only when both $p \to q$ and $q \to p$ are true. "p if and only if q" means (p if q) $\wedge$ (p only if q); that is,

$$(q \to p) \wedge (p \to q).$$

The other two propositions related to a given implication, the inverse and the contrapositive, will be defined next.

Definition 1-5: The *inverse* of a given implication is the implication that results when the p and the q are replaced by their negations (i.e., the inverse of $p \rightarrow q$ is $\sim p \rightarrow \sim q$).

The inverse of "If a polygon is a rectangle, then it is a parallelogram" is "If a polygon is not a rectangle, then it is not a parallelogram." The truth of an implication in no way ensures the truth of the inverse. Thus the inverse of a true implication may be either true or false, and you should construct some examples to see that this is so.

Definition 1-6: The *contrapositive* of the implication "If p, then q" is "If not q, then not p" (i.e., the contrapositive of $p \rightarrow q$ is $\sim q \rightarrow \sim p$).

Thus, to form the contrapositive of an implication, one may interchange the p and the q and then negate them; or one may negate the p and the q and then interchange them. The contrapositive of "If a polygon is a rectangle, then it is a parallelogram" is "If a polygon is not a parallelogram, then it is not a rectangle." From your high school geometry, you will remember that both statements are true. In fact, the truth of the contrapositive *does ensure* the truth of the associated implication, as will be shown in the next section.

To summarize the definitions of converse, inverse, and contrapositive, consider the following examples:

(a) *Implication:* If Mr. Jones buys a new automobile, he will select a red convertible.

(b) *Converse:* If Mr. Jones selects a red convertible, he will buy a new automobile.

(c) *Inverse:* If Mr. Jones does not buy a new automobile, he will not select a red convertible.

(d) *Contrapositive:* If Mr. Jones does not select a red convertible, he will not buy a new automobile.

Exercise Set 1-2

1. Determine which of the following are propositions and classify each proposition as either true or false.

(a) Abraham Lincoln was born in Texas.
(b) Good morning.
(c) All athletes over seven feet tall play basketball.
(d) $5 + 4 = 9$.
(e) $x + 5 = 8$.
(f) Help stop inflation.
(g) Some students work hard at their studies.
(h) All industries pollute the atmosphere.
(i) Some professors are intelligent.
(j) All men weigh more than 150 pounds.
(k) $2 \cdot 3 = 7$.
(l) $3x = 6$.

2. Write in two ways the negations of the parts of Exercise 1 that are propositions.
3. Translate the following propositions into symbolic form using A, B, C, D, $\wedge$, $\vee$, $\sim$, and $\rightarrow$ where A, B, C, and D denote the following statements:

A, it is snowing;
B, the roofs are white;
C, the streets are not slick;
D, the trees are beautiful.

(a) It is snowing, and the trees are beautiful.
(b) The trees are beautiful, or it is snowing.
(c) If it is not snowing, then the roofs are not white.
(d) If the streets are not slick, then it is not snowing.
(e) The streets are not slick, and the roofs are not white.
(f) The trees are not beautiful, and it is not snowing.

4. Using the statements of Exercise 3, translate the following into English sentences.

(a) $A \wedge \sim B$ (b) $\sim B \vee \sim C$ (c) $A \wedge (B \vee C)$
(d) $(A \vee \sim C) \wedge D$ (e) $\sim (A \wedge \sim D)$ (f) $\sim (A \wedge \sim C)$

5. State the converse, inverse, and contrapositive of each of the following implications:

(a) If a triangle is a right triangle, then one angle is 90°.
(b) If a number is a prime, then it is odd.
(c) If two lines are parallel, then alternate interior angles are equal.
(d) If Joyce is smiling, then she is happy.
(e) If x is divisible by 10, then x is divisible by 5.
(f) If Tom is John's father, then John is Tom's son.

6. Write each of the following statements in the form "If . . . then."

(a) Triangles are not squares.
(b) There is a pot of gold at the end of the rainbow.

(c) Honest politicians will not accept bribes.
(d) A rolling stone gathers no moss.

7. For the implication "If I use leaded gasoline, I will pollute the atmosphere," find the following.
(a) Contrapositive of the inverse
(b) Inverse of the converse
(c) Converse of the contrapositive
(d) Contrapositive of the converse

*8. Answer true or false where p represents "A car is a Chevrolet" and q represents "A car is a General Motors product."
(a) $q \rightarrow p$
(b) p is necessary and sufficient for q
(c) p if and only if q
(d) If q, then p
(e) p is a sufficient condition for q

*9. Write the following statements in symbolic form, using p, q, and r as indicated below. Then, assuming p to be T, q to be F, and r to be F, find the truth value of each statement.

p represents "*ABC* is a triangle."
q represents "*ABC* is a polygon."
r represents "*ABC* is an equilateral triangle."

(a) If *ABC* is an equilateral triangle, then *ABC* is a polygon and *ABC* is a triangle.
(b) If *ABC* is a polygon, then *ABC* is a triangle; and if *ABC* is not a triangle, then *ABC* is not an equilateral triangle.
(c) If *ABC* is not a triangle, then *ABC* is an equilateral triangle or *ABC* is not a polygon.
(d) *ABC* is not an equilateral triangle, or *ABC* is a polygon.

*4

Introduction to Deductive Reasoning

Consider the validity of the argument "If one is a Communist, then he reads *The Daily Worker*. Edward reads *The Daily Worker*. Therefore, Edward is a Communist." Did you consider this argument to be invalid deductive reasoning? If so, the material in this section will confirm your intuition.

In contrast to the inductive general conclusions drawn from particular

* The asterisk on the section number indicates that the section may be omitted at the discretion of the teacher. A section marked with an asterisk is not necessary for the remainder of the book.

observations, deductive reasoning presents a proof of a general statement. We introduce the concept of deductive logic by use of truth tables. The following examples give additional practice in the construction of such tables and introduce the concept of *logically equivalent.*

Examples: Construct truth tables for

(a) $\sim(\sim p \vee q)$ (Table 1-5),
(b) $\sim p \rightarrow \sim q$ (Table 1-6),
(c) $(p \wedge q) \rightarrow r$ (Table 1-7).

(a)

p	q	$\sim p$	$\sim p \vee q$	$\sim(\sim p \vee q)$
T	T	F	T	F
T	F	F	F	T
F	T	T	T	F
F	F	T	T	F

Table 1-5

(b)

p	q	$\sim p$	$\sim q$	$\sim p \rightarrow \sim q$
T	T	F	F	T
T	F	F	T	T
F	T	T	F	F
F	F	T	T	T

Table 1-6

There are eight possibilities involving p, q, and r in the truth table for part (c) of this example.

(c)

p	q	r	$p \wedge q$	$(p \wedge q) \rightarrow r$
T	T	T	T	T
T	T	F	T	F
T	F	T	F	T
T	F	F	F	T
F	T	T	F	T
F	T	F	F	T
F	F	T	F	T
F	F	F	F	T

Table 1-7

Example: By means of a truth table, find the relationship between the truth values of $p \leftrightarrow q$ and $(p \rightarrow q) \wedge (q \rightarrow p)$.

p	q	$p \leftrightarrow q$	$p \rightarrow q$	$q \rightarrow p$	$(p \rightarrow q) \wedge (q \rightarrow p)$
T	T	T	T	T	T
T	F	F	F	T	F
F	T	F	T	F	F
F	F	T	T	T	T

Table 1-8

Note in Table 1-8 that the truth values for $p \leftrightarrow q$ and $(p \rightarrow q) \wedge (q \rightarrow p)$ are the same. In other words these two propositions are said to be *logically equivalent.*

Definition 1-7: Two propositions are said to be *logically equivalent* if and only if they have the same truth values.

To illustrate this definition, consider the equivalence of an implication and its contrapositive, as shown in Table 1-9.

p	q	$\sim p$	$\sim q$	$p \rightarrow q$	$\sim q \rightarrow \sim p$
T	T	F	F	T	T
T	F	F	T	F	F
F	T	T	F	T	T
F	F	T	T	T	T

Table 1-9

We observe that, for all possible truth values of p and q, the truth values for $p \rightarrow q$ and $\sim q \rightarrow \sim p$ are exactly the same. Thus, an implication and its contrapositive are logically equivalent.

A proposition is said to be *logically true* if it is always true regardless of the truth values of the statements that compose it. Such propositions are called *tautologies.*

Definition 1-8: Propositions that are logically true are called *tautologies.*

It is easy to show that certain statements are tautologies by means of truth tables. If a statement is a tautology, the column for this statement will contain all T's.

Example: p or not p [denoted by $p \vee (\sim p)$] is a tautology. That this compound statement is always true is verified by the following truth table.

p	$\sim p$	$p \vee (\sim p)$	*Examples*
T	F	T	$2 + 3$ equals 5, or $2 + 3$ is unequal to 5.
F	T	T	Your name is Rusey, or your name is not Rusey.

Table 1-10

Example: $\sim[p \wedge (\sim p)]$ is a tautology. This is verified in Table 1-11.

p	$\sim p$	$p \wedge (\sim p)$	$\sim[p \wedge (\sim p)]$
T	F	F	T
F	T	F	T

Table 1-11

The fact that $\sim[p \wedge (\sim p)]$ is always true is sometimes called the *Law of the Excluded Middle.*

Example: $\sim(p \vee q) \rightarrow \sim p$ is a tautology. This is verified by Table 1-12.

p	q	$\sim p$	$p \vee q$	$\sim(p \vee q)$	$\sim(p \vee q) \rightarrow \sim p$
T	T	F	T	F	T
T	F	F	T	F	T
F	T	T	T	F	T
F	F	T	F	T	T

Table 1-12

In mathematics, we use deductive or *valid reasoning* when we use propositions, which are called *premises*, to arrive without error at other propositions known as *conclusions*. If there is an error in the reasoning process, the reasoning is said to be *invalid*. For illustrative purposes, we shall consider a number of examples that may at first seem far removed from mathematics. In these examples, the first two statements, or premises, are known as the *hypothesis* and the third statement as a *conclusion*. We say the third statement is *implied* by the first two.

Examples:

(a) All residents of Dallas are Texans.
All Texans brag about Texas.
Therefore, all residents of Dallas brag about Texas.

(b) All college professors are absent-minded.
Mr. Smith is a college professor.
Therefore, Mr. Smith is absent-minded.

(c) If Joyce is smiling, she is happy.
If Joyce is happy, she is polite.
Therefore, if Joyce is smiling, she is polite.

Thus, we say that reasoning is *valid* if a conclusion is a logical outcome of the assumed hypothesis. One way to show that reasoning is valid is to show that the argument expressed as an implication is logically true or is a tautology. Many times one uses known tautologies as rules in deducing valid conclusions. The most commonly used rules of deductive reasoning are the *Rule of Detachment* and the *Chain Rule*, which we will state here as theorems.

Theorem 1-1: *Rule of Detachment.* If $p \to q$ is true and if p is true, then q is true.

Proof: The theorem written in terms of connectives becomes $(p \to q) \wedge p$ implies q, or $[(p \to q) \wedge p] \to q$. Table 1-13 indicates that there is only one case where $(p \to q) \wedge p$ is true; in this case, the conclusion q is likewise true. Thus, $[(p \to q) \wedge p] \to q$ is true for all possibilities.

p	q	$p \to q$	$(p \to q) \wedge p$	$[(p \to q) \wedge p] \to q$
T	T	T	T	T
T	F	F	F	T
F	T	T	F	T
F	F	T	F	T

Table 1-13

This theorem states that when $p \to q$ is accepted as true and p is accepted as true, then we must also accept q as true. Suppose p represents "It is snowing" and q "The streets are slick." If one accepts as true "If it is snowing, then the streets are slick" and if it is in fact snowing, then by the Rule of Detachment, one knows that the streets are slick.

Theorem 1-2: *Chain Rule.* If the implications $p \to q$ and $q \to r$ are true, then the implication $p \to r$ is true.

Proof: Written in terms of connectives, the theorem may be stated as $(p \to r)$ is true when $(p \to q) \wedge (q \to r)$ is true. Table 1-14 verifies the fact that $[(p \to q) \wedge (q \to r)] \to (p \to r)$ is always true.

p	q	r	$p \to q$	$q \to r$	$(p \to q) \wedge (q \to r)$	$p \to r$	$[(p \to q) \wedge (q \to r)] \to (p \to r)$
T	T	T	T	T	T	T	T
T	T	F	T	F	F	F	T
T	F	T	F	T	F	T	T
T	F	F	F	T	F	F	T
F	T	T	T	T	T	T	T
F	T	F	T	F	F	T	T
F	F	T	T	T	T	T	T
F	F	F	T	T	T	T	T

Table 1-14

Consider one set of premises and a conclusion as given on a previous page:

If Joyce is smiling, she is happy.
If Joyce is happy, she is polite.
Therefore, if Joyce is smiling, she is polite.

This can be written as $p \rightarrow q$, and $q \rightarrow r$; therefore, $p \rightarrow r$. Thus, the acceptance of the premises that "If Joyce is smiling, she is happy" and "If Joyce is happy, she is polite" requires by the Chain Rule that one consider as true the conclusion "If Joyce is smiling, she is polite."

Of course, the Chain Rule can be extended to include any number of premises in the hypothesis—for example, if $p \rightarrow q$, $q \rightarrow r$, and $r \rightarrow s$, then $p \rightarrow s$.

The testing of the validity of an argument may be simplified by using tautologies or equivalent statements as indicated by the following examples.

Example: If Earl is not guilty, then John is telling the truth.
John is not telling the truth.
Therefore, Earl is guilty.

Symbolically this can be written as $\sim p \rightarrow q$ and $\sim q$; therefore, p (where p represents "Earl is guilty" and q represents "John is telling the truth"). Since an implication and its contrapositive are equivalent, $\sim p \rightarrow q$ may be replaced by $\sim q \rightarrow \sim(\sim p)$. But $\sim(\sim p)$ is equivalent to p. Thus, the first premise may be replaced by $\sim q \rightarrow p$; so the argument becomes $\sim q \rightarrow p$, $\sim q$; therefore, p. This argument is valid by the Rule of Detachment.

Consider now a set of premises that does not lead to a valid conclusion.

Example: If one is a college graduate, then he is intelligent.
Bill Jones is intelligent.
Therefore, Bill Jones is a college graduate.

This can be written as $p \rightarrow q$, and q; therefore, p. Of course, this is not valid because it would imply that the truth of the conclusion of a true implication leads to the truth of the hypothesis.

Example: If I go bowling, I will not study.
If I do not study, I will take a nap.
I will not take a nap.
Therefore, I do not go bowling.

We write this as $p \rightarrow \sim q$, $\sim q \rightarrow r$, and $\sim r$; therefore, $\sim p$. Now $p \rightarrow \sim q$ and $\sim q \rightarrow r$ can be replaced by $p \rightarrow r$ by the Chain Rule. Likewise, $p \rightarrow r$ can be replaced by its contrapositive $\sim r \rightarrow \sim p$. Thus, the argument can be written as $\sim r \to \sim p$ and $\sim r$, which leads to the conclusion $\sim p$ by the Rule of Detachment.

Exercise Set 1-3

1. Determine which of the following are tautologies.
   (a) $p \to p$ (b) $[p \wedge (p \to q)] \to q$
   (c) $(p \to q) \to p$ (d) $[(\sim p \vee q) \wedge p] \to q$
   (e) $(p \to q) \to (p \vee q)$ (f) $(p \wedge q) \to q$
   (g) $(\sim p \vee q) \to (p \to q)$ (h) $(p \to q) \to q$
2. By means of truth tables, prove that the following pairs of statements are logically equivalent.
   (a) $\sim(\sim p)$ and p (b) $\sim(p \wedge q)$ and $\sim p \vee \sim q$
   (c) $\sim(p \vee q)$ and $\sim p \wedge \sim q$ (d) $\sim(p \to q)$ and $p \wedge \sim q$
   (e) $p \to q$ and $\sim p \vee q$

For Exercises 3 through 15, tell which of the arguments are valid and which are invalid. Specify the rules of logic that are used in each step.

3. If two sides of a triangle are equal, the angles opposite these sides are equal.
   Side BC equals side AB in triangle ABC.
   Therefore, the angles opposite BC and AB are equal.
4. If he is a thief, he is a lawbreaker.
   He is not a thief.
   Therefore, he is not a lawbreaker.
5. If a man is a good speaker, he will make a good teacher.
   Mr. Bishop is a good teacher.
   Therefore, he must be a good speaker.
6. If I can't go to town, I will go to the shopping center.
   I can go to town.
   Therefore, I will not go to the shopping center.
7. You will fail this course, if you do not study enough.
   You are not studying enough.
   Therefore, you will fail this course.
8. If R, then $\sim S$.
   If $\sim S$, then T.
   Therefore, if R then T.
9. Every football player has good coordination.
   William has good coordination.
   Therefore, William is a football player.
10. If a geometric figure is a square, then its diagonals are equal.
    Geometric figure $ABCD$ is a square.

Therefore, the diagonals of $ABCD$ are equal.

11. If a quadrilateral is a parallelogram, then opposite sides are parallel. Opposite sides of quadrilateral $ABCD$ are parallel. Therefore, $ABCD$ is a parallelogram.
12. If globs are nobs, then gleeks are sleeks. If gleeks are sleeks, then globs are fobs. Globs are nobs. Therefore, globs are fobs.
13. All nobs are globs. Some globs are gleeks. All gleeks are fobs. Therefore, all nobs are fobs.
14. If globs are nobs, then gleeks are sleeks. Globs are not nobs. Therefore, gleeks are not sleeks.
15. If gleeks are sleeks, then globs are fobs. Globs are not fobs. Therefore, gleeks are not sleeks.

*16. Determine a valid conclusion for the following arguments.
   (a) If $ABCD$ is a square, then $ABCD$ is a quadrilateral. $ABCD$ is not a quadrilateral. If $ABCD$ is not a square, then $ABCD$ is a triangle. Therefore, $ABCD$ is ______________.
   (b) If x is an integer, then x is a rational number. If x is not a real number, then x is not a rational number. x is an integer. Therefore, x is ______________.
   (c) If a person's chest X-ray gives a negative result, he probably does not have tuberculosis. If a person's skin test gives a positive result, he probably has tuberculosis. Dick's chest X-ray gives a negative result. Therefore, Dick's skin test is ______________.

*17. Under what conditions are the following four implications simultaneously true?
   (a) $A \vee B$ (b) $A \rightarrow B$
   (c) $(\sim B \rightarrow A) \rightarrow \sim A$ (d) $(A \wedge \sim B) \rightarrow A$

*5

Venn Diagrams

If a premise contains words such as "some," "all," and "no," one of the best ways to test whether or not a conclusion is a necessary consequence of the premises is by means of Venn diagrams. An English logician named John Venn used closed figures, usually circles, to indicate relationships. The following example illustrates a deductive argument using Venn diagrams.

Example: All residents of Dallas are Texans.
All Texans brag about Texas.
Therefore, all residents of Dallas brag about Texas.

In Figure 1-1 let the small circle and its interior (which we call D) represent all residents of Dallas. According to the first premise, all residents of Dallas D are a part of, or are contained in, a larger circle T composed of Texans. The second premise states that all Texans T are a part of a larger group B, namely, those who brag about Texas. Since circle D lies wholly within the circle B representing those who brag about Texas, it is valid to draw the conclusion that all residents of Dallas brag about Texas. If it were possible for some part of D to lie outside of B, the conclusion would not be valid.

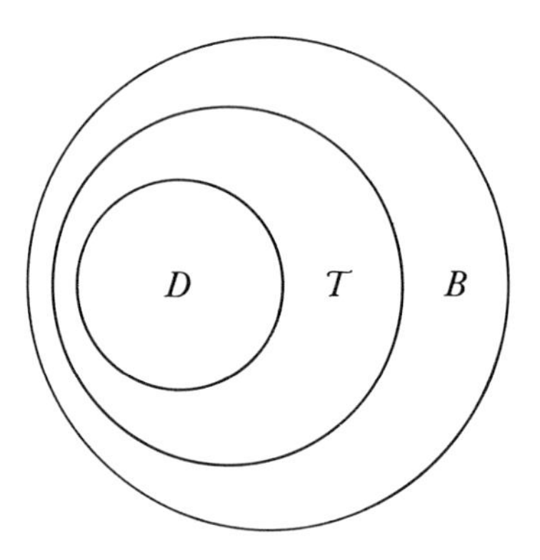

Figure 1-1

The statement "If p, then q" can often be represented by "All p are q." Consequently the circle representing the condition p would be contained within the circle representing q. That is, if you have p, you must have q. This fact is demonstrated by the following example.

Example: If Joyce is smiling, she is happy.
If Joyce is happy, she is polite.
Therefore, if Joyce is smiling, she is polite.

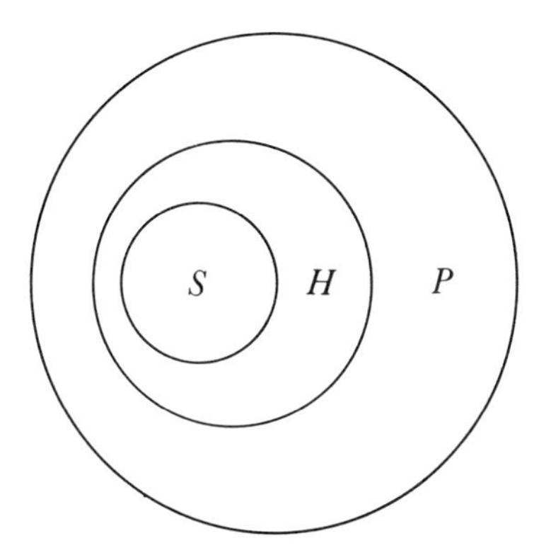

Figure 1-2

In Figure 1-2, the circle representing the situation "when Joyce is smiling" is included as a part of the region representing "happy situations." Similarly, "when she is happy" is contained within "when she is polite." Thus, Figure 1-2 indicates that every time Joyce is smiling, she is also polite. This, of course, means that the conclusion is a valid consequence of the hypothesis.

Example: All college graduates are intelligent.
Bill Jones is intelligent.
Therefore, Bill Jones is a college graduate.

The set of college graduates C is a part of the group of intelligent people I. Bill Jones is intelligent, and so he is contained as a point within the circle representing intelligent people. Now look at the two Venn diagrams in Figure 1-3. Both diagrams satisfy the premises of the problem.

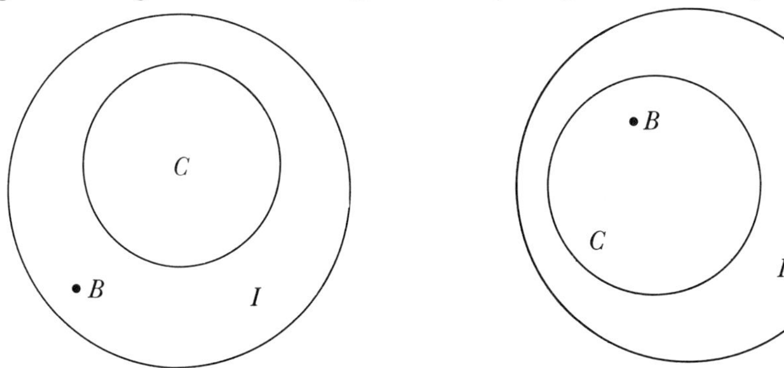

Figure 1-3

Thus, we have one possibility which states that Bill is not a college graduate and one which states that he is a college graduate. However, only one example that gives a result not in agreement with a conclusion is needed to disprove a conclusion. Thus, it is invalid to draw the given conclusion for this example. *This example illustrates the fact that just one figure in which the conclusion is not true is sufficient to prove the reasoning invalid.*

Examples using the words *some* and *no* (sometimes called quantifiers) are easily illustrated by Venn diagrams. "No fat people are interesting" would be represented by two circles that are disjoint (have nothing in common). *Some* implies *at least one.* "Some fat people are humorous" means that at least one fat person is humorous. However, it does not preclude the possibility that all fat people are humorous. For example, consider the following argument:

Example: Some college professors are absent-minded.
Mr. Smith is a college professor.
Therefore, Mr. Smith is absent-minded.

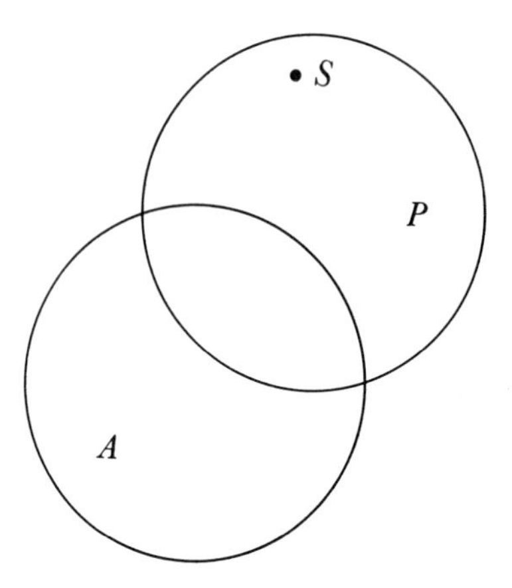

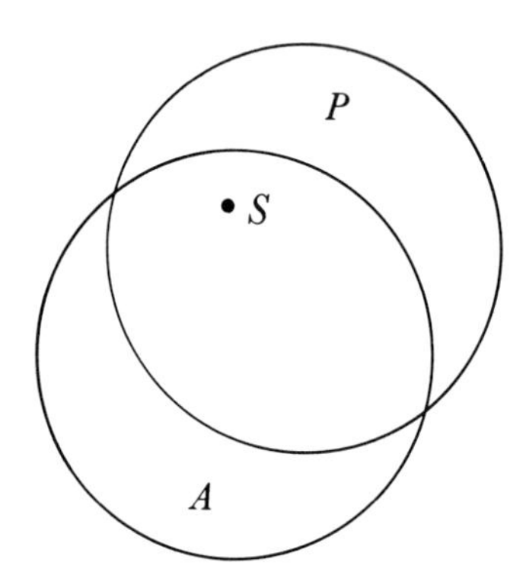

Figure 1-4

The premises of this problem can be illustrated in two ways by Venn diagrams. In the right diagram of Figure 1-4, Mr. Smith is a college professor and also absent-minded. However, the left diagram demonstrates that Mr. Smith can be a college professor without being absent-minded. Thus, the truth of the premises does not require the truth of the conclusion, and so Mr. Smith may or may not be absent-minded. The argument, then, contains a conclusion that is not a valid result of the premises.

Exercise Set 1-4

Using appropriate Venn diagrams, determine which conclusions in Exercises 1–8 are valid and which are invalid.

1. All fast runners are athletes.
   No people with long legs are athletes.
   Therefore, no people with long legs are fast runners.
2. All spheres are round.
   All circles are round.
   Therefore, all circles are spheres.
3. Carelessness leads to accidents.
   Mrs. Yeager had an accident.
   Therefore, Mrs. Yeager was careless.
4. All cats are intelligent animals.
   Fido is an intelligent animal.
   Therefore, Fido is a cat.
5. Some em's are red.
   All im's are em's.
   Therefore, some im's are red.
6. All athletes are fast runners.
   Some people with short legs are not fast runners.
   Therefore, no athletes have short legs.
7. All wise boys are interested in girls.
   George is interested in girls.
   Therefore, George is wise.
8. All women are fickle.
   Some women are friendly.
   Therefore, some friendly people are fickle.

Find valid conclusions, where possible, for the sets of premises given in Exercises 9 through 13.

9. All college students are clever.
   Paul is clever.
10. All triangles are polygons.
    All polygons are plane figures.
11. If Larry is not on time, he will be fired.
    Larry was not fired.
12. All right angles are equal.
    Angles A and B are right angles.
13. Not all students are radicals.
    Some students are Republicans.
    Henry is a student.

*14. (a)

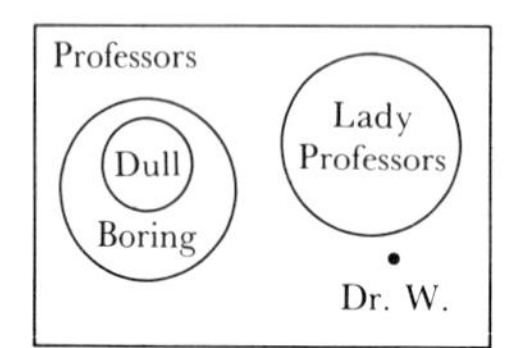

What conclusion can be drawn about Dr. W?

(b)

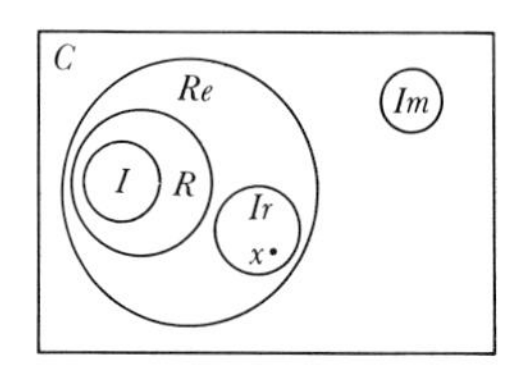

I represents integers; *R*, rational numbers; *Ir*, irrational numbers; *Re*, real numbers; *Im*, imaginary numbers; and *C*, complex numbers. What conclusion can be drawn about x?

(c)

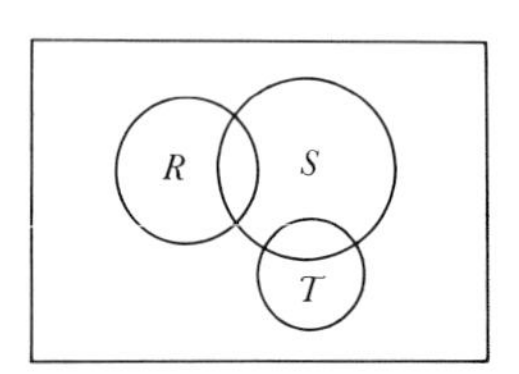

What conclusions can be drawn concerning the members of set R?

Using Venn diagrams, select valid conclusions from the given possibilities in Exercises 15 through 17.

*15. All x's are y's.
Some x's are z's.
Some x's are w's.
Therefore, (a) all w's are y's.
(b) all z's are y's.
(c) some y's are w's.
(d) some x's are not y's.

*16. All whole numbers are integers.
All integers are rationals.
No irrationals are rational.

All irrationals are reals.
x is an integer.
Therefore, (a) x is a whole number.
(b) x is a real number.
(c) x is an irrational number.
(d) x is not a rational number.

*17. All authors are intelligent.
Some men from New York are authors.
Some men from Philadelphia are not intelligent.
Therefore, (a) no man from Philadelphia is an author.
(b) no authors are from Philadelphia.
(c) no authors are from New York.
(d) some New York men are intelligent.

*6

Mathematical Systems

In the first section of this chapter, mathematics was described as an art, a science, a language, a tool, and a game. However, we are now ready to present a more formal answer to the question "What is mathematics?" *Mathematics will be considered as the grand total of all mathematical systems, along with the application of these systems to all forms of human endeavour.*

A mathematical system is built up deductively from a set of elements, a relation (at least one), an operation (at least one), and a set of axioms or postulates. Throughout this book we shall study the structure of mathematical systems—that is, we shall study the nature of the assumptions, the properties of the elements, the operations, and the relations that are derived from the postulates. The logical development of such mathematical systems starts with undefined terms.

It may seem strange to you that mathematics contains some terms deliberately *undefined.* Yet if you spend a few minutes using a dictionary, it will become immediately obvious why this is necessary. Look up any word in the dictionary. It will be defined in terms of another word. Look up the second word, which is defined in terms of a third word. Repeating the process eventually leads to a word defined in terms of the first word. And thus you have completed a circle. In a recent issue of a standard American dictionary, "sad" is defined in terms of unhappy, whereas "unhappy" is defined in terms of sad. Where do you start in the definition process? One ought to start with certain terms so fundamental that they do not need a formal definition. Thus a mathematical system begins with a few words which must of necessity remain *undefined.* Words such as "point," "line," "rotation," "between," "operation," "number," "equality," "length," "magnitude," and "set" are many times left undefined. In different mathematical systems, one finds that different words are left as undefined terms. However, all other terms in the system must be clearly defined relative to these basic undefined terms.

Bertrand Russell once remarked that formal *definitions* in mathematics are simply statements of symbolic abbreviations. Thus, a formal definition may be regarded as a pattern defining how a symbol is to be used. The statement of the definition should be motivated by a desire to clarify and explain. It should use undefined terms or previously defined terms or both to classify and describe the basic properties and characteristics of the idea or symbol being defined. However, definitions are not altogether arbitrary. That which is defined must be consistent with previous knowledge in the system; it must not lead to redundant answers.

When we have compiled an adequate vocabulary of undefined and defined terms, we are ready to form sentences using these words. We have defined a *proposition* as an expression which is meaningful and which can be regarded as either true or false. Just as with definitions, we choose some propositions and arbitrarily designate them as true. That is, we agree that in our particular mathematical system these propositions are acceptable for purposes of our discourse. Such propositions are called axioms, postulates, or assumptions. That is, an *axiom* or *postulate* is a proposition that is assumed to be true without proof. The acceptance of the axiom has nothing whatsoever to do with absolute truth or with empirical facts. The choice of axioms is creative rather than logical. We may try to choose axioms to reflect observed properties of nature, or we may base our choice on some patterns of experiences which suggest the form that the postulates should take.

A *theorem* is an assertion that may be proved to be true as a conclusion in a logical argument. In the next section, we study how one would mathematically prove a theorem. When a theorem has been proved, it may be used to prove other theorems. In this manner a mathematical system is constructed.

*7

Mathematical Proofs

A proof of a theorem is a valid argument that a conclusion is a result of given premises which take the form of certain basic facts, theorems, definitions, and assumptions. To prove a theorem, one shows that the hypothesis, along with the given premises, implies the conclusion. Ordinarily, proofs are classified as either *direct* or *indirect*. Direct proofs start in a variety of ways, but the following outline is typical. First of all, a proof starts with a known axiom, definition, or previously proved theorem, or with a combination of these. The problem of constructing the proof is to arrange a logically valid sequence of steps leading to the desired conclusion. There is no fixed way to accomplish this result; however, one develops skill through experience and study. Whenever alternative proofs are available, one is as valid as another.

Suppose H represents the hypothesis of a theorem and C represents the conclusion. We wish to prove the theorem "If H, then C" or "H implies C." A typical procedure for the proof involves the Chain Rule. Suppose $H \rightarrow D$, $D \rightarrow E$, $E \rightarrow F$, $F \rightarrow G$, ..., $R \rightarrow S$ and $S \rightarrow C$ are all axioms or valid theorems where $D, E, F, G, \ldots,$ and S are different conclusions. We know D is true by the Rule of Detachment, which states that if H is true and $H \rightarrow D$ is true, then D is true. Likewise by the Chain Rule $E, F, G, \ldots, R, S,$ and finally C are all true. Thus the theorem is proved.

Example: Prove that "If $x = 2$, then $x + 3 = 5$."

Proof: We prove this theorem by a series of implications, each of which is true because of assumed properties of the number system under consideration. For example, we assume we can add a number to both sides of an equation in the first step and that addition is possible in the second step.

If $x = 2$, then $x + 3 = 2 + 3$. $\quad A \rightarrow B$
If $x + 3 = 2 + 3$, then $x + 3 = 5$. $\quad B \rightarrow C$
Therefore, if $x = 2$, then $x + 3 = 5$. $\quad A \rightarrow C$ by Chain Rule

However, in mathematical proofs of this nature we usually assume the Rule of Detachment and Chain Rule and simply list the properties or reasons for each implication or statement.

We again prove that if $x = 2$, then $x + 3 = 5$.

Proof:

$x = 2$	Given
$x + 3 = 2 + 3$	Assumed axiom that a number may be added to both sides of an equality to produce a second equality
$x + 3 = 5$	Addition

Similarly, with other assumed axioms we can prove that if $x + 3 = 5$, then $x = 2$.

Proof:

$x + 3 = 5$	Given
$(x + 3) - 3 = 5 - 3$	Assumed axiom that one can subtract a number from both sides of an equality
$x + (3 - 3) = 5 - 3$	Assumed axiom that one can group the $3 - 3$ together
$x + 0 = 2$	Subtraction
$x = 2$	Addition of 0

These two results may be stated as "$x + 3 = 5$ if and only if $x = 2$" or "A necessary and sufficient condition that $x + 3 = 5$ is that $x = 2$." The statement "$x + 3 = 5$ if and only if $x = 2$" means (a) if $x + 3 = 5$, then $x = 2$ and (b) if $x = 2$, then $x + 3 = 5$. If $x + 3 = 5$, it is necessary that $x = 2$; and a sufficient condition for $x = 2$ is $x + 3 = 5$.

A good method of direct proof for problems containing unknowns involves *substitution.* For example, consider the theorem, "If $x = 2$, then $x + 3 = 5$." Replace or substitute for x the value given in the hypothesis, namely, $x = 2$. Then $x + 3 = 5$ becomes $2 + 3 = 5$. Therefore when $x = 2$, $x + 3 = 5$, and the proof is complete.

Since it is not always easy to establish a list of implications and then apply the Chain Rule, the direct method of proof is sometimes difficult. Many times it is easier to prove a theorem by using what is called an *indirect proof.* One method of indirect proof is to assume the conclusion is false (or $\sim q$ is true where the theorem is $p \rightarrow q$) and then arrive at a contradiction of the hypothesis or some other truth.

An indirect proof that $p \rightarrow q$ may take on the following pattern:

(a) Assume $\sim q$ is true.

(b) Using $\sim q$, deduce that some statement r, which is known to be false, is true or that p, which is given as a true statement, is false.

(c) But r is known to be false; likewise p is known to be true.

(d) Our assumption that $\sim q$ is true has led to a contradiction whether it involves statement r or hypothesis p.

(e) Therefore, $\sim q$ is false and hence q is true.

Example: Prove that if $x + 1 \neq 7$, then $x \neq 6$ (where $\neq$ means "is unequal to").

Proof:

Assume $x = 6$	Allowable assumption in an indirect proof
$x + 1 = 6 + 1$	Assumed axiom that one can add a number to both sides of an equality
$x + 1 = 7$	Addition
But $x + 1 \neq 7$	Hypothesis
Therefore, we have a contradiction	Last two statements
Assumption $(x = 6)$ is false	Because of the contradiction
Therefore, $x \neq 6$	A negation of a false proposition is true

Another method of indirect proof is to prove a theorem by proving the *contrapositive* of the theorem. We have already stated that an implication and its contrapositive are logically equivalent. Thus when you prove the contrapositive of a theorem, you are at the same time proving the theorem.

If you have tried unsuccessfully to prove what you think is a theorem, it may well be that your theorem is not true. Thus you would like a method to disprove the proposed theorem. There are two standard methods for disproving theorems. The easiest way to disprove a theorem is by *counterexample.* In other words, you look for an example that satisfies all of the conditions of the hypothesis but does not satisfy the conclusion of the proposed theorem. If you find one, the theorem is not true.

Example: Prove that if $x = 5$, then $x + 3 = 11$. The conclusion is false because, by substituting $x = 5$ into the equation of the conclusion, one obtains $5 + 3 = 8 \neq 11$. $x = 5$ satisfies the conditions of the hypothesis but does not satisfy the conclusion. Thus we have disproved this statement as a theorem by means of a counterexample.

A second method of disproving a theorem is by *contradiction*. In this case, you assume that the given theorem is true, then proceed to derive additional relationships from it. If you can derive a relationship that contradicts a definition, axiom, or known theorem, then the given statement is false.

We close this chapter with this *warning*. Although a theorem may be disproved by an example, you cannot prove a theorem with examples unless you consider *all* possible examples, and such a procedure is often impossible.

Exercise Set 1-5

1. Why must some of the terms in mathematics remain undefined?
2. What is the principal difference between an axiom and a theorem?
3. How may a theorem be proved directly? Indirectly?
4. What is a counterexample? How may a theorem be proved false?
5. Identify each of the following statements as true or as false.
   (a) Even if the contrapositive of a theorem is proven to be true, the theorem itself is not necessarily true.
   (b) A postulate is equivalent to an axiom.
   (c) Proving that a contrapositive of a theorem is true and using the fact that if $\sim p$ is false then p is true are two forms of an indirect proof.
   (d) Most theorems can be proved by examples.
   (e) All false implications have true contradictions.
   (f) A conjecture is derived using definitions and assumed axioms.
   (g) One method of direct proof makes use of the fact that if $\sim q$ is false then q is true.
6. In the following "if ... then" statements, select the hypothesis and the conclusion. Does the conclusion follow as a logical consequence of the hypothesis?
   (a) If $x + 5 = 7$, then $x = 2$.
   (b) If $x = 3$, then $x + 4 = 7$.
   (c) If $x = y + 1$, and if $y = 4$, then $x = 5$.
   (d) If $x + 4 = 5$, then $x + 6 = 7$.
7. Find a counterexample to disprove each of the following:

(a) If a person lives in the United States, then he lives in Florida.
(b) For all x, $2 \cdot x = 4$.
(c) If $x + 1 = 5$, then $x = 6$.
(d) If line A is perpendicular to line B and line B is perpendicular to line C, then line A is perpendicular to line C.

8. Suppose you are given the axioms that all numbers are either even or odd and that the product of two odd numbers is odd. Prove that if the product of two numbers is even, at least one of the numbers must be even.

*9. Prove or disprove each of the following theorems or propositions. List reasons for each step in the proofs.
   (a) For all values of x, $x + 6 = 9$.
   (b) If $x + 7 = 15$, then $x \neq 6$.
   (c) $x + 1 = 4$ if and only if $x = 3$.
   (d) If $x + 2 \neq 7$, then $x \neq 5$.
   (e) If $x + 3 = y + 4$, then $x \neq y + 1$.
   (f) There are no two words in the English language whose first two letters are each *a*.

Review Exercise Set 1-6

1. Write the converse, inverse, and contrapositive of each statement.
   (a) If it does not rain, we will play tennis.
   (b) If I study diligently, I will pass the course.
2. Specify whether the following are examples of inductive or deductive reasoning.
   (a) The 7:30 A.M. bus on which Martha rides to work has been late every morning for two weeks.
   Therefore, the bus will be late today.
   (b) All warm-blooded animals are mammals.
   A dog is warm-blooded.
   Therefore, a dog is a mammal.
   (c) The product of $3 \cdot 5$ is 15, and the product of $7 \cdot 9$ is 63.
   3, 5, 7, 9, 15, and 63 are odd numbers.
   Therefore, one can conclude that the product of two odd numbers is an odd number.
   (d) The biscuits baked for the past five meals have been burned.
   Therefore, the biscuits tonight will be burned.

3. Classify each of the following arguments as valid or invalid.
   (a) If you like mathematics, then you will like this course.
   You do not like mathematics.
   Therefore, you will not like this course.
   (b) Some freshmen are mathematics majors.
   Some seniors are mathematics majors.
   All mathematics majors are intelligent.
   Therefore, some intelligent juniors are mathematics majors.
   (c) All girls are beautiful.
   All beautiful people seldom study.
   Therefore, if you seldom study you are a girl.
   (d) All a's are b's.
   Some b's are c's.
   Therefore, some a's are c's.
4. Determine which of the following are logically equivalent to (a). Is any one of them a tautology?
   (a) $(p \to q) \wedge \sim q$ (b) $[\sim p \to (q \wedge p)] \wedge \sim q$
   (c) $(p \vee q) \to (\sim q \to p)$ (d) $\sim(\sim q \to \sim p)$
   (e) $[q \to (p \wedge \sim q)] \wedge (p \vee q)$ (f) $\sim(p \to q) \to \sim(\sim p \vee q)$
5. In each of the following, two premises are given. Can you find a conclusion that is a logical result of the premises?
   (a) All college students are clever.
   Glenda is a college student.
   (b) All college students are sharp.
   Joyce is sharp.
   (c) All right angles are equal.
   Angles A and B are equal.
   (d) Some college girls are beautiful.
   Linda is a college girl.

*6. Answer the following questions by constructing truth tables.
   (a) Is the contrapositive of $p \to (\sim q \wedge r)$ a tautology?
   (b) Is the inverse of $[\sim p \wedge \sim r] \to (r \vee p)$ a tautology?
   (c) Is the converse of $[p \to \sim r] \to (r \wedge p)$ a tautology?
   (d) Is the contrapositive of $[p \vee \sim q] \to (\sim q \vee r)$ a tautology?

*7. If you are willing to accept that $3 = 5$, can you prove that $13 = 15$? Explain.

*8. Prove or disprove.
   (a) If $x = 1$, then $x + 3 = 4$.
   (b) If $x \neq 5$, then $x + 2 \neq 9$.
   (c) If $x + 5 = 9$, then $x = 4$.
   (d) $(x - 2)(x + 2) = x^2 - 4x + 4$ for all x.

Suggested Reading

Deductive Reasoning: Boyer, pp. 308, 342. Campbell, pp. 9–14. Copeland, pp. 240–241. Keedy, pp. 67, 70, 138. Meserve, Sobel, pp. 300–307. Nichols, Swain, pp. 36–47. Ohmer, Aucoin, Cortez, pp. 23–38. Spector, pp. 53–73, 86–102. Willerding, pp. 35–39.

Inductive Reasoning: Boyer, pp. 305, 318. Campbell, pp. 1–4. Copeland, pp. 240–241. Graham, p. 86. Keedy, pp. 67, 194. Spector, pp. 105–107.

Mathematical Proofs: Garstens, Jackson, pp. 20–23. Graham, pp. 85–87, 285. Ward, pp. 45–48, 54–55, 373–374. Willerding, pp. 43–45.

Propositions, Negations, Implications: Allendoerfer, pp. 13–47. Armstrong, pp. 6–14. Bouwsma, Corle, Clemson, pp. 142–148. Byrne, p. 61. Garstens, Jackson, pp. 9, 13–16. Ohmer, Aucoin, Cortez, pp. 1–22, 29–32. Nichols, Swain, pp. 29–34. Spector, pp. 40–50. Willerding, pp. 29–34. Wren, pp. 70–73.

2

Sets and Relations

1

Introduction

In Chapter 1 we noted that mathematics rests on logic. In addition to logic, certain other concepts are so basic that they enter into the development of all mathematics. We shall study one such concept in this chapter.

During the latter part of the nineteenth century, Georg Cantor, while working with mathematical entities called infinite series, found it helpful to borrow a concept in common usage to describe a mathematical idea. The idea he borrowed was that of a *set*. We often say "set of books," "collection of shells," "congregation of people," "class of children," and "ball team." Cantor used the idea to mean that a collection or group of objects could be considered as a whole. In addition, we are indebted to George Boole, an English mathematician, for developing the *algebra* of sets. Boolean algebra has been named in his honor.

We shall not attempt to give a formal definition of the word "set." Instead, we choose to declare it a concept so basic that it should be considered an undefined term. Our purposes will be served if we describe a set as a *collection* of objects or symbols possessing a certain property that enables one to determine whether a given object is or is not in the set.

The individual objects of a set are called *elements* of the set. They are said to *belong to* or to *be members of* or to *be in* the set. The relationship between objects of the set and the set itself is expressed in the form "is an element of" or "is a member of." The symbol $\in$ is used to denote this relationship; thus, $x \in A$ means that x is an element of the set A. Nonmembership is indicated by $\notin$, and the symbol is read "is not an element of." In this book, we adopt the notation that a set is designated by a capital letter; its elements, when represented by letters, will be lower-case letters.

Often it is possible to specify a set simply by listing the members of the set. This method of listing the elements of a set is called *tabulation.* A set indicated by braces $\{\quad\}$ should be read "the set whose members are." The set of counting numbers less than 10 can be written

$$\{1, 2, 3, 4, 5, 6, 7, 8, 9\}.$$

Patty, Gail, and Shirley comprise the set {Patty, Gail, Shirley}. The order of tabulating the elements is immaterial; $\{1, 2, 3\}$ is the same as $\{2, 1, 3\}$, $\{3, 2, 1\}$, $\{3, 1, 2\}$, $\{1, 3, 2\}$, and $\{2, 3, 1\}$.

Sometimes, sets have so many elements that it is difficult or even impossible to tabulate them. Sets of this nature may be indicated by a descriptive statement or rule. For example, the following sets are well specified without a tabulation of members: the counting numbers, the even numbers less than 1000, the Presidents of the United States, and all the football teams in Pennsylvania.

The difficulty of tabulating sets is often overcome by enclosing within braces a letter representing an element of a set followed by a qualifying description of the element. The letter or symbol that is used to denote any element of a specified set is called a *variable* over the set, and the set is referred to as the *domain* of the variable. This method for representing sets is sometimes called the *set-builder* notation. In addition to being convenient, it has the added advantage of describing a defining property of the set. For example, let A represent the set of counting numbers less than 10; then $A = \{n \mid n \text{ is a counting number less than } 10\}$. The vertical line may be read "such that" or "satisfying the condition that." Set A may be read "The set of all n such that n is a counting number less than 10."

Example: Use set-builder notation to describe the set of counting numbers less than 140.

$$\{x \mid x \text{ is a counting number less than } 140\}$$

Many times in mathematics, three dots (sometimes called an ellipsis) are used to indicate the omission of terms. The set of even counting numbers less than or equal to 100 may be written as $\{2, 4, 6, \ldots, 100\}$.

This notation saves time in tabulating the elements of complicated sets; but unless the set has been specified completely by another description, this practice may lead to ambiguity. Exercise 13 of Exercise Set 2-1 illustrates that dots representing omitted elements must be used carefully.

A set may consist of no elements, a limited collection of elements, or an unlimited collection of elements. A set that contains no elements is called an *empty* or *null* set and is denoted by the symbols $\varnothing$ or $\{\ \ \}$. Can you think of examples of empty sets? What about the set of three-headed professors? The set of elephants in your classroom? The set of letters preceding *a* in our alphabet? A set containing either no elements or a definite number of elements is called a *finite* set. {Doris, Ruth, Betty} is a finite set with three elements. A set that contains an unlimited collection of elements is called an *infinite* set. (Both finite and infinite sets are defined and discussed in Chapter 3.) An infinite set is denoted either by a descriptive statement or by the use of braces with an ellipsis after the last-mentioned member. The set of all counting numbers is an infinite set and may be written in the following ways: $\{y \mid y$ is a counting number$\}$, or $\{1, 2, 3, \ldots\}$, or simply by the statement "the set of counting numbers."

Definition 2-1: Set A is said to be a *subset* of set B, denoted by $A \subseteq B$, if and only if each element of A is an element of B.

For example, the set of all male mice is a subset of the set of all mice. The set {Gail, Patty, Shirley} is a subset of {Gail, Patty, Shirley}. Note that $\varnothing$ is a subset of every set. (Why?) The set $\{a, b, c\}$ has eight subsets, namely $\{a, b, c\}$, $\{a, b\}$, $\{a, c\}$, $\{b, c\}$, $\{a\}$, $\{b\}$, $\{c\}$, and $\varnothing$.

Example: The set $\{y \mid y$ is the set of all industries guilty of water pollution$\}$ is a subset of $\{z \mid z$ is the set of all industries guilty of pollution$\}$.

Definition 2-2: Set A is said to be a *proper subset* of set B, denoted by $A \subset B$, if and only if each element of A is an element of B and there is at least one element of B that is not an element of A.

Consider the set of all animals. Since all dogs are animals, the set of dogs is a subset of the set of animals. Notice that the relationship between the set and subset is that of inclusion. The set of animals includes the set of dogs. You should recognize that the set of dogs is also a proper subset of the set of animals, since there are animals that are not dogs.

Let $A = \{m, n, o, p\}$ and $B = \{n, m, p, o\}$. By Definition 2-1, A is a subset of B, but B is also a subset of A. This suggests a definition of equality of sets.

Definition 2-3: Two sets A and B are said to be equal, denoted by $A = B$, if and only if $A \subseteq B$ and $B \subseteq A$.

By this definition, every element of A is an element of B, and every element of B is an element of A. That is, two sets are equal if and only if they contain precisely the same elements. Thus, if $A = \{1, 2, 3, 4\}$ and $B = \{1, 4, 2, 3\}$, then $A \subseteq B$ and $B \subseteq A$; hence, $A = B$. Notice again that the order in which the elements are listed is unimportant.

You need to be careful not to confuse the meaning of

$$x \in A \quad \text{and} \quad X \subset A.$$

$x \in A$ states that x is an element of set A, and $X \subset A$ states that set X is a subset of set A.

Example: For

$$A = \{a, b, c\},$$

let

$$x = a,$$

and

$$X = \{a\}.$$

Note that $x \in \{a, b, c\}$ but $X \subset \{a, b, c\}$.

Exercise Set 2-1

1. If A is the set of all counting numbers less than or equal to 16, indicate which of the following statements are true and which are false.
   (a) $11 \in A$ (b) $16 \in A$ (c) $\{5\} \in A$
   (d) $\{1, 2, 3, \ldots, 15\} \subset A$ (e) $^-4 \in A$ (f) $0 \in A$
   (g) $81 \in A$ (h) $\{\ \} \subseteq A$ (i) $A \subset \varnothing$
2. (a) Give three examples, other than those in the text, specifying a set by the set-builder notation.

(b) Give three other examples specifying a set by tabulation.
(c) Give three examples of infinite sets.

3. Indicate which of the following sets are empty sets.
   (a) All men who can run a mile in three seconds
   (b) The counting numbers greater than 7 and less than 8
   (c) All cats that can fly
   (d) The set of vowels in the English language
   (e) The set of even counting numbers not greater than 2
   (f) The set of women presidents of the United States
4. Indicate which of the following sets are infinite sets.
   (a) The even counting numbers
   (b) The citizens of Alabama
   (c) All fractions with denominators of 3
   (d) The male citizens of India
   (e) The cities of the world
5. Write $\subset$, $\not\subset$, $\notin$, or $\in$ between the following pairs to give a correct statement.

(a)	$\{2, 4, 6, 8\}$	$\{1, 2, 3, 4, 5, 6, 7, 8, 9, 10\}$
(b)	x	$\{x, y, z\}$
(c)	$\{a, o\}$	$\{a, e, i, o, u\}$
(d)	$\{\triangle, \star, \bigcirc, \square\}$	$\{\square, \star, \triangle, \bigcirc\}$
(e)	h	$\{l, m, n\}$
(f)	$\{y\}$	$\{x, y, z\}$

6. Let $A = \{1, 2, 3, 4\}$, $B = \{1, 2\}$, $C = \{3, 4\}$, and $D = \{3, 4, 5\}$. Classify each as true or as false.

(a) $B \subset A$	(b) $C \subseteq D$	(c) $D \subset \varnothing$
(d) $D \subset A$	(e) $\varnothing \subseteq A$	(f) $\{3\} \subset C$
(g) $3 \in D$	(h) $\{3\} \subseteq D$	(i) $\varnothing \subseteq C$

7. For each set in the left column, choose the sets from the right column that are subsets of it.

(a) $\{a, b, c, d\}$	(a) $\{\ \}$
(b) $\{o, p, k\}$	(b) $\{1, 4, 8, 9\}$
(c) Set of letters in the word "book"	(c) $\{1/10\}$
(d) $\{2, 4, 6, 8, 10, 12\}$	(d) $\{o, k\}$
(e) $\{1/4, 1/6, 1/8, 1/10\}$	(e) $\{12\}$
(f) Set of counting numbers less than 10	(f) $\{b\}$

8. We are given sets A, B, and C as follows:

$$A = \{x \mid x \text{ is a counting number between 1 and 7}\}$$
$$B = \{1, 2, 3, 4, 5, 6\}$$
$$C = \{2, 3, 4, 5, 6, 7\}.$$

Which of the following sets are equal to either A, B, or C?

(a) $\{2, 4, 5, 3, 6\}$ (b) $\{5, 6, 7, 2, 3, 4\}$
(c) $\{1, 3, 5, 2, 3, 6\}$ (d) $\{2, 4, 6, 3, 5, 7\}$
(e) $\{6, 5, 4, 3, 2\}$ (f) $\{4, 5, 6, 2, 3, 7\}$

9. Verbally describe sets A, B, C, D, and E that satisfy the following conditions.
(a) A is a proper subset of B (b) $C = D$ (c) $E = \{\ \}$

*10. (a) Is there any difference between $\{\ \}$ and $\varnothing$?
(b) Is $\varnothing \in \{\ \}$?
(c) Is $0 \in \{\ \}$?
(d) If A is any set, is $A \subset \varnothing$?
(e) Is $\{0\}$ the same as $\varnothing$?
(f) Is $\varnothing \subset \{\ \}$?

*11. The set of odd counting numbers consists of $\{1, 3, 5, \ldots\}$, while $\{2, 4, 6, \ldots\}$ is the set of even counting numbers. For each set given below, define the number of subsets of each set.
(a) $\{x \mid x$ is an even counting number less than $8\}$
(b) $\{2x + 1 \mid x$ is an odd counting number greater than 1 and less than $10\}$

*12. In each of the following, state whether or not the sets C and D are equal. Explain your answers.
(a) $C = \{2n + 1 \mid n$ is an odd counting number$\}$,
$D = \{2n - 1 \mid n$ is an even counting number$\}$
(b) $C = \{4n + 1 \mid n$ is a counting number$\}$,
$D = \{2n + 1 \mid n$ is an odd counting number greater than $1\}$

*13. In each of the following, give two examples of different sets that might be described by the bracketed members.
(a) $\{2, 4, \ldots, 64\}$
(b) $\{3, 5, 7, \ldots, 19\}$ (*Hint:* Find out what a prime number is—see page 4.)
(c) $\{A, B, \ldots, F\}$
(d) $\{B, C, D, \ldots, 0\}$
(e) $\{1, 2, 3, \ldots\}$

2

The Intersection of Sets

A very useful device for visualizing and discussing sets is the Venn diagram, named for the English logician John Venn and introduced in the preceding chapter. By using such diagrams, one may represent the members of a set under discussion by a set of points within a region formed by a closed curve. When representing set A by a circular region, the statement that $x \in A$ means that x is represented by a point within the circle, as illustrated by Figure 2-1.

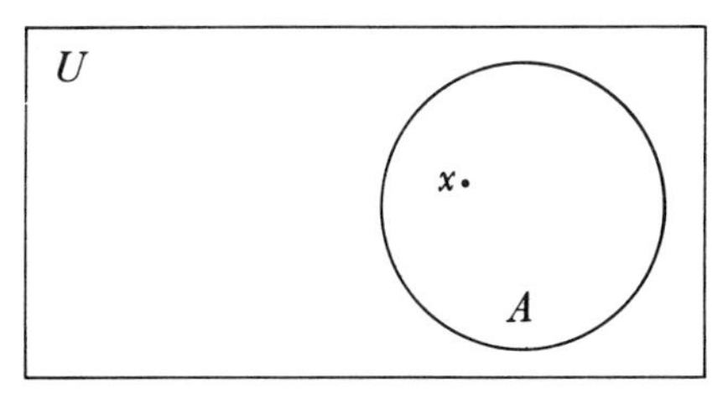

Figure 2-1

If one has a fixed set of objects to which the discussion will be limited, and if all the sets to be discussed are subsets of this class, then one may take this overall class to be the *universal set*, or simply *the universe*. We will learn as we continue that the counting numbers, the odd numbers, the prime numbers, the negative integers, and the even numbers are all subsets of the universal set of integers. Generally, in mathematics, one knows in advance the universal set to which a discussion belongs. This set may vary from situation to situation, but it is fixed for the duration of a particular discussion.

Many mathematicians represent the universal set as points within a rectangle and the set under consideration as the interior of a circle within the rectangle. For example, for the Venn diagram in Figure 2-1, let the interior of the rectangle represent a universe consisting of all students at Diploma College and let the interior of the circle represent freshmen at Diploma College; x then is a freshman student at Diploma College. Venn diagrams are also used to give a representation of special relations that occur among sets. The complement of set A, denoted by $\overline{A}$, is one such relation. In Figure 2-1, the region outside of the circle and inside the rectangle represents what is called the *complement* of set A. This set concept is defined in the following manner.

Definition 2-4: The *complement* of set A is the set of elements in the universe that are not in set A.

If A is a subset of the universe U, then the complement of A can be written in set-builder notation as $\overline{A} = \{x \mid x \in U \text{ and } x \notin A\}$.

If the universe consists of all counting numbers less than 10 and $A = \{1, 2, \ldots, 7\}$, then $\overline{A}$ is the set $\{8, 9\}$. If the universe is "all college students," and if A is the set of college students who have made all A's, then "all college students who have made at least one grade lower than an A" is the complement of A.

One may also discuss the complement of set A relative to set B. The *complement of A relative to B* is the set of all elements in B that are not in A. This set is denoted by $B - A$. Thus,

$$B - A = \{x \mid x \in B \text{ and } x \notin A\}.$$

Example: Let A represent all cities with either air pollution, water pollution, or industrial odors. Let B represent all cities with either air pollution, water pollution, or traffic problems. Then $B - A$ consists of all cities with traffic problems.

Sometimes, two sets have elements in common. The shaded region of Figure 2-2(a) represents the elements that are common to two sets, each of which is represented by the interior of a circle.

Definition 2-5: The *intersection* of two sets A and B is the set of all elements common to both A and B and is denoted by $A \cap B$.

Symbolically, $A \cap B = \{x \mid x \in A \text{ and } x \in B\}$, and $A \cap B$ is read "A intersection B."

Definition 2-6: Two sets A and B are said to be *disjoint* if and only if $A \cap B = \varnothing$.

That is, two sets are disjoint if they have no common elements, as illustrated in Figure 2-2(b).

The four diagrams in Figure 2-2 represent $A \cap B$ under different conditions. If A and B overlap or have elements in common, then the intersection is shown as the shaded area in (a). If A and B are disjoint, or have no elements in common, then the intersection, $A \cap B$, is the null set as shown by the lack of shading in (b). If $A \subset B$, A is a proper subset of B, then the elements common to A and B are the elements in set A, so $A \cap B = A$, as shown in (c). In a similar manner, if $B \subset A$, then $A \cap B = B$, as demonstrated by (d). If, by chance, A and B are equal, then $A \cap B = A = B$.

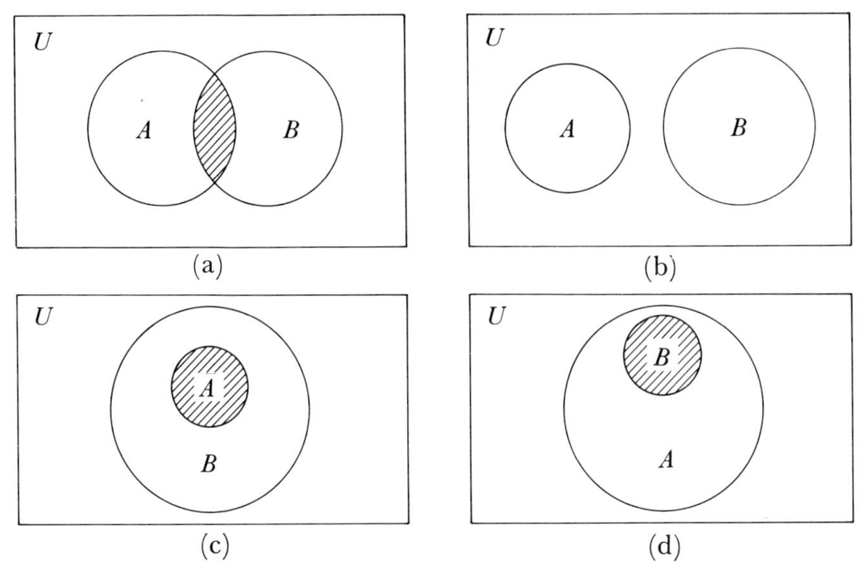

Figure 2-2

The following three examples illustrate the intersection of sets.

Example: If $A=\{1, 2, 3, \ldots, 100\}$ and $B=\{60, 61, \ldots, 1000\}$, then

$$A \cap B=\{60, 61, \ldots, 100\}.$$

Example: If $A=\{1, 2, \ldots, 10\}$ and $B=\{2, 3, 4\}$, then $A \cap B=B=\{2, 3, 4\}$, as in the shaded region of Figure 2-2(d). Note that the third set obtained from the intersection of two sets is not necessarily distinct from the other two.

Example: If $A=\{1, 2\}$ and $B=\{1, 2\}$, then $A \cap B=\{1, 2\}$.

Notice that in the definition of intersection, we start with two sets and perform on them some *operation* to get a third set. This operation of "forming the intersection" defines the process by which the third set is obtained.

Since $\cap$ is an operation, we will discuss for it the usual properties of operations. First of all, sets are *closed* relative to the operation $\cap$. That is, $\cap$ has the property that when it is applied to two sets, it produces a set.

The definition of the intersection of two sets implies that the order in which the sets are considered is immaterial in forming an intersection. Thus, the statement $A \cap B = B \cap A$ is true for any two sets A and B. When the order in which we take two things in performing an operation has no effect on the result, the operation is said to be *commutative.*

Theorem 2-1: *The Commutative Property of Intersection.* The process of forming the intersection of two sets is a commutative operation: $A \cap B = B \cap A$.

Example: Let $A = \{1, 2, \ldots, 6\}$ and $B = \{4, 5, \ldots, 10\}$; then

$$A \cap B = \{1, 2, \ldots, 6\} \cap \{4, 5, \ldots, 10\} = \{4, 5, 6\}.$$

Similarly, $B \cap A = \{4, 5, \ldots, 10\} \cap \{1, 2, \ldots, 6\} = \{4, 5, 6\}$.

The truth table introduced in Chapter 1 is most useful in illustrating intersection. $x \in A$ is either true or false and $x \in B$ is either true or false. Thus, there are four combinations of truth values.

$x \in A$	$x \in B$	$x \in (A \cap B)$
T	T	T
T	F	F
F	T	F
F	F	F

(a)

$x \in A$	$x \in B$	$x \in (B \cap A)$
T	T	T
T	F	F
F	T	F
F	F	F

(b)

Table 2-1

To prove Theorem 2-1, we examine the truth values of Table 2-1. Use the definition of intersection to verify the third column of Table 2-1(a) and Table 2-1(b). Note that the third columns are identical. Thus, in every possible situation, $A \cap B = B \cap A$.

The discussion above defined the operation of forming the intersection of two sets only, so we must define what we mean by the intersection of three sets. Let us agree to the following: for any sets A, B, and C, $A \cap B \cap C$ denotes the set $(A \cap B) \cap C$. Thus, we consider $A \cap B \cap C$ as the set formed by first forming the intersection of A and B (indicated by the parentheses) and then forming the intersection of this set with C.

For example, if $A = \{1, 3\}$, $B = \{3, 5, 7\}$, and $C = \{3, 7, 11\}$, then $A \cap B = \{3\}$. Thus,

$$A \cap B \cap C = (A \cap B) \cap C = \{3\} \cap \{3, 7, 11\} = \{3\}.$$

$A \cap B \cap C$ could also be defined as $A \cap (B \cap C)$. For this case, the parentheses indicate that $B \cap C$ is formed first and then the intersection of the set with A is formed. For the preceding example,

$$A \cap (B \cap C) = \{1, 3\} \cap (\{3, 5, 7\} \cap \{3, 7, 11\}) = \{1, 3\} \cap \{3, 7\} = \{3\}.$$

For this example, it is seen that $A \cap (B \cap C) = (A \cap B) \cap C$.

In Figure 2-3, (a) and (b) represent the formation of $(A \cap B) \cap C$. Similarly, in Figure 2-4, (a) and (b) represent the formation of $A \cap (B \cap C)$. Thus, you can see that the shaded region representing

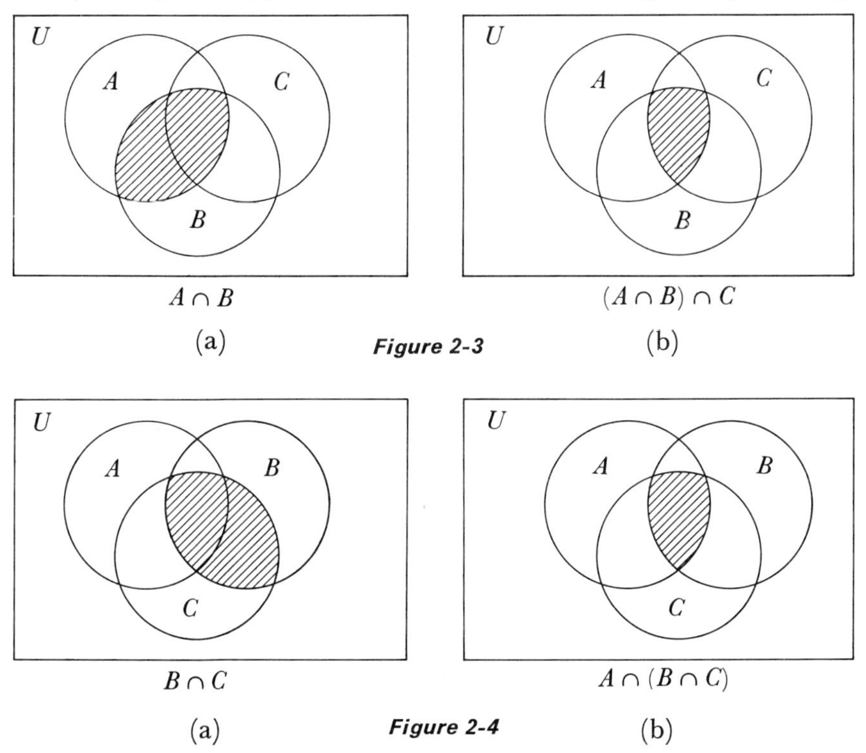

$A \cap B$ (a) **Figure 2-3** $(A \cap B) \cap C$ (b)

$B \cap C$ (a) **Figure 2-4** $A \cap (B \cap C)$ (b)

$(A \cap B) \cap C$ is the same as the region representing $A \cap (B \cap C)$. Although this reasoning does not constitute a proof, it does make plausible the following property.

Theorem 2-2: *The Associative Property of Intersection.* If A, B, and C are any three sets, then $A \cap (B \cap C) = (A \cap B) \cap C$.

This theorem will be proved by truth tables in Exercise 13.

Example: Let

$$A = \{a, b, c, d, e, f\},\ B = \{d, e, f, g, h, i\}, \text{ and } C = \{b, d, f, h, m, n\}.$$

Then

$A \cap B = \{d, e, f\}$ and $(A \cap B) \cap C = \{d, e, f\} \cap \{b, d, f, h, m, n\} = \{d, f\}$.
$B \cap C = \{d, f, h\}$ and $A \cap (B \cap C) = \{a, b, c, d, e, f\} \cap \{d, f, h\} = \{d, f\}$.

Thus, $(A \cap B) \cap C = A \cap (B \cap C) = \{d, f\}$.

Example: If the universal set U is the set $\{a, b, c, d, e, f\}$ and sets A, B, and C are defined as $A = \{a, b, c\}$, $B = \{a, c, e, f\}$, and $C = \{d, e, f\}$, find $(\overline{A \cap B}) \cap \overline{C}$.

Now, $\overline{C} = \{a, b, c\}$ and $A \cap B = \{a, c\}$. Consequently, $\overline{A \cap B} = \{b, d, e, f\}$. Thus $(\overline{A \cap B}) \cap \overline{C} = \{b\}$.

Example: Draw a Venn diagram representing $X \cap (\overline{Y \cap Z})$.
(a) Shade $Y \cap Z$. (b) Shade $\overline{Y \cap Z}$. (c) Shade $X \cap (\overline{Y \cap Z})$.

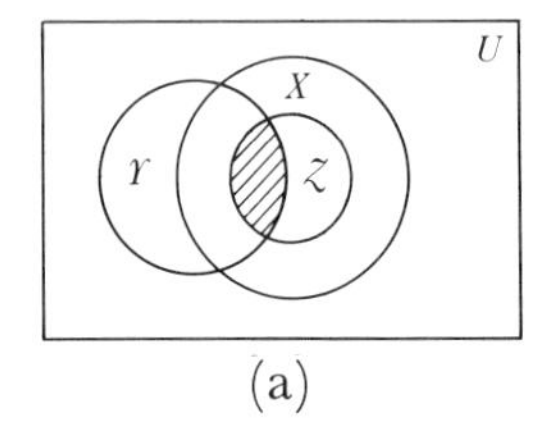

(a)

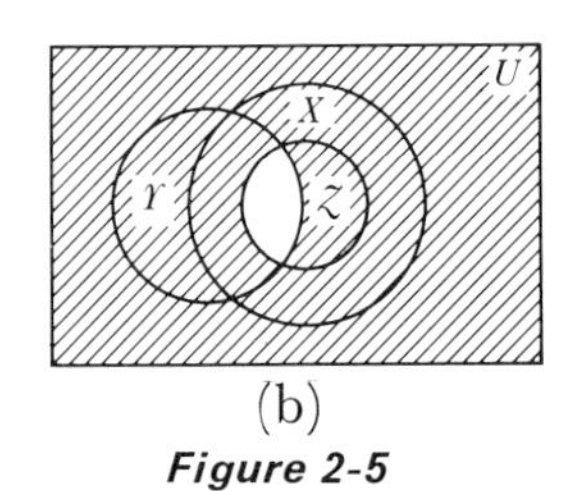

(b)

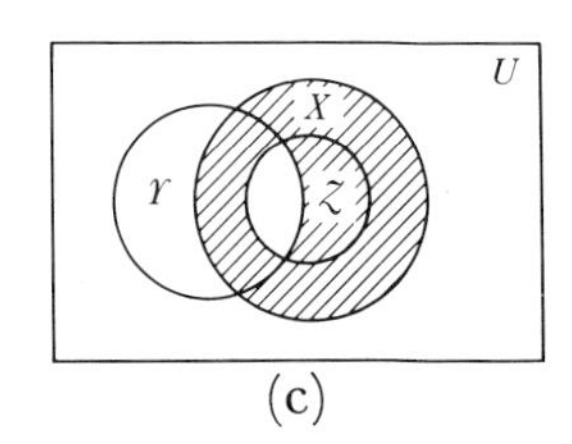

(c)

Figure 2-5

Exercise Set 2-2

1. Form the intersection of the following pairs of sets.
   (a) $R = \{5, 10, 15\}$ $\quad T = \{15, 20\}$
   (b) $L = \{a, e, m, n\}$ $\quad S = \{m, e, a, n\}$
   (c) $A = \{0, 10, 100, 1000\}$ $\quad B = \{10, 100\}$

(d) $G = \{\text{odd counting numbers less than } 100\}$
$H = \{\text{even counting numbers between 1 and } 31\}$
(e) $A = \{x, y, z, t\}$ $B = \{x, y, s, r\}$

2. Let $A = \{1, 2, 3, 4, 5\}$, $B = \{4, 5, 6, 7\}$, $C = \{2, 3, 4, 5, 6\}$.
   (a) Show that $(A \cap B) \cap C = A \cap (B \cap C)$.
   (b) Illustrate with Venn diagrams.
3. Shade that portion of the diagram that will illustrate each of the following sets.

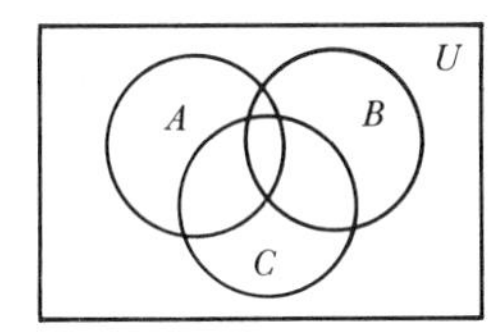

   (a) $\overline{A} \cap B$ (b) $\overline{A} \cap C$
   (c) $\overline{A} \cap (B \cap C)$ (d) $\overline{(A \cap B)} \cap C$
4. If the universal set is the set $\{1, 2, 3, \ldots, 12\}$ of the first dozen counting numbers and sets A, B, C are defined as $A = \{1, 3, 5, 7\}$, $B = \{2, 7, 10\}$, and $C = \{1, 3, 8\}$, find the following.
   (a) $A \cap B$ (b) $\overline{B \cap C}$
   (c) $\overline{A}$ (d) $A \cap C$
   (e) $\overline{C}$ (f) $A \cap (B \cap C)$
   (g) $C \cap (A \cap B)$ (h) $\overline{(B \cap C) \cap A}$
5. If the universal set consists of all automobiles, if C is the set of all red convertibles, and if B is the set of all automobiles built in 1973, describe in words the sets specified by each of the following.
   (a) $\overline{C}$ (b) $C \cap B$ (c) $\overline{C} \cap B$
   (d) $\overline{C} \cap \overline{B}$ (e) $C \cap \overline{B}$ (f) $\overline{C \cap B}$
6. Draw Venn diagrams such that sets A, B, and C satisfy the following conditions.
   (a) $A \subset B, C \cap B \neq \varnothing, A \cap C = \varnothing$
   (b) $A \cap B \neq \varnothing, C \subset (A \cap B)$
   (c) $B \cap C = \varnothing, \overline{A} \cap C = C, A \cap B \neq \varnothing$
   (d) $B \subset A, C \subset A, B \cap C \neq \varnothing$
7. Given U is the universe of all drivers of automobiles; let $A = \{\text{all female drivers under } 18\}$ and $B = \{\text{all male drivers under } 21\}$. Describe each of the following sets.
   (a) $(A \cap B) \cap \varnothing$ (b) $\overline{A} \cap \varnothing$ (c) $(A \cap B) \cap U$
   (d) $A \cap U$ (e) $\overline{A} \cap U$ (f) $\overline{(A \cap B) \cap \varnothing}$
8. Let A and B be subsets of a universal set U. Under what conditions

on the sets A and B will the following statements be true?

(a) $A \cap B = \varnothing$ (b) $B \cap A = A$ (c) $A \cap B = B$

(d) $(A \cap B) \subseteq A$ (e) $B \cap \overline{B} = \varnothing$ (f) $A \cap \varnothing = \varnothing$

(g) $A \cap U = \varnothing$ (h) $B \cap \overline{A} = U$

9. If $A \subseteq B$, describe

(a) $A \cap B$; (b) $A \cap \overline{B}$.

*10. Describe the shaded portion of each of the following Venn diagrams.

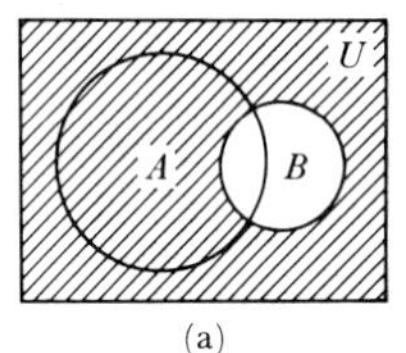

(a)

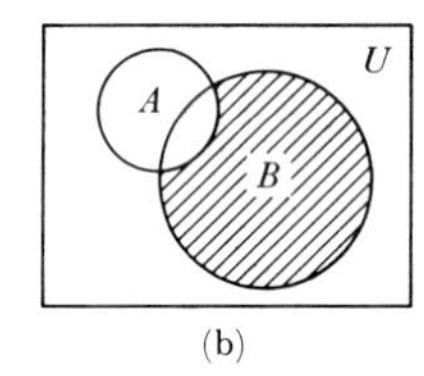

(b)

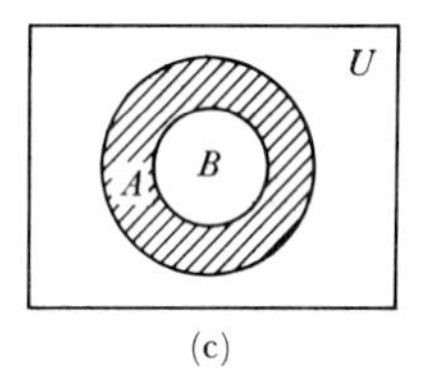

(c)

*11. An investigator found in a poll of 100 people that 40 used only Brand A and 25 used both Brand A and Brand B. Everyone contacted used either Brand A or Brand B. But to his sorrow, the poor investigator lost his tabulation of those who used only Brand B. Can you help this investigator find his answer?

*12. Use a truth table to prove that $\left(\overline{\overline{A}}\right) = A$.

*13. Prove the associative property of intersection, $(A \cap B) \cap C = A \cap (B \cap C)$, by means of truth tables using the eight possibilities for the truth values of A, B, and C.

3

The Union of Sets

In the preceding section, we learned how a new set could be formed from two sets by the operation of "forming the intersection." In this section, we form a third set from two given sets by an operation called *union*.

Definition 2-7: If A and B are any two sets, the *union* of A and B, denoted by $A \cup B$, is the set consisting of all the elements in set A or in set B or in both A and B. Symbolically,

$$A \cup B = \{x \mid x \in A \text{ or } x \in B\}.$$

Thus, we have a new operation on sets. If A and B are any two sets, we can form a new set $A \cup B$, which is read "A union B." The three Venn diagrams in Figure 2-6 illustrate the union of two sets A and B under different conditions.

The union of A and B, if A and B overlap or have elements in common, is shown by the shaded area in (a). Recall that $A \cup B$ is the set of all elements which belong to A or B or to both A and B. It should be noted that if there are elements common to both sets, they are listed only once in the union of the set. Given

$$A = \{a, b, c, d, e\} \quad \text{and} \quad B = \{c, d, e, f, g\},$$

then

$$A \cup B = \{a, b, c, d, e, f, g\}.$$

The c, d, and e, which are common to A and B, are listed only once.

In Figure 2-6(b), A and B are disjoint, and the union of A and B is the set represented by the shaded area which consists of the interior of both circles. In (c), $B \subset A$, or B is a proper subset of A. In this example, the union of A and B is the interior of circle A.

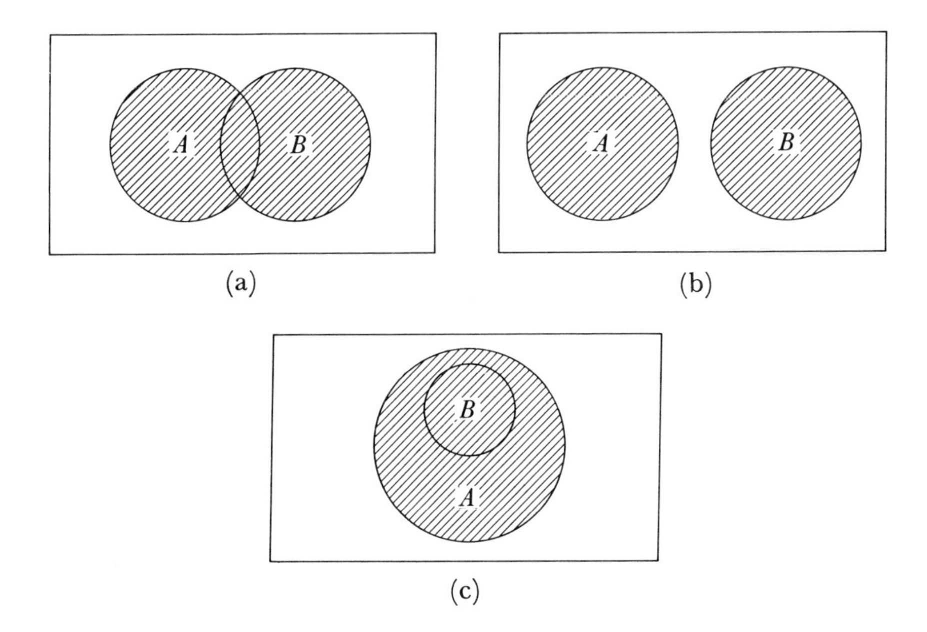

Figure 2-6

The following six examples illustrate the definition of the union of two sets.

Example: Let $A = \{\text{Patty, Gail, Shirley}\}$ and $B = \{\text{Doris, Gail}\}$; then

$$A \cup B = \{\text{Patty, Gail, Shirley, Doris}\}$$

because each of the four girls belongs either to A or to B or to both.

Example: The union of $A = \{4, 5, \ldots, 11\}$ and $B = \{1, 2, \ldots, 9\}$ is the set $\{1, 2, \ldots, 11\}$.

Example: For the disjoint sets $A = \{\text{Birmingham, Atlanta, Nashville}\}$ and $B = \{\text{New York, Chicago}\}$, the union of A and B is $\{\text{Birmingham, Atlanta, Nashville, New York, Chicago}\}$.

Example: If $A = \{1, 2, \ldots, 10\}$ and $B = \{2, 3, 4\}$, then

$$A \cup B = A = \{1, 2, \ldots, 10\}.$$

Example: If $A = \{1, 2\}$, then $A \cup A = \{1, 2\}$. The fact that $A \cup A = A$ for any set A is called the *idempotent* property for union.

Example: If $A = \{1, 2, 3, 4\}$ and $B = \varnothing$, then $A \cup B = A$.

In the preceding section, forming the intersection was considered an operation, and certain properties for operations were discussed. Intuitively, it seems reasonable to think that the union could have the same basic properties as the intersection. Suppose we examine this conjecture.

The definition of union implies that this operation has the closure property. This means that the union of any two sets is a set. That is, if A and B are any two sets, then $A \cup B$ is a set.

Earlier, we noted that the intersection of sets is both commutative and associative. These properties also hold for the union of sets.

Theorem 2-3: *The Commutative Property of Union.* If A and B are any two sets, then $A \cup B = B \cup A$.

Let us consider the following example:

Example: Consider $A = \{1, 3\}$ and $B = \{3, 5, 7\}$. Then

$$A \cup B = \{1, 3, 5, 7\} \text{ and } B \cup A = \{1, 3, 5, 7\},$$

so $A \cup B = B \cup A$.

We will use a truth table to prove that $A \cup B = B \cup A$. Again, in the first two columns of the table are listed all possibilities for an element x. Since the third and fourth columns are identical in Table 2-2, $A \cup B = B \cup A$.

$x \in A$	$x \in B$	$x \in (A \cup B)$	$x \in (B \cup A)$
T	T	T	T
T	F	T	T
F	T	T	T
F	F	F	F

Table 2-2

Although the process of forming the union of two sets, like that of forming their intersection, is a binary operation, we can extend it to more than two sets by means of the following agreement. For sets A, B, and C, let $A \cup B \cup C$ denote the set $(A \cup B) \cup C$. Again, the parentheses indicate that $A \cup B$ is formed first. In a like manner, we could consider $A \cup B \cup C$ to be $A \cup (B \cup C)$.

Example: Draw Venn diagrams such that $A \cap B \neq \varnothing$, $B \cap C \neq \varnothing$, but $A \cap C = \varnothing$; by shading illustrate that $(A \cup B) \cup C = A \cup (B \cup C)$.

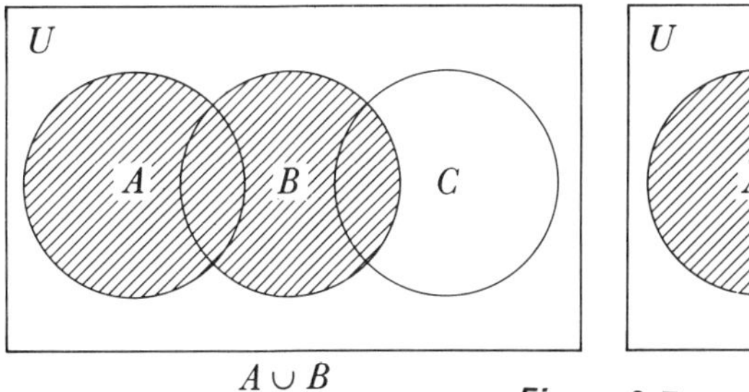

$A \cup B$

U A B C

$(A \cup B) \cup C$

Figure 2-7

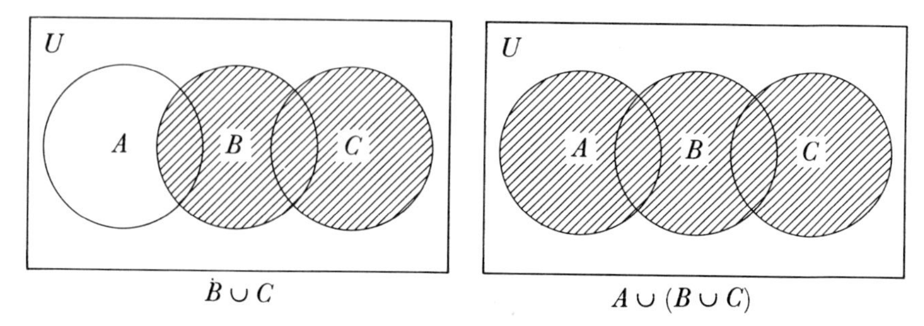

Figure 2-8

Theorem 2-4: *The Associative Property of Union.* If A, B, and C are any three sets, then $A \cup (B \cup C) = (A \cup B) \cup C$.

Example: Suppose $A = \{1, 3\}$, $B = \{3, 5, 7\}$, and $C = \{3, 7, 11\}$. Then

$$A \cup B = \{1, 3, 5, 7\} \text{ and } (A \cup B) \cup C = \{1, 3, 5, 7, 11\}.$$

$$B \cup C = \{3, 5, 7, 11\} \text{ and } A \cup (B \cup C) = \{1, 3, 5, 7, 11\}.$$

Thus, $(A \cup B) \cup C = A \cup (B \cup C)$.

Theorem 2-4 will be proved in Exercise 14.

The commutative and associative properties of the union of three sets imply that, to find the union of sets A, B, and C, one can group the sets in twelve different ways, all of which represent the same set. Four arrangements are $A \cup (B \cup C)$, $A \cup (C \cup B)$, $C \cup (A \cup B)$, and $(C \cup A) \cup B$. Can you list the others?

The operations of union and intersection have another interesting and important property. For each of these operations, there is a set called the *identity*. If the identity set and any set A are involved in either union or intersection, the set A is produced by the operation.

What is the identity set for the operation of union? The null set, which contains no elements, is the identity for union because the union of any set A with the null set would give set A. That is, $A \cup \varnothing = \varnothing \cup A = A$.

Similarly, the operation of intersection has an identity set. The universal set U contains every element under consideration. Thus, it contains all the elements of A. Consequently, $U \cap A = A \cap U = A$.

Theorem 2-5: *Identities for Union and Intersection.*

(a) The set $\varnothing$ is an identity for the operation of finding the union of sets. For any set A, $\varnothing \cup A = A \cup \varnothing = A$.

(b) The universal set U is an identity for the operation of finding the intersection of two sets. For any set A, $U \cap A = A \cap U = A$.

Seventeen interesting properties of sets are given at the end of this section. All these properties can be geometrically verified by means of Venn diagrams. Note that such a check does not constitute a proof, but it does provide intuitive satisfaction.

In Figure 2-9(a), $B \cup C$ is shaded with horizontal lines, and the intersection of $B \cup C$ with A is indicated by vertical lines to give $A \cap (B \cup C)$. In a like manner, $A \cap B$ is indicated in (b) by horizontal lines and $A \cap C$ by vertical lines. The union of $A \cap B$ and $A \cap C$ is the shaded portion of the diagram in Figure 2-9(b). The double shading in Figure 2-9(a) covers the same region as the shading on Figure 2-9(b). Thus, we have verified geometrically the distributive property of intersection over union; that is,

$$A \cap (B \cup C) = (A \cap B) \cup (A \cap C).$$

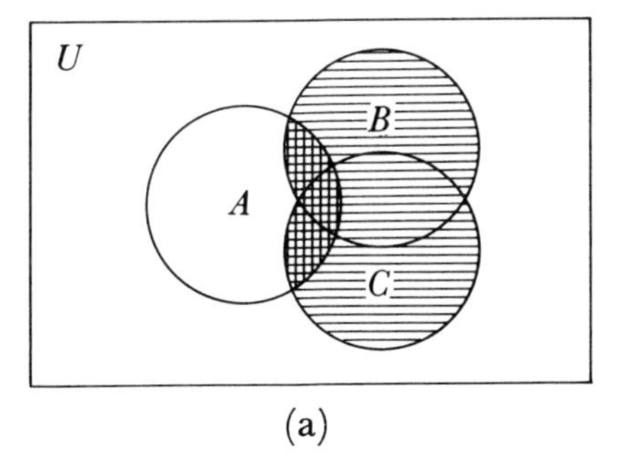

(a)

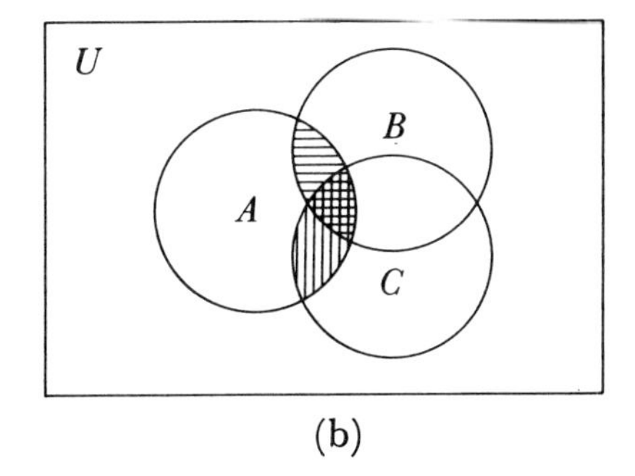

(b)

Figure 2-9

The distributive property of intersection over union may be proved using the truth table in Table 2-3.

$x \in A$	$x \in B$	$x \in C$	$x \in (B \cup C)$	$x \in [A \cap (B \cup C)]$	$x \in (A \cap B)$	$x \in (A \cap C)$	$x \in [(A \cap B) \cup (A \cap C)]$
T	T	T	T	T	T	T	T
T	T	F	T	T	T	F	T
T	F	T	T	T	F	T	T
T	F	F	F	F	F	F	F
F	T	T	T	F	F	F	F
F	T	F	T	F	F	F	F
F	F	T	T	F	F	F	F
F	F	F	F	F	F	F	F

Table 2-3

Since the fifth and eighth columns are identical, then $A \cap (B \cup C) = (A \cap B) \cup (A \cap C)$. The distributive property of union over intersection also holds true. See Exercise 15.

In Figure 2-10(a), $A \cup B$ is everything not shaded, and $\overline{A \cup B}$ is everything that is shaded.

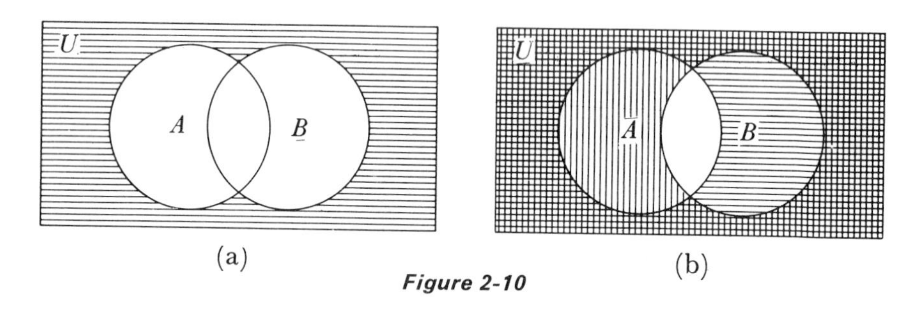

(a) (b)

Figure 2-10

In Figure 2-10(b), $\overline{A}$ is shaded with horizontal lines and $\overline{B}$ with vertical lines. Thus, $\overline{A} \cap \overline{B}$ is everything with the double shading. Note that the double-shaded region in (b) is the same as the shaded region in (a). Thus, by Venn diagrams, we have verified what is called *De Morgan's Law*, the rule that $\overline{A \cup B} = \overline{A} \cap \overline{B}$. Similarly, $\overline{A \cap B} = \overline{A} \cup \overline{B}$.

The basic properties of sets may be listed as follows. If A, B, and C are any sets, then

(a) $A \cap B = B \cap A$ — Commutativity for intersection

(b)	$A \cup B = B \cup A$	Commutativity for union
(c)	$A \cap (B \cap C) = (A \cap B) \cap C$	Associativity for intersection
(d)	$A \cup (B \cup C) = (A \cup B) \cup C$	Associativity for union
(e)	$A \cap (B \cup C) = (A \cap B) \cup (A \cap C)$	Distributivity of intersection over union
(f)	$A \cup (B \cap C) = (A \cup B) \cap (A \cup C)$	Distributivity of union over intersection
(g)	$A \cup \overline{A} = U$	Complementation of union
(h)	$A \cap \overline{A} = \varnothing$	Complementation of intersection
(i)	$A \cup A = A$	Idempotent property for union
(j)	$A \cap A = A$	Idempotent property for intersection
(k)	$\overline{A \cup B} = \overline{A} \cap \overline{B}$	De Morgan's law
(l)	$\overline{A \cap B} = \overline{A} \cup \overline{B}$	De Morgan's law
(m)	$A \cup \varnothing = A$	Identity for union
(n)	$A \cap \varnothing = \varnothing$	Property of empty set
(o)	$\overline{\overline{A}} = A$	Complement of a complement
(p)	$A \cap U = A$	Identity for intersection
(q)	$A \cup U = U$	Property of the universal set

Exercise Set 2-3

1. Find the union of the following pairs of sets.
   (a) $R = \{5, 10, 15\}$ $\quad T = \{5, 20\}$
   (b) $W = \{x \mid x \text{ is a counting number larger than } 5\}$
   $Z = \{x \mid x \text{ is a counting number between 3 and } 10\}$
   (c) $G = \{\text{odd counting numbers less than } 100\}$
   $H = \{\text{even counting numbers less than } 100\}$
   (d) $A = \{x, y, z, t\}$ $\quad B = \{x, y, r, s\}$
2. If the universal set S is the set $\{1, 2, \ldots, 12\}$ and set $A = \{1, 3, 5, 7\}$, $B = \{2, 7, 10\}$, and $C = \{1, 2, 3, 5\}$, find
   (a) $A \cup \overline{B}$ (b) $\overline{A} \cup \overline{B}$ (c) $C \cup (A \cup B)$
   (d) $\overline{B \cup C}$ (e) $\overline{A \cup (B \cup C)}$ (f) $\overline{A} \cup (B \cup \overline{C})$

3. Shade the following using Venn diagrams for sets A, B, and C such that $A \cap B = \varnothing$, $A \cap C \neq \varnothing$, and $B \cap C \neq \varnothing$, where the symbol $\neq$ is read "unequal to."

   (a) $A \cup \overline{B}$ (b) $\overline{A} \cup (\overline{B} \cup C)$

   (c) $\overline{A} \cap (\overline{B} \cup C)$ (d) $A \cup \overline{(B \cup C)}$

4. Given $U = \{a, b, c, d, e, f, g\}$, $X = \{a, b, d, c\}$, and $Y = \{c, d, e, f\}$. Enumerate the elements of the following sets.

   (a) $X \cup Y$ (b) $\overline{X}$ (c) $\overline{X \cup Y}$

   (d) $X \cap Y$ (e) $\overline{Y}$ (f) $\overline{X \cap Y}$

5. Given that $M = \{v, w, x, s\}$, $N = \{v, w, x, y\}$, $P = \{r, w, x\}$. Show that each of the following is true.

   (a) $(M \cup P) \cup N = M \cup (P \cup N)$

   (b) $(M \cap P) \cap N = M \cap (P \cap N)$

   (c) $M \cup (P \cap N) = (M \cup P) \cap (M \cup N)$

   (d) $M \cap (P \cup N) = (M \cap P) \cup (M \cap N)$

6. Illustrate the results of Exercise 5 with Venn diagrams.

7. Draw a Venn diagram representing the relationships between the sets $U = \{x \mid x$ is a counting number less than 25$\}$; $A = \{x \mid x$ is the set of counting numbers larger than 6 and less than 20$\}$; $B = \{x \mid x$ is a counting number less than 25 and greater than 19$\}$; and $C = \{x \mid x$ is a counting number less than 8$\}$. On this diagram, first shade each of the following and then indicate the members of that set.

   (a) $A \cup B$ (b) $A \cup (B \cup C)$ (c) $\overline{(A \cup B)} \cup C$

   (d) $\overline{A} \cup C$ (e) $A \cap (B \cup C)$ (f) $A \cup \overline{(B \cap C)}$

8. If set M represents the set of all girls with dates on this Saturday night and N represents the set of all girls with dates last Saturday night, then describe in words each of the following sets.

   (a) $M \cup N$ (b) $\overline{M} \cup N$ (c) $M \cup \overline{N}$ (d) $\overline{M \cup N}$

   (e) $\overline{M} \cup \overline{N}$

9. Let $U = \{1, 2, 3, \ldots, 10\}$, $A = \{1, 2, 3, 4, 5\}$, $B = \{1, 3, 7, 9\}$, and $C = \{1, 2, 9, 10\}$. Verify each of the following.

   (a) $A \cap \overline{(B \cap C)} = (A \cap \overline{B}) \cup (A \cap \overline{C})$

   (b) $A \cup \overline{(B \cup C)} = (A \cup \overline{B}) \cap (A \cup \overline{C})$

   (c) $\overline{A \cup B} = \overline{A} \cap \overline{B}$ (d) $\overline{A \cap B} = \overline{A} \cup \overline{B}$

10. Verify with Venn diagrams the following. Use A, B, and C as given in Exercise 3.

   (a) $A \cap \overline{(B \cap C)} = (A \cap \overline{B}) \cup (A \cap \overline{C})$

   (b) $A \cup \overline{(B \cup C)} = (A \cup \overline{B}) \cap (A \cup \overline{C})$

   (c) $\overline{A \cup B} = \overline{A} \cap \overline{B}$ (d) $\overline{A \cap B} = \overline{A} \cup \overline{B}$

11. Indicate whether the following statements are true or false.
    (a) $\overline{A \cap B} = \overline{A} \cap \overline{B}$
    (b) If $A \cap B = A \cup B$, then $A \subseteq B$
    (c) If $A \cap B = A \cup B = \{\ \}$, then $A = B = \{\ \}$
    (d) If $A \subseteq B$ and $A \cap B = B$, then $A = B$
    (e) If $A \cap (B \cap C) = \varnothing$ and $B \cap C = \varnothing$, then $A \cap B = \{\ \}$
    (f) $A \cap (B \cap C) \equiv (A \cap \overline{B}) \cup (A \cap \overline{C})$
12. Give a verbal statement indicating the relationship between A and B or between A and U in each of the following.
    (a) $A \subseteq (A \cup B)$ (b) $A \cup B = \varnothing$
    (c) $A \cup B = A$ (d) $A \cup U = U$
*13. What is the simplest possible description of the following sets?
    (a) $(A \cap \overline{B}) \cup (\overline{A \cup B})$ (b) $[A \cup (\overline{B} \cap \overline{C})] \cap [A \cup (B \cap C)]$
*14. Using a truth table, prove Theorem 2-4.
*15. Using truth tables, prove
    (a) $\overline{A \cup B} = \overline{A} \cap \overline{B}$ (b) $\overline{A \cap B} = \overline{A} \cup \overline{B}$
    (c) $A \cup (B \cap C) = (A \cup B) \cap (A \cup C)$ (d) $A \cap \overline{A} = \varnothing$
*16. A pollster on a college campus interviewed the coeds as to the shampoo they used. Of the 7900 girls on campus, he reported that 4500 used Shiny Shampoo, 4850 used Dandruff Doo, and 4750 used Glow and Gleam. Out of these girls, 3000 used both Shiny Shampoo and Dandruff Doo, 2750 used Dandruff Doo and Glow and Gleam, 2250 used Shiny Shampoo and Glow and Gleam, and 1750 used all three. Was this poll accurate? Why or why not?
*17. Prove each of the following for all A and B where $B \subset A$.
    (a) $B \cup (A - B) = A$ (b) $B \cap (A - B) = \varnothing$
    (c) $A - B = A \cap \overline{B}$

4

The Cartesian Product

In this section, we define a new set called the cartesian product of two sets. This set is named for the philosopher and mathematician René Descartes (1596–1650), who invented the coordinate system for graphing in mathematics.

Forming the *cartesian product* of two sets is quite different from forming the union or intersection of two sets. The elements of the cartesian product are not elements from either set but are what we shall call *ordered pairs*.

First of all, we introduce the concept of an ordered pair as an outgrowth of everyday experience. On an ordinary road map, one is often directed to find a location by scanning the area designated by K and 4. This means you can locate your position by following the horizontal boundary until you locate section K; then follow the vertical boundary until you find section 4. Thus, the direction K and 4 locates a position on the map. This (K, 4) is an example of an ordered pair.

Definition 2-8: An *ordered pair* is an element (x, y) formed by taking x from a set and y from a set in such a way that x is designated as the "first" element and y as the "second" element. In formal language, the x is termed the *abscissa* of the ordered pair while the y is called the *ordinate.*

If x and y are any elements of any two sets (or the same set), one can put them together to form an ordered pair denoted by (x, y). This pair is ordered in the sense that (x, y) and (y, x) are not equal unless $x = y$. We observe that $(x, y) \neq (y, x)$ although $\{x, y\} = \{y, x\}$. We define equality for ordered pairs as follows:

$$(a, b) = (c, d) \text{ if and only if } a = c \text{ and } b = d.$$

Example: One can travel from Chicago to Miami by auto, plane, train, or bus, and from Miami to Nassau by plane or ship. The different ways a person can travel from Chicago to Nassau through Miami can be described in terms of ordered pairs such as

{(auto, plane), (auto, ship), (plane, plane),
(plane, ship), (train, plane), (train, ship),
(bus, plane), (bus, ship)}.

If A represents {auto, plane, train, bus} and B represents {train, plane}, then the set of ordered pairs representing the different means of travel is denoted by $A \times B$ and is called the *cartesian product* of the sets A and B.

Definition 2-9: The *cartesian product*, denoted by $A \times B$, of two sets A and B is the set of all ordered pairs (a, b) with the first element chosen from A and the second element chosen from B. Symbolically,

$$A \times B = \{(a, b) \mid a \in A \text{ and } b \in B\}.$$

For example, if $A = \{a, b, c\}$ and $B = \{d, e\}$, then

$$A \times B = \{(a, d), (a, e), (b, d), (b, e), (c, d), (c, e)\}.$$

Note that $(x, y) \in A \times B$ if and only if $x \in A$ and $y \in B$.

Example: Suppose you specify that the interior of your new car is to be one of the colors white, black, or yellow, denoted by $\{w, b, y\}$. The exterior is to be painted tangerine, chartreuse, baby blue, or scarlet, denoted by $\{t, c, bb, s\}$. Let the first and second elements of an ordered pair represent the exterior and interior colors, respectively, of your new car. The set of possible color combinations is given by

$$\begin{array}{lll} \{(t, w), & (t, b), & (t, y) \\ (c, w), & (c, b), & (c, y) \\ (bb, w), & (bb, b), & (bb, y) \\ (s, w), & (s, b), & (s, y)\}. \end{array}$$

Figure 2-11

This example suggests a grouping of the elements in a cartesian product in a systematic way that facilitates enumerating large and cumbersome Cartesian product sets. By grouping the elements so that the first element of an ordered pair is the same along any (horizontal) row and the second element is constant in a (vertical) column, you may obtain the arrangement in Figure 2-11. This notation allows some of the elements to be omitted and represented by dots with no ambiguity to the reader.

Note that the set of ordered pairs in a cartesian product includes every pair that can be constructed by choosing a left partner from the elements of A and a right partner from the elements of B. $A \times B$ is sometimes called the *cross product* of A and B and can be read "A cross B." The definition does not specify whether the sets used in forming the cross product have to be different or can be the same. Some examples will illustrate that they can be either.

Example: If $A = \{1, 2, 3\}$ and $B = \{4, 5, 6\}$, then

$$A \times B = \{(1, 4), (1, 5), (1, 6), (2, 4), (2, 5), (2, 6), (3, 4), (3, 5), (3, 6)\}.$$

Example: If $A = \{a, b\}$, form $A \times A$. By definition, the ordered pairs (x, y) must be such that $x \in A$ and $y \in A$. Thus,

$$A \times A = \{(a, a), (a, b), (b, a), (b, b)\}.$$

Consider the cartesian product of $A \times \varnothing$. Now, $A \times \varnothing$ consists of ordered pairs of elements where the first element is from A and the second element is from $\varnothing$. Since $\varnothing$ has no elements, it is impossible to form any ordered pairs of elements for $A \times \varnothing$. Thus,

$$A \times \varnothing = \varnothing.$$

In a like manner, it is easy to see that $\varnothing \times A = \varnothing$.

Example: If $S = \{1, 2, 3\}$, find $\varnothing \times S$. From the preceding discussion, $\varnothing \times S = \varnothing$.

We define $A \times (B \times C)$ to consist of ordered pairs of elements in which the first element is from A and the second element is from $B \times C$. Thus, in forming $A \times (B \times C)$, you first consider $B \times C$. Likewise $(A \times B) \times C$ is easily found after $A \times B$ has been formed.

Example: If $A = \{1, 2\}$, $B = \{4, 5\}$, and $C = \{6, 7, 8\}$, then we consider $A \times (B \times C)$ to be the following set of number pairs:

$$\{[1, (4, 6)], [1, (4, 7)], [1, (4, 8)], [1, (5, 6)], [1, (5, 7)], [1, (5, 8)], [2, (4, 6)], [2, (4, 7)], [2, (4, 8)], [2, (5, 6)], [2, (5, 7)], [2, (5, 8)]\}.$$

In a like manner,

$$(A \times B) \times C = \{[(1, 4), 6], [(1, 5), 6], [(2, 4), 6], [(2, 5), 6], [(1, 4), 7], [(1, 5), 7], [(2, 4), 7], [(2, 5), 7], [(1, 4), 8], [(1, 5), 8], [(2, 4), 8], [(2, 5), 8]\}.$$

Since $(A \times B) \times C \neq A \times (B \times C)$ for this example, the operation of cross product for sets is not associative. However, note that $(A \times B) \times C$ and $A \times (B \times C)$ have the same number of elements in this example and that each element of one set can be associated in a natural way with an element of the other.

Theorem 2-6: *The Distributive Property of Cartesian Product over Union.* If A, B, and C are any sets, then $(A \cup B) \times C = (A \times C) \cup (B \times C)$ and $C \times (A \cup B) = (C \times A) \cup (C \times B)$.

To demonstrate the truth of this theorem, consider $A = \{1, 2\}$, $B = \{2, 3\}$, and $C = \{1, 5\}$. Then

$$\begin{aligned}(A \cup B) \times C &= \{1, 2, 3\} \times \{1, 5\} \\ &= \{(1, 1), (2, 1), (3, 1), (1, 5), (2, 5), (3, 5)\}.\end{aligned}$$

Now

$$A \times C = \{1, 2\} \times \{1, 5\} = \{(1, 1), (2, 1), (1, 5), (2, 5)\}$$

and

$$B \times C = \{2, 3\} \times \{1, 5\} = \{(2, 1), (3, 1), (2, 5), (3, 5)\};$$

so

$$(A \times C) \cup (B \times C) \\ = \{(1, 1), (2, 1), (3, 1), (1, 5), (2, 5), (3, 5)\} = (A \cup B) \times C.$$

Likewise,

$$C \times (A \cup B) = \{1, 5\} \times \{1, 2, 3\} \\ = \{(1, 1), (1, 2), (1, 3), (5, 1), (5, 2), (5, 3)\}.$$

$$C \times A = \{1, 5\} \times \{1, 2\} = \{(1, 1), (1, 2), (5, 1), (5, 2)\}$$

and

$$C \times B = \{1, 5\} \times \{2, 3\} = \{(1, 2), (1, 3), (5, 2), (5, 3)\}.$$

Hence,

$$(C \times A) \cup (C \times B) \\ = \{(1, 1), (1, 2), (1, 3), (5, 1), (5, 2), (5, 3)\} = C \times (A \cup B).$$

Proof: To prove Theorem 2-6, we list in Table 2-4 all possible truth values for x as an element of A and B and for y as an element of C. Note that the truth values for $(x, y) \in [(A \cup B) \times C]$ are identical to those for $(x, y) \in [(A \times C) \cup (B \times C)]$. Thus, $(A \cup B) \times C = (A \times C) \cup (B \times C)$.

$x \in A$	$x \in B$	$y \in C$	$x \in A \cup B$	$(x,y) \in [(A \cup B) \times C]$	$(x,y) \in (A \times C)$	$(x,y) \in (B \times C)$	$(x,y) \in [(A \times C) \cup (B \times C)]$
T	T	T	T	T	T	T	T
T	T	F	T	F	F	F	F
T	F	T	T	T	T	F	T
T	F	F	T	F	F	F	F
F	T	T	T	T	F	T	T
F	T	F	T	F	F	F	F
F	F	T	F	F	F	F	F
F	F	F	F	F	F	F	F

Table 2-4

Exercise Set 2-4

1. If $A = \{a, b, c\}$ and $B = \{r, s, t\}$, then tabulate the elements of the indicated cartesian product sets.
   (a) $A \times B$ (b) $A \times A$ (c) $B \times A$ (d) $B \times B$
2. Tabulate the cartesian product set $A \times A$ for $A = \{1, 2, 3, 4\}$.
3. Let $A = \{a, b, c\}$, $B = \{c, d, e\}$, $C = \{a, c, x\}$. Tabulate the following cartesian product sets.
   (a) $A \times (B \cap C)$ (b) $A \times (B \cup C)$
   (c) $(A \cap B) \times C$ (d) $(A \cup C) \times B$
4. Find $B \times C$ for each of the following pairs of sets.
   (a) $B = \{3\}$, $C = \{0\}$
   (b) $B = \{3\}$, $C = \{\ \}$
   (c) $B = \{3\}$, $C = \{3, 4\}$
5. The cartesian product $B \times C$ is given in each of the following. Find B and C for each.
   (a) $\{(1, 1), (1, 2), (1, 3), (4, 1), (4, 2), (4, 3)\}$
   (b) $\{(1, 4), (1, 5), (0, 4), (0, 5)\}$
   (c) $\{(6, 6), (6, 7), (6, 8)\}$
   (d) $\{(x, x), (x, y), (x, z), (y, x), (y, y), (y, z), (z, x), (z, y), (z, z)\}$
6. On Sundays, the cafeteria serves beef, fish, and chicken as its meats, and carrots, potatoes, peas, and corn as its vegetables. If Jim orders one meat and one vegetable, what are all of the possible food combinations from which he can select his Sunday dinner?
7. If $A = \{a_1, a_2, a_3, a_4\}$ and $B = \{b_1, b_2, b_3\}$, classify each of the following either as true or as false.
   (a) $(a_1, b_2) \in A \times B$
   (b) $\{(a_2, b_1), (a_2, b_2), (a_2, b_3)\} \subset B \times A$
   (c) $(b_3, a_2) \in B \times A$
   (d) $A \times B = B \times A$
   (e) $(b_1, b_2) \in B \times B$
   (f) $[a_1, (b_1, a_2)] \in (A \times B) \times A$
8. If $A = \{1, 2\}$, $B = \{3, 4, 5\}$, and $C = \{4, 5\}$, tabulate the elements of the following sets.
   (a) $B \times (C \times A)$ (b) $A \times (B \times C)$ (c) $C \times (B \times A)$
   (d) $(A \times B) \times C$ (e) $B \times (C \times C)$ (f) $(B \times C) \times C$

*9. (a) Does $A \times B = B \times A$ imply that $A = B$?
   (b) If $A \neq \varnothing$ and $B \neq \varnothing$ and $A \times B = B \times A$, does $A = B$?
   (c) If $A \neq \varnothing$ and $B \neq \varnothing$ and $A \times B = C \times D$, does $A = C$ and $B = D$?

*10. An experiment in psychology involves three white mice running a maze. The three mice, Do, Re, and Mi, must initially select one of four entrances to the maze, designated by a, b, c, and d. Find the total number of possible combinations of mouse and path by finding the cartesian product.

*11. Outline a proof for $C \times (A \cup B) = [(C \times A) \cup (C \times B)]$.

5

Relations and Functions

We use the idea of ordered pairs to introduce one of the most useful concepts in mathematics—the concept of a function. To introduce functions, we first consider the idea of a *relation*.

In anthropology and sociology, one is often concerned with kinship relations, such as "being the son of" or "being an ancestor of." In mathematics, we use relations such as "is less than," "is perpendicular to," and "is an element of."

What do all these relations have in common, and how can these common qualities be usefully employed to abstract these ideas into a single precise concept? To answer these questions, consider some ordinary statements involving relations: "Terry is engaged to Patty," "Lee is a daughter of Richard," "Alaska is larger than Texas," "Today is hotter than yesterday," and "Gail is more beautiful than Jane." Each statement involves a relation between persons or things. The combining expressions, "is engaged to," "is a daughter of," "is larger than," and so on, are connectives that express the relations. Sometimes these connectives are themselves called relations. A relation is something that "holds" between certain objects x and y and fails to hold between others. Using a Biblical example, the connective "is the father of" gives a relation between Adam and Abel but not between Cain and Abel.

Consider the cartesian product of the set $A = \{$Atlanta, New York, Chicago, Buffalo$\}$ and the set $B = \{$Georgia, New York$\}$. $A \times B$ is $\{$(Atlanta, Georgia), (New York, Georgia), (Chicago, Georgia), (Buffalo, Georgia), (Atlanta, New York), (New York, New York), (Chicago, New York), (Buffalo, New York)$\}$. If x is an element of A and y is an element of B, the relation "x is a city in y" would define the set $\{$(Atlanta, Georgia), (New York, New York), (Buffalo, New York)$\}$. This new set is a subset of $A \times B$. This example illustrates intuitively why a relation is defined as a subset of a cartesian product.

Definition 2-10: Given any two sets A and B, a *relation* from set A to set B is a subset of the cartesian product $A \times B$. If R is a relation from A to B, then $R \subseteq (A \times B)$. If $(c, d) \in R$, then we write ${}_cR_d$.

Example: Let A represent all men and B represent all women. Then R, the set of all married couples, is a subset of $A \times B$. ${}_cR_d$ or $(c, d) \in R$ would mean that c is married to d (in particular, that c is the husband of d).

If R is a relation from A to B, then the *domain* of R is $\{x \mid x \in A$ and $(x, y) \in R$ for some $y \in B\}$. The set of y's is called the *range*. For example, in the relation above, "x is a city in y," the domain is {Atlanta, New York, Buffalo}, and the range is {Georgia, New York}.

Of course, in the definition of a relation, it is implied that if $A = B$, then R is a subset of $A \times A$. For this case, the domain and range are both subsets of the same set A, and the relation is said to be "defined on A." If $A = \{1, 2, 3, 4\}$ and "x is greater than y" defines a relation on A, then $R = \{(2, 1), (3, 1), (4, 1), (3, 2), (4, 2), (4, 3)\}$.

Definition 2-11: A relation R is *reflexive* if $(a, b) \in R$ implies $(a, a) \in R$ and $(b, b) \in R$.

For example, the relation "is the same species as" is reflexive. A dog is the same species as a dog. In mathematics, equality is a reflexive relation because $1 = 1$, $2 = 2$, and, in general, $a = a$. The relation "is the mother of," however, is not reflexive because one is not the mother of himself. While geometric congruence is reflexive, perpendicularity, less than, and greater than are not. For example, it is not true that "a is perpendicular to a" or "2 is greater than 2."

Example: Determine if the relation $\{(1, 1), (1, 2), (1, 3), (2, 1), (2, 2), (2, 3), (3, 1), (3, 2), (3, 3)\}$ is reflexive. This relation is reflexive because all three elements $(1, 1)$, $(2, 2)$, and $(3, 3)$ are within the relation.

Definition 2-12: A relation R is said to be symmetric if

$$(a, b) \in R \rightarrow (b, a) \in R.$$

For example, the relation "is the brother of" is symmetric. If Bob is the brother of Bill, then Bill is also the brother of Bob. In a symmetric relation, whenever a is related to b, b is related to a. Perpendicularity is symmetric even though it is not reflexive. However, "is the mother of" is not symmetric because, while Amy is the mother of Patty, Patty cannot be the mother of Amy. Likewise, "is less than" and "is greater than" are not symmetric.

Example: Determine if the relation $\{(1, 1), (1, 2), (1, 3), (2, 1), (2, 2), (2, 3), (3, 1), (3, 2), (3, 3)\}$ is symmetric. This relation is symmetric because for each (a, b) there exists an element (b, a). For example, $(1, 2)$ and $(2, 1)$, $(1, 3)$ and $(3, 1)$, $(2, 3)$ and $(3, 2)$ exist.

Definition 2-13: A relation R is *transitive* if

$$(a, b) \in R \text{ and } (b, c) \in R \rightarrow (a, c) \in R.$$

For example, the relation "is the same age as" is transitive. If Burt is the same age as Rick and Rick is the same age as Bob, then Burt is the same age as Bob. The transitive property for relations, then, states that whenever a is related to b and b is related to c, then a is related to c. Note that we are not excluding the fact that a and c can be the same. "Is equal to," "is greater than," and "is less than" are good examples of relations for which the transitive property holds. However, a relation such as "is the mother of" is not transitive. Elsie is the mother of Amy, and Amy is the mother of Patty; but Elsie is not the mother of Patty. Perpendicularity, likewise, is not a transitive relation.

The transitive property of equality can be generalized to involve any finite number of equalities; for example,

$$\text{if } a = b,\ b = c,\ c = d,\ d = e, \text{ and } e = f, \text{ then } a = f.$$

Definition 2-14: A relation that is reflexive, symmetric, and transitive is called an *equivalence relation.*

"Is the same age as," "is related to," and "is equal to" are examples of equivalence relations. "Is greater than" is not an equivalence relation. Why not?

An equivalence relation partitions a cross product into sets called *equivalence classes.*

Example: Separate {Sue, Jim, John, Pete, Laura} into equivalence classes with the equivalence relation "is the same sex as." The two equivalence classes are {Sue, Laura} and {Jim, John, Pete}.

Note the following characteristics of the equivalence classes.

(a) The sets or classes are disjoint.
(b) Every element belongs to one class.
(c) If a and b are elements of the same equivalence class, then $_aR_b$ where R is an equivalence relation.

Example: Tom, Richard, Sue, and Jane all have black hair; Joyce and Joe have blond hair; and Simon has red hair. Partition this set of people into equivalence classes where the relation is "has the same color hair as." The answer is: {Tom, Richard, Sue, Jane}, {Joyce, Joe}, and {Simon}.

Functions are frequently introduced in elementary school mathematics with what is called a function machine. A function machine is a machine that produces a *unique* output number for a given input number. If we enter the machine with a number x, we obtain an output which we denote by $f(x)$. The machine actually represents, mathematically, a formula or rule. Suppose the rule is to add 4 to a number x. If $x = 3$, then $f(x) = 3 + 4 = 7$. If $x = 10$, then $f(x) = 14$. Thus, a function f from set A to set B is a rule or procedure that associates with each element of A one and only one element of set B.

Example: Consider the function machine or the rule $y = f(x) = x + 2$. An input of $x = 1$ yields an output of $y = 3$. We sometimes denote the output by $f(1) = 1 + 2 = 3$. Likewise, $f(3) = 3 + 2 = 5$ and $f(50) = 50 + 2 = 52$.

When the notation $y = f(x)$ is used, x is said to be the *independent variable* and y is said to be the *dependent variable.*

Example: We say $y = x + 1$ is a function on the set of real numbers because, for every real number we assign to x, there is one and only one value for y. x is the independent variable and y is the dependent variable.

A rule that defines a function may be expressed in a number of ways, some of which are

(a) sets of ordered pairs,
(b) tables of data,
(c) charts or graphs,
(d) equations or formulas,
(e) verbal statements of principles.

When $y = f(x)$, we say $(x, y) \in f$. If $f(1) = 3$, then $(1, 3) \in f$. Suppose $f(x) = x + 7$. When $x = 2, f(2) = 9$; when $x = 4, f(4) = 11$; and when $x = 5, f(5) = 12$. The ordered pairs (2, 9), (4, 11), and (5, 12) are said to be elements of the function. Thus, a function from set A to set B is a relation whose domain is A with the property that each element of A is paired with exactly one element of B. This result may be formally stated as follows:

Definition 2-15: f is a *function* when f is a relation such that if $(x, y) \in f$ and $(x, z) \in f$, then $y = z$.

Example: Consider the following relations from A to B, where $A = \{2, 3, 4\}$ and $B = \{3, 4, 5, 6\}$. Which of these relations are functions?

(a) $R = \{(2, 3), (3, 4), (4, 5)\}$
(b) $R = \{(3, 3), (3, 4)\}$
(c) $R = \{(2, 3), (3, 4), (4, 5), (2, 6)\}$
(d) $R = \{(2, 5), (3, 5), (4, 5)\}$

(a) and (d) are functions. (c) is not a function because 2 is paired with both 3 and 6. (b) is not a function because 3 is paired with both 3 and 4.

6

One-to-One Correspondence

Before we can define the concept of a number, we must have an understanding of *one-to-one correspondence*. It is possible that early man used the concept of one-to-one correspondence to keep count of his possessions. He probably used a pebble or a mark to represent each animal in his herd. To him, each pebble or mark represented one and only one animal, and for each animal, he had one and only one pebble. Today, one-to-one correspondence is defined in just about the same way it must have been used thousands of years ago.

Definition 2-16: Two sets A and B are said to be in *one-to-one correspondence* if the elements of A and B are so paired that for every element $a \in A$ there corresponds exactly one element $b \in B$ and for every element $b \in B$ there corresponds exactly one element $a \in A$. When an element of A is paired with an element of B, that element of B is automatically paired with the element of A.

Thus, two sets are in one-to-one correspondence when each member of the first set is paired with one and only one member of the second set, and each member of the second set is paired with one and only one element of the first set.

Consider the sets $A = \{1, 2, 3\}$ and $B = \{\text{John, Jane, Jill}\}$. One way of placing these two sets in a one-to-one correspondence is to use the double-headed arrow $\leftrightarrow$ as shown below. Of course, these two sets

$$\begin{array}{ccc} 1 & 2 & 3 \\ \updownarrow & \updownarrow & \updownarrow \\ \text{John} & \text{Jane} & \text{Jill,} \end{array}$$

could be placed in a one-to-one correspondence in several other ways. For example, 1 could correspond to Jill, 2 to John, and 3 to Jane. However, try to set up a one-to-one correspondence between

$$A = \{1, 2, 3, 4\} \quad \text{and} \quad B = \{\text{John, Jane, Jill}\}.$$

If we let John correspond to 1, Jane to 2, and Jill to 3, no element is left to correspond to 4. Thus, one cannot set up a one-to-one correspondence between these two sets. Why is this impossible? It should be intuitively evident from the definition of a one-to-one correspondence between two finite sets that the sets must contain the same number of elements.

Example: Let $H = \{a, c, e, g, i\}$ and $K = \{2, 4, 6, 8, 10\}$. One possible one-to-one correspondence is $a \leftrightarrow 2$, $c \leftrightarrow 4$, $e \leftrightarrow 6$, $g \leftrightarrow 8$, $i \leftrightarrow 10$. Another is $a \leftrightarrow 6$, $c \leftrightarrow 8$, $e \leftrightarrow 10$, $g \leftrightarrow 2$, $i \leftrightarrow 4$. Can you guess how many different one-to-one correspondences can be established between set H and set K?

In the elementary grades, the word "matching" is sometimes used in place of "one-to-one correspondence." Thus, two sets A and B are *matched* if they can be placed in a one-to-one correspondence. Of course, it is not necessary that a correspondence be one-to-one. For example, the set of students in mathematics classes can be made to correspond to the set of teachers, but this correspondence is not one-to-one.

Definition 2-17: If two sets are such that there exists a one-to-one correspondence between the two sets, the sets are said to be *equivalent.*

Suppose a room contains 70 seats and one person is sitting in each of the seats with no one standing. In this case, the set of seats and the set of people are equivalent.

The term *equivalent* should not be confused with equality as used earlier in this chapter. If two sets are equal, then each element of one set must equal a corresponding element of the other, and conversely. However, sets are equivalent if a one-to-one correspondence exists between elements. Thus, two sets may be equivalent but not necessarily equal; however, if two sets are equal, they are equivalent. Likewise, equivalent sets should not be confused with equivalence relations.

Exercise Set 2-5

1. Which of the following pairs of sets can be put into one-to-one correspondence?
   (a) $\{d, e, f, g, h\}$ and $\{3, 6, 7, 8, 5\}$
   (b) {the set of odd counting numbers below 20} and {the set of prime numbers below 20}
   (c) {Mary, Sue, Becky} and {Joe, Richard, Sanders, William}
   (d) $\{1, 2, 3, 4\}$ and $\{11, 12, \ldots, 14\}$
2. Let $A = \{1, 2, 3, 4\}$ and let R be a relation on A.
   (a) Tabulate the set R which describes "a is less than $b + 1$," where a and b are contained in A.
   (b) Tabulate the set R which describes "$a - b = 2$," where a and b are contained in A.
   (c) If R is the set $\{(1, 3), (2, 4)\}$, what is a verbal description for the relation?
   (d) If R is the set $\{(1, 1), (2, 2), (3, 3), (4, 4)\}$, describe the relation in other terms.

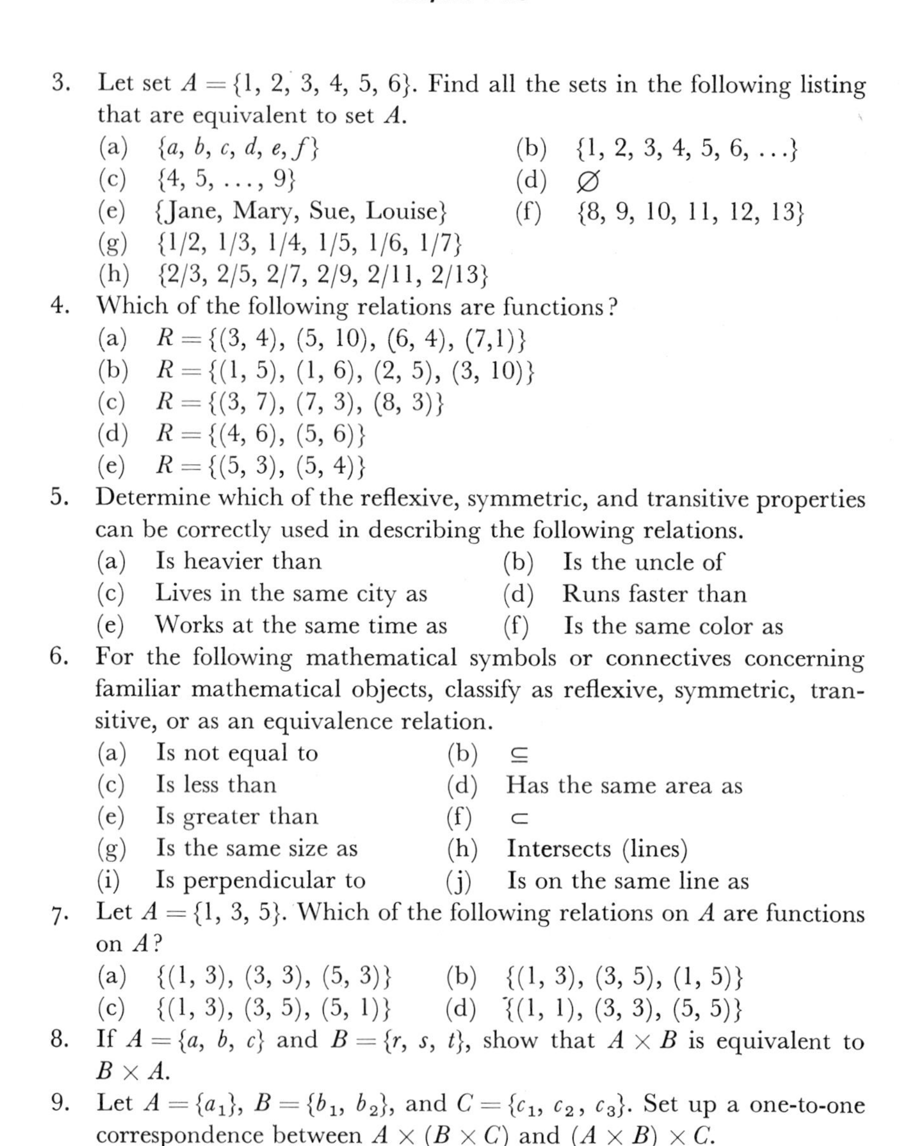

3. Let set $A = \{1, 2, 3, 4, 5, 6\}$. Find all the sets in the following listing that are equivalent to set A.
   (a) $\{a, b, c, d, e, f\}$ (b) $\{1, 2, 3, 4, 5, 6, \ldots\}$
   (c) $\{4, 5, \ldots, 9\}$ (d) $\varnothing$
   (e) {Jane, Mary, Sue, Louise} (f) $\{8, 9, 10, 11, 12, 13\}$
   (g) $\{1/2, 1/3, 1/4, 1/5, 1/6, 1/7\}$
   (h) $\{2/3, 2/5, 2/7, 2/9, 2/11, 2/13\}$
4. Which of the following relations are functions?
   (a) $R = \{(3, 4), (5, 10), (6, 4), (7,1)\}$
   (b) $R = \{(1, 5), (1, 6), (2, 5), (3, 10)\}$
   (c) $R = \{(3, 7), (7, 3), (8, 3)\}$
   (d) $R = \{(4, 6), (5, 6)\}$
   (e) $R = \{(5, 3), (5, 4)\}$
5. Determine which of the reflexive, symmetric, and transitive properties can be correctly used in describing the following relations.
   (a) Is heavier than (b) Is the uncle of
   (c) Lives in the same city as (d) Runs faster than
   (e) Works at the same time as (f) Is the same color as
6. For the following mathematical symbols or connectives concerning familiar mathematical objects, classify as reflexive, symmetric, transitive, or as an equivalence relation.
   (a) Is not equal to (b) $\subseteq$
   (c) Is less than (d) Has the same area as
   (e) Is greater than (f) $\subset$
   (g) Is the same size as (h) Intersects (lines)
   (i) Is perpendicular to (j) Is on the same line as
7. Let $A = \{1, 3, 5\}$. Which of the following relations on A are functions on A?
   (a) $\{(1, 3), (3, 3), (5, 3)\}$ (b) $\{(1, 3), (3, 5), (1, 5)\}$
   (c) $\{(1, 3), (3, 5), (5, 1)\}$ (d) $\{(1, 1), (3, 3), (5, 5)\}$
8. If $A = \{a, b, c\}$ and $B = \{r, s, t\}$, show that $A \times B$ is equivalent to $B \times A$.
9. Let $A = \{a_1\}$, $B = \{b_1, b_2\}$, and $C = \{c_1, c_2, c_3\}$. Set up a one-to-one correspondence between $A \times (B \times C)$ and $(A \times B) \times C$.
10. Are $(A \times B) \times C$ and $A \times (B \times C)$ equivalent in Exercise 9?
*11. Suppose that a relation R is defined on the set $\{1, 3, 5, 7\}$. Give an R which satisfies the following conditions.
   (a) R is transitive, but not reflexive and symmetric.
   (b) R is transitive, reflexive, and symmetric.
   (c) R is reflexive and transitive, but not symmetric.

*12. If $A = \{1, 2, 3, 4, 5, 6\}$, $B = \{2, 4, 6\}$, and $C = \{6\}$, classify each of the following as equivalent or not equivalent.
(a) $A \times B, A \cup B$ (b) $A \cap B, B \times C$
(c) $A \cup C, A \cup B$ (d) $C \times A, A \cap C$

*13. In how many ways can $\{a, b, c, d\}$ be placed in a one-to-one correspondence with itself?

*14. If S and T are transitive relations, discuss whether or not the following are transitive.
(a) $S \cap T$ (b) $S \cup T$ (c) $\bar{S}$ (d) $S \cup \overline{T}$

*15. Prove in general that $A \times B$ is equivalent to $B \times A$.

*16. Prove $(A \times B) \times C$ is equivalent to $A \times (B \times C)$.

Review Exercise Set 2-6

1. Given sets $A = \{1, 2, \ldots, 5\}$ and $B = \{4, 5, \ldots, 10\}$, find each of the following.
(a) $A \cup B$ (b) $A \cap B$
(c) A subset of B that can be put into a one-to-one correspondence with A
(d) $A \cap (B \cup A)$

2. Indicate whether the following statements are true or false. If false, explain why the statement is false.
(a) The transitive property holds true for the set $\{(a, b), (b, c), (c, d), (a, c), (b, c), (a, d)\}$.
(b) If set A is reflexive, then it is also symmetric.
(c) The set $\{1, 2, 3, \ldots, 1{,}000{,}000\}$ is a set including a number of elements that cannot be tabulated.
(d) A set must be empty, finite, or infinite.
(e) The complement of set A, which is the set of vowels within the universe of the English alphabet, is the set of the consonants of the alphabet.
(f) The commutative, associative, and identity properties hold for both union and intersection.
(g) The commutative, associative, and distributive properties hold for the cartesian product and intersection.
(h) If two sets are equivalent, they are equal.
(i) If sets A and B are equal, they are equivalent.

3. Shade each set listed below in the given Venn diagram.
   (a) $(A \cap B) \cup (D \cup E)$
   (b) $(E \cap C) \cap B$
   (c) $E \cup [\overline{(A \cup B) \cup D}]$
   (d) $(B \cap C) \cup D$
   (e) $[A \cup (B \cap C)] \cup [D \cap E]$

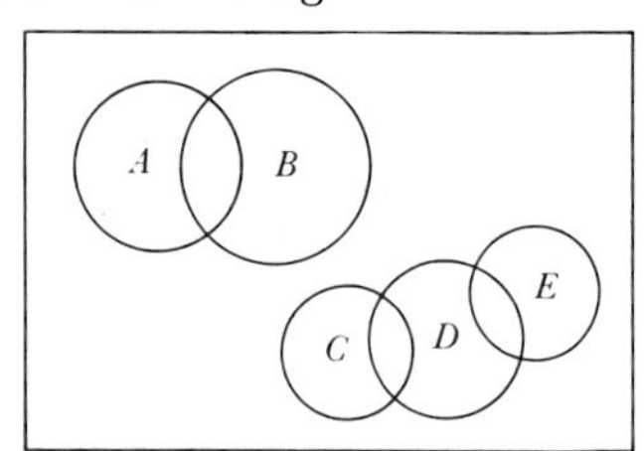

4. Consider set $C = \{6, 7, 8, \ldots, 15\}$ along with A and B as defined in Exercise 1; then find each of the following.
   (a) $A \cap (B \cup C)$ (b) $A \cap (B \cap C)$
   (c) $C \cup (B \cup A)$ (d) $B \cup (A \cap C)$
5. A survey was taken of two hundred males and two hundred females to determine those who prefer football, F, to three other sports: baseball, B; basketball, BB; and hockey, H.

	F	B	BB	H
Males	80	40	20	60
Females	70	20	30	80

   (a) How many males are in set $A = B \cup F$? Verbally describe this set.
   (b) How many males are in set $D = (B \cup BB) \cup H$? Verbally describe this set.
   (c) How many females are in set $E = BB \cup H$? Verbally describe this set.
   (d) How many females are in set $G = F \cup BB$? Verbally describe this set.
   (e) Verbally describe $A \cap D$, where A and D are defined as in (a) and (b) above. How many are in this set?
   (f) Verbally describe $E \cap A$. How many are in this set?
   (g) Verbally describe $(D \cup E) \cap A$. How many are in this set?
6. For the sets $A = \{1, 2, 3, 4, 5\}$ and $B = \{1, 2, 3, 4\}$, find a subset of $A \times B$ such that
   (a) the first component of each ordered pair is 2.
   (b) the sum of the two components is 5.
   (c) the components are equal.
   (d) the sum of the two components is 10.
7. Indicate whether $A \subset B$, $B \subset A$, or $A = B$ (or none of these) for each of the following.
   (a) $A = \{m, n, o\}$, $B = \{m, o\}$
   (b) $A = \{a, b, c\}$, $B = \{a, b, c\}$

(c) $A = \{x\}$, $B = \{x, y, z\}$
(d) $A = \{a, b, c\}$, $B = \{a, c, d\}$
(e) $A = \{c, d\}$ $B = \{d, c\}$

8. Let U be the set of all living humans. Some subsets are: M: All married people; B: All Baptists; T: All teachers; A: All Alabamians. Translate the following to set notation.
   (a) All teachers who are not married
   (b) All teachers and all who are not Alabamians
   (c) All single people and all teachers who are not Baptist
9. Using symbolic notation, describe the sets indicated by the shaded part of each diagram.

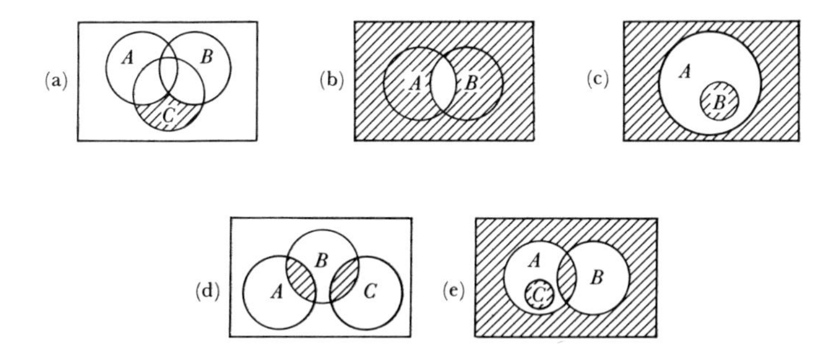

10. Given that set $A = \{1, 2, 3, 4\}$ and set $B = \{5, 6, 7, 8\}$, define a relation R where x and y are contained in A if and only if $x + y \in B$ where $x \neq y$.

*11. Suppose A, B and C are sets such that
   (a) $A \cup B = A$. What can you conclude?
   (b) $A \times B = A$. What can you conclude?
   (c) $A \cap B = A$. What can you conclude?
   (d) $A \times B = B \times A$. What can you conclude?
   (e) $A \times B = A \times C$. What can you conclude?
   (f) $A \times B = C \times A$. What can you conclude?

*12. Determine if and why any relation below is an equivalence relation.
   (a) Congruence
   (b) Is the father of
   (c) $\{(1, 1), (2, 2), (3, 3), (1, 2), (2, 1)\}$
   (d) $A \times B$, where $A = \{x | x \text{ is a vowel of the English alphabet}\}$ and $B = \{y | y \text{ is a counting number such that } 6 < y < 8\}$

Suggested Reading

Identifying Sets: Allendoerfer, pp. 49–71. Armstrong, pp. 84–94, 103–105. Bouwsma, Corle, Clemson, pp. 3–6, 10–14. Brumfiel, Krause, pp. 6–14. Byrne, pp. 1–8. Campbell, pp. 21–24, 27–28. Copeland, pp. 11–13. Garner, pp. 3–9, 12, 15–19. Garstens, Jackson, pp. 98–103. Graham, pp. 1–11, 53–54. Hutton, pp. 1–16, 22–24. Meserve, Sobel, pp. 93–99, 104. McFarland, Lewis, pp. 7–15. Nichols, Swain, pp. 1–10. Ohmer, Aucoin, Cortez, pp. 41–48. Peterson, Hashisaki, pp. 22–31. Podraza, Blevins, Hanson, Prall, pp. 1–4, 8–11. Scandura, pp. 67–74. Smith, pp. 24–29, 36–39. Spector, pp. 119–121. Willerding, pp. 1–9, 15–16. Wren, pp. 3–5, 8–9, 13. Zwier, Nyhoff, pp. 1–11.

Intersection and Union: Allendoerfer, pp. 111–113, 169–171. Armstrong, pp. 94–98, 107–111. Bouwsma, Corle, Clemson, pp. 26–28, 131–141. Brumfiel, Krause, pp. 17–20. Byrne, pp. 5–16. Campbell, pp. 24–31. Copeland, pp. 49–50, 157. Garner, pp. 10–15, 23–29. Garstens, Jackson, p. 104. Graham, pp. 13–25. Hutton, pp. 78–85. Meserve, Sobel, pp. 97–101, 106–116. Nichols, Swain, pp. 11–13. Ohmer, Aucoin, Cortez, pp. 48–64. Peterson, Hashisaki, pp. 29–37. Podraza, Blevins, Hanson, Prall, pp. 8–9, 12–17. Scandura, pp. 118–121. Smith, pp. 30–33. Spector, pp. 120–130. Willerding, pp. 9–12, 17–19. Wren, pp. 6–12. Zwier, Nyhoff, pp. 12–23.

Cartesian Product: Allendoerfer, pp. 137–140. Bouwsma, Corle, Clemson, pp. 85–86. Byrne, pp. 17–22, 27–31. Campbell, pp. 32–33. Copeland, pp. 102–106. Garner, pp. 30–33. Graham, pp. 27–29. Hutton, pp. 96–102. Meserve, Sobel, pp. 218–222. Ohmer, Aucoin, Cortez, pp. 64–67. Peterson, Hashisaki, pp. 37–39. Podraza, Blevins, Hanson, Prall, pp. 15–16. Scandura, pp. 188–191. Smith, pp. 35–36. Willerding, pp. 24–26. Wren, pp. 59–60. Zwier, Nyhoff, pp. 48–51.

Equivalence Relations and One-to-One Correspondence: Allendoerfer, pp. 69–72. Armstrong, pp. 24–33, 98–100. Bouwsma, Corle, Clemson, pp. 204–220. Byrne, pp. 23–27. Campbell, pp. 33–38. Garner, pp. 30–33. Garstens, Jackson, pp. 58–59, 110–113. Graham, pp. 31–38, 48–50. Hutton, pp. 18–20. Meserve, Sobel, pp. 231–236. Nichols, Swain, pp. 100–104. Ohmer, Aucoin, Cortez, pp. 45–57, 67–72. Peterson, Hashisaki, pp. 41–52, 55–60. Podraza, Belvins, Hanson, Prall, pp. 4–5. Scandura, pp. 68–69, 75–83. Spector, pp. 136–138. Willerding, pp. 3–5, 35–36. Wren, pp. 12–13, 308–317. Zwier, Nyhoff, pp. 29–35, 319–323.

3

The System of Whole Numbers

1

Introduction

In our everyday use of numbers, we give little thought to the symbols being used. Most of the time we classify numerals as numbers. What is a number? What is a numeral? Is it important to make a distinction? If you told someone that the even counting numbers consist of the set {2, 4, 6, 8, 10, 12, 14, 16, 18, ...}, he might well conclude that "An even number is one which ends in 0, 2, 4, 6, or 8." The inadequacy of this definition can be easily demonstrated, for there are systems of numeration in which the even numbers are represented by sequences of digits that do not necessarily end in a 0, 2, 4, 6, or 8. We should define mathematical terms explicitly to prevent misunderstanding; thus, we must distinguish between a number and a numeral.

The distinction between a number and a numeral can be described as the distinction between an object and the name associated with that object. At the bottom of this page, you see a folio representing the page number. What you see is not a number; it is a numeral. Thus, a numeral is a symbol which represents a number. The symbol for a number can take many different forms. For example, 4, $2+2$, and $1+3$ all represent the same number. In various languages, the *name* of a counting number is expressed in different ways, both written and verbal, but the *concept* of a counting number does not change. However, in the teaching of modern mathematics there is a disagreement concerning the best way to introduce the number concept simply, yet correctly.

One way to develop a number system is to set up a system of postulates and definitions and, from these, to develop the properties of the system in the form of theorems. Thus, the system of numbers satisfying the set of postulates and theorems is the specified number system. This approach usually postulates the existence of numbers and leaves the term *number* undefined. In the latter part of the nineteenth century, three mathematicians, Richard Dedekind (1831–1916), Georg Cantor (1845–1918), and Giuseppe Peano (1858–1932), added significantly to the foundations of our basic system of numbers. These men contributed ideas on how one could develop a complete theory of numbers by starting with certain undefined terms and selected postulates.

The approach in this book will be to examine the number concept from a historical standpoint, to formulate definitions intuitively, and to develop these intuitive concepts. We begin by setting up a correspondence between the elements of a set and a selected subset of counting numbers. We develop what is known as the whole number system, a system in which the operations will be defined in terms of operations for sets. The lack of closure for certain inverse operations will provide motivation for extending or enlarging the whole number system.

In the following chapters, we will be reaching toward a two-operation system ($+$ and $\cdot$) with the properties of closure, associativity, commutativity, identities, inverses, and the distributive property of $\cdot$ over $+$. The final development will give what we call the *real number system.* If a, b, and c are any real numbers, then the basic arithmetic properties of the number system are as follows.

Closure	$a + b$ is a real number	$a \cdot b$ is a real number
Commutativity	$a + b = b + a$	$a \cdot b = b \cdot a$
Associativity	$a + (b + c) = (a + b) + c$	$a \cdot (b \cdot c) = (a \cdot b) \cdot c$
Identities	$a + 0 = a$	$a \cdot 1 = a$
Inverses	$a + {}^{-}a = 0$	$a \cdot \frac{1}{a} = 1$
Distributivity of $\cdot$ over $+$		$a \cdot (b + c) = (a \cdot b) + (a \cdot c)$

2

Historical Development of Numbers

Leopold Kronecker, a German mathematician of the nineteenth century, is supposed to have said "God created the natural numbers; all else is the work of man." Some people have interpreted Kronecker's statement to mean that man's mind is naturally endowed with the power to comprehend the concept of the counting numbers, while other numbers are a result of the inventiveness of man.

The number concept as it is considered by mathematicians today is the result of many centuries of study. Numbers undoubtedly originated in man's need for counting. Primitive man probably developed the idea of number by the practice of matching objects in one set with objects in another set. In order to keep a record of the number of animals killed in a hunt, he often made marks on the wall of his cave. If he were asked how many animals he had killed during the hunt, he could point to the marks on the wall. Actually, this primitive man was answering "How many?" by establishing what we now call a one-to-one correspondence between the animals killed and the marks on the wall. Thus, the first concepts of number seem to involve what we have called equivalent sets.

In Chapter 2, equivalent sets were defined as sets that could be put into one-to-one correspondence. Suppose we consider four equivalent sets:

$$\{1, 2, 3\}, \quad \{a, b, c\}, \quad \{x, y, z\}, \quad \text{and} \quad \{\bigcirc, \square, \lambda\}.$$

What do these sets have in common? The answer is that they have the same number of elements, namely, three. Thus, the sets have a common property of *threeness*. In a similar manner, equivalent sets with two elements each have a common property of *twoness*, and equivalent sets with only one element each have a property of *oneness*. Thus, every class of equivalent sets has a unifying numerical idea associated with it.

This approach to numbers, which is based on the general notion of a class of sets, was discussed mathematically by Gottlob Frege in 1893. In our intuitive approach, we say that equivalent sets have a common property, which we shall call the *cardinal number* of the set. The names and symbols that are used to represent this abstract conception of cardinal number should always be classified as numerals.

3

The Natural Numbers

The *set of natural numbers* (or counting numbers) is the set $\{1, 2, 3, \ldots\}$. When one counts the elements of a set, he is essentially setting up a one-to-one correspondence between the elements of the set and a certain subset of the natural numbers. The subset needs to be arranged *in order*, such as $\{1, 2, 3, \ldots, m\}$. Here, we illustrate the establishment of a one-to-one correspondence between set A to be counted and the subset of natural numbers:

$$\begin{array}{rl} A = & \{x, y, z, p, q\} \\ & \ \ \updownarrow\ \updownarrow\ \updownarrow\ \updownarrow\ \updownarrow \\ N = & \{1, 2, 3, 4, 5\}. \end{array}$$

Since the subset of counting numbers is arranged *in order*, the last element, 5, gives the *cardinal number* of the set A. The setting up of the one-to-one correspondence is not unique; but no matter how the correspondence is arranged, there will be an element corresponding to the last element in N. For example, the set {John, Jane, Jim} can be placed into a one-to-one correspondence with {1, 2, 3} in six different ways, two of which are shown.

$$\begin{array}{ccc} \{\text{John}, & \text{Jane}, & \text{Jim}\} \\ \updownarrow & \updownarrow & \updownarrow \\ \{1, & 2, & 3\} \end{array} \qquad \begin{array}{ccc} \{\text{John}, & \text{Jane}, & \text{Jim}\} \\ \updownarrow & \times & \\ \{1, & 2, & 3\} \end{array}$$

The symbol $n(A)$ will represent the number of elements in set A. It will be read "the number of A." This number is the cardinal number of all sets equivalent to A. Suppose there are 30 students in a classroom. If S stands for the set of students, then $n(S) = 30$. Similarly, if

$$A = \{u, v, w, x, y, z\},$$

then

$$n(A) = 6.$$

It should be noted that the number of a set is another property of a set that does not depend in any way on the kind of things represented as elements or on the order in which the elements are listed.

Sometimes we are not interested in how many are in a set but in the position that a given member of the set may occupy. You are now reading the "fourth" page of this chapter. The "fourth" is an example of an *ordinal* number as it refers to the relative position or order. The following examples illustrate the two number concepts cardinal and ordinal.

Example: The administration building is five stories high (cardinal). The president's office is on the second floor (ordinal).

Example: Johnny weighs 210 pounds (cardinal). In weight, however, he ranks second (ordinal) on the football team.

Some sets exist with which one cannot easily associate a number. How many grains of sand are there on the beach at Miami? Even though we have no practical way of determining the number of the set, we know that such a number exists. A set is said to be *finite* if and only if it is the null set or there is a counting number n such that the set can be put into a one-to-one correspondence with the set $\{1, 2, 3, \ldots, n\}$. Otherwise we classify a set as *infinite*. By this definition, the set $\{x, y, z, \ldots\}$ is an infinite set, as is the set of all counting numbers.

In the preceding discussion, our numbers were restricted to the *natural numbers*, or *counting numbers*, $\{1, 2, 3, 4, \ldots\}$. We now consider the natural numbers along with a new number, zero, denoted by the symbol 0. This new set of numbers will be called the *whole numbers*. Thus, the *whole numbers* will be the natural numbers and zero. *Zero* will be defined as the cardinal number of the null set; that is, $n(\varnothing) = 0$. You may have heard the statement, "Zero is nothing." This is not correct. 0 is a symbol that represents the cardinal number of the null set and answers such questions as "How many elephants are in your classroom?"

Exercise Set 3-1

1. Place a C by the phrases involving cardinal number concepts and an O by those involving the ordinal number concepts.
   (a) 16 sheep (b) page 9 (c) 82nd Congress
   (d) third grade (e) five people (f) fifth session
   (g) 7 days in a week (h) 10th round (i) 12 months in a year
   (j) \$15 (k) a dozen eggs (l) 2nd period
2. Exhibit a one-to-one correspondence between one of the sets $\{1\}$, $\{1, 2\}$, $\{1, 2, 3\}$, $\{1, 2, 3, 4\}$, or $\{1, 2, 3, 4, 5\}$, and the following sets.
   (a) $\{b\}$ (b) $\{\varnothing, \lambda\}$
   (c) $\{a, b, c, d, e\}$ (d) $\{4, 6, 8\}$
   (e) $\{10, 40, 30, 50, 70\}$ (f) $\{2\}$
3. Use the results of Exercise 2 to find the number of elements in each set given in Exercise 2.
4. Compute $n(A)$ through $n(F)$ for the following sets.
   (a) $A = \{201, 1, 2, 3\}$

(b) $B = \{x, y, z\}$
(c) $C = \{n \mid n$ is a natural number less than $8\}$
(d) $D = \{n \mid n$ is a natural number greater than 5 and less than $6\}$
(e) $E = \{n \mid n$ is a natural number less than $3\}$
(f) $F = \{x \mid x$ is a whole number less than $10\}$

5. For the sets in Exercise 4, compute $n(S)$ where S is given by each of the following.
   (a) $A \cup B$ (b) $A \cap B$ (c) $A \cap D$
   (d) $A \times B$ (e) $D \times F$ (f) $B \cup C$
   (g) $C \times A$ (h) $E \times A$ (i) $D \times E$
6. Indicate whether the following statements are true or false.
   (a) Sets $A = \{$○, △, ▭, □, ⬠$\}$, $B = \{1, 6, 8, 3, 0\}$, and $C = \{$Rick, Steve, Allan, Paul, Ed$\}$ are equivalent because they share the property of oneness.
   (b) Sets A and B in part (a) have the same cardinal number because they are equal.
   (c) Although sets A, B, and C in part (a) are not equal, they still have the same cardinal number and are equivalent.
   (d) The smallest element in the set of natural numbers is 0.
   (e) A subset of a finite set may be infinite.
7. Answer true or false.
   (a) $n\{0\} = 0$ (b) $n(\varnothing) = 0$
   (c) $n\{\ \} = 1$ (d) $n\{0, 1\} = 1$
   (e) $n\{0, 1, 2\} = 3$ (f) $n\{0, 1, 2, 3\} = 4$
8. For $A = \{2, 3, \ldots, 6\}$ and $B = \{6, 7, \ldots, 10\}$, verify that
   (a) $n(A \cup B) = n(A) + n(B) - n(A \cap B)$
   (b) $n(A \times B) = n(A) \cdot n(B)$
   (c) $n(A \times A) = n(A) \cdot n(A)$
9. Let P be the set of letters used in spelling the word "teach" and Q be the set of letters in "cheat." Is it correct to write $P = Q$? Why? Is $n(P) = n(Q)$? Why? Compute $n(P \cup Q)$ and $n(P \cap Q)$.
10. For each pair of sets given, compute $n(A)$, $n(B)$, $n(A \cup B)$, $n(A \cap B)$, and $n(A \times B)$.
   (a) $A = \{a, b, c\}$, $B = \{x, y\}$ (b) $A = \{a, b, c\}$, $B = \{c, d\}$
   (c) $A = \{a, b, c\}$, $B = \{b, c\}$ (d) $A = \{a, b, c\}$, $B = \varnothing$
   (e) $A = \{a, b, c\}$, $B = \{x\}$ (f) $A = \{a, b, c\}$, $B = A$
11. For each pair of sets given in Exercise 10, compare
   (a) $n(A) + n(B)$ with $n(A \cup B)$,
   (b) $n(A) \cdot n(B)$ with $n(A \times B)$.
12. Using the set N of natural numbers and the set W of whole numbers, determine each of the following.

(a) $N \cap \{0\}$ (b) $N \cup \{0\}$ (c) $W \cap \{0\}$
(d) $W \cup \{0\}$ (e) $N \cap W$ (f) Is $N \subset W$?

*13. If $n(R) = 10$, $n(S) = 6$, and $n(R \cup S) = 12$, calculate $n(R \cap S)$.

*14. In surveying a small group of pre-school children, a psychologist determined that fourteen of the children liked the television program "At Home," seventeen liked the program "Back and Forth," and eleven liked the program "Catch It!" Of these children, eleven liked both A ("At Home") and B ("Back and Forth"), six liked both B and C ("Catch It!"), and five liked both A and C. Only two of the children liked all three. Determine from this information each of the following.
(a) $n(A \cup B)$ (b) $n(B \cup C)$
(c) $n(A \cup C)$ (d) $n(A \cup B \cup C)$

4

The Addition of Whole Numbers

How well do you understand addition? What do you mean by the operation $+$? Why is $2+3=3+2$? Do you get the same answer if you compute both $2+3+5$ and $5+2+3$? Perhaps some new ideas are needed to help us to understand and to explain fully the operation of addition.

Just as a number is defined in terms of sets in the preceding section, the operation of addition will be considered with respect to sets here. Consider the two sets

$$A = \{a, b, c\} \quad \text{and} \quad B = \{\triangle, \lambda, \bigcirc, \square\}.$$

Now $n(A) = 3$ and $n(B) = 4$. How many elements are there in the union of these two sets? Assume that we know how to add to obtain the answer $3 + 4 = 7$. How many members are contained in $A \cup B$? Did you get 7? Thus, it seems that

$$n(A) + n(B) = n(A \cup B).$$

Now consider a second example. Let $A = \{a, b, c\}$ and $B = \{c, d, e, h\}$.

$$n(A) + n(B) = 3 + 4 = 7.$$

But $n(A \cup B) = 6$. In this example,

$$n(A) + n(B) \neq n(A \cup B).$$

($\neq$ is the symbol for "is not equal to.") What is the difference? Did you notice that in the first example the sets were disjoint, while in the second example the letter c was common to both sets?

Consider the problem of adding two whole numbers, say 3 and 5. Assume this time that we do not know how to add. Consider A to be any set of the equivalent class of sets that has 3 for a cardinal number; A could be $\{x, y, z\}$. Let B be a set disjoint from A that has 5 for a cardinal number.

$$B = \{a, b, c, d, e\}$$

is an example. Then

$$3 + 5 = n(A) + n(B).$$

Since A and B are disjoint, we might guess from the discussion above that $n(A) + n(B) = n(A \cup B)$. But $A \cup B = \{x, y, z, a, b, c, d, e)$ and

$$n(A \cup B) = 8,$$

since $A \cup B$ can be put into one-to-one correspondence with the set $\{1, 2, 3, 4, 5, 6, 7, 8\}$. Thus, $3 + 5$ seems to be 8.

Let A be the set of birds perched on the edge of a birdbath. Set A consists of a sparrow and a robin. On a fence nearby, a bluejay, a redbird, and a mockingbird make up set B. If the birds on the fence *join* the birds at the birdbath, there are 5 birds in all. $2 + 3 = 5$. Therefore, addition is associated with the *union* or *joining together* of two sets. These intuitive ideas are stated in the following definition.

Definition 3-1: Let a and b represent any two whole numbers, and choose A and B to be disjoint finite sets so that $n(A) = a$ and $n(B) = b$. Then $a + b = n(A \cup B)$.

Example: If $A = \{a, b, c\}$ and $B = \{x, y, z, w\}$, then $n(A) = 3$ and $n(B) = 4$. Since

$$A \cup B = \{a, b, c, x, y, z, w\},$$

$n(A \cup B) = 7$. Also, $A \cap B = \varnothing$; thus

$$n(A \cup B) = n(A) + n(B),$$

or

$$7 = 3 + 4.$$

Here we found that $7 = 3 + 4$. The 3 and 4 can be termed the *addends* of the problem while the $3 + 4$, or 7, is called the *sum.*

Example: If $A = \{a, b, c, d, e\}$ and $B = \{a, j, i\}$, then

$$A \cup B = \{a, b, c, d, e, j, i\} \quad \text{and} \quad n(A \cup B) \neq n(A) + n(B).$$

Why?

We should note that addition as defined is a binary operation. A binary operation, denoted by $*$, assigns to each ordered pair of elements (a, b) a third (but not necessarily different) element denoted by $a * b$. Thus, when we say that addition is a binary operation we mean that the operation $+$ replaces $*$ to give $a + b$.

The closure property for addition for the whole numbers means that for any two whole numbers a and b, there exists a unique whole number $a + b$. Note that closure has two properties: uniqueness and existence. Existence stipulates that an answer exists for the addition of any two whole numbers. Uniqueness requires that only one answer exists. See Exercises 11, 12, and 13 of Exercise Set 3-2.

Suppose we add the whole numbers 3 and 4. Does it matter in which order we add them? In other words, is $3 + 4$ the same as $4 + 3$? In either case the answer is 7. This property for the addition of whole numbers is called *commutativity*.

A commuter is a person who goes back and forth. In the same sense, when whole numbers are commuted with respect to an operation, they simply change places.

Theorem 3-1: *The Commutative Property of Addition for Whole Numbers.* For all whole numbers a and b, $a + b = b + a$.

Proof: Let A and B be any sets such that $a = n(A)$, $b = n(B)$, and $A \cap B = \varnothing$.

$A \cup B = B \cup A$	Commutative property of the union of sets
$A \cup B$ is equivalent to $B \cup A$	Equal sets are equivalent
$a + b = n(A \cup B)$	Definition 3-1
$= n(B \cup A)$	Cardinal number of equivalent sets
$= b + a$	Definition 3-1
Therefore,	
$a + b = b + a$	Transitive property of equalities

The commutative property of addition states that the order in which we add two whole numbers is immaterial; that is, it does not matter whether we add 2 to 3 or 3 to 2. Of course, you have known since kindergarten that $2 + 3 = 3 + 2$, but now you can state the reason why the order of addition can be interchanged. Remember this name—the commutative property of addition.

When we wish to add three or more numbers, we must agree on the order in which the operations are to be performed. Suppose we wish to add 2, 3, and 7. Shall we add 2 and 3 and then 7 or shall we add the sum of 3 and 7 to 2? Usually we indicate repeated additions by the use of parentheses ( ) or brackets [ ]. $(2 + 3) + 7$ means that 2 and 3 are added and 7 is added to the answer. $2 + (3 + 7)$ means that the sum of 3 and 7 is added to 2. The parentheses serve as mathematical punctuation. Perform these additions and see if you get the same answer. If you do, then you have verified for this example another important property of the addition of whole numbers, the *associative property of addition for whole numbers.*

Thus to add three whole numbers 5, 7, and 6, one may group them in two ways; and, by the associative property, the answers will be the same.

$$(5 + 7) + 6 = 12 + 6 = 18 \quad \text{or} \quad 5 + (7 + 6) = 5 + 13 = 18$$

Theorem 3-2: *The Associative Property of Addition for Whole Numbers.* For all whole numbers a, b, and c, $(a + b) + c = a + (b + c)$.

Proof: Let A, B, and C be finite sets such that $a = n(A)$, $b = n(B)$, and $c = n(C)$, where $A \cap B = \varnothing$, $A \cap C = \varnothing$, and $B \cap C = \varnothing$.

$a + b = n(A \cup B)$	Definition 3-1
$b + c = n(B \cup C)$	Definition 3-1
$(A \cup B) \cup C = A \cup (B \cup C)$	Associative property of the union of sets
$(A \cup B) \cup C$ is equivalent to $A \cup (B \cup C)$	Equal sets are equivalent
$(a + b) + c = n(A \cup B) + n(C)$	Definition 3-1
$= n[(A \cup B) \cup C]$	Definition 3-1
$= n[A \cup (B \cup C)]$	Cardinal number of equivalent sets
$= n(A) + n(B \cup C)$	Definition 3-1
$= a + (b + c)$	Definition 3-1

Therefore,

$(a + b) + c = a + (b + c)$	Transitive property of equalities

Example:

$$3 + 4 + 7 = (3 + 4) + 7 = 7 + 7 = 14$$

or

$$3 + 4 + 7 = 3 + (4 + 7) = 3 + 11 = 14$$

When there are four numbers to be added, one may associate the numbers in several ways, such as $(3 + 2) + (5 + 6)$ or $3 + [(2 + 5) + 6]$. That these will all yield the same number is a generalization or extension of the associative property of addition.

Example: Using the commutative and associative properties for addition, prove that $(2 + 3) + 5 = (5 + 2) + 3$.

Proof:

$(2 + 3) + 5 = 5 + (2 + 3)$	Commutative property of addition
$5 + (2 + 3) = (5 + 2) + 3$	Associative property of addition
Therefore,	
$(2 + 3) + 5 = (5 + 2) + 3$	Transitive property of equalities

Example: Prove that $(c + a) + b$ is equal to $a + (b + c)$.

$(c + a) + b = (a + c) + b$	Commutative property of addition
$(a + c) + b = a + (c + b)$	Associative property of addition
$a + (c + b) = a + (b + c)$	Commutative property of addition
Therefore,	
$(c + a) + b = a + (b + c)$	Transitive property of equalities

Consider the addition $4 + 0$. We may associate with 4 the set $A = \{a, b, c, d\}$ and with 0 the null set, $\varnothing$. The union of A and $\varnothing$ is A. Thus, $n(A) + n(\varnothing) = n(A \cup \varnothing) = n(A)$. Hence, $4 + 0 = 4$. In a like manner, $0 + 4 = 4$. Thus, it seems that $a + 0 = 0 + a = a$ for any whole number a. If a happens to be 0, then $0 + 0 = 0$. This result is stated in the following theorem.

Theorem 3-3: *Additive Identity.* If a is any whole number, then $0 + a = a$, and $a + 0 = a$.

Proof: Since $0 + a = n(\varnothing) + n(A)$, by Definition 3-1,

$$0 + a = n(\varnothing \cup A).$$

But $\varnothing \cup A = A$. Thus $0 + a = n(\varnothing \cup A) = n(A) = a$. By the commutative property of addition, $a + 0 = 0 + a = a$.

Because an additive identity is a number which when added to any number a within the set gives a number identical to a, 0 is the additive identity ($a + 0 = a$) for the set of whole numbers.

Historically, 0 has played a very special role in the development of numbers and the use of numbers. We have defined the number 0 to be the cardinal number of the null set. Now we have introduced 0 as the additive identity for the binary operation of addition.

The following addition property is very useful in our study of whole numbers.

Theorem 3-4: (a) If a, b, and c are whole numbers with $a = b$, then $a + c = b + c$.

(b) If a, b, c, and d are whole numbers with $a = b$ and $c = d$, then $a + c = b + d$.

Proof: Part (a) of this theorem is implied by the definition of addition. Since $a = b$, let D be some set such that $n(D) = a = b$. Take C to be a set disjoint to D, ($D \cap C = \varnothing$), such that $n(C) = c$. Then

$$a + c = n(D \cup C) \quad \text{and} \quad b + c = n(D \cup C).$$

Thus by the transitive property of equalities,

$$a + c = b + c.$$

Part (b): $a + c = b + c$ by part (a) of this theorem. Similarly, $b + c = b + d$. Thus, $a + c = b + d$ by the transitive property of equalities.

Exercise Set 3-2

1. For the following pairs of sets, state whether or not

$$n(R) + n(S) = n(R \cup S).$$

(a) $R = \{3, 2, 7, 1\}$, $S = \{4, 6, 5\}$

(b) $R = \{3, 2, 4, 7\}$, $S = \{1, 4, 6, 5\}$

(c) $R = \{x \mid x \text{ is greater than 1 and less than 6}\}$
$S = \{x \mid x \text{ is greater than 4 and less than 9}\}$

(d) $R = \{x | x \text{ is greater than 1 and less than 5}\}$
$S = \{x | x \text{ is greater than 4 and less than 5}\}$

2. Explain in detail how you would associate sets to obtain the answers to the following addition problems.
(a) $2 + 4$ (b) $4 + 0$ (c) $5 + 3$
(d) $8 + 2$ (e) $0 + 6$ (f) $0 + 0$
3. The following problems are examples of the application of properties described in this section. State which properties are being applied in each example.
(a) $2 + 3 = 3 + 2$ (b) $4 + (2 + 3) = (4 + 2) + 3$
(c) $6 + 0 = 6$ (d) $(a + c) + d = a + (c + d)$
(e) $[2 + (4 + 6)] + 8 = (2 + 4) + (6 + 8)$ (f) $2 + (5 + 7) = 2 + (7 + 5)$
(g) $5 + 0 = 0 + 5$ (h) $(8 + 2) + 3 = 3 + (8 + 2)$
(i) $(4 + 2) + 3 = 3 + (4 + 2)$ (j) $(2 + 5) + 3 = (5 + 2) + 3$
4. Use inductive reasoning to determine whether the following sets are closed with respect to addition. Does the set contain an identity element?
(a) $\{0, 2, 3, 4, \ldots\}$ (b) $\{2, 4, 6, 8, 10, \ldots\}$
(c) $\{2, 3, 4, 5\}$ (d) $\{0\}$
(e) $\{x | x \text{ is a whole number greater than 15}\}$
(f) $\{x | x \text{ is a natural number less than 10}\}$
(g) The set of odd counting numbers
(h) The set of counting numbers greater than 10
5. Prove the following addition facts. State the reason for each step of a proof.
(a) $(2 + 4) + 3 = (3 + 2) + 4$ (b) $(2 + 0) + 1 = 1 + 2$
(c) $8 + (5 + 2) = 2 + (8 + 5)$ (d) $6 + (9 + 1) = 9 + (6 + 1)$
(e) $(a + b) + (c + d) = (b + d) + (a + c)$ (f) $(a + d) + (b + c) = (b + d) + (a + c)$
6. Prove that each of the following is equal to $a + (b + c)$.
(a) $a + (c + b)$ (b) $(a + b) + c$ (c) $(b + c) + a$
(d) $(a + c) + b$ (e) $c + (a + b)$ (f) $(c + b) + a$
7. Verify the commutative law of addition by partitioning the set $\{l, m, n, o, p, q\}$ and adding the cardinal numbers of each part of the partition. Show how you can also verify the associative law of addition by partitioning this same set.
8. Sophomore Sandy had a difficult time understanding this section on addition. After solving two exercises, she looked in the back of the book only to discover that she had found the wrong answers! Study Sandy's proof below and find her mistake (or mistakes).

$(6+7)+(0+1)=(6+7)+0$ Additive identity
$(6+7)+0=6+7$ Commutative property
$6+7=13$ Addition

Therefore,

$(6+7)+(0+1)=13$ Transitive property of equalities

9. For the natural numbers, suppose we invent a new operation which states that a sum is the first number added to twice the second. Is this operation commutative?

*10. What is wrong with the following?
 (a) {Glenda, Joyce, Kathy} + {Linda, Joyce} = 5
 (b) 5 peaches + 4 apples = 9 pieces of fruit
 (c) $4 \cup 7 = 11$

*11. What is meant by the statement that the closure property has two requisites: uniqueness and existence?

*12. If one person computes the sum $a+b$ by using disjoint sets A and B and another person computes the same sum using disjoint sets C and D, different from A and B, make an argument showing that both get the same answer.

*13. Prove that addition is unique.

*14. Exhibit three sets A, B, and C with the property that $A \cap B \cap C = \varnothing$ but for which it is not the case that each of the statements $A \cap B = \varnothing$, $B \cap C = \varnothing$, and $C \cap A = \varnothing$ is true.

5

The Multiplication of Whole Numbers

In the preceding section, we defined and developed the properties of the addition of whole numbers. In this section, our attention will be focused on the operation of *multiplication*. Multiplication is the operation of finding the product of two numbers. Generally, it is defined as repeated addition. For example, $3 \cdot 2$ could mean $2+2+2$ or 6. However, in this chapter the definition of multiplication will be given in terms of sets. This form of the definition facilitates the development of the properties of multiplication.

Paul keeps his money separated into boxes. The other day, Paul had three boxes, each of which contained two one-dollar bills. How much money did Paul possess? The answer, of course, is $3 \cdot 2 = 6$ one-dollar bills. We attempt now to obtain this answer by a reasoning process that will lead to our definition of multiplication.

Suppose Paul names his boxes a, b, c and the two one-dollar bills in each box he marks 1 and 2. Now he can list his one-dollar bills in the following way: $(a, 1)$ denotes the one-dollar bill in box a marked with 1; $(a, 2)$ denotes the one-dollar bill in box a marked with 2, and so on. His complete set of one-dollar bills can be listed as

$$\{(a, 1), (a, 2), (b, 1), (b, 2), (c, 1), (c, 2)\}.$$

By the process of counting, we find that this is indeed six one-dollar bills. Now, does the tabulation above look familiar? Does it look anything like the elements in the cartesian product $\{a, b, c\} \times \{1, 2\}$? Thus, with the set of three boxes, $\{a, b, c)$, and the set of two one-dollar bills in each box $\{1, 2\}$, the number of elements in the cartesian product

$$\{a, b, c\} \times \{1, 2\}$$

gives the number of one-dollar bills.

Example: Set $A = \{1, 2, 3\}$ has the cardinal number 3, and set $B = \{0, 1, 2, 3\}$ has the cardinal number 4. The number of the set

$$A \times B = \{(1, 0), (1, 1), (1, 2), (1, 3), (2, 0), (2, 1), (2, 2), (2, 3), (3, 0), (3, 1), (3, 2), (3, 3)\}$$

is 12. Previously we have used $3 \cdot 4$ to be equal to 12. This discussion leads to the following definition.

Definition 3-2: Let a and b represent any two whole numbers. Select finite sets A and B such that $n(A) = a$ and $n(B) = b$. Then

$$a \cdot b = n(A \times B).$$

To multiply two numbers a and b, one merely investigates the number of terms in the cartesian product $A \times B$. Notice that the definition implies that the cartesian product involved is independent of the nature of the elements of the sets. This means that the sets need not be disjoint. In fact, they could be identical. In an expression indicating multiplication, such as $4 \cdot 3 = 12$, the 3 and 4 are called *factors* and 12 is called the *product*. We also write $a \cdot b$ as ab, as $(a)(b)$, or as $a(b)$.

The following example illustrates Definition 3-2.

Example: $2 \cdot 3 = n(A \times B)$, where $n(A) = 2$ and $n(B) = 3$. If

$$A = \{a, b\} \quad \text{and} \quad B = \{k, l, m\},$$

then

$$2 \cdot 3 = n\{(a, k), (b, k), (a, l), (b, l), (a, m), (b, m)\} = 6.$$

Multiplication is clearly a binary operation since it is defined for every ordered pair (a, b) of whole numbers. One might question whether $a \cdot b$ is a whole number. Since $a \cdot b = n(A \times B)$, where A and B are finite sets, it is intuitively evident that $A \times B$ is finite; thus, $a \cdot b$ is a whole number.

Similarly, we can argue that $a \cdot b$ is unique. Suppose one selects sets C and D instead of A and B. That is, $n(A) = a = n(C)$ and $n(B) = b = n(D)$. Again our intuition suggests that $n(A \times B) = n(C \times D)$ (see Exercise 10, Exercise Set 3-3), so $a \cdot b$ is uniquely defined.

Thus we have the *closure property for the multiplication* of whole numbers. If a and b are any two whole numbers, then $a \cdot b$ is a whole number.

The following illustration presents a second concept of multiplication. Suppose three boys have four pennies each and they wish to know how many pennies they have or if they have enough money to buy a ten-cent candy bar. Now these youngsters do not know how to multiply $3 \cdot 4$, but instead they add together the pennies and count them. From this standpoint, multiplication is repeated addition, or to multiply 3 times 4, add $4 + 4 + 4$. This special property of multiplication is further illustrated by the following example.

If a classroom has four rows of desks with six desks in each row, how many desks are in the room? Designate the desks in each row by a, b, c, d, e, f, and number the rows 1, 2, 3, 4. Then the desks can be identified by the following sets of ordered pairs:

$$\begin{aligned}
R_1 &= \{(1, a), (1, b), (1, c), (1, d), (1, e), (1, f)\},\\
R_2 &= \{(2, a), (2, b), (2, c), (2, d), (2, e), (2, f)\},\\
R_3 &= \{(3, a), (3, b), (3, c), (3, d), (3, e), (3, f)\},\\
R_4 &= \{(4, a), (4, b), (4, c), (4, d), (4, e), (4, f)\}.
\end{aligned}$$

But if we let $R = \{1, 2, 3, 4\}$ and $D = \{a, b, c, d, e, f\}$, then $R \times D$ is the same set of ordered pairs as $R_1 \cup R_2 \cup R_3 \cup R_4$. Since

$$R \times D = R_1 \cup R_2 \cup R_3 \cup R_4,$$

then

$$n(R \times D) = n[R_1 \cup R_2 \cup R_3 \cup R_4],$$

or

$$n(R) \cdot n(D) = n(R_1) + n(R_2) + n(R_3) + n(R_4).$$

There are $4 \cdot 6 = 6 + 6 + 6 + 6 = 24$ desks in the room.

Theorem 3-5: If a is a counting number and b is a whole number, then $a \cdot b = b + b + b + \cdots + b$, where b is used (appears) a times.

Proof: Let A and B be finite sets such that $n(A) = a$ and $n(B) = b$. Consider A as the union of disjoint subsets consisting of one element each.

$$A = A_1 \cup A_2 \cup A_3 \cup \cdots \cup A_a$$

$a \cdot b = n[A \times B]$	Definition 3-2
$= n[(A_1 \cup A_2 \cup A_3 \cup \cdots \cup A_a) \times B]$	Substitution for A
$= n[(A_1 \times B) \cup (A_2 \times B) \cup (A_3 \times B) \cup \cdots \cup (A_a \times B)]$	Repeated application of Theorem 2-6
$= n(A_1 \times B) + n(A_2 \times B) + \cdots + n(A_a \times B)$	Repeated application of Definition 3-1
$= n(A_1) \cdot n(B) + n(A_2) \cdot n(B) + \cdots + n(A_a) \cdot n(B)$	Definition 3-2
$= (1 \cdot b) + (1 \cdot b) + \cdots + (1 \cdot b)$	Since each A_i has only one element ($i = 1, 2, \ldots, a$)
$= b + b + \cdots + b$ (b appears a times)	Since i varies in A_i from 1 to a

Examples:

$$5 \cdot 4 = 4 + 4 + 4 + 4 + 4 = 20$$
$$2 \cdot 5 = 5 + 5 = 10$$
$$7 \cdot 0 = 0 + 0 + 0 + 0 + 0 + 0 + 0 = 0$$

The process of pairing each element in set A with each element in set B and counting the total number of ordered pairs can be accomplished easily by arranging the elements of $A \times B$ in an array of rows and columns. Such a procedure also demonstrates the "repeated addition" aspect of multiplication. Figure 3-1 is an array showing the number of ordered pairs obtained from $A = \{a, b, c\}$ and $B = \{k, m, n, o\}$. If $3 = n(A)$ and $4 = n(B)$, then $3 \cdot 4 = 12$.

	k	m	n	o
a	$\{(a, k)$	(a, m)	(a, n)	(a, o)
b	(b, k)	(b, m)	(b, n)	(b, o)
c	(c, k)	(c, m)	(c, n)	$(c, o)\}$

Figure 3-1

Sometimes the elements of an array are represented by dots, such as in Figure 3-2(a). This array demonstrates again that $3 \cdot 4 = 12$.

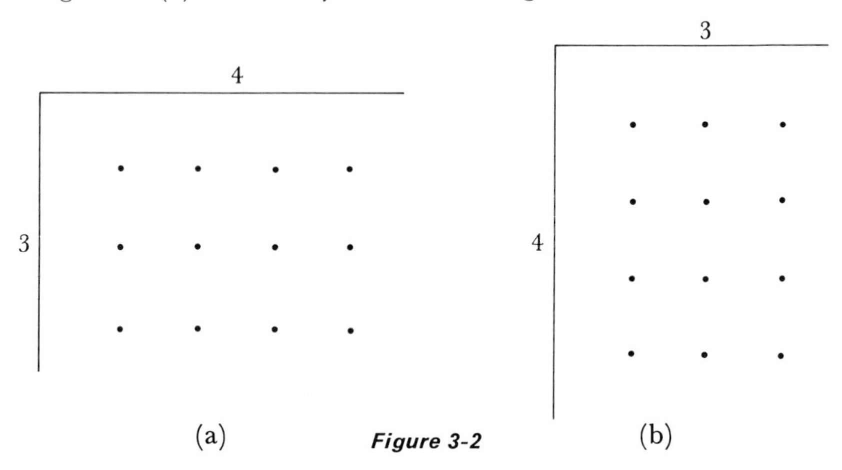

Figure 3-2

If three students each check four books from the library, then the number of books borrowed is twelve. However, if four students each take three books, the total number of books borrowed is still twelve. To see this visually, consider the arrays in Figure 3-2(a) and (b). The number of elements in each array is the same. This is one way of demonstrating that $3 \cdot 4 = 4 \cdot 3$. This property is given as follows:

Theorem 3-6: *The Commutative Property of Multiplication of Whole Numbers.* If a and b are whole numbers, then $a \cdot b = b \cdot a$.

Proof: Choose sets A and B so that $n(A) = a$ and $n(B) = b$. Then

$ab = n(A \times B)$ Definition 3-2
$\quad = n(B \times A)$ Since $A \times B$ is equivalent to $B \times A$ as proved in Exercise 15, Exercise Set 2-5
$\quad = ba$ Definition 3-2

Therefore,

$ab = ba$ Transitive property of equalities

To multiply 2, 3, and 5, one may multiply 2 and 3 to get an answer of 6, and then multiply 6 and 5 to obtain 30. In a like manner, 2 multiplied by the product of 3 and 5 is 30; thus $(2 \cdot 3) \cdot 5 = 2 \cdot (3 \cdot 5)$.

Theorem 3-7: *The Associative Property of Multiplication of Whole Numbers.* For all whole numbers a, b, and c, $(ab)c = a(bc)$.

Proof: Let A, B, and C be sets such that $a = n(A)$, $b = n(B)$, and $c = n(C)$.

$(ab)c = n(A \times B) \cdot n(C)$	Definition 3-2
$= n[(A \times B) \times C]$	Definition 3-2
$= n[A \times (B \times C)]$	In Exercise 16, Exercise Set 2-5, it was proved that $(A \times B) \times C$ is equivalent to $A \times (B \times C)$
$= n(A) \cdot n(B \times C)$	Definition 3-2
$= a(bc)$	Definition 3-2

Therefore,

$(ab)c = a(bc)$	Transitive property of equalities

Example: $7 \cdot (4 \cdot 6) = (7 \cdot 4) \cdot 6$ because

$$7 \cdot (4 \cdot 6) = 7 \cdot 24 = 168 \quad \text{and} \quad (7 \cdot 4) \cdot 6 = 28 \cdot 6 = 168.$$

Example: Use the properties of multiplication to show that

$$b \cdot (4 \cdot 3) = 3 \cdot (4 \cdot b).$$

Now

$b \cdot (4 \cdot 3) = (4 \cdot 3) \cdot b$	Commutative property of multiplication
$(4 \cdot 3) \cdot b = (3 \cdot 4) \cdot b$	Commutative property of multiplication
$(3 \cdot 4) \cdot b = 3 \cdot (4 \cdot b)$	Associative property of multiplication

Therefore,

$b \cdot (4 \cdot 3) = 3 \cdot (4 \cdot b)$	Transitive property of equalities

The associative property not only gives meaning to the product of three factors, it can also be interpreted to give meaning to the product of any number of factors. Consider the multiplication of four factors such as $abcd$. The associative property states that this problem can be considered as $(ab) \cdot (cd)$, or $a[b(cd)]$, or $[a(bc)]d$, all of which give the same answer.

The next property formalizes a very obvious characteristic of the number 1 in the system of whole numbers.

Theorem 3-8: *Multiplicative Identity.* If a is any whole number, then $1 \cdot a = a \cdot 1 = a$.

For example, consider the multiplication $3 \cdot 1 = 3$. Let $Q = \{a, b, c\}$ and $R = \{a\}$.

$$Q \times R = \{(a, a), (b, a), (c, a)\};$$

by counting, we see that the number of elements in $Q \times R$ is the same as the number of elements in Q. Therefore, $n(Q) = n(Q \times R) = 3$. $R \times Q = \{(a, a), (a, b), (a, c)\}$ and $n(R \times Q) = n(Q) = 3$. So

$$3 \cdot 1 = 1 \cdot 3 = 3.$$

Intuitively, we see that

$$x \cdot 1 = 1 \cdot x = x,$$

where x is any whole number.

Proof: Let 1 equal $n(B)$ and $a = n(A)$. Then

$$1 \cdot a = n(B) \cdot n(A) = n(B \times A).$$

But $B \times A$ consists of number pairs whose first elements are the single element of B and whose second elements are the different elements of A. Thus $B \times A$ and A have a elements each. Hence $1 \cdot a = n(B \times A) = a$. By the commutative property of multiplication, $1 \cdot a = a \cdot 1 = a$.

Thus, for the system of whole numbers, there exists a number, namely 1, having the property that $(1)(a) = (a)(1) = a$. This existence of the multiplicative identity is more important than it seems; actually, some mathematical systems do not have a multiplicative identity.

Exercise Set 3-3

1. Justify each of the following statements.
   (a) $2 \cdot 3 = 3 \cdot 2$
   (b) $5(4 + 7) = (4 + 7) \cdot 5$
   (c) $3(2 + 4) = 3(4 + 2)$
   (d) $2(3 \cdot 4) = (2 \cdot 3) \cdot 4$
   (e) $5 + (2 + 7) = (5 + 2) + 7$
   (f) $6(5 \cdot 8) = 6(8 \cdot 5)$
   (g) $5 + (2 + 7) = (2 + 7) + 5$
   (h) $6(5 \cdot 4) = (5 \cdot 4) \cdot 6$
2. Write out a statement of the meaning of "the product of two whole numbers is uniquely defined."

3. Group 2, 3, 5, and 6 for multiplication in at least three different ways; for example, one way is $2 \cdot [3(5 \cdot 6)]$. Perform the multiplication to show that all the products are the same. Thus, the associative property of multiplication of three whole numbers has been extended to four whole numbers.
4. If $A = \{c, d\}$ and $B = \{e, f, g\}$, set up an equivalence between $A \times B$ and $B \times A$ to show that $2 \cdot 3 = 3 \cdot 2$.
5. Identify the property or properties that justify each of the following statements.
   (a) $(6 + 5) \cdot (4 + 8) = (5 + 6) \cdot (8 + 4)$
   (b) $(5 \cdot 4) \cdot 7 = (7 \cdot 4) \cdot 5$
   (c) $(6 + 5) \cdot (4 + 8) = (8 + 4) \cdot (6 + 5)$
   (d) $(5 + 4) \cdot (4 + 8) = (8 + 4) \cdot (4 + 5)$
   (e) $4 + 3 = (4 + 3) + 0$
   (f) $(7 \cdot 3) \cdot 2 = 7 \cdot (3 \cdot 2)$
   (g) $1 \cdot (4 + 2) = (2 + 4)$
6. Determine whether the following sets are closed with respect to multiplication. Does the set contain an identity element for multiplication?
   (a) $\{1, 2, 3, 4, 5\}$ (b) $\{1\}$
   (c) The even whole numbers
   (d) The odd whole numbers
   (e) $\{x \mid x$ is a whole number greater than $10\}$
   (f) $\{x \mid x$ is a whole number less than $2\}$
   (g) $\{1, 4, 7, 10, 13, 16, \ldots\}$
   (h) $\{1, 4, 9, 16, 25, 36, 49, 64, \ldots\}$
7. Rick has eight bars of candy to sell for his club at \$.50 each. He also agreed to take 25 tickets for the club booth at the campus carnival to sell at \$.15 each. If Rick sells the candy and tickets, how much money will he have to return to the club?
8. Sophomore Sandy is still unsure of herself as she studies this section. Find her mistakes, if any, in the proof below.

$8(13 \cdot 0) + 4 \cdot 1 + (0 \cdot 2) = 8 \cdot 0 + 4 \cdot 1 + 0$	Multiplicative identity
$8 \cdot 0 + 4 \cdot 1 + 0 = 8(4 \cdot 1) + 0$	Additive identity
$8(4 \cdot 1) + 0 = 8(4) + 0$	Multiplicative identity
$8(4) + 0 = 32 + 0$	Multiplication
$32 + 0 = 32$	Additive identity

Therefore,

$8(13 \cdot 0) + 4 \cdot 1 + (0 \cdot 2) = 32$	Transitive property of equalities

*9. In the following table the operation $\otimes$ is a binary operation similar to our addition and multiplication. Here it is defined for the set $\{0, 1, 2, 3, 4\}$. The entry 3 on the third row inside the table is $2 \otimes 4 = 3$. Note also that $3 \otimes 2$ is 1.

$\otimes$	0	1	2	3	4
0	0	0	0	0	0
1	0	1	2	3	4
2	0	2	4	1	3
3	0	3	1	0	2
4	0	4	3	2	1

(a) Is the set closed with respect to $\otimes$?
(b) Is the operation $\otimes$ commutative on this set?
(c) Is the operation $\otimes$ associative on this set?
(d) Does an identity element exist for the operation $\otimes$ on this set? If so, what is it?

*10. If $n(A) = n(C) = a$ and $n(B) = n(D) = b$, prove that $n(A \times B) = n(C \times D)$; and then reason that $a \cdot b$ is unique.

*11. For the whole numbers $\{0, 1, 2, 3, 4, \ldots\}$ define $x \odot y$ as follows:

$$x \odot y = 0 \text{ if } x \text{ or } y \text{ or both } x \text{ and } y \text{ are even,}$$
$$x \odot y = 1 \text{ if both } x \text{ and } y \text{ are odd.}$$

Discuss $\odot$ as a candidate for a binary operation. Is it commutative and associative?

6

Connectives between + and ·

So far, all the problems in this chapter have involved either addition or multiplication, but not both. Now we combine these two operations in a single expression, such as $2(6 + 4)$ or $(3 \cdot 9) + 2$.

If we were asked to find the answer to $(3 \cdot 9) + 2$, we would perform the operations as multiplication followed by addition; that is, multiply 3 by 9 and add 2 to get 29. In $2(6 + 4)$, the parentheses indicate that the addition of 6 and 4 should be performed first; then the result should be multiplied by 2 to get $2 \cdot 10 = 20$. Note that $2(6 + 4) \neq (2 \cdot 6) + 4$, because $(2 \cdot 6) + 4 = 12 + 4 = 16$.

Example:

$$(3 + 2) \cdot 4 = 5 \cdot 4 = 20, \quad \text{and} \quad (3 \cdot 4) + (2 \cdot 4) = 12 + 8 = 20,$$

so

$$(3 + 2) \cdot 4 = (3 \cdot 4) + (2 \cdot 4).$$

These examples illustrate an important property of whole numbers; it is called the *distributive property*.

Theorem 3-9: *The Distributive Property of Multiplication over Addition.* For all a, b, and c in the set of whole numbers,

$$a(b + c) = ab + ac$$

and

$$(b + c)a = ba + ca.$$

Proof: Consider sets A, B, and C such that $n(A) = a$, $n(B) = b$, and $n(C) = c$, where $B \cap C = \emptyset$. Since $B \cap C = \emptyset$, then

$$(B \times A) \cap (C \times A) = \emptyset. \quad \text{Why?}$$

$(b + c)a = n(B \cup C) \cdot n(A)$	Definition 3-1
$= n[(B \cup C) \times A]$	Definition 3-2
$= n[(B \times A) \cup (C \times A)]$	Cardinal number of equivalent sets since, by Theorem 2-6, $(B \cup C) \times A = (B \times A) \cup (C \times A)$
$= n(B \times A) + n(C \times A)$	Definition 3-1
$= ba + ca$	Definition 3-2

Therefore,

$(b + c)a = ba + ca$	Transitive property of equalities

To illustrate this property, look at the 3 by 8 array in Figure 3-3(a). This array shows that $3 \cdot 8 = 24$.

Since $8 = 6 + 2$, then

$$3 \cdot 8 = (3 \cdot 6) + (3 \cdot 2) = 18 + 6 = 24,$$

as shown by Figure 3-3(b).

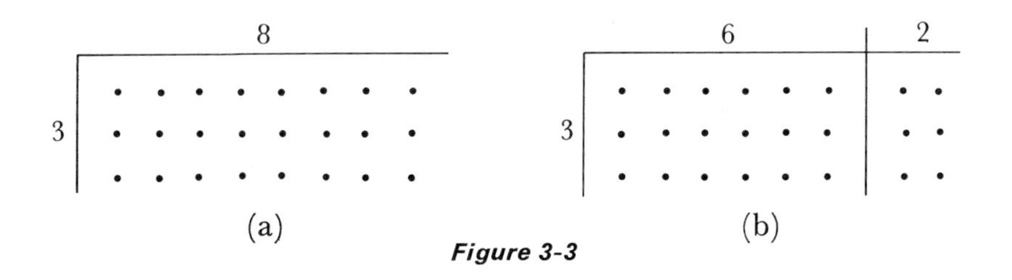

Figure 3-3

Let us illustrate the distributive property with the following examples.

Example:

$$8(6+11) = 8 \cdot 17 = 136$$
$$(8 \cdot 6) + (8 \cdot 11) = 48 + 88 = 136$$

Therefore,

$$8(6+11) = (8 \cdot 6) + (8 \cdot 11).$$

Example:

$$(5+4)(8) = 9 \cdot 8 = 72$$
$$(5 \cdot 8) + (4 \cdot 8) = 40 + 32 = 72$$

Therefore,

$$(5+4)(8) = (5 \cdot 8) + (4 \cdot 8).$$

Example: The distributive property is useful in the process of multiplying mentally. For example, (34)(41) could be solved without pencil and paper if considered as $34(40+1)$. The distributive property gives

$$(34)(41) = (34)(40) + (34)(1) = 1360 + 34 = 1394.$$

Example:

$$3x^3 + 4x^2 = (3x+4)x^2 \quad \text{and} \quad ax^3 + bx = (ax^2 + b)x$$

Example:

$$a(b+c+d) = a([b+c]+d) = a[b+c] + ad = ab + ac + ad$$

This example serves as a first step in obtaining by inductive reasoning the generalized distributive property of multiplication over addition:

$$a(b_1 + b_2 + \cdots + b_n) = ab_1 + ab_2 + \cdots + ab_n.$$

Another interesting multiplication fact involves the additive identity. Many times operations with 0 are confusing, but a little thought will show clearly how to operate with 0.

Theorem 3-10: If a is any whole number, then $0 \cdot a = a \cdot 0 = 0$.

Examples:

$$0 \cdot 3 = 3 \cdot 0 = 0$$
$$0 \cdot 7 = 7 \cdot 0 = 0$$
$$0 \cdot 0 = 0$$

Proof: Associate with the number a the set A, where $n(A) = a$. Now $\varnothing$ is the only set such that $n(\varnothing) = 0$. Thus $0 \cdot a = n(\varnothing \times A)$ by the definition of multiplication. But $\varnothing \times A$ is the set of number pairs whose first elements are elements from $\varnothing$ and whose second elements are elements from A. However, there are no elements in $\varnothing$. Thus, there are no elements in $\varnothing \times A$, so $\varnothing \times A = \varnothing$. Hence

$$0 \cdot a = n(\varnothing \times A) = n(\varnothing) = 0.$$

By the commutative property of multiplication $0 \cdot a = a \cdot 0 = 0$.

Theorem 3-11: (a) If a, b, and c are whole numbers and $a = b$, then $ac = bc$.

(b) If a, b, c, and d are whole numbers, and if $a = b$ and $c = d$, then $ac = bd$.

Proof: Part (a): This theorem is implied by the definition of multiplication. Since $a = b$, sets A and B have the same number of elements. Let C be a set such that $n(C) = c$. The number of first elements in the number pairs of $A \times C$ is the same as the number of first elements in $B \times C$, since A and B have the same number of elements. Because the second elements in the number pairs from $A \times C$ and $B \times C$ come from C in both cases, the number of elements in $A \times C$ is the same as the number of elements in $B \times C$. Thus $ac = bc$.

Part (b): Since $a = b$, by Part (a) of this theorem, $ac = bc$. In a like manner, since $c = d$, $bc = bd$. Thus, by the transitive property of equalities, $ac = bd$.

Example: Since $3 = (2 + 1)$, then $4 \cdot 3 = 4(2 + 1)$.

Example: Since $8 = (5 + 3)$ and $(7 + 2) = 9$, then

$$8(7 + 2) = (5 + 3)9.$$

Exercise Set 3-4

1. Classify each of the following as true or as false.
   (a) $8 \cdot 0 = 8$ (b) $(16 + 1) + 0 = 17$
   (c) $7 + 0 = 0$ (d) $(ab)(0) = 0$
   (e) $(16 + 1)0 = 17$ (f) $(1 + 1)1 = 1$
2. Compute the following expressions in two different ways.
   (a) $3(5 + 4)$ (b) $11(6 + 3)$
   (c) $(2 + 7)9$ (d) $(3 + 1)4$
3. Evaluate.
   (a) $(2 + 3) \cdot 4$ (b) $2 + (3 \cdot 4)$ (c) $(2 \cdot 3) + 4$
   (d) $(2 + 3) + 4$ (e) $2 \cdot (3 + 4)$ (f) $2(3 \cdot 4)$
4. Rename the following numbers using the distributive property.
   (a) $2(3 + 4)$ (b) $5a + 3a$ (c) $(4 \cdot 2) + (4 \cdot 3)$
   (d) $ax + a$ (e) $4xay + 2xz$ (f) $ax^2 + bx^2 + cx$
5. State which properties justify the following equalities.
   (a) $6 + 8 = 8 + 6$ (b) $(a)(c) + b = (c)(a) + b$
   (c) $c(a + b) = ca + cb$ (d) $(a + c)d + (a + c)e = (a + c)(d + e)$
6. Use the generalized distributive property to prove Theorem 3-5 for the following cases.
   (a) $3 \cdot 4$ (b) $4 \cdot 7$ (c) $5 \cdot 6$
7. We have illustrated that multiplication is distributive over addition. Is addition distributive over multiplication? That is, if a, b, and c are any whole numbers, is it always true that $a + (b \cdot c) = (a + b) \cdot (a + c)$? Back up your answer with an example.
8. Sophomore Sandy is still unsure of herself. Find her mistakes, if any, in the proof below.

   $42(a + b) + (42c \cdot 1) = 42(a + b) + 42c$ Additive identity
   $42(a + b) + 42c = 42a + 42b + 42c$ Distributive property
   $42a + 42b + 42c = 42(a + b + c)$ Associative property

   Therefore,

   $42(a + b) + (42c \cdot 1) = 42(a + b + c)$ Transitive property of equalities
9. You can see that Sandy needs some help. She is discouraged, so you can help by filling in the reasons in the steps below.

   $(3 + a)b + 2c + (1 + 0)c = (3 + a)b + 2c + (1)c$ (a) ________
   $(3 + a)b + 2c + (1)c = 3b + ab + 2c + (1)c$ (b) ________
   $3b + ab + 2c + (1)c = 3b + ab + (2 + 1)c$ (c) ________
   $3b + ab + (2 + 1)c = 3b + ab + 3c$ (d) ________

   Therefore,

   $(3 + a)b + 2c + (1 + 0)c = 3b + ab + 3c$ (e) ________

10. Give explicit reasons for each step in showing that each of the following statements is true.
    (a) $(2 \cdot 3)4 = (4 \cdot 2)3$ (b) $(8 + 2)7 = 7(2 + 8)$
    (c) $(6 + 1)5 = (1)(5) + (5)(6)$ (d) $3(4 + 7) = (4)(3) + (3)(7)$
    (e) $2(3 \cdot 1) = (1 \cdot 2)3$ (f) $(ab)c = (ca)b$

*11. Using the fact that $(b + c)a = ba + ca$, prove that $a(b + c) = ba + ca$.

*12. Let $\odot$ be defined by $x \odot y = x$ and $\triangle$ be defined by $x \triangle y = y$. State the distributive property of $\odot$ over $\triangle$ using three whole numbers a, b, and c. Demonstrate the answer when $a = 3$, $b = 2$, and $c = 4$; when $a = 1$, $b = 2$, and $c = 3$.

7

The Ordering of Whole Numbers

Recall that to count the number of elements in a set, we put the set in a one-to-one correspondence with $\{1, 2, 3, \ldots, n\}$. Without realizing it, we were really establishing an order relation. For example, if $A = \{1, 2\}$, the number associated with A is 2. If $B = \{1, 2, 3\}$, the number associated with B is 3. Since $\{1, 2\}$ is a proper subset of $\{1, 2, 3\}$, we can conclude that 2 is less than 3. Generally, if $\{1, 2, \ldots, d\}$ is a proper subset of $\{1, 2, \ldots, e)$, then d is less than e. That is, if A and B are any finite sets and A can be matched in a one-to-one correspondence to a proper subset of B, then $n(A)$ is defined to be less than $n(B)$.

Another way of looking at the order relation is summarized as follows. If a and b are different whole numbers, then there is a natural number n so that either $a + n = b$ or $b + n = a$. If we consider 2 and 6, then 2 is less than 6 because $2 + 4 = 6$; but 6 is greater than 5 because $6 = 5 + 1$.

Definition 3-3: If a and b are any whole numbers, then a is said to be less than b, denoted by $a < b$, if and only if there exists a natural number c such that $a + c = b$.

For example, $6 < 14$ because there exists a natural number 8 such that $6 + 8 = 14$. In a like manner, $4 < 10$ because $4 + 6 = 10$. $3 < 2$ is not true because there is no natural number x such that $3 + x = 2$.

Definition 3-4: If a and b are any two whole numbers, then a is said to be greater than b, written $a > b$, if and only if there exists a natural number d such that $a = b + d$.

If a is greater than b, then b is less than a; and if a is less than b, then b is greater than a. Thus, the expressions $x < y$ and $y > x$ express the same idea. Sometimes the equality relation is combined with these *inequalities* to give a "less than or equal to" relation. When one writes $a \leq b$, it means that $a < b$ or $a = b$.

The application of these definitions leads to a very important property of whole numbers called the *trichotomy property*. The word "trichotomy" refers to a situation in which each possibility must fall into one of three mutually exclusive classes. This property is now stated as an axiom.

Axiom 3-1: *The Trichotomy Property for Whole Numbers.* If a and b are any two whole numbers, then *precisely* one of the three following possibilities holds: (1) $a < b$, (2) $a = b$, or (3) $a > b$.

This axiom states that two whole numbers are equal or the first is less than the second or the first is greater than the second. No two of these cases can occur at the same time.

Closely associated with ordering the whole numbers is the geometric representation of the whole numbers on a number line. A *number line* is a representation of a geometric line extending endlessly in two opposite directions. The line is marked with two fundamental points, one representing 0 and one representing 1. We have marked what we will call in Chapter 10 a line segment of length 1, sometimes called a unit segment (Figure 3-4). Now that a unit length has been established, other points are marked on the number line (Figure 3-5).

. . . ———•——•——————— . . .
0 1

Figure 3-4

Thus the number line is divided into equal divisions by points, each point being placed in a one-to-one correspondence with a whole number; so there is a one-to-one correspondence between a set of points on the number line and the set of whole numbers. No matter how large a whole number may be, it can be matched in a one-to-one correspondence with a point on the number line.

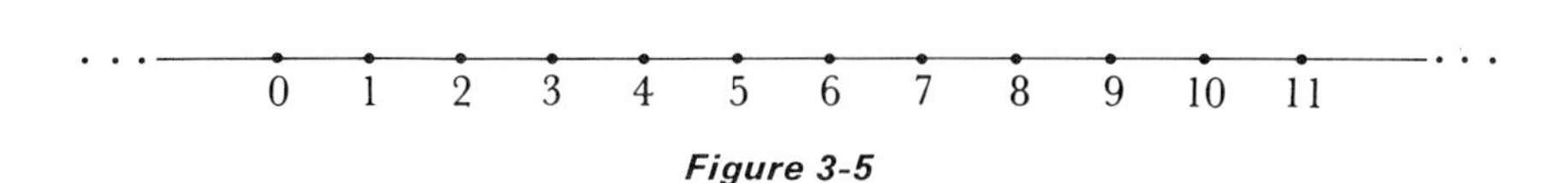

Figure 3-5

Note that many points on the number line are not labeled. In fact, between any two labeled points on the number line there are infinitely many unlabeled points. This fact will be discussed in detail in Chapter 8.

When we represent the whole numbers on a number line, we are actually using the order relation. If we did not have the concept of order, we would not know how to label the points on the number line. In using the number line, we agree that numbers of greater value are represented by points to the right of points representing numbers of lesser value. Some evident relations are $1 < 2$, $2 < 3$, $3 < 4$, and so forth.

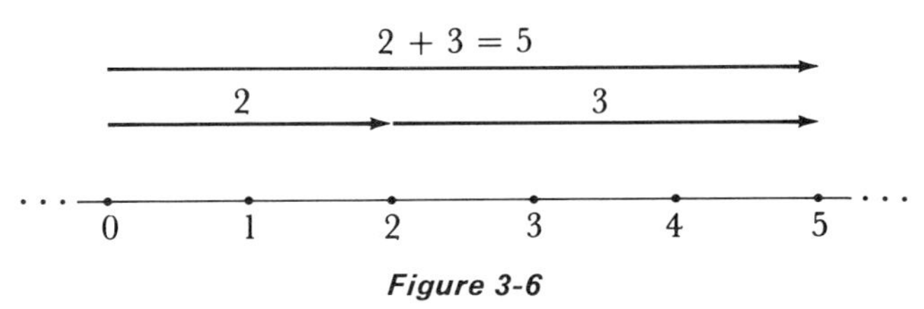

Figure 3-6

We are now ready to represent on the number line the operation of addition involving two natural numbers. Consider the statement $2 + 3 =$? Starting at 0, draw a line 2 units long and place an arrow on the end of this line. At the end of the first line, measure another line (with an arrow on the end) 3 units long and note that the arrow of the second is at the point labeled 5 (Figure 3-6). Therefore, $2 + 3 = 5$.

The lines with arrows on the end are called *arrow diagrams*, or directed line segments, and are used to represent the numbers in the addition process. This geometric representation of numbers can be used to demonstrate the commutative property of addition. We will use the preceding example and show on the number line that $2 + 3 = 3 + 2$ (Figure 3-7).

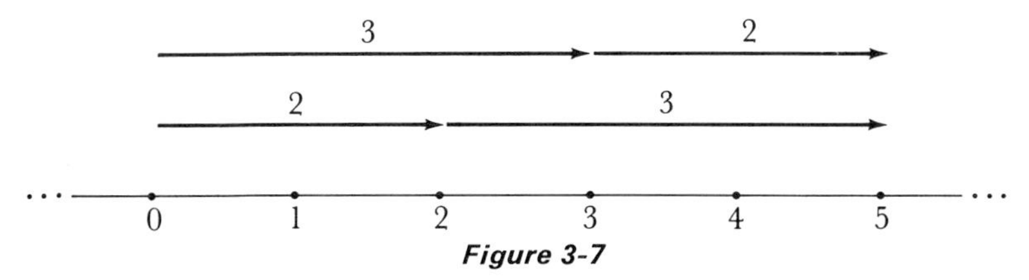

Figure 3-7

An illustration of the multiplication of two whole numbers on the number line should make clear the concepts given in Theorem 3-5. The multiplication of 3 times 4 can be demonstrated on the number line by adding 4 three times (Figure 3-8).

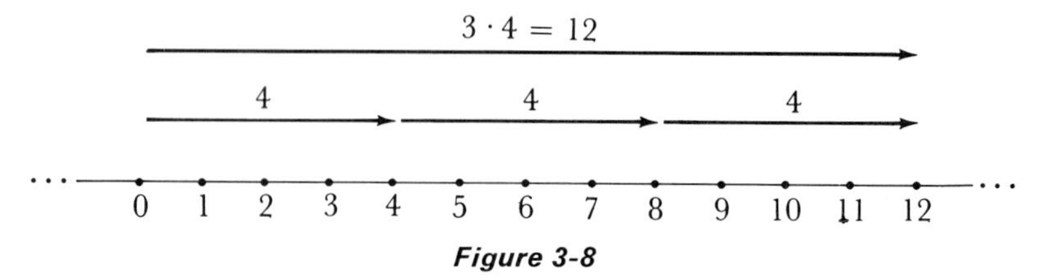

Figure 3-8

A property of order is illustrated in Figure 3-9 for the whole numbers 2, 3, and 5. Note that 2 is to the left of 3 and 3 is to the left of 5; and thus 2 is to the left of 5, or $2 < 3$, $3 < 5$; therefore $2 < 5$. Later on, this property will be called the *transitive property of inequalities.*

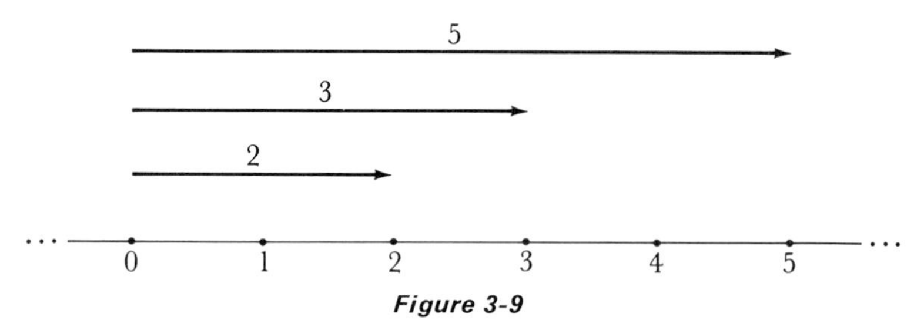

Figure 3-9

We conclude this section with a discussion of two fundamental order properties of the natural numbers, which we state as axioms.

Axiom 3-2: *Archimedean Property.* If a and b are any natural numbers such that $a < b$, then there exists a natural number k such that $b < ka$.

This property of order is illustrated in Figure 3-10 for the natural

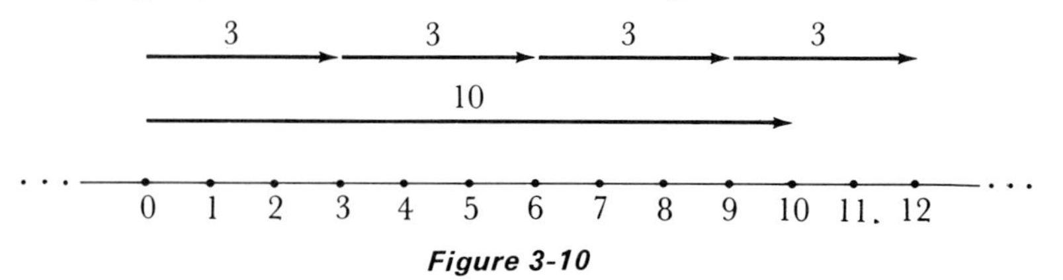

Figure 3-10

numbers $a = 3$, $b = 10$. The diagram shows that at least 4 arrow diagrams of length 3 are necessary for $b < ka$; $10 < 4 \cdot 3$.

Axiom 3-3: *Well-Ordering Property.* If C is any nonempty subset of the whole numbers, there is one and only one element of C which is less than or equal to the elements of C.

This element is called the least element, and the well-ordering property guarantees the existence of such an element for every nonempty subset of the whole numbers. For example, if A is the subset of whole numbers $\{4, 5, 6, 7, 8, 9\}$, on the number line we see that 4 is the least element of this subset.

8

Subtraction and Division of Whole Numbers

Many times each day, you do something and then undo it. You put on your coat, then take it off; open the door, then close it; pull down the window shade, then raise it. Pairs of opposites are also encountered in mathematics. These opposites are called inverses—inverse operations, inverse elements, and inverse functions.

In this section, you will study the inverse of the operation of addition, called *subtraction.* If you add 8 and 3 to get 11, then by subtracting 3 from 11 you should get 8, and by subtracting 8 from 11 you should get 3. When we subtract b from a, we are looking for a whole number c called the *difference* such that when we add b to c we get a.

Definition 3-5: The difference $a - b$ in the subtraction of a whole number b from a whole number a is equal to a whole number c if and only if $a = b + c$.

Examples:

$$16 - 5 = 11 \quad \text{because} \quad 5 + 11 = 16$$
$$8 - 3 = 5 \quad \text{because} \quad 3 + 5 = 8$$
$$10 - 0 = 10 \quad \text{because} \quad 0 + 10 = 10$$
$$0 - 0 = 0 \quad \text{because} \quad 0 + 0 = 0$$

This difference, $a - b$, may be read as "a minus b," "b subtracted from a," or "a take away b." The definition does not imply that the whole number c exists for all pairs a and b. In fact, the whole number c is defined to exist only when a is greater than or equal to b. What happens when $a < b$? Then the definition does not have meaning. For example, subtract 5 from 3 $(3 - 5 = ?)$. What whole number added to 5 will give an answer of 3? There is no such number in the system of whole numbers. Thus, the set of whole numbers is *not closed* under the operation of subtraction.

Let us consider the question "What number added to 3 will equal 8?" To obtain the answer to this question, one solves the simple *equation* $x + 3 = 8$ or $3 + x = 8$. The answer is obtained from Definition 3-5, which states that $8 - 3 = x$ if and only if $8 = x + 3$. Thus, $x = 8 - 3 = 5$.

Example: Solve $x + 7 = 9$.

By Definition 3-5,

$$x = 9 - 7 \quad \text{or} \quad x = 2.$$

Example: Solve $n + 10 = 21$.

$$n = 21 - 10 = 11.$$

The solution of equations will be discussed in more detail in Chapter 5.

Let us now consider subtraction on the number line. Addition has been shown as moving to the right on the number line, and subtraction for whole numbers will be shown as moving to the left. Thus, since subtraction is the inverse of addition, the arrow for subtraction and the arrow for addition will point in opposite directions. Consider on the number line (Figure 3-11) the example $8 - 3 = ?$ Starting with 0, move to the right to the point labeled 8. To subtract 3, we move 3 units to the left. The arrow at the end of the arrow diagram for 3 is at the point labeled 5. Thus $8 - 3 = 5$.

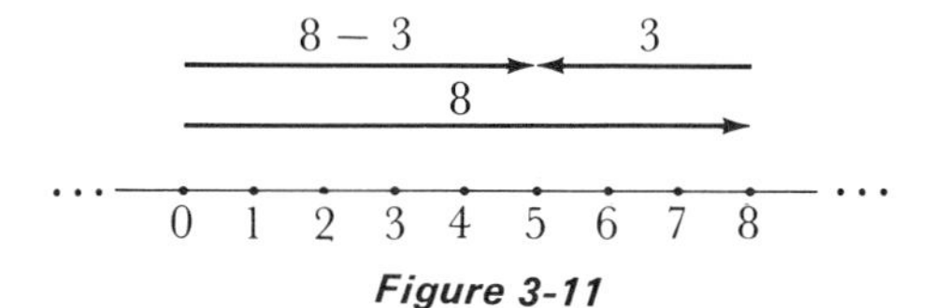

Figure 3-11

Does multiplication distribute itself over subtraction? This question will be answered by the next theorem, which, along with Theorem 3-13, will be proved in Chapter 5.

Theorem 3-12: *The Distributive Property of Multiplication over Subtraction.* If a, b, and c are whole numbers such that $a \geq b$, then

$$c(a - b) = ca - cb.$$

Example: $6(8 - 5) = (6)(8) - (6)(5) = 48 - 30 = 18$
$(6)(3) = 18$

Example: $(71)(143) - (71)(142) = 71(143 - 142) = (71)(1) = 71$

Theorem 3-13: If $b \geq c$, then $(a + b) - c = a + (b - c)$.

Example:
$$\begin{aligned}(10 + 7) - 3 &= 10 + (7 - 3)\\ &= 10 + 4\\ &= 14\end{aligned}$$

Theorem 3-14: If $a \geq c$, $b \geq d$, then

$$(a + b) - (c + d) = (a - c) + (b - d).$$

Proof: Let $a - c = f$ and $b - d = g$. Then

$a = c + f$, and $b = d + g$	Definition 3-5
$a + b = (c + f) + (d + g)$	Theorem 3-4
$= c + [f + (d + g)]$	Associative property of addition
$= c + [(f + d) + g]$	Associative property of addition
$= c + [(d + f) + g]$	Commutative property of addition
$= c + [d + (f + g)]$	Associative property of addition
$= (c + d) + (f + g)$	Associative property of addition
$(a + b) - (c + d) = f + g$	Definition 3-5
$(a - c) + (b - d) = f + g$	Theorem 3-4

Therefore,

$(a + b) - (c + d) = (a - c) + (b - d)$ Transitive property of equalities

Example: $(8+6)-(3+4)=(8-3)+(6-4)$, since

$$14-7=5+2.$$

Multiplication also has an inverse operation. We call the inverse of multiplication the operation of *division*. For instance, since $(2)(8)=16$, then 16 divided by 8 equals 2, and 16 divided by 2 equals 8.

Definition 3-6: If a is a whole number and b is a natural number, then a divided by b is equal to a whole number d if and only if $a=bd$.

The operation of division is commonly denoted symbolically in one of three ways:

$$a \div b = d, \quad a/b = d, \quad \text{or} \quad \frac{a}{b} = d.$$

In any of these cases, a is called the *dividend*, b is called the *divisor*, and d is called the answer or *quotient*.

The division concept as introduced in this definition is sometimes called *exact division* when compared with the results of the *Division Algorithm* in Chapter 4. From Definition 3-6, there is one number d such that $a=bd$; whereas in the *Division Algorithm* there are two numbers d and r such that $a=bd+r$. That is, the product of b and d is not exactly equal to a; it differs by a remainder r.

Examples:

$$\begin{aligned} 16 \div 8 &= 2 \quad \text{because} \quad (8)(2) = 16 \\ 21 \div 7 &= 3 \quad \text{because} \quad (7)(3) = 21 \\ 7 \div 4 &= ? \quad \text{because} \quad (4)(?) = 7 \end{aligned}$$

Obviously, 4 does not divide 7, because there is no whole number which when multiplied by 4 gives a result of 7. Thus $7 \div 4$ is not possible on the set of whole numbers.

Example: Solve $3x=15$.

By Definition 3-6, since $15=3x=x \cdot 3$, then $15/3=x$. $15/3=5$. Thus, the solution of $3x=15$ is $x=5$.

The number 0 plays a very special role in division. Consider the problem $0 \div 9 = ?$. Written as a multiplication problem, $9(?)=0$;

the answer is clearly 0. Since the answer to this problem is 0 no matter what numbers are used instead of 9, we can generalize and say that 0 divided by any whole number unequal to 0 is always 0.

Now consider the problem $4 \div 0 = ?$. As a multiplication problem, this can be written as $4 = 0(?)$. There is no solution to this problem, since 0 multiplied by any number is 0 and not 4. Consider the definition $a \div b = d$ where d is defined such that $a = bd$. From $a = bd$, it is obvious that when $b = 0$, then $a = 0$; thus $a \div b$, where $a \neq 0$ and $b = 0$ is impossible. For the case $a = 0$ and $b = 0$, d can have any value, and the equality $a = bd$ still holds. Thus, $0 \div 0$ has no unique answer. Division by 0 ($a \div 0$) where $a \neq 0$ is impossible, and division by 0 ($a \div 0$) where $a = 0$ has no unique answer. We avoid both situations by *excluding all division by 0 by saying that division by 0 is undefined.*

To summarize:

> If $a \neq 0$, then $0 \div a = 0$, because $a \cdot 0 = 0$.
> If $a \neq 0$, then $a \div 0$ is undefined, because $a \neq 0 \cdot (?)$.
> $0 \div 0$ is undefined, because there is no unique answer.

Exercises Set 3-5

1. Illustrate on the number line the following operations.

(a) $5 + 7 = 12$	(b) $8 + 5 = 5 + 8$
(c) $4 \cdot 3 = 12$	(d) $6 + 8 = 14$
(e) $12 + 7 = 7 + 12$	(f) $2 \cdot 5 = 10$
(g) $4 + 3 = 3 + 4$	(h) $5 \cdot 4 = 4 \cdot 5$
(i) $(5 + 2) + 8 = 5 + (2 + 8)$	(j) $(2 + 3) + 4 = 2 + (3 + 4)$
(k) $2(3 + 4) = 2 \cdot 3 + 2 \cdot 4$	(l) $2(3 \cdot 1) = (2 \cdot 3) \cdot 1$

2. The Archimedean property is illustrated for different numbers on four number lines below. Find a, b, and k in each case.

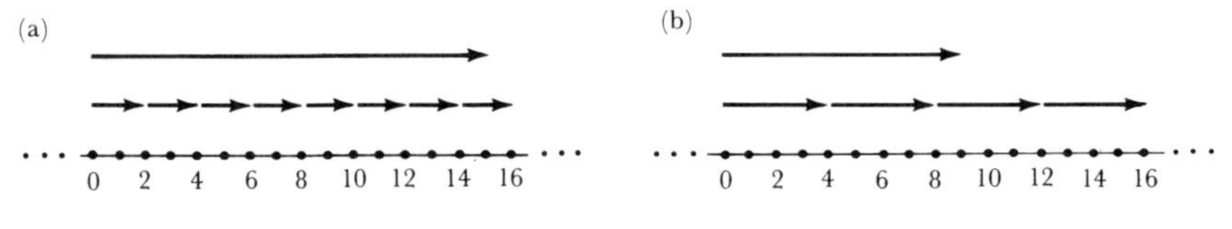

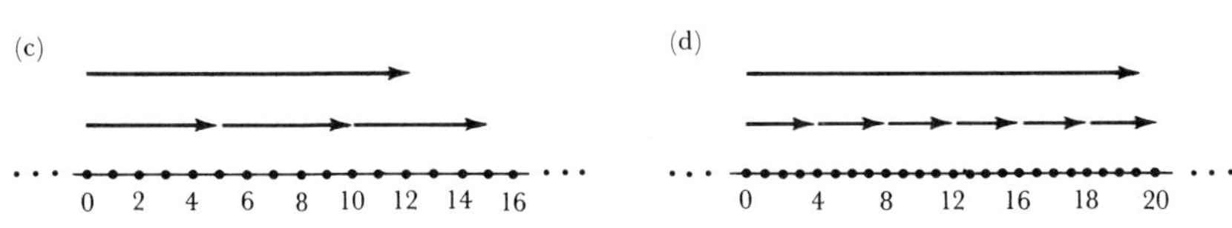

3. By means of Definition 3-3, classify the inequalities either as true or as false.

(a) $6 < 9$ (b) $8 < 8$ (c) $1 < 1$
(d) $5 < 7$ (e) $7 + 3 \geq 10$ (f) $8 + 9 > 13$
(g) $3 + 3 \leq 6$ (h) $4(6 + 2) \leq 27$

4. Complete the following sentences.
   (a) $(10 \div 2) =$ __ because $10 =$ ___ ___.
   (b) $(0 \div 7) =$ __ because $0 =$ ___ ___.
   (c) $(\_\_ \div 3) = 4$ because __ $= 3 \cdot$ ___.
   (d) $(18 \div \_) = 3$ because __ $=$ ___ ___.
5. Solve for x in the following equations.
   (a) $x + 6 = 11$ (b) $x + 7 = 9$
   (c) $3x = 12$ (d) $2x = 16$
   (e) $5 + x = 12$ (f) $17 + x = 20$
   (g) $6 = (2 \cdot 3) + x$ (h) $6x = 0$
   (i) $48 - 24 = x \cdot 6$ (j) $6 + x = 6$
6. Perform the following operations if possible; otherwise state that no answer exists for whole numbers.
   (a) $6 \div 0$ (b) $0 \div 7$ (c) $(3 - 3) \div 4$
   (d) $14 \div (2 - 2)$ (e) $0 \cdot 4$ (f) $(0 \cdot 4) \div 2$
7. Perform the following operations if possible; otherwise state that no answer exists for whole numbers.
   (a) $6 - 4$ (b) $2 - 7$ (c) $4 - (3 + 2)$
   (d) $8 + 4 - 7$ (e) $6 - (8 - 5)$ (f) $6 - (12 - 3)$
8. Let $A = \{a \mid a \text{ is a whole number } < 3\}$ and $B = \{b \mid b \text{ is a whole number } \leq 6\}$.
   (a) List the elements of $A \cup B$.
   (b) List the elements of $A \cap B$.
9. Given that $4 < 6$.
   (a) What can you say about $4 + 3$ and $6 + 3$?
   (b) What can you say about $2(4)$ and $2(6)$?
   (c) What can you say about $4 + b$ and $6 + b$ where b is any whole number?
   (d) What can you say about $4b$ and $6b$ when b is any natural number?
10. What would be a transitive property of less than? Illustrate with three whole numbers.
11. Construct examples to prove that division and subtraction are not associative or commutative.
12. For each of the following statements, write two other statements that can be derived from the given statement.
    (a) $5 + 16 = 21$ (b) $(7 + 4) - 3 = 8$
    (c) $a - c = f$ (d) $2(8 + 5) - 4 = 22$
13. The following statements are counterexamples disproving certain properties. State the property disproved in each case.

(a) $(18 \div 6) \div 3 \neq 18 \div (6 \div 3)$
(b) $(15 - 4) - (3 - 2) \neq (15 - 4) - (2 - 3)$
(c) $(6 \cdot 15) \div (6 \cdot 5) \neq 6(15 \div 5)$
(d) $(24 - 20) - 4 \neq 24 - (20 - 4)$
(e) $(48 \div 12) \div 4 \neq 4 \div (48 \div 12)$

*14. Consider the subtraction $6 - 2$. Write

$$6 = n\{1, 2, 3, 4, 5, 6\} = n(A)$$

and

$$2 = n\{1, 2\} = n(B).$$

Define subtraction to be $n(A - B) = n\{3, 4, 5, 6\} = 4$. Make up several examples to illustrate this definition. What happens when $A \subset B$? When $A = B$?

*15. Using $12 \div 3$, consider division as repeated subtraction. Then define division in terms of repeated subtraction.

9

Summary

By the system of whole numbers, we mean the set

$$W = \{0, 1, 2, 3, \ldots\},$$

the binary operations, addition $(+)$, multiplication $(\cdot)$, and the following properties. If a, b, and c are elements of W, then

Closure Properties

1. $a + b$ is a whole number.
2. $a \cdot b$ is a whole number.

Commutative Properties

3. $a + b = b + a$.
4. $(a)(b) = (b)(a)$.

Associative Properties

5. $a + (b + c) = (a + b) + c$.
6. $a(b \cdot c) = (a \cdot b)c$.

Distributive Property of Multiplication over Addition

7. $a(b + c) = ab + ac$.

Identities

8. Additive identity, $0 + a = a$.
9. Multiplicative identity, $1 \cdot a = a$.

Properties of Zero

10. $0 \cdot a = 0$.
11. $0 \div a = 0$ if $a \neq 0$.
12. $a \div 0$ is meaningless.

Trichotomy Property

13. Either $a < b$, $a = b$, or $a > b$.

Additional Properties

14. $a < b$ if and only if $a + c = b$ where $c > 0$.
15. If $a = b$, then $a + c = b + c$ and $ac = bc$.
16. If $a = b$ and $c = d$, then $a + c = b + d$ and $ac = bd$.
17. Inverse operations of addition and multiplication are so defined that

$$a - b = c \quad \text{if and only if } a = b + c.$$

$$\frac{a}{b} = c \quad (b \neq 0) \quad \text{if and only if } a = bc.$$

18. $a(b - c) = ab - ac$ if $b \geq c$.
19. $(a + b) - c = a + (b - c)$ where $b \geq c$.
20. $(a + b) - (c + d) = (a - c) + (b - d)$ where $a \geq c$ and $b \geq d$.

In arithmetical expressions without parentheses such as $2 \cdot 4 + 5$, it is conventional to do the multiplication and then the addition. As a general rule, first do multiplications and divisions, then additions and subtractions.

Example:

$$8 - 12 \div 4 = 8 - 3 = 5$$
$$8 - 2 \cdot 3 = 8 - 6 = 2$$

Review Exercise Set 3-6

1. Select the symbols that represent 0.
   (a) $5 - 5$ (b) $0 \div 4$

(c) $4 \div (3 - 3)$ (d) $4(2 - 2)$
(e) $(5 - 5) \div 2$ (f) $(2 \cdot 3) \div 0$

2. What properties are used in the following steps to rename the expression $y(m + n) + my$?
(a) $y(m + n) + my = (ym + yn) + my$
(b) $(ym + yn) + my = ym + (yn + my)$
(c) $ym + (yn + my) = ym + (my + yn)$
(d) $ym + (my + yn) = (ym + my) + yn$
(e) $(ym + my) + yn = (my + my) + ny$
(f) $(my + my) + ny = (1 \cdot my + 1 \cdot my) + ny$
(g) $(1 \cdot my + 1 \cdot my) + ny = (1 + 1)my + ny$
(h) $(1 + 1)my + ny = 2my + ny$
(i) $y(m + n) + my = 2my + ny$

3. Verify that the following are true. Give the reason for each step of the discussion.
(a) $(110)(9) - (9)(6) = 9(110 - 6)$
(b) $3(a + 0) = (a)(3)$
(c) $4(23 \cdot 25) = 25(23 \cdot 4)$
(d) $(108)(6) - (6)(108) = 0$
(e) $c(d \cdot e) = e(d \cdot c)$
(f) $a + (b + c) = c + (b + a)$

4. Show that $(a + b)(c + d) = (ac + ad) + (bc + bd)$ and justify each step.

5. Show that $(a + b)(a + b) = (a \cdot a) + (2 \cdot a \cdot b) + (b \cdot b)$ and justify each step.

6. Group the numbers 2, 4, 5, and 8 as a sum in several different ways. Carry out the additions and compare the results. Do the same for multiplication.

7. Select the true statements from below.
(a) If $n + 6 = 11$, then $n \geq 5$.
(b) If $n + 89 = 91$, then $n < 2$.
(c) If $n + 13 = 12$, then n is not a whole number.
(d) If $n + 16 = 14$, then $1 < n < 2$.
(e) If $0 - 6 = n$, then n is a whole number.
(f) If $0 \div 6 = n$, then n is a whole number.
(g) If a is a natural number, then $\dfrac{a - 0}{a} = 1$.
(h) If a is a natural number, then $\dfrac{a}{a - a} = 1$.

*8. Solve each of the following problems.
   (a) Julie brought a cake cut into 16 pieces to work. If she shares the cake with seven other friends (plus herself), how much cake will each person have?
   (b) If five girls weigh a total of 560 pounds, what is the average weight of each girl?
   (c) Mary Francis perked coffee in an urn holding 20 cups. Joyce drank three cups. How many cups are left for the other secretaries?
   (d) Rick and Joyce traveled 160 miles to Atlanta. On their way back home, they stopped to eat 32 miles outside of Atlanta. How far were they from home?

*9. Assume that $\boxdot$ and $\oplus$ represent operations defined on set S, when x, y, and z are elements of S. Describe what the following mean.
   (a) $\boxdot$ is commutative.
   (b) $\oplus$ is commutative.
   (c) $\boxdot$ is associative.
   (d) $\oplus$ is associative.
   (e) $\boxdot$ is distributive over $\oplus$.

Suggested Reading

Addition: Allendoerfer, pp. 159–197. Armstrong, pp. 33–41. Bouwsma, Corle, Clemson, pp. 13, 26–33. Byrne, pp. 2, 33–41. Campbell, pp. 42–50. Copeland, pp. 44–49. Garner, pp. 47, 130–136. Graham, pp. 88–94. McFarland, Lewis, pp. 84–90. Ohmer, Aucoin, Cortez, pp. 78–86. Nichols, Swain, pp. 117–120. Peterson, Hashisaki, pp. 82–85. Podraza, Blevins, Hanson, Prall, pp. 36–45. Scandura, pp. 179–184. Smith, pp. 64–74. Spector, pp. 176–187. Willerding, pp. 50–54. Wren, pp. 45–55. Zwier, Nyhoff, pp. 40–45.

Multiplication: Allendoerfer, pp. 213–222. Armstrong, pp. 41–49. Bouwsma, Corle, Clemson, pp. 84–92. Byrne, pp. 42–50. Campbell, pp. 42–43. Copeland, pp. 97–126. Garner, pp. 49–63, 66–68. Graham, pp. 95–107. McFarland, Lewis, pp. 96–100. Nichols, Swain, p. 122. Ohmer, Aucoin, Cortez, pp. 86–103. Peterson, Hashisaki, pp. 86–92. Podraza, Blevins, Hanson, Prall, pp. 51–57. Scandura, pp. 188–192. Spector, pp. 174–183, 190–199. Smith, pp. 92–105. Willerding, pp. 54–62. Wren, pp. 58–66. Zwier, Nyhoff, pp. 49–60.

Ordering, Number Line: Allendoerfer, pp. 197–202. Armstrong, pp. 102–103. Bouwsma, Corle, Clemson, pp. 14–16, 37. Byrne, pp. 100–101, Campbell, pp. 50–52. Garner, pp. 43–46, 68–75. Graham, pp. 117–123. Hutton, p. 31. Meserve, Sobel, pp. 165–166. Ohmer, Aucoin, Cortez, pp. 103–110, 174. Peterson, Hashisaki, pp. 100–102. Podraza, Blevins, Hanson, Prall, pp. 36–37, 46, 52. Scandura, pp. 87–91, 283–284. Spector, pp. 248, 250, 290. Willerding, pp. 46–48. Zwier, Nyhoff, pp. 127, 220–221.

Subtraction and Division: Allendoerfer, pp. 203–212, 223–234. Armstrong, pp. 57–74. Bouwsma, Corle, Clemson, pp. 49–56, 149–162. Byrne, pp. 104–121. Campbell, pp. 58–61. Garner, pp. 76–82, 145–154. Graham, pp. 107–112. Hutton, p. 107. Meserve, Sobel, p. 156. Nichols, Swain, pp. 128–131. Ohmer, Aucoin, Cortez, pp. 150–152. Podraza, Blevins, Hanson, Prall, pp. 45–51, 57–60. Scandura, pp. 185–187, 192–195. Spector, pp. 255–256, 267. Smith, pp. 80–85, 111–118. Willerding, pp. 64–69. Wren, pp. 80, 89–93. Zwier, Nyhoff, pp. 91, 94, 97–100, 116–117.

4

Numeration Systems

1

History of Numeration Systems

In Chapter 3 we learned that the name or symbol for a number is called a *numeral.* A set of symbols for representing numbers and the rules or principles for combining the symbols to name numbers is called a *system of numeration.*

Our modern numeration system did not suddenly appear in its completed form. Like many great inventions of the human mind, it has developed over the centuries. Since many early attempts to represent numbers are lost in the dimness of the past, we discuss in this section only a part of the development of numerals. Archaeologists have uncovered interesting facts about numeration systems of various civilizations. In order to appreciate our own system, it helps to examine several early systems of numeration.

One of the earliest systems of numeration was the *tally* system. Marks or strokes were used to represent numbers. For example, *//////* could represent a herd of six animals.

Early systems of numeration may be characterized as additive systems, multiplicative systems, and place-value systems. Additive systems rely primarily on the addition of numbers represented by symbols. That is, a number represented by a set of symbols is the sum of the numbers represented by the individual symbols.

One of the earliest additive systems (of which we have a record) was the Egyptian system. The oldest known mathematical book was written on papyrus by an Egyptian scribe named Ahmes more than 35 centuries ago. This *Ahmes Papyrus* shows how ingenious the Egyptians were in inventing ways to deal with numbers. The early Egyptian numerals were hieroglyphic, or picture symbols. A stroke was used to represent 1; the heel bone ∩ represented 10; a coiled rope or scroll 𓍢 represented 100; a lotus flower 𓆼, 1000; a pointed finger 𓂭, 10,000; a burbot fish 𓆐, 100,000; and, finally, an astonished man 𓁨 represented 1 million. A great number of astonished men would thus be necessary to represent numbers used in government circles today. The symbols in the Egyptian numeration system could be placed in any order to represent the same number. For example, 43 might have been written in several ways: ∩∩∩∩|||, or ∩|∩|∩|∩ , or ∩∩|||∩∩. Thus, there was no "place value" in this system (whereas in ours, 916 is different from 196).

The age of the Greek influence on numeration systems extended roughly from 600 B.C. to A.D. 100. The Greek system was very simple; it was based on ten but used neither zero nor place value. Actually, the Ionic Greek numeration system used the letters of the Greek alphabet to represent numbers: α for 1, β for 2, γ for 3, δ for 4, and so on.

About 3000 B.C., the great Babylonian civilization flourished in that part of the world known today as the Middle East. This civilization wrote numerals on clay tablets with a piece of wood. The vertical wedge ▼ was used to represent 1, and the symbol ◀ represented 10. Symbols for numbers from 1 to 59 were formed by repeated use of these two symbols. For example, ◀◀▼▼▼ represented 23. Sixty was represented by the same symbol that represented 1. Thus, the Babylonian numeration system seemed to use the idea of place value even though the Babylonians did not have a symbol for zero. To distinguish between the symbol for 60 and the symbol for 1, a wider space was left between characters.

Examples:

	▼	◀◀▼	or	$60 + 21 = 81$
	◀▼	◀▼▼▼	or	$11(60) + 13 = 673$
▼▼	◀◀	◀◀◀	or	$2(60)^2 + 20(60) + 30 = 8430$

At the peak of its civilization, around A.D. 100, the Roman Empire needed an elaborate system of numeration for keeping records and accounting, a situation created by the collection of taxes and by commerce in the vast empire. Essentially, the Roman system was an *additive* system with *subtractive* and *multiplicative* features. If symbols decrease in value from left to right, their values are to be added; however, if a symbol has a smaller value than a symbol on the right, it is to be subtracted. For example, $CX = 100 + 10 = 110$, but $XC = 100 - 10 = 90$.

Roman numerals are still in use today. You recognize that the letters I, V, X, L, C, D, and M represent 1, 5, 10, 50, 100, 500, and 1000, respectively. Not more than three identical symbols are used in succession. For example, IV was used for 4 instead of IIII. Likewise, there were never more than two symbols involved in the subtractive feature.

Examples: LXXVI = 76; XLIV = 44

To write large numbers using Roman numerals, one utilizes a multiplicative feature involving bars above the symbols. For example, $\overline{\text{V}}$ indicates 5 multiplied by 1000, or 5000. $\overline{\overline{\text{V}}}$ represents 5,000,000. Thus, a symbol with a bar above it indicates a number represented by the symbol multiplied by 1000; a double bar means multiplication by 1,000,000.

Examples:

(a) $\overline{\text{IV}}\text{DCXLVII} = 4{,}647$

(b) $\overline{\overline{\text{L}}}\text{MDXXI} = 50{,}001{,}521$

Early Chinese and Japanese numeration systems used both the additive and multiplicative principles to represent a number. Probably the best-known contribution of the early Chinese civilization was the invention of the abacus. Even today, the abacus is widely used in Asia. It is also a good teaching aid in the elementary school, and children generally seem eager to learn how to use it.

The Chinese abacus contained a bar that separated the beads on each rod into two sets of two and five beads each, as illustrated in Figure 4-1. Each bead above the bar had associated with it a value five times that of the bead below the bar on the same rod. Numbers were indicated by moving the beads toward the dividing bar. Figure 4-1 demonstrates an arrangement of the beads representing the number 3248.

The Japanese abacus is very similar to the Chinese abacus except that the dividing bar separates the beads on a rod into sets of four and one. The one bead represents 5; thus, one can count up to 9 on a given rod.

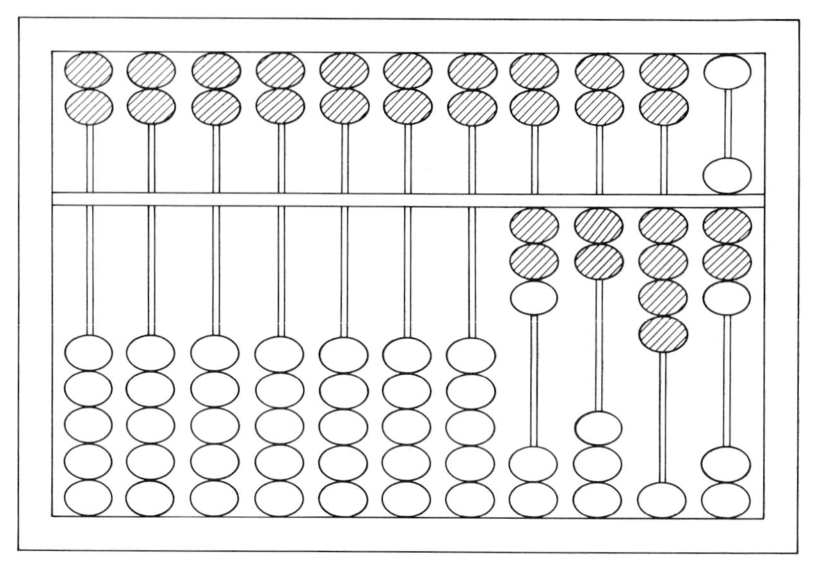

Figure 4-1

Another interesting numeration system comes from the early civilizations of Guatemala and Honduras. Although no one has found a clear-cut way to decipher the thousands of characters and symbols left by these people, it is evident that an advanced civilization lived in the area. Recent archaeological discoveries indicate that some other civilization preceded that of the Mayan Indians, who have received credit for developing an advanced numeration system and an unusual calendar. Although there is evidence that the numeration system was developed before the advent of Mayan Indians, most people still describe the numeration system they used as the Mayan numeration system; we will use that terminology in this book. At one time, this numeration system seems to have been based on 5; however, it was developed to the degree that it was finally based on 20. Instead of using figures to represent a number, this system used a set of dots and horizontal lines. Each dot was a unit, and each line represented 5. The largest number that was recorded in one place with these symbols was 19, three bars and four dots: ····. Numbers larger than 19 were written in terms of base 20, making use of a symbol for zero that looked somewhat like a football,.

Their system used $18(20)^2$ instead of $(20)^3$, $18(20)$ instead of $(20)^2$, but in spite of this discrepancy, their place-value system of numeration was very ingenious. The Mayans (as well as the Chinese and Japanese) wrote their numerals in vertical form.

Examples:

(a)
$$1(20) = 20$$
$$0 = 0$$
$$\overline{20}$$

(b)
$$2(20) = 40$$
$$4 = 4$$
$$\overline{44}$$

(c)
$$1(18)(20) = 360$$
$$0(20) = 0$$
$$5 = 5$$
$$\overline{365}$$

(d)
$$6(18)(20) = 2160$$
$$6(20) = 120$$
$$0 = 0$$
$$\overline{2280}$$

(e)
$$2(18)(20)^2 = 14{,}400$$
$$6(18)(20) = 2{,}160$$
$$0(20) = 0$$
$$8 = 8$$
$$\overline{16{,}568}$$

The numerals we use today to represent numbers were invented in India and brought to Europe by the Arabs. Consequently, our numeration system is called the Hindu-Arabic system of numeration. Its important characteristics are as follows:

(a) Each symbol represents a number.
(b) The position of the symbol in the numeral has a fixed meaning.
(c) Different symbols represent different numbers.
(d) All numbers are constructed from a small number (ten) of basic symbols.

The Hindu-Arabic system probably had its beginnings about the third century B.C. The date of the invention of zero is unknown; yet we do know that it was in existence before the ninth century A.D. The introduction of zero as a symbol denoting the absence of units or of certain powers of ten has been rated as one of the greatest inventions of all time. It completely revolutionized the formation of numeration systems.

The Hindu-Arabic numerals reached Europe through traders. At first, the European people clung to the familiar Roman numerals and ignored the new system. For about four hundred years, there was a battle between those who favored the Roman system and the advocates of the Hindu-Arabic system; nevertheless, by the thirteenth century, the Hindu-Arabic system was well established throughout Europe.

It seems remarkable that so many thousands of years passed before man thought of a really simple way of representing numbers. Today, it all seems so obvious that it is difficult for us to realize the years of struggle and effort that lie behind our present arithmetic processes. As we follow the development of mathematics through the ages, we see again and again that concepts and processes that appear difficult to one generation become clear and matter-of-fact to a later generation.

Exercise Set 4-1

1. What numbers are represented by the following Egyptian symbols?
   (a) IIII (b) ∩∩II (c) ∩∩∩IIIII
2. Write the Hindu-Arabic symbols that represent the same numbers as the following Babylonian numerals.
   (a) ▼▼ (b) ◀◀ (c) ◀▼
   (d) ◀ ◀ ▼▼ (e) ▼ ◀◀▼ ▼▼ (f) ◀▼ ◀◀ ◀
3. Write the following Roman numerals in Hindu-Arabic notation.
   (a) XXI (b) XIV (c) XLIX
   (d) CXXI (e) CLVI (f) LXVI
   (g) $\overline{\text{III}}$DCCCLXXXVIII (h) $\overline{\text{XIII}}$CDXLIV
   (i) $\overline{\text{XX}}$DCLXVI (j) $\overline{\text{L}}$DCX
4. What numbers are represented by the following Mayan symbols?

(a)	(b)	(c)	(d)	(e)	(f)
• —	• ═	•• —	• ≡	•••• —	••• ≡

(g)	(h)	(i)	(j)
•• • — •••	• — ═ ••• —	•• •• — •• ═ ≡	• — •• ═ • ≡ ••• ═

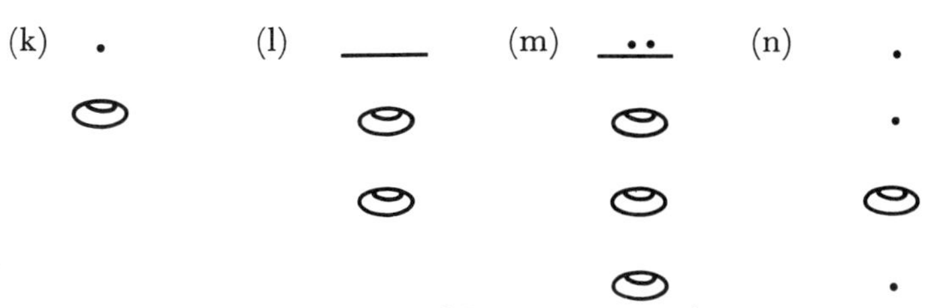

5. Write the following in terms of Roman numerals.
   (a) 76 (b) 101 (c) 189
   (d) 44 (e) 148 (f) 948
6. As a basis of comparison for the various systems we have discussed, complete the following table. If one system does not utilize a single symbol, use a combination of symbols.

Hindu-Arabic	0	1	5	10	50	100
Egyptian						
Roman						
Babylonian						
Mayan						

*7. Consider the five numeration systems described above. Give at least one advantage and at least one disadvantage of each system.
*8. Find the number (in the same system) that is 10 less than the given number.
   (a) CIV (b) 𓍢∩∩∩∩I∩∩I
   (c) ◀ ◀◀ ▼ (d) 1669
   (e) • / —— / ⬭

2

Understanding Our Numeration System

One should not overlook the fact that numeration systems have meaning only by agreement. We agree on ten as the base of the numeration system we use today. That is, we do our counting by grouping repeatedly by ten. Thirty-six apples can be grouped as three groups of ten apples and one group of six apples. Because our system uses groups of ten, it is called a *decimal* system. (The word "decimal" is derived from the Latin word *decem*, which means "ten.")

Long ago, man learned the advantages of grouping objects for counting. We use the same idea today in our monetary system by letting a dime represent ten pennies and a dollar, ten dimes. By counting in sets of ten, we need but ten symbols in our numeration system. These symbols, called *digits*, are written 0, 1, 2, 3, 4, 5, 6, 7, 8, and 9. With these ten digits, we can represent any whole number whatsoever. (The word *digit* means "a finger or toe," and it is natural to guess that our number system is based on ten digits *because* we have ten fingers.)

The decimal system uses the idea of *place value* to represent the size of a group in a grouping process. The size of the group represented by a digit depends on the position of the digit in the numeral. The digit also tells us how many we have of a particular group. In the numeral 234, the 2 represents two groups of one hundred (200), the 3 represents three groups of ten (30), and the 4 represents four ones (4). The idea of place value makes the decimal system very convenient in many ways.

Example: The digit 4 is found in each of the numerals 654, 456, and 546. In each instance, because of its place, it has a different value. In 654, the 4 represents four ones; in 456, it represents four hundreds; and in 546, it represents four tens.

Each successive place to the left in a base ten numeral represents a group ten times that of the preceding group. If one begins at the right and moves toward the left, the first place digit tells how many groups of one; the second place indicates the number of groups of 10; the third place tells how many groups of 10 times 10; the next place gives the number of groups of 10 times 10 times 10, and so on. Thus, 2346 is an abbreviation for

$$2(10 \cdot 10 \cdot 10) + 3(10 \cdot 10) + 4(10) + 6(1).$$

In order to simplify the preceding ideas, we give meaning to what are called *powers* of numbers. Let a be a whole number. When we write the symbol a^2, we mean $a \cdot a$; by a^3 we mean $a \cdot a \cdot a$. These concepts lead to the following definition.

Definition 4-1: If a and n are any natural numbers, then a^n is the product obtained by using "a" as a factor n times.

In this definition, the superscript n is called the *exponent*; the number a is called the *base*; and the complete symbol is read "the nth power of a" or "a to the nth power." A number expressed in the form a^n is said to be written in *exponential form*, a notation introduced by a French mathematician, Descartes, in the sixteenth century. For example, when considering the exponential 6^3, we call 6 the base and 3 the exponent; and 6^3 is read "the third power of 6," or "the cube of 6," or "6 cubed," and it means $6 \cdot 6 \cdot 6$.

Since $a^3 = a \cdot a \cdot a$ and $a^4 = a \cdot a \cdot a \cdot a$, then

$$(a^3)(a^4) = (a \cdot a \cdot a) \cdot (a \cdot a \cdot a \cdot a) = a^7.$$

More generally, if m and n are natural numbers,

$$a^m \cdot a^n = \underbrace{(a \cdot a \cdot a \ldots \cdot a)}_{m \text{ factors}} \cdot \underbrace{(a \cdot a \cdot a \ldots \cdot a)}_{n \text{ factors}} = a^{m+n}.$$

In order to divide a^5 by a^3 $(a \neq 0)$, we use Definition 3-6, and a whole number must exist such that, when the whole number is multiplied by a^3, the multiplication yields an answer of a^5. We recognize the answer to be a^2, since $a^3 \cdot a^2 = a^5$. Thus

$$a^5 \div a^3 = a^2 = a^{5-3}.$$

Similarly, $a^m \div a^n$, where $a \neq 0$ and $m \geq n$ is a^{m-n} because

$$\begin{aligned} a^n \cdot a^{m-n} &= a^{n+(m-n)} \\ &= a^{(n+m)-n} \\ &= a^{(m+n)-n} \\ &= a^{m+(n-n)} \\ &= a^m. \end{aligned}$$

Theorem 4-1: *Properties of Whole Numbers in Exponential Form.* If a is any whole number unequal to zero and m and n are whole numbers, then

(a) $a^m \cdot a^n = a^{m+n}$, (b) $a^m \div a^n = a^{m-n}$ if $m \geq n$,

(c) $(a^m)^n = a^{mn}$.

Example: $(5^2)^3 = 5^2 \cdot 5^2 \cdot 5^2 = (5 \cdot 5)(5 \cdot 5)(5 \cdot 5) = 5^6$
or $(5^2)^3 = 5^{2(3)} = 5^6$

In order for the preceding concepts to hold for all whole-number exponents, a^0 $(a \neq 0)$ must be defined to be 1. This will be illustrated by the following example.

$$a^3 \div a^3 = a^{3-3} = a^0$$

by property (b) above; also, $a^3 \div a^3 = 1$, because $a^3 = a^3 \cdot 1$. Thus, a^0 should be defined to be 1 if $a \neq 0$.

Definition 4-2: If a is any natural number, then $a^0 = 1$.

Since 10 has been selected as the base of our numeration system, let us examine successive powers of 10.

$$
\begin{aligned}
10^0 &= 1 \\
10^1 &= 10 \\
10^2 &= (10)(10) = 100 \\
10^3 &= (10)(10)(10) = 1000 \\
10^4 &= (10)(10)(10)(10) = 10{,}000 \\
10^5 &= (10)(10)(10)(10)(10) = 100{,}000
\end{aligned}
$$

Thus, 2346 can now be written in terms of decreasing powers of 10 as

$$2(10)^3 + 3(10)^2 + 4(10)^1 + 6(10)^0.$$

Since 10^3 is 1000, 10^2 is 100, 10^1 is 10, and 10^0 is 1, this symbol is read "two thousand three hundred forty-six." The 6 is called the *units digit*; 4 is the *tens digit*; 3 is the *hundreds digit*; and 2 is the *thousands digit*. In the same way, 386,472 means

$$3(10)^5 + 8(10)^4 + 6(10)^3 + 4(10)^2 + 7(10)^1 + 2(10)^0.$$

Note that the digits represent or "count" powers of 10 in a strictly decreasing order.

Examples:

$$
\begin{aligned}
4{,}321 &= 4(10)^3 + 3(10)^2 + 2(10)^1 + 1(10)^0 \\
15 &= 1(10)^1 + 5(10)^0 \\
12{,}000 &= 1(10)^4 + 2(10)^3 + 0(10)^2 + 0(10)^1 + 0(10)^0
\end{aligned}
$$

These numbers are said to be written in *expanded form*.

The associative and distributive properties of whole numbers studied in Chapter 3 are useful when working with exponents, as indicated by the following examples.

Examples:

$$
\begin{aligned}
5b^2 + 3b^2 &= (5 + 3)b^2 = 8b^2 \\
5x^2 + 3x^2 + 4x + 2x &= (5 + 3)x^2 + (4 + 2)x = 8x^2 + 6x \\
7(10)^2 + 2(10)^2 &= (7 + 2)(10)^2 = 9(10)^2 \\
5(10)^2 + 3(10)^2 + 4(10)^1 + 2(10)^1 & \\
&= (5 + 3)(10)^2 + (4 + 2)(10)^1 = 8(10)^2 + 6(10)^1 \\
3(10)^3 + 2(10)^3 + 4(10)^3 &= (3 + 2 + 4)(10)^3 = 9(10)^3
\end{aligned}
$$

Man has always looked for ways and methods of making his work easier. The procedures developed to facilitate the addition of numbers are a result of man's efforts to find an easier way. First of all, he made use of the associative property of addition to find a process for placing an addition problem such as $2 + 3 + 7 + 5$ in a column:

$$\begin{array}{r} 2 \\ 3 \\ 7 \\ 5 \\ \hline \end{array}$$

By the repeated use of the associative property of addition one can add $5 + 7$; and then to that sum add 3, $[(5 + 7) + 3]$, and to that sum add 2, as

$$[(5 + 7) + 3] + 2.$$

In the same way, one can add down the column by first adding $2 + 3$ and to that sum adding 7, $[(2 + 3) + 7]$, and then adding 5 to the answer,

$$[(2 + 3) + 7] + 5.$$

The use of place value also enabled man to simplify computations in addition, multiplication, subtraction, and division. It is important to realize that the short-cut procedures we ordinarily use in arithmetic computations produce the correct answers through numerical manipulation based on the properties of addition, multiplication, subtraction, and division discussed in the preceding chapter. The reasons for each step in the numerical manipulation will be studied in the next section. However, we need to understand the importance of place value before analyzing each step of the manipulation. In the examples that follow, we present the usual short-cut methods of computing on the left and compare them with the steps involving place value, shown on the right. In the next section it will be shown that a combination of the associative and commutative properties, along with the distributive property of multiplication over addition (abbreviated as ACD), are utilized in writing the problem in column form for the computations shown on the right.

Examples:

Addition (a)

$$\begin{array}{r} 46 \\ 32 \\ \hline 78 \end{array} \qquad \begin{array}{r} 4(10) + 6 \\ 3(10) + 2 \\ \hline 7(10) + 8 \end{array}$$

(b)

$$\begin{array}{r} 421 \\ 176 \\ \hline 597 \end{array} \qquad \begin{array}{r} 4(10)^2 + 2(10) + 1 \\ 1(10)^2 + 7(10) + 6 \\ \hline 5(10)^2 + 9(10) + 7 \end{array}$$

Subtraction

$$\begin{array}{r} 87 \\ 35 \\ \hline 52 \end{array} \qquad \begin{array}{r} 8(10) + 7 \\ 3(10) + 5 \\ \hline 5(10) + 2 \end{array}$$

Multiplication

$$\begin{array}{r} 23 \\ 21 \\ \hline 23 \\ 46 \\ \hline 483 \end{array} \qquad \begin{array}{r} 2(10) + 3 \\ 2(10) + 1 \\ \hline 2(10) + 3 \\ 4(10)^2 + 6(10)\phantom{{}+3} \\ \hline 4(10)^2 + 8(10) + 3 \end{array}$$

Division

$$3\overline{)69} = 23 \qquad 3\overline{)6(10) + 9} = 2(10) + 3$$

Exercise Set 4-2

1. Using exponents, write in simple form the following numerals.
   (a) (2)(2)(2)(2)(2) (b) (6)(36)
   (c) (64)(16)(4) (d) 10,000 ÷ 100
   (e) (3)(3)(3)(3)(3)(3)(3) (f) 64 ÷ 16
2. Perform the indicated operations.
   (a) 3^4 (b) $(x^4)(x^7)$ (c) 9^0
   (d) $x^8 \div x^5$ (e) $(2^3)(2^0)$ (f) $3^4 \div 3^4$
   (g) $5^4 \div 5$ (h) $(49)^2 \div 7$ (i) $(27)^2 \div 9$
   (j) $x^3(x^4 \cdot x^2)$ (k) $(k^3 \cdot k^8) \div k^6$ (l) $4^3 \cdot (136)^0$
3. Using exponents, write the following numbers in expanded form.
   (a) 768 (b) 902 (c) 23,105
   (d) 147,258 (e) 7604 (f) 80,001
4. Find the value of n that makes the following equalities true.
   (a) $2^2 \cdot 2^n = 2^5$
   (b) $\dfrac{3^n}{3^4} = 3^7$
   (c) $5(10)^2 + 6(10)^2 = n(10)^2$

5. Compute each of the following sums using the associative and distributive properties.
   (a) $3c^2 + 5c^2$ (b) $6c^3 + c^3 + 2c^3$
   (c) $4(10)^3 + 2(10)^3$ (d) $3(10)^2 + 4(10)^2 + 5(10) + 3(10)$
   (e) $8(10)^3 + 7(10)^2 + 2(10)^2 + 6(10) + 10$
6. Suppose you are given the digits 3, 5, and 7.
   (a) What is the largest number that can be represented in base ten by these three digits?
   (b) What is the smallest number that can be represented in base ten by these three digits?
7. Compute the following by first writing in expanded form.
   (a) $24 + 15$ (b) $(23)(3)$
   (c) $53 - 21$ (d) $514 + 273$
   (e) $(21)(32)$ (f) $(12)(23)$
   (g) $768 - 254$ (h) $684 \div 2$
   (i) $724 + 235$ (j) $(214)(12)$

*8. Is it true that $(a^x)(a^y) = (a^y)(a^x)$, where a, x, and y are natural numbers? Why?

*9. Arrange in increasing order: 10^2, 7^3, 4^0, 2^{10}, 3^7, 9^4, 4^9.

*10. Show that it is not true in general that $a^x + a^y = a^{x+y}$. (*Hint:* You need only to find an exception such as $a = 10$, $x = 1$, $y = 2$.) Also, show that it is not generally true that $a^x + b^x = (a + b)^x$.

*11. Suppose that $a^x = b^y$ for whole numbers a, b, x, and y.
   (a) If $a = b$, is x necessarily equal to y?
   (b) If $x = y$, is $a = b$ always true?

3

Some Fundamental Algorithms

An *algorithm* is a systematic procedure for performing some mathematical operation such as addition or multiplication. Whenever we perform operations with single-digit numerals, we write down the answer from memory. However, when problems involve larger numbers, it is convenient to have a pattern to follow.

The term *algorithm* comes from the name of a Persian mathematician, Mohammed ben Musa al-Khowarizmi, who published a book on rules of computation in about the year 820 A.D.

The algorithms used today are generally refined versions of algorithms that have been passed down from generation to generation. Actually, most of the algorithms we use are not unique. As an example, there are several good patterns for performing multiplications. For instance, have you tried the *lattice method of multiplication?*

In Figure 4-2, the problem is to find the product of 476 and 753.

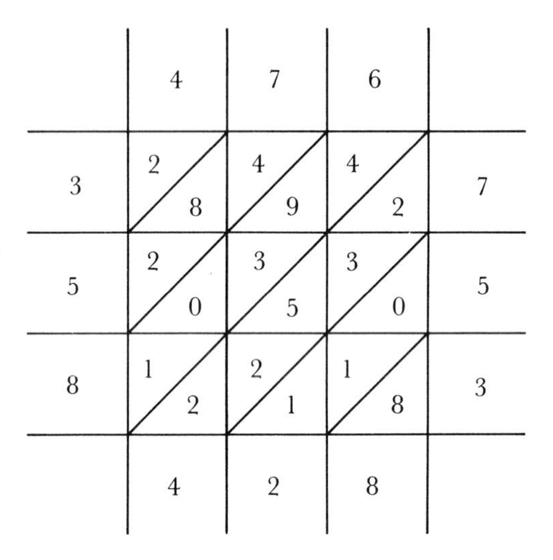

Figure 4-2

Since the two numbers to be multiplied are represented by numerals with three digits each, a square is drawn containing nine small squares of equal size. The numerals 4, 7, 6 are written across the top of the square and 7, 5, 3 are written along the right side from top down. Diagonals (from lower left to upper right) are drawn in each of the squares to form a *lattice* design. Products of pairs of digits taken from the top and the right side of the rectangle are now entered in the squares. The tens digit of the product is written above the diagonal and the units digit below the diagonal. Thus, in the first square, $6 \cdot 7 = 42$, so 4 is above the diagonal and 2 is below it. Now add the elements between adjacent diagonals, beginning at the lower right corner. There is only one element below the first diagonal. This is an 8, so write down an 8 in the space at the bottom of this diagonal. The sum of the elements between the next two diagonals is $1 + 1 + 0$, and the answer 2 is recorded below the area between diagonals. If the sum is more than 9, record the units digit as before and regroup the tens digit to be added to the elements between the next two adjacent diagonals. When all diagonal elements are totaled, the answer is read down the left side and along the bottom. The answer is 358,428.

In the preceding example, the two factors contained the same number of digits. If the two numbers to be multiplied contain a different number of digits, a rectangle rather than a square is drawn. Like the square, the rectangle is then divided into small squares of equal size, and the lattice is formed by drawing the diagonals. The multiplication of the numbers is performed as in the example.

This example of the lattice method of multiplication indicates that our usual algorithms are not the only methods for performing mathematical operations. However, it is believed that the algorithms in use today are the most effective. First, we verify the addition pattern we have used for many years to compute $17 + 8$.

$17 + 8 = (10 + 7) + 8$	Decimal system of numeration
$= 10 + (7 + 8)$	Associative property of addition
$= 10 + 15$	Addition
$= 10 + (10 + 5)$	Decimal system of numeration
$= (10 + 10) + 5$	Associative property of addition
$= (1 + 1)10 + 5$	Distributive property of multiplication over addition
$= 2 \cdot 10 + 5$	Addition
$= 25$	Decimal system of numeration

The preceding steps justify the conventional algorithm for computing $17 + 8$.

$$\begin{array}{r} {\scriptstyle 1} \\ 17 \\ 8 \\ \hline 25 \end{array}$$

Rename the sum of 7 and 8 as $10 + 5$ and place the 1 for the one ten above the 1 in 17. Then add the numbers in the tens column. We use the word *rename* or *regroup* in place of the term "carrying" since these words seem to describe more accurately the computational procedures involved.

In the next example, we extend the preceding algorithm to the addition of two digit numerals, such as 76 and 59. To shorten our work, we assume the transitive property of equality in each step and use the following abbreviations here and in the remainder of this chapter.

S	By our system of numeration
A_a	Associative property of addition.
A_m	Associative property of multiplication
C_a	Commutative property of addition
C_m	Commutative property of multiplication

D_a Distributive property of multiplication over addition
D_s Distributive property of multiplication over subtraction

Example: $76 + 59$

(a)	$7 \cdot 10 + 6$	$+ 5 \cdot 10 + 9$	S
(b)	$7 \cdot 10$	$+ [6 + (5 \cdot 10 + 9)]$	A_a
(c)	$7 \cdot 10$	$+ [(5 \cdot 10 + 9) + 6]$	C_a
(d)	$7 \cdot 10$	$+ [5 \cdot 10 + (9+6)]$	A_a
(e)	$7 \cdot 10 + 5 \cdot 10$	$+ 9 + 6$	A_a
(f)	$(7 + 5)10$	$+ 9 + 6$	D_a
(g)	$(7 + 5)10$	$+ 6 + 9$	C_a
(h)	$12(10)$	$+ 15$	Addition

Note that the first seven steps in this algorithm verify that the problem can be written for column addition. Step (h),

76	$7(10) + 6$	
59	$5(10) + 9$	
135	$12(10) + 15$	by ACD

performs this addition.

Step (g) was not necessary except to demonstrate that the problem could be written for column addition. In the preceding section, we denoted by ACD (a combination of the associative and commutative properties of addition and multiplication along with the distributive property of multiplication over addition) the reasons one could write the problem in column form as indicated above. However, both $6 + 9$ and $7 + 5$ give answers greater than 9. Steps (i) through (l) justify the renaming or regrouping from the units column to the tens column and from the tens column to the hundreds column. The remaining steps complete the computation.

(i)	$(1 \cdot 10 + 2)10$	$+ 1 \cdot 10 + 5$	S
(j)	$(1 \cdot 10)10 + 2 \cdot 10$	$+ 1 \cdot 10 + 5$	D_a
(k)	$1(10 \cdot 10) + 2 \cdot 10$	$+ 1 \cdot 10 + 5$	A_m
(l)	$1 \cdot 10^2 + 2 \cdot 10$	$+ 1 \cdot 10 + 5$	Renaming
(m)	$1 \cdot 10^2$	$+ [2 \cdot 10 + (1 \cdot 10 + 5)]$	A_a
(n)	$1 \cdot 10^2 + (2 \cdot 10 + 1 \cdot 10) + 5$		A_a
(o)	$1 \cdot 10^2 + (2 + 1)10 + 5$		D_a
(p)	$1 \cdot 10^2 + 3 \cdot 10$	$+ 5$	Addition
(q)	135		S

These examples do not, of course, *prove* the validity of the algorithm that we use for adding counting numbers. They do, however, indicate a method of proof and justify the theory for the addition of the two numbers considered.

Of course, this process can be extended to cover the sum of three or more numbers represented by numerals with two digits. It can also be extended to cover the addition of numerals with any number of digits. Thus, the addition algorithm provides for the addition of all numbers with any number of digits.

Before considering an algorithm for multiplication, we need to consider examples involving repeated application of the distributive property. In the next example, consider $a+5$ as a single counting number in the first application.

Example:

$$\begin{aligned} (a+5)(b+3) &= (a+5)b + (a+5)3 && \text{D}_\text{a} \\ &= (a+5)b + 3(a+5) && \text{C}_\text{m} \\ &= (ab+5b) + (3a+15) && \text{D}_\text{a} \\ &= ab + 5b + 3a + 15 && \text{A}_\text{a} \end{aligned}$$

Example:

$$(a+b)(c+d) = (a+b)c + (a+b)d = ac + bc + ad + bd$$

Example:

$$\begin{aligned} (a+b)(c+d+e) &= (a+b)c + (a+b)d + (a+b)e \\ &= ac + bc + ad + bd + ae + be \end{aligned}$$

We consider now what happens when we multiply a number by 10.

Example: Find $10 \cdot 246$.

$$\begin{aligned} 10 \cdot 246 &= 10[2(10)^2 + 4(10) + 6] && \text{S} \\ &= [2(10)^2 + 4(10) + 6] \cdot 10 && \text{C}_\text{m} \\ &= [2(10)^2]10 + [4(10)](10) + 6(10) && \text{D}_\text{a} \\ &= 2[(10)^2 \cdot 10] + 4[(10)(10)] + 6(10) && \text{A}_\text{m} \\ &= 2(10)^3 + 4(10)^2 + 6(10) && \text{Renaming} \\ &= 2(10)^3 + 4(10)^2 + 6(10) + 0 && \text{Additive identity} \\ &= 2460 && \text{S} \end{aligned}$$

Thus, the multiplication of a number by 10 can generally be accomplished by annexing a 0 to the numeral to the right of the other digits. In a similar manner, multiplication by 10^2 annexes two 0's and by 10^3, three 0's.

$$10^2 \cdot 246 = 24{,}600$$
$$10^3 \cdot 246 = 246{,}000$$

In a manner similar to the algorithm used for addition, the definitions and properties of the preceding chapter will be used to verify a well-known multiplication algorithm.

In an algorithm similar to the one used for addition, we use the definitions and properties of the preceding chapter to multiply 4 times 136.

Example: $4 \cdot 136$

(a)	$4(1 \cdot 10^2 + 3 \cdot 10 + 6)$	S
(b)	$4(1 \cdot 10^2) + 4(3 \cdot 10) + 4 \cdot 6$	D_a
(c)	$(4 \cdot 1)10^2 + (4 \cdot 3)10 + 4 \cdot 6$	A_m
(d)	$4(10)^2 + 12(10) + 24$	Multiplication

The preceding steps justify

$$\begin{array}{r} 136 \\ 4 \\ \hline 24 \\ 120 \\ 400 \\ \hline 544. \end{array}$$

(e)	$4(10)^2 + 12 \cdot 10 + (2 \cdot 10 + 4)$	S
(f)	$4(10)^2 + (12 \cdot 10 + 2 \cdot 10) + 4$	A_a

These two steps are indicated by the regrouping technique in our usual procedure for multiplication.

$$\begin{array}{r} {\scriptstyle 2} \\ 136 \\ 4 \\ \hline 4 \end{array}$$

(g)	$4(10)^2 + (12 + 2)10 + 4$	D_a
(h)	$4(10)^2 + 14 \cdot 10 + 4$	Addition
(i)	$4(10)^2 + (1 \cdot 10 + 4)10 + 4$	S
(j)	$4(10)^2 + (1 \cdot 10)10 + 4 \cdot 10 + 4$	D_a
(k)	$4(10)^2 + 1(10 \cdot 10) + 4 \cdot 10 + 4$	A_m

(l) $4(10)^2 + 1(10)^2 + 4 \cdot 10 + 4$ Multiplication
(m) $[4(10)^2 + 1(10)^2] + 4 \cdot 10 + 4$ A_a

The preceding steps validate the following:

$$\begin{array}{r} 12 \\ 136 \\ 4 \\ \hline 44. \end{array}$$

(n) $(4 + 1)(10)^2 + 4 \cdot 10 + 4$ D_a
(o) $5 \cdot 10^2 + 4 \cdot 10 + 4$ Addition
(p) 544 S

Thus,

$$\begin{array}{r} 12 \\ 136 \\ 4 \\ \hline 544. \end{array}$$

The step-by-step validation of our multiplication algorithm is easily extended to larger numbers.

$42 \cdot 36 = 42(30 + 6)$	S
$= 42(30) + 42(6)$	D_a
$= (42 \cdot 3)10 + 42(6)$	Previous algorithm on multiplication by multiples of 10
$= 126(10) + 252$	Algorithm for single-digit multiplication
$= 1260 + 252$	Multiplication by 10
$= 1512$	Addition algorithm

In the three forms of the algorithm that follow, the form on the left includes the validation of single-digit multiplication. The form in the middle is a summary of the preceding validation, and the form on the right is our usual shorthand notation.

$$\begin{array}{r} 42 \\ 36 \\ \hline 12 \\ 240 \\ 60 \\ 1200 \\ \hline 1512 \end{array} \quad \text{or} \quad \begin{array}{r} 42 \\ 36 \\ \hline 252 \\ 1260 \\ \hline 1512 \end{array} \quad \text{or} \quad \begin{array}{r} 42 \\ 36 \\ \hline 252 \\ 126 \\ \hline 1512 \end{array}$$

By combining the associative and commutative properties of addition and multiplication and the distributive property of multiplication over addition into what we again call "combination of associative, commutative, and distributive properties," we obtain the following results:

$$\begin{array}{r} 4(10) + \ 2 \\ 3(10) + \ 6 \\ \hline 24(10) + 12 \\ 12(10)^2 + \ 6(10) \qquad \\ \hline 12(10)^2 + 30(10) + 12 \end{array} \quad \text{by ACD.}$$

When 12 is written as $1 \cdot 10 + 2$ and the terms are regrouped, the answer becomes $12(10)^2 + 31(10) + 2$. Since

$$31(10) = [3(10) + 1](10) = 3(10)^2 + 1(10),$$

the answer is $15(10)^2 + 1(10) + 2$. But

$$15(10)^2 = (1 \cdot 10 + 5)(10)^2 = 1(10)^3 + 5(10)^2.$$

Hence, the answer is

$$1(10)^3 + 5(10)^2 + 1(10) + 2,$$

or 1,512.

Exercise Set 4-3

1. Consider the following justification of the algorithm used in the addition of 28 and 9. List the properties of whole numbers that are applied in each step.
   (a) $28 + 9 = (2 \cdot 10 + 8) + 9$
   (b) $= 2 \cdot 10 + (8 + 9)$
   (c) $= 2 \cdot 10 + 17$
   (d) $= 2 \cdot 10 + (1 \cdot 10 + 7)$
   (e) $= (2 \cdot 10 + 1 \cdot 10) + 7$
   (f) $= (2 + 1) \cdot 10 + 7$
   (g) $= 3 \cdot 10 + 7$
   (h) $= 37$
2. (a) In Exercise 1, which steps validate the following procedure?

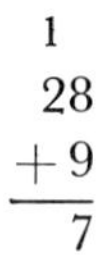

(b) Which steps validate the addition of $1+2$ and the final answer of 37?

$$\begin{array}{r} 1 \\ 28 \\ +9 \\ \hline 37 \end{array}$$

3. Multiply 496 by 2468 using the lattice method of multiplication.
4. Multiply 479 by 320 using the lattice method of multiplication.
5. Consider the multiplication of 46 by 3. List the properties of whole numbers that are applied in each step. Does this verify the algorithm usually used for this multiplication?

(a) $3(46) = 3(4 \cdot 10 + 6)$
(b) $ = 3(4 \cdot 10) + 3 \cdot 6$
(c) $ = (3 \cdot 4)10 + 3 \cdot 6$
(d) $ = 12(10) + 18$
(e) $ = 120 + 18$
(f) $ = 138$

6. State the properties that allow each step of the following computation.

(a) $46 \cdot 23 = 46[2(10) + 3]$
(b) $ = 46 \cdot 2(10) + 46 \cdot 3$
(c) $ = (46 \cdot 2)10 + 46 \cdot 3$
(d) $ = 92 \cdot 10 + 138$
(e) $ = 920 + 138$
(f) $ = 1058$

7. By writing each part of the problem in expanded form in terms of powers of 10, perform the following operations. The repeated use of the associative, commutative, and distributive properties in the first step may be classified as ACD. For example,

$$\begin{array}{rll} 35 + 28 = & 3(10) + 5 & \\ & \underline{2(10) + 8} & \\ & 5(10) + 13 & \text{By ACD} \\ & 5(10) + (1 \cdot 10 + 3) & \text{S} \\ & [5(10) + 1 \cdot 10] + 3 & \text{A}_\text{a} \\ & 6(10) + 3 & \text{D}_\text{a} \text{ and renaming} \\ & 63 & \text{S} \end{array}$$

(a) $27 + 15$ (b) $23 \cdot 37$
(c) $53 + 24$ (d) $584 + 273$
(e) $28 \cdot 32$ (f) $65 \cdot 23$

8. Simplify $(a + 3)(a + 4)(a + 5)$ by the repeated use of the distributive property of multiplication over addition.
9. In the product of $364 \cdot 182$, why can we ignore the zeros?

$$\begin{array}{r} 364 \\ \underline{182} \\ 728 \\ 29120 \\ \underline{36400} \\ 66248 \end{array}$$

*10. In the following lattice, find the digits represented by U, V, W, X, Y, and Z.

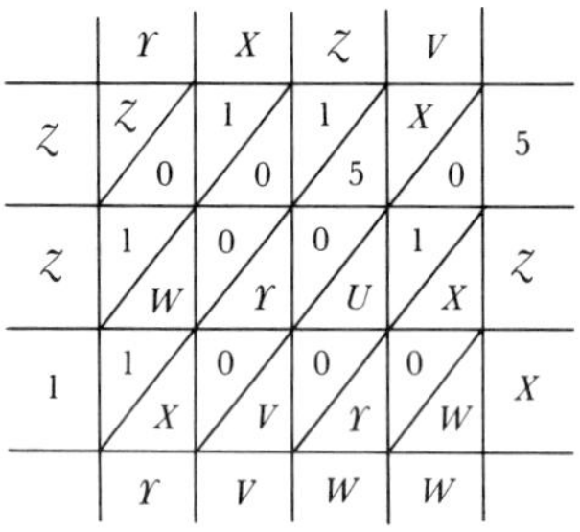

	Y	X	Z	V	
Z	Z / 0	1 / 0	1 / 5	X / 0	5
Z	1 / W	0 / Y	0 / U	1 / X	Z
1	1 / X	0 / V	0 / Y	0 / W	X
	Y	V	W	W	

*11. Use your mathematics library to determine what the following multiplication algorithms involve, and illustrate each with an example.
(a) Napier's bones (b) Russian peasant multiplication
(c) Repiego method (d) Scacchera (or chessboard) method

*12. In a manner similar to the examples given at the beginning of this section, verify the algorithms usually employed to perform the following operations.
(a) $61 + 74$ (b) $146 + 28$
(c) $61 \cdot 74$ (d) $56 \cdot 71$

4

Algorithms for Subtraction and Division

In Chapter 3, the operation of addition assigns to a pair of numbers (called addends) a specified number called the *sum*. The operation of finding one addend when the sum and the other addend are known is the operation of *subtraction.* Thus, subtraction is the inverse operation of addition.

Before considering an algorithm for subtraction, we should consider a property of subtraction suggested by Theorem 3-14.

Examples:

(a) $(10 + 7) - (4 + 3) = (10 - 4) + (7 - 3) = 6 + 4 = 10$
(b) $(28 + 9) - (21 + 2) = (28 - 21) + (9 - 2) = 7 + 7 = 14$
(c) $(20 + 9) - (3) = (20 + 9) - (0 + 3) = (20 - 0) + (9 - 3) = 26$

This property decreases the number of steps necessary to verify a subtraction algorithm.

Example: Compute 89 — 43.

$$\begin{aligned} 89 - 43 &= [8(10) + 9] - [4(10) + 3] && \text{S} \\ &= [8(10) - 4(10)] + [9 - 3] && \text{Theorem 3-14} \\ &= (8 - 4)10 + (9 - 3) && D_s \\ &= 4(10) + 6 && \text{Subtraction} \\ &= 46 && \text{S} \end{aligned}$$

Notice how this compares with our usual procedure.

$$\begin{array}{r} 8(10) + 9 \\ -[4(10) + 3] \\ \hline 4(10) + 6 \end{array} \quad \begin{array}{c} \\ \text{or} \\ \text{by ACD} \end{array} \quad \begin{array}{r} 89 \\ -43 \\ \hline 46 \end{array}$$

We will now discuss an example in which we *rename* or *regroup* before we complete the subtraction; sometimes this procedure is called *borrowing*.

Example: Compute 74 — 36.

If we attempt to use the method of the last example, we will get, as the first step, 4 — 6, which is not defined in the system of whole numbers. Rewrite 74 in the following manner.

$$74 = 7(10) + 4 = (6 + 1)10 + 4 = 6(10) + 1(10) + 4 = 6(10) + 14.$$

Then

$$\begin{aligned} 74 - 36 &= [6(10) + 14] - [3(10) + 6] \\ &= [6(10) - 3(10)] + [14 - 6] \\ &= 3(10) + 8 \\ &= 38. \end{aligned}$$

Compare with

$$\begin{array}{r} 74 \\ -36 \\ \hline \end{array} \quad \text{or} \quad \begin{array}{r} 7(10) + 4 \\ -[3(10) + 6] \\ \hline \end{array} \quad \text{or} \quad \begin{array}{r} 6(10) + 14 \\ -[3(10) + \ \ 6] \\ \hline 3(10) + \ \ 8 = 38. \end{array}$$

In the preceding chapter, division was defined as the inverse of multiplication. If a and b ($b \neq 0$) are any whole numbers, then $a \div b$ is some number c (if it exists) such that $a = bc$. It was emphasized that division is not possible for every pair of whole numbers. For example, it is not possible to compute $8 \div 5$ in the system of whole numbers. It is possible, however, in every division problem involving whole numbers (where the divisor is unequal to zero) to find an answer consisting of a *quotient* and a *remainder*, both of which are whole numbers. This property is stated in the form of a theorem.

Theorem 4-2: *Division Algorithm.* If a and b ($b \neq 0$) are whole numbers, then there exist unique whole numbers q and r such that $a = bq + r$, where $0 \leq r < b$.

In the definition of division in Chapter 3, for division to exist there was one number q such that $a = bq$. This theorem states that for any two whole numbers a and b with $b \neq 0$ one can obtain two numbers, q (called the quotient) and r (the remainder) such that $a = bq + r$ where $0 \leq r < b$. In other words, dividend = divisor · quotient + remainder. The proof of this theorem will be given in Chapter 6.

Example: If $a = 47$ and $b = 7$, write a in the form $bq + r$, where $0 \leq r < b$. Answer: $47 = 7(6) + 5$; $q = 6$ and $r = 5$; $0 \leq 5 < 7$.

Example: If $a = 54$ and $b = 9$, write a in the form $bq + r$, where $0 \leq r < b$. Answer: $54 = 9(6) + 0$; $q = 6$ and $r = 0$.

Example: If $a = 5$ and $b = 8$, write a in the form $bq + r$, where $0 \leq r < b$. Answer: $5 = 8(0) + 5$; q is 0 and r is 5.

Many procedures are known for performing the operation to find q and r. Generally, these procedures are algorithms given in such a way that the properties of whole numbers relative to the operations involved are completely obscured. For most people, this process consists of guessing, multiplying, subtracting, and then guessing again; and if an initial guess was wrong, it is replaced by a new guess. This process, a series of repeated subtractions, is often called *long division.* The Division Algorithm states that the answer exists; thus, after a sufficient number of guesses, one should always find the correct answer. Let us consider the division indicated by $8132 \div 38$. What whole number multiplied by 38 can be subtracted from 8132? The answer is any whole number such that the product is not greater than 8132. We usually consider products that involve 10, 100, 1000, and so on, as factors. We will now perform this division a long way to illustrate the procedure. Now $38 \cdot 100 = 3800$, which is less than 8132; hence 3800 can be subtracted from 8132. Similarly, $38 \cdot 100$ can be subtracted again. (This is equivalent to subtracting $2 \cdot (38 \cdot 100) = 38 \cdot 200$.) Next we use the product $38 \cdot 10$, and then $38 \cdot 1$, which must be subtracted four times. This last subtraction is equivalent to subtracting $38 \cdot 4$.

38)8132	
3800	$38 \cdot 100$
4332	
3800	$38 \cdot 100$
532	
380	$38 \cdot 10$
152	
38	$38 \cdot 1$
114	
38	$38 \cdot 1$
76	
38	$38 \cdot 1$
38	
38	$38 \cdot 1$

Now, how many 38's have been subtracted from 8132? The answer is $100 + 100 + 10 + 1 + 1 + 1 + 1 = 214$. Since $38(214) = 8132$, then $8132 \div 38 = 214$.

Now let us look at a more familiar form, one suggested in the preceding discussion. Instead of subtracting $38 \cdot 100$, we will subtract $38 \cdot 200$.

$$\begin{array}{r|l} 38\overline{)8132} & \\ 7600 & 38 \cdot 200 \\ \hline 532 & \\ 380 & 38 \cdot 10 \\ \hline 152 & \\ 152 & 38 \cdot 4 \\ \hline \end{array}$$

This process is now listed in the usual form for comparison.

$$\begin{array}{r} 2(10)^2 + 1(10) + 4 \\ 3(10) + 8\overline{)8(10)^3 + 1(10)^2 + 3(10) + 2} \\ 7(10)^3 + 6(10)^2 + 0(10) + 0 \\ \hline 5(10)^2 + 3(10) + 2 \\ 3(10)^2 + 8(10) + 0 \\ \hline 1(10)^2 + 5(10) + 2 \\ 1(10)^2 + 5(10) + 2 \\ \hline \end{array} \quad \textit{or} \quad \begin{array}{r} 214 \\ 38\overline{)8132} \\ 7600 \\ \hline 532 \\ 380 \\ \hline 152 \\ 152 \\ \hline \end{array}$$

Example: Divide 153,999 by 723, using the three algorithms discussed in this section.

(a)
$$\begin{array}{rl} 723\overline{)153999} & \\ 72300 & 723 \cdot 100 \\ \hline 81699 & \\ 72300 & 723 \cdot 100 \\ \hline 9399 & \\ 7230 & 723 \cdot 10 \\ \hline 2169 & \\ 723 & 723 \cdot 1 \\ \hline 1446 & \\ 723 & 723 \cdot 1 \\ \hline 723 & \\ 723 & 723 \cdot 1 \\ \hline \end{array}$$

$$100 + 100 + 10 + 1 + 1 + 1 = 213$$

(b)
$$\begin{array}{rl} 723\overline{)153999} & \\ 144600 & 723 \cdot 200 \\ \hline 9399 & \\ 7230 & 723 \cdot 10 \\ \hline 2169 & \\ 2169 & 723 \cdot 3 \\ \hline \end{array}$$

$$200 + 10 + 3 = 213$$

(c)

$$\begin{array}{r|l} & 2(10)^2+1(10)+3=213 \\ \hline 7(10)^2+2(10)+3 & 1(10)^5+5(10)^4+3(10)^3+9(10)^2+9(10)+9 \\ & 1(10)^5+4(10)^4+4(10)^3+6(10)^2+0(10)+0 \\ \hline & 9(10)^3+3(10)^2+9(10)+9 \\ & 7(10)^3+2(10)^2+3(10)+0 \\ \hline & 2(10)^3+1(10)^2+6(10)+9 \\ & 2(10)^3+1(10)^2+6(10)+9 \\ \hline \end{array}$$

It is not necessary to go through the process of verifying each step by the properties of whole numbers to show that this algorithm gives the correct answer. Since division is defined as the inverse of multiplication, the answer can be verified by the operation of multiplication. Thus, $8132 \div 38$, by the process discussed in the preceding paragraph, yields the correct answer 214 because $(214)(38) = 8132$, or $153{,}999 \div 723 = 213$ because $(213)(723) = 153{,}999$.

Exercise Set 4-4

1. For the following pairs of numbers, let a be the first number of the pair and b the second number. Find whole numbers q and r for each pair such that $a = bq + r$, where $0 \leq r < b$.
   (a) 72, 11 (b) 16, 9 (c) 11, 18 (d) 106, 13
   (e) 51, 14 (f) 25, 39 (g) 54, 9 (h) 176, 21
2. Perform the following divisions in three ways as indicated by the examples of this section. Verify that your answer is correct by using the definition of division.
   (a) $27\overline{)1075}$ (b) $314\overline{)35{,}304}$
   (c) $203\overline{)12{,}061}$ (d) $873\overline{)35{,}714}$
3. Consider the subtraction of 9 from 28. List the properties of whole numbers that allow each of the steps in this operation.
   (a) $28 - 9 = (2 \cdot 10 + 8) - (0 \cdot 10 + 9)$
   (b) $= [(1 + 1)10 + 8] - (0 \cdot 10 + 9)$
   (c) $= [(1 \cdot 10 + 1 \cdot 10) + 8] - (0 \cdot 10 + 9)$
   (d) $= [(1 \cdot 10 + 10) + 8] - (0 \cdot 10 + 9)$
   (e) $= [1 \cdot 10 + (10 + 8)] - (0 \cdot 10 + 9)$
   (f) $= (1 \cdot 10 + 18) - (0 \cdot 10 + 9)$

(g) $= (1 \cdot 10 - 0 \cdot 10) + (18 - 9)$
(h) $= (1 - 0)10 + (18 - 9)$
(i) $= 1 \cdot 10 + 9$
(j) $= 19$

4. According to the Division Algorithm, $a = bq + r$. Find the missing values in the following divisions.
   (a) $a = 11$, $b = 2$ (b) $b = q = 5$, $r = 2$
   (c) $a = 57$, $q = 8$, $r = 1$ (d) $b = 8$, $q = 16$, $r = 6$
5. Use the incomplete example

$$\begin{array}{r} 4 \\ 33\overline{)1605} \\ 1320 \\ \hline 285 \end{array}$$

   as a basis for (a), (b), and (c) below.
   (a) Discuss the meaning of 4 in the quotient.
   (b) Discuss why the 4 is placed above the 0.
   (c) Some people omit the 0 in 1320 and say "Bring down the 5." Discuss the meaning of this procedure.
6. In the statement of the Division Algorithm, what is the necessity for stating that $b \neq 0$?
7. Describe how you would explain to a fourth grader why "the differences of big numbers are not always big."
8. Compute the q and r of Theorem 4-2 for the division of
   (a) 438 by 1000. (b) 438 by 100.
   (c) 438 by 10. (d) 438 by 1.
9. Repeat Exercise 8, but replace 438 with 8104.
10. The pairs given below consist of the dividend a and remainder r. Find a quotient $q \neq 1$ and the largest possible divisor.
   (a) 6, 2 (b) 333, 11 (c) 729, 18 (d) 2418, 27
11. Study each of the following subtraction algorithms and the example given and then give another example demonstrating each algorithm.
   (a) The Austrian method of subtraction consists of the following steps. Write $764 - 348$ as $(700 + 60 + 4) - (300 + 40 + 8)$. Then add 10 in each parenthesis: $(700 + 60 + 10 + 4) - (300 + 40 + 10 + 8) = (700 + 60 + 14) - (300 + 50 + 8) = 400 + 10 + 6 = 416$.
   (b) Subtracting by taking complements consists of adding to both numbers the same sum. This sum is selected so that the number to subtract is a multiple of 10.

$$\begin{array}{r} 1734 \\ -468 \\ \hline \end{array} \qquad \begin{array}{r} 1734 \\ +532 \\ -532 \\ -468 \\ \hline \end{array} \qquad \begin{array}{r} 2266 \\ -1000 \\ \hline 1266 \end{array}$$

(c) A complement of a digit a is $9 - a$. The method of using complements consists of the following. Find the complement of each digit in the subtrahend. Add the resulting numeral to the minuend. Then subtract 1 from the left digit and add 1 to the right digit, giving the difference for a subtraction problem.

$$\begin{array}{r} 5316 \\ -3927 \\ \hline \end{array} \qquad \begin{array}{r} 5316 \\ +6072 \\ \hline 11388 \\ -1 \;\; +1 \\ \hline 1389 \end{array}$$

5

Numeration Systems with Bases Other Than Ten

The preceding sections dealt with the Babylonian system of base 60 and the Mayan system of base 20, as well as with the Hindu-Arabic system, which we called the decimal, or base ten, system. Obviously, numbers other than ten can be used for bases of numeration systems. Can one find a base for a numeration system that would be, for some purposes, better than base ten?

To answer this question, we now consider numeration systems in bases other than ten. This study of systems with different bases should increase your understanding of decimal symbols. Additionally, the difficulties and challenges of learning a new system are comparable to the problems encountered by a child as he learns the base ten system. Finally, a knowledge of operations in bases other than ten is important in order to understand the operation of a digital computer.

We begin by considering a base seven numeration system; in this notation, groups of seven are selected. Examine the x's listed in groups of seven in Figure 4-3. In (a), the numeral representing the x's could be written 12_{seven}, representing one group of seven with two remaining.

If the x's should correspond to days of the week, then the 12_{seven} would represent one week and two days. The numeral representing the x's in

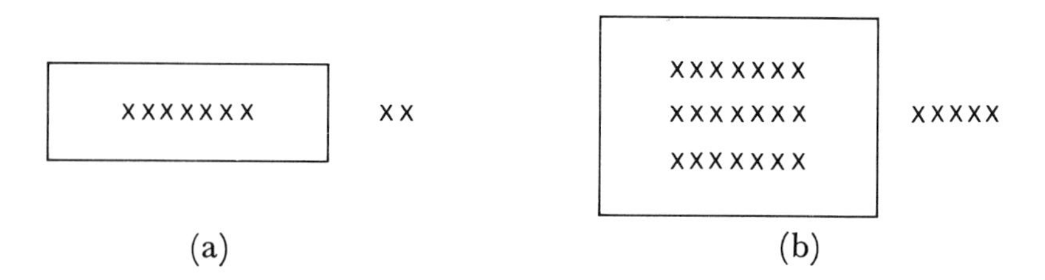

Figure 4-3

(b) could be written as 35_{seven} to mean three groups of seven and five more. We write the subscript "seven" to show that the numeral is not the "thirty-five" we usually think of in base ten.

Having seven for the base of a numeration system limits the number of symbols for counting to seven—0, 1, 2, 3, 4, 5, and 6. For our work in this chapter, 0 will be a necessary member of sets of digits for all numeration systems. So, a set of digits for a system in base b contains the digits 0, 1, 2, . . ., $(b - 1)$. In the counting process, the next counting number after 6 in base seven would be 10. However, this is not the "10" you are accustomed to seeing. In the base seven numeration system, 10 represents one group of seven and no more. The next counting number, 11, represents one seven and one. The counting process continues with 12, 13, 14, 15, 16, and then 20; 20 represents two sevens and no more. If you continue counting in this numeration system, you will soon reach the numeral 66. What is the next numeral? Be sure you understand that it is 100.

Consider now a base seven numeral written in expanded notation. Remember that $(10)^2$ represents base squared and 10 represents one base.

$$243_{\text{seven}} = [2(10)^2 + 4(10) + 3]_{\text{seven}}$$

The base squared, $(10)^2$, is equivalent to 7^2 in base ten and 10 is equivalent to the base 7. Thus, in base ten we have

$$243_{\text{seven}} = [2(7)^2 + 4(7) + 3]_{\text{ten}}.$$

This provides an easy method for changing 243 to a base ten numeral. Thus,

$$243_{\text{seven}} = [98 + 28 + 3]_{\text{ten}} = 129_{\text{ten}}.$$

Hence, it is seen that to change a numeral expressed in terms of a given base to a base ten numeral requires that one write out the meaning of the numeral in terms of the base. Then a series of multiplications followed by additions gives the equivalent expression in base ten.

Example:

$$
\begin{aligned}
3462_{\text{seven}} &= [3(10)^3 + 4(10)^2 + 6(10) + 2]_{\text{seven}} \\
&= [3(7)^3 + 4(7)^2 + 6(7) + 2]_{\text{ten}} \\
&= [1029 + 196 + 42 + 2]_{\text{ten}} \\
&= 1269_{\text{ten}}
\end{aligned}
$$

A seventeenth-century mathematician, Gottfried Wilhelm von Leibnitz, is reported to have been an advocate of the *binary numeration system*, which has two for a base; this system uses only the digits 0 and 1. It may seem strange that such an eminent mathematician would advocate such a system; however, we note that the binary system has become very popular in the last few years. This system is of importance today because many high-speed computers use base two for computations. An electric switch needs only two digits (0 and 1) to represent *off* and *on*, and so a switch position can be represented by a binary symbol.

Binary numerals are based on groups of two, just as the preceding arithmetic was based on groups of seven and the decimal system is based on groups of ten. The groups in Figure 4-4 are set up for counting in the binary system. Figure 4-4(a) represents one group of (two times two) elements, one group of two elements, and one element, or

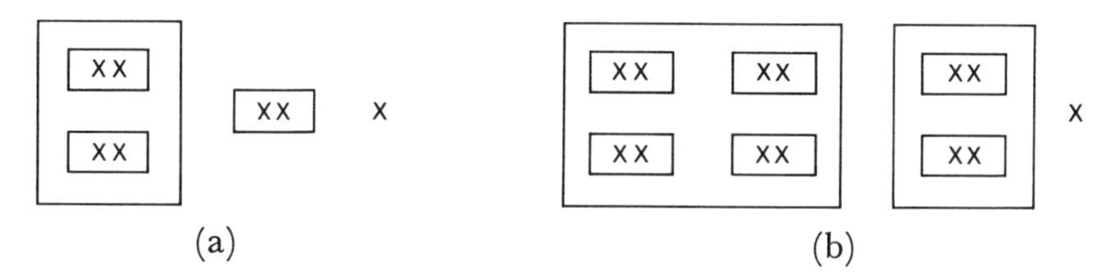

Figure 4-4

111_{two}. In (b), there are no groups of two elements, so the symbol becomes 1101_{two}. The first ten counting symbols in the base two system are 1, 10, 11, 100, 101, 110, 111, 1000, 1001, 1010. What is the next symbol?

It is very easy to convert a binary numeral to a base ten numeral, as the following examples illustrate.

Example: What decimal numeral is equivalent to 1011_{two}?

$$\begin{aligned} 1011_{\text{two}} &= [1(10)^3 + 0(10)^2 + 1(10) + 1]_{\text{two}} \\ &= [1(2)^3 + 0(2)^2 + 1(2) + 1]_{\text{ten}} \\ &= 11_{\text{ten}} \end{aligned}$$

Example: Change 10111_{two} to a base ten numeral.

$$\begin{aligned} 10111_{\text{two}} &= [1(10)^4 + 0(10)^3 + 1(10)^2 + 1(10) + 1]_{\text{two}} \\ &= [1(2)^4 + 0(2)^3 + 1(2)^2 + 1(2) + 1]_{\text{ten}} \\ &= 23_{\text{ten}} \end{aligned}$$

A system that has received wide attention and that has even been suggested as a replacement for the decimal system is the *duodecimal system*, with a base of twelve. An argument for twelve as a number base is that it has more divisors than ten; the only counting numbers that divide 10 are 1, 2, 5, 10, whereas the divisors of 12 are 1, 2, 3, 4, 6, 12. Georges Buffon, a French naturalist, suggested over two hundred years ago that the base twelve system be universally adopted, a suggestion that has been carried into this century. Even today, the Duodecimal Society of America lobbies for the base twelve system as a replacement for our present system.

Figure 4-5 illustrates how we group by dozens. The marks in (a) are

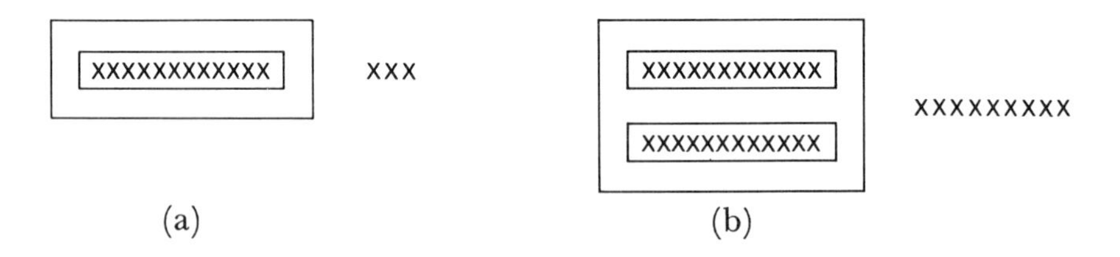

Figure 4-5

denoted by 13_{twelve}, because there is one group of a dozen and three ones. In Figure 4-5(b), there are two groups of a dozen and nine ones. This is written 29_{twelve}.

To group and count by twelves requires that two new symbols be created. T and E are the symbols commonly used for what we usually call ten and eleven, and so the base twelve numerals will be as follows: 1, 2, 3, ..., 8, 9, T, E, 10, 11, 12, ..., 19, 1T, 1E, 20, 21, ...

One needs, of course, to compute powers of twelve in order to translate numbers from base twelve to base ten. Using $(12)^2 = 144$, $(12)^3 = 1728$, and $(12)^4 = 20{,}736$, verify the changes from base twelve to base ten in the following examples.

Examples:

$$
\begin{aligned}
32_{\text{twelve}} &= [3(10) + 2]_{\text{twelve}} \\
&= [3(12) + 2]_{\text{ten}} \\
&= 38_{\text{ten}}
\end{aligned}
$$

$$
\begin{aligned}
209_{\text{twelve}} &= [2(10)^2 + 0(10) + 9]_{\text{twelve}} \\
&= [2(12)^2 + 0(12) + 9]_{\text{ten}} \\
&= 297_{\text{ten}}
\end{aligned}
$$

$$
\begin{aligned}
12E5_{\text{twelve}} &= [1(10)^3 + 2(10)^2 + E(10) + 5]_{\text{twelve}} \\
&= [1(12)^3 + 2(12)^2 + E(12) + 5]_{\text{ten}} \\
&= 2153_{\text{ten}}
\end{aligned}
$$

You have learned in the preceding paragraphs to change numerals from different bases to base ten. Now consider the inverse process—changing from base ten to other bases. We begin this discussion by considering problems in base seven.

Suppose we wish to change 23 in base ten to a base seven numeral. What is the largest power of 7 contained in 23? 7^2 is larger than 23, so 7^1 is the largest power of 7 contained in 23. Since there are three 7's and a remainder of 2 in 23, then $23 = 3(7) + 2 = 32_{\text{seven}}$. Thus, to change a numeral in base ten to an equivalent numeral in a given base requires division, using the powers of the given base as divisors.

Example: Express 59 as a numeral in base two. Since the numeral is to be expressed in base two, consider in base ten the powers of 2, which are 1, 2, 4, 8, 16, 32, 64, etc. Since 59 is less than 64, it is only necessary to find the number of 32's in 59.

$$
\begin{array}{r|l|l}
32 & 59 & 1 \\
 & 32 & \\
\hline
 & 27 &
\end{array}
$$

The remainder is 27. The next step is to determine how many 16's are in 27.

32	59	1	
	32		
16	27	1	
	16		
8	11	1	How many 8's in 11?
	8		
4	3	0	How many 4's in 3?
	0		
2	3	1	How many 2's in 3?
	2		
1	1	1	How many 1's in 1?
	1		
	0		

Hence,

$$\begin{aligned} 59 &= [1(32) + 1(16) + 1(8) + 0(4) + 1(2) + 1(1)]_{\text{ten}} \\ &= [1(2)^5 + 1(2)^4 + 1(2)^3 + 0(2)^2 + 1(2) + 1]_{\text{ten}} \\ &= [1(10)^5 + 1(10)^4 + 1(10)^3 + 0(10)^2 + 1(10) + 1]_{\text{two}}. \end{aligned}$$

Thus,

59 is expressed in base two as $59 = 111011_{\text{two}}$.

Example: Express 289 as a numeral in base five. Since the numeral is to be expressed in base five, consider in base ten the successive powers of 5, which are 1, 5, 25, 125, 625, etc. Since 289 is less than 625, determine how many 125's are in 289. Continue with the same steps used in the preceding example.

125	289	2	
	250		
25	39	1	How many 25's in 39?
	25		
5	14	2	How many 5's in 14?
	10		
1	4	4	How many 1's in 4?
	4		
	0		

Hence,

$$\begin{aligned} 289 &= [2(125) + 1(25) + 2(5) + 4(1)]_{ten} \\ &= [2(5)^3 + 1(5)^2 + 2(5) + 4]_{ten} \\ &= [2(10)^3 + 1(10)^2 + 2(10)^1 + 4]_{five}. \end{aligned}$$

Thus,

$$289 = 2124_{five}.$$

Exercise Set 4-5

1. Write the first fifteen counting numbers using each of the following as a number base.
   (a) five (b) two (c) four (d) twelve
2. Write each of the following as a base ten numeral.
   (a) 157_{nine} (b) 504_{six}
   (c) 101101_{two} (d) $T02E_{twelve}$
   (e) 430_{eight} (f) 2010_{four}
   (g) $E6_{twelve}$ (h) 11110_{two}
3. Write in the indicated base the next consecutive counting number for each of the following.
   (a) 16_{seven} (b) 505_{six} (c) 111_{two}
   (d) EEE_{twelve} (e) 607_{eight} (f) 3033_{four}
4. Change each of the following base ten numerals to an expression in base two.
   (a) 9 (b) 35 (c) 55
   (d) 285 (e) 1000 (f) 5280
5. Repeat Exercise 4 by changing each of the numerals to an expression in base seven.
6. Repeat Exercise 4 by changing each of the numerals to an expression in base twelve.
7. Choose the largest number from each of the following sets.
   (a) 5_{six}, 11_{four}, 101_{three}
   (b) 122_{three}, 112_{four}, 76_{eight}
   (c) ET_{twelve}, 101_{eight}, 11110_{two}
   (d) 325_{twelve}, 523_{six}, 10122_{three}
8. Find the unknown base indicated by the letter *b*.
   (a) $67_{ten} = 61_b$ (b) $12_{ten} = 1100_b$ (c) $234_{ten} = 176_b$
9. Change the following numerals to the base indicated.
   (a) 231_{four} to base twelve (b) $27TE_{twelve}$ to base six

*10. Can you characterize even and odd numbers merely by looking at the units digit of a given number when it is expressed in base two? In base three? In base four? In base five?

6

Addition and Subtraction in Different Bases

To add and subtract numbers in base ten, you have memorized about one hundred combinations, such as $7 + 8 = 15$. However, Table 4-1, an addition table in base ten, is included for review. Note that, because the commutative property of addition holds true for this system, the entries within the table are symmetric about a line drawn diagonally from the upper left corner to the lower right corner.

Addition Table, Base Ten

+	0	1	2	3	4	5	6	7	8	9
0	0	1	2	3	4	5	6	7	8	9
1	1	2	3	4	5	6	7	8	9	10
2	2	3	4	5	6	7	8	9	10	11
3	3	4	5	6	7	8	9	10	11	12
4	4	5	6	7	8	9	10	11	12	13
5	5	6	7	8	9	10	11	12	13	14
6	6	7	8	9	10	11	12	13	14	15
7	7	8	9	10	11	12	13	14	15	16
8	8	9	10	11	12	13	14	15	16	17
9	9	10	11	12	13	14	15	16	17	18

Table 4-1

If you were planning to use repeatedly any numeration system, you would probably memorize all the addition combinations for the given base. However, for the purposes of this course, such memorization is unnecessary, and addition tables will be constructed for several different bases.

From your knowledge of the counting process, verify that the answers for the following base seven additions, which are represented in Table 4-2, are correct.

$$\begin{bmatrix} 1+6=10 & & & \\ 2+5=1+(1+5) & =1+ & 6 & =10 \\ 2+6=1+(1+6) & =1+ & 10 & =11 \\ 3+4=2+(1+4) & =2+ & 5 & =10 \\ 3+5=1+(2+5) & =1+ & 10 & =11 \\ 3+6=1+(2+6) & =1+ & 11 & =12 \end{bmatrix}_{\text{seven}}$$

Now verify the remaining entries in Table 4-2. You can use a number line to assist you with this problem.

Example: What is $5+4$ in base seven?

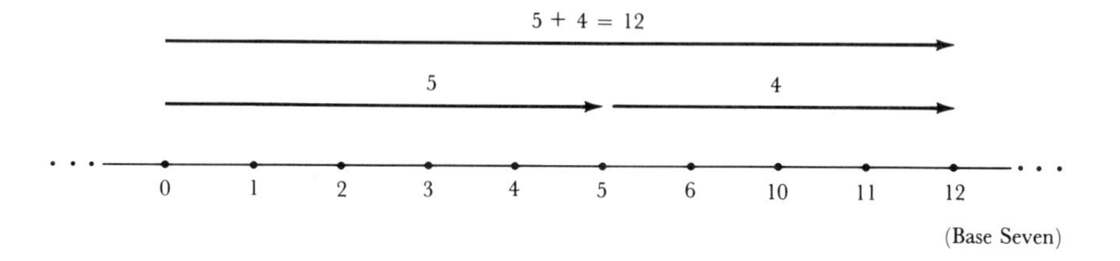

Figure 4-6

Addition Table, Base Seven

+	0	1	2	3	4	5	6
0	0	1	2	3	4	5	6
1	1	2	3	4	5	6	10
2	2	3	4	5	6	10	11
3	3	4	5	6	10	11	12
4	4	5	6	10	11	12	13
5	5	6	10	11	12	13	14
6	6	10	11	12	13	14	15

Table 4-2

Table 4-2 shows the addition of pairs of numbers from 0 to 6. To add larger numbers, one needs an algorithm similar to those discussed in the third section of this chapter. It is easy to see that the algorithm for the addition of numbers in base ten is applicable to the addition of numbers expressed in other bases.

$$\begin{aligned} 26_{\text{seven}} &= 2(10) + 6 \\ +34_{\text{seven}} &= 3(10) + 4 \\ &= 5(10) + 13 \\ &= 5(10) + 1(10) + 3 \\ &= 6(10) + 3 \\ &= 63_{\text{seven}} \end{aligned}$$

Some people are more successful in working this type of problem when they "think" base ten (that is, they perform all operations in base ten) and write the answers in terms of the given base. Consider the previous example.

$$\begin{aligned} 26_{\text{seven}} &= 2(7) + 6 \\ 34_{\text{seven}} &= 3(7) + 4 \\ &\quad 5(7) + [1(7) + 3] = 6(7) + 3 = 63_{\text{seven}} \end{aligned}$$

The *renaming* process is performed in the next set of examples without writing the numerals in expanded form. If you have difficulty understanding the computations, write each numeral in expanded form.

Examples:

Addition	Suggested Help
$\begin{bmatrix} 435 \\ 342 \\ \hline 1221 \end{bmatrix}_{\text{six}}$	$\begin{bmatrix} 5 + 2 = 7 = 1(6) + 1 \\ 1 + 3 + 4 = 8 = 1(6) + 2 \\ 1 + 4 + 3 = 8 = 1(6) + 2 \end{bmatrix}_{\text{ten}}$
$\begin{bmatrix} 233 \\ 312 \\ \hline 1211 \end{bmatrix}_{\text{four}}$	$\begin{bmatrix} 3 + 2 = 5 = 1(4) + 1 \\ 1 + 3 + 1 = 5 = 1(4) + 1 \\ 1 + 2 + 3 = 6 = 1(4) + 2 \end{bmatrix}_{\text{ten}}$
$\begin{bmatrix} 111 \\ 101 \\ \hline 1100 \end{bmatrix}_{\text{two}}$	$\begin{bmatrix} 1 + 1 = 2 = 1(2) + 0 \\ 1 + 1 + 0 = 2 = 1(2) + 0 \\ 1 + 1 + 1 = 3 = 1(2) + 1 \end{bmatrix}_{\text{ten}}$

What is $11_{\text{seven}} - 2_{\text{seven}}$? According to the definition of subtraction, the problem could be stated "What number when added to 2 gives an answer of 11_{seven}?" By searching the addition table for base seven addition, we find that 6 when added to 2 gives 11_{seven}. Thus,

$$11_{\text{seven}} - 2_{\text{seven}} = 6_{\text{seven}}.$$

This same idea can be used in every subtraction problem; however, it is sometimes necessary to *regroup* or *rename* the sums of the powers of the base so that subtraction can be performed.

Examples: Subtract.

$$\begin{array}{rl} 465_{\text{seven}} & = 4(10)^2 + 6(10) + 5 \\ -132_{\text{seven}} & = 1(10)^2 + 3(10) + 2 \\ \hline & = 3(10)^2 + 3(10) + 3 = 333_{\text{seven}} \end{array}$$

$$\begin{array}{rlll} 624_{\text{seven}} & = 6(10)^2 + 2(10) + 4 & = 5(10)^2 + 11(10) + 14 & \\ -246_{\text{seven}} & = 2(10)^2 + 4(10) + 6 & = 2(10)^2 + 4(10) + 6 & \\ \hline & & 3(10)^2 + 4(10) + 5 & = 345_{\text{seven}} \end{array}$$

We work the same problem performing the operations in base ten:

$$\begin{array}{rlll} 624_{\text{seven}} & = 6(7)^2 + 2(7) + 4 & = 5(7)^2 + 8(7) + 11 & \\ -246_{\text{seven}} & = 2(7)^2 + 4(7) + 6 & = 2(7)^2 + 4(7) + 6 & \\ \hline & & = 3(7)^2 + 4(7) + 5 & = 345_{\text{seven}} \end{array}$$

Example:

$$\begin{bmatrix} 514 \\ -145 \\ \hline 336 \end{bmatrix}_{\text{seven}} \text{ or } \begin{bmatrix} 5(7)^2 + 1(7) + 4 \\ 1(7)^2 + 4(7) + 5 \\ \hline \end{bmatrix}_{\text{ten}} \text{ or } \begin{bmatrix} 4(7)^2 + 7(7) + 11 \\ 1(7)^2 + 4(7) + 5 \\ \hline 3(7)^2 + 3(7) + 6 \end{bmatrix}_{\text{ten}}$$

Check | Suggested Help

$$\begin{bmatrix} 336 \\ +145 \\ \hline 514 \end{bmatrix}_{\text{seven}} \text{ or } \begin{bmatrix} 6 + 5 = 11 = 1(7) + 4 \\ 1 + 3 + 4 = 8 = 1(7) + 1 \\ 1 + 3 + 1 = 5 \end{bmatrix}_{\text{ten}}$$

Some examples of subtraction in bases two, four, and six will now be given. If you have trouble understanding the *regrouping* process, you will want to write the numerals in expanded form involving the base.

Example: Subtract.

$$\begin{array}{r} 1010_{two} \\ 101_{two} \\ \hline 101_{two} \end{array} \qquad \begin{array}{r} 312_{four} \\ 123_{four} \\ \hline 123_{four} \end{array} \qquad \begin{array}{r} 523_{six} \\ 245_{six} \\ \hline 234_{six} \end{array}$$

A good way to check your work in both addition and subtraction is to translate each part of the problem to base ten. Then compare the answers obtained by performing the operations in base ten with the answers obtained in the given base translated to base ten.

Example: Subtract.

$$\begin{array}{rl} 11101_{two} & = 1(2)^4 + 1(2)^3 + 1(2)^2 + 0(2) + 1 = 29 \\ 1011_{two} & = \qquad\qquad 1(2)^3 + 0(2)^2 + 1(2) + 1 = 11 \\ \hline 10010_{two} & \qquad\qquad\qquad\qquad\qquad\qquad\qquad 18 \end{array}$$

$10010_{two} = 1(2)^4 + 0(2)^3 + 0(2)^2 + 1(2) + 0 = 18$; thus, the answer checks.

Exercise Set 4-6

1. Perform the following additions.
   (a) $304_{seven} + 356_{seven}$ (b) $233_{four} + 32_{four}$
   (c) $1101_{two} + 111_{two}$ (d) $552_{six} + 345_{six}$
   (e) $211_{four} + 232_{four}$ (f) $1001_{two} + 1001_{two}$
2. Perform the following subtractions.
   (a) $1000_{two} - 11_{two}$ (b) $403_{seven} - 54_{seven}$
   (c) $331_{four} - 132_{four}$ (d) $1111_{six} - 555_{six}$
   (e) $110101_{two} - 1011_{two}$ (f) $321_{four} - 103_{four}$
3. Make an addition table for the following bases.
   (a) three (b) five (c) twelve
4. Use the addition tables of Exercise 3 to perform the following operations.
   (a) $35_{twelve} + 15_{twelve}$ (b) $2304_{five} + 121_{five}$
   (c) $12_{twelve} - 9_{twelve}$ (d) $121_{three} + 22_{three}$
   (e) $8T2_{twelve} + 26E_{twelve}$ (f) $321_{five} - 143_{five}$
   (g) $503_{twelve} - 2TE_{twelve}$ (h) $714_{twelve} - ET_{twelve}$
5. Perform the indicated operations. Check by translating to base ten.
   (a) $320_{five} - 43_{five}$ (b) $3E8_{twelve} + 3TT_{twelve}$
   (c) $11011_{two} + 1111_{two}$ (d) $2003_{five} - 342_{five}$

6. For each of the following bases, name the largest and then the smallest number represented by a numeral of the form *abc* where $a \neq 0$. Find the difference of these two numbers.
   (a) five (b) nine
   (c) three (d) six

*7. Perform the following operations. Leave the answer in terms of the base of the numeral on the left.
   (a) $563_{\text{seven}} - \text{ET}_{\text{twelve}}$ (b) $11011_{\text{two}} + 323_{\text{four}} + \text{TE}_{\text{twelve}}$
   (c) $10011_{\text{two}} + 423_{\text{five}}$

*8. Arrange in increasing order: 1101000110_{two}, 2132_{four}, 2132_{five}, 22212_{three}, 2237_{eight}, 2237_{nine}, $\text{ETE}_{\text{twelve}}$.

*9. The creatures of Long Lost Land have a number system based on 4 numerals because they have 3 fingers on each of their 3 hands. But instead of using the Hindu-Arabic numerals, they use $\triangle$, $\mathscr{L}$, $\square$, and / for numerals. Given below is an addition table for this system. Determine the following sums and differences.

$+$	$\triangle$	$\mathscr{L}$	$\square$	/
$\triangle$	$\triangle$	$\mathscr{L}$	$\square$	/
$\mathscr{L}$	$\mathscr{L}$	$\square$	/	$\mathscr{L}\triangle$
$\square$	$\square$	/	$\mathscr{L}\triangle$	$\mathscr{L}\mathscr{L}$
/	/	$\mathscr{L}\triangle$	$\mathscr{L}\mathscr{L}$	$\mathscr{L}\square$

   (a) $\square - \triangle$ (b) $(/ + /)$
   (c) $(\mathscr{L}\mathscr{L} - \square) + (\mathscr{L}\mathscr{L} - /)$ (d) $(\square + \square) - \mathscr{L}$
   (e) What is the additive identity?

*10. (a) If $243 + 461 = 724$, what base is being used?
   (b) If $576 + 288 = 842$, what base is being used?

7

Multiplication and Division in Various Bases

In this section, multiplication and division will be considered for bases other than ten. By knowing all the products of the digits in a numeration system, one can perform multiplication in much the same manner as in base ten. Instead of memorizing the different number combinations, it is recommended that multiplication tables such as Table 4-3 be constructed.

Multiplication Table, Base Seven

·	0	1	2	3	4	5	6
0	0	0	0	0	0	0	0
1	0	1	2	3	4	5	6
2	0	2	4	6	11	13	15
3	0	3	6	12	15	21	24
4	0	4	11	15	22	26	33
5	0	5	13	21	26	34	42
6	0	6	15	24	33	42	51

Table 4-3

The algorithm for the multiplication of a number by 10 in any base is simply to annex a zero to the end of the numeral. For example,

$$\begin{aligned} 24_{\text{seven}} \cdot 10_{\text{seven}} &= \left[\begin{aligned} &[2(7)+4][1(7)+0] \\ &[2(7)+4](7) \\ &[2(7)](7)+4(7) \\ &2(7)^2+4(7) \\ &2(7)^2+4(7)+0 \end{aligned}\right]_{\text{ten}} \\ &= 240_{\text{seven}} \end{aligned}$$

Example: Find the product of 216_{seven} and 14_{seven}. To illustrate this multiplication, each factor will be expressed as sums of powers of the base seven. The multiplication may be performed using the same procedure or multiplication algorithm as used in base ten.

$$\left[\begin{array}{r} 2(7)^2+1(7)+6 \\ 1(7)+4 \\ \hline 8(7)^2+4(7)+24 \\ 2(7)^3+1(7)^2+6(7) \quad\quad \\ \hline \end{array}\right]_{\text{ten}} \quad \text{or} \quad \left[\begin{array}{r} 2(7)^2+1(7)+6 \\ 1(7)+4 \\ \hline 1(7)^3+2(7)^2+0(7)+3 \\ 2(7)^3+1(7)^2+6(7) \quad\quad \\ \hline 3(7)^3+3(7)^2+6(7)+3 \end{array}\right]_{\text{ten}}$$

$$216_{\text{seven}} \cdot 14_{\text{seven}} = 3363_{\text{seven}}.$$

Consider the following calculation with accompanying suggested helps:

Suggested Help

$$\left[\begin{array}{r} 216 \\ 14 \\ \hline 1203 \\ 216 \\ \hline 3363 \end{array}\right]_{\text{seven}} \qquad \left[\begin{array}{l} 4 \cdot 6 = 24 = 3(7) + 3 \\ 4 \cdot 1 + 3 = 7 = 1(7) + 0 \\ 4 \cdot 2 + 1 = 9 = 1(7) + 2 \\ \text{first line } 1203 \\ 1 \cdot 6 = 6,\ 1 \cdot 1 = 1 \text{ and } 1 \cdot 2 = 2 \\ \text{second line } 216 \end{array}\right]_{\text{ten}}$$

Additional examples of multiplications will be given using bases two, four, and seven. Any step you cannot understand should be checked by writing the factors in expanded form.

Examples: Multiply.

$$\left[\begin{array}{r} 111 \\ 11 \\ \hline 111 \\ 111 \\ \hline 10101 \end{array}\right]_{\text{two}} \quad \left[\begin{array}{r} 1011 \\ 101 \\ \hline 1011 \\ 1011 \\ \hline 110111 \end{array}\right]_{\text{two}} \quad \left[\begin{array}{r} 123 \\ 32 \\ \hline 312 \\ 1101 \\ \hline 11322 \end{array}\right]_{\text{four}} \quad \left[\begin{array}{r} 123 \\ 56 \\ \hline 1104 \\ 651 \\ \hline 10614 \end{array}\right]_{\text{seven}}$$

The Division Algorithm states that if a and b are whole numbers and $b \neq 0$, then a can be written as $a = b \cdot q + r$, where q and r are whole numbers such that $0 \leq r < b$. A discussion was given on division in the decimal system in Section 4 of this chapter. Let us illustrate the Division Algorithm by an example in base two.

Example: Divide 11010_{two} by 1011_{two}.

To solve this problem, we organize our work in the usual schematic form for division.

$$\left[\begin{array}{r|r} & 1(2) + 0 \\ \hline 1(2)^3 + 0(2)^2 + 1(2) + 1 & 1(2)^4 + 1(2)^3 + 0(2)^2 + 1(2) + 0 \\ & 1(2)^4 + 0(2)^3 + 1(2)^2 + 1(2) + 0 \\ \hline & 1(2)^2 + 0(2) + 0 \\ & 0(2)^2 + 0(2) + 0 \\ \hline & 1(2)^2 + 0(2) + 0 \end{array}\right]_{\text{ten}}$$

Since there are only two digits in the binary system, the first digit of the answer will be either $0(2)$ or $1(2)$. Suppose we try $1(2)$. $1(2)$ multiplied by the divisor and subtracted from the dividend leaves a remainder of $1(2)^2 + 0(2) + 0$. The divisor will divide this 0 times. Thus, the answer is 10_{two} with a remainder of 100_{two}.

A somewhat more difficult example will involve base seven. The trial and error process of selecting a digit of the quotient is not indicated; however, your first selection of a trial divisor may often give an answer either too large or too small. After working several problems, you should become much more proficient at selecting trial divisors.

Example: Divide 1662_{seven} by 24_{seven}.

$$\left[\begin{array}{r|r} & 5(7) + 2 \\ 2(7) + 4 & 1(7)^3 + 6(7)^2 + 6(7) + 2 \\ & 1(7)^3 + 5(7)^2 + 6(7) + 0 \\ \hline & 1(7)^2 + 0(7) + 2 \\ & 5(7) + 1 \\ \hline & 2(7) + 1 \end{array}\right]_{\text{ten}}$$

Exercise Set 4-7

1. Perform the following operations in the bases indicated.
   (a) $(1101_{\text{two}})(11_{\text{two}})$ (b) $(213_{\text{four}})(32_{\text{four}})$
   (c) $(1011_{\text{two}}) \div (11_{\text{two}})$ (d) $(312_{\text{four}}) \div (2_{\text{four}})$
   (e) $(146_{\text{seven}})(31_{\text{seven}})$ (f) $(266_{\text{seven}})(146_{\text{seven}})$
   (g) $(1642_{\text{seven}}) \div (6_{\text{seven}})$ (h) $(6242_{\text{seven}}) \div (14_{\text{seven}})$
2. Check Exercises 1(a), (b), (c), (d), and (e) by translating each factor to a base ten numeral before performing the operation. Then translate the answer to a base ten number to see if the two answers agree.
3. Make a multiplication table for the following bases.
   (a) three (b) five (c) twelve
4. Perform the indicated operations, and write your answers in the indicated bases.
   (a) $(44_{\text{five}})(4_{\text{five}})$ (b) $(22_{\text{twelve}})(5_{\text{twelve}})$
   (c) $(323_{\text{five}})(43_{\text{five}})$ (d) $(40\text{E}_{\text{twelve}})(3\text{T}_{\text{twelve}})$
   (e) $(155_{\text{twelve}}) \div (4_{\text{twelve}})$ (f) $(3240_{\text{five}}) \div (23_{\text{five}})$
   (g) $(432_{\text{twelve}}) \div (\text{T}_{\text{twelve}})$ (h) $(64\text{TE}_{\text{twelve}}) \div (45_{\text{twelve}})$

5. Work 4(a), (b), and (c) with the numerals in base nine; base seven.

*6. Perform the following divisions, leaving the answer in the same base as the a in $a \div b$.
   (a) $(4T0E_{twelve}) \div (1101_{two})$ (b) $(1204_{five}) \div (212_{three})$

*7. The following multiplication table is developed from the numeration system of Long Lost Land. Solve each of the following multiplication and division problems. Use this table and the addition table in Exercise 9, Exercise Set 4-6.

$\cdot$	$\triangle$	$\mathscr{L}$	$\square$	/
$\triangle$	$\triangle$	$\triangle$	$\triangle$	$\triangle$
$\mathscr{L}$	$\triangle$	$\mathscr{L}$	$\square$	/
$\square$	$\triangle$	$\square$	$\mathscr{L}\triangle$	$\mathscr{L}\mathscr{L}$
/	$\triangle$	/	$\mathscr{L}\mathscr{L}$	$\mathscr{L}\square$

   (a) $(\square \cdot /) \div \square$ (b) $\mathscr{L}\square \div /$
   (c) $(\mathscr{L}\mathscr{L} \div \square) \div /$ (d) $(\mathscr{L}\mathscr{L} \div \square) \div \mathscr{L}$
   (e) $(\square \cdot /) \div /$ (f) What is the multiplicative identity?

*8. (a) If $43 \cdot 25 = 1501$, what base is being used?
   (b) If $1021 \cdot 12 = 20022$, what base is being used?

Review Exercise Set 4-8

1. A time machine took Adventurer back to visit many ancient civilizations. Adventurer found that his work in elementary mathematics helped him understand these civilizations. What answers did he find to the following problems? (Write the answer in base ten.)
   (a) CXLIV + DCLXVI (b) ▼▼▼▼▼ ◀ + ◀ ▼▼▼▼
   (c) 𓆼 𓍢|𓍢𓍢 ∩| ∩∩ − 𓍢|𓍢∩𓍢|𓍢|𓍢𓍢|
   (d) Mayan numerals: ——, ⊜ ∓ ••••, ——, ——, ••••, ——

2. (a) Write 157 in base twelve. (b) Write 31 in base two.
   (c) Write 534 in base seven. (d) Write 86 in base three.
   (e) Write 234 in base six. (f) Write 999 in base nine.

3. Simplify.
   (a) $(3)(3)(3)^4$ (b) $(8)^2 \div (2)^5$

(c) $(9)^0(4)$ (d) $\frac{(a)^3(a)^5}{a^4}$

(e) $\frac{(2)^4(2)^3}{2^0}$ (f) $\frac{2^0}{3^0}$

4. Translate each of the following to a base ten numeral.
   (a) $T3E_{\text{twelve}}$ (b) 10110_{two}
   (c) 1231_{five} (d) 461_{seven}
   (e) 101011_{two} (f) 154_{six}
5. Perform the indicated additions and subtractions.
   (a) $3123_{\text{five}} + 2314_{\text{five}}$ (b) $930E_{\text{twelve}} - 3T4_{\text{twelve}}$
   (c) $11011_{\text{two}} - 1101_{\text{two}}$ (d) $465_{\text{seven}} + 356_{\text{seven}}$
6. Perform the following operations after expressing each numeral in terms of powers of ten.
   (a) $2346 + 984$ (b) $23{,}410 \div 37$
   (c) $1728 \div 32$ (d) $2435 - 867$

*7. Name the number base for the following solved problems.
   (a) $54 \div 8 = 8$ (b) $403 - 134 = 214$
   (c) $452 - 263 = 156$ (d) $1111 + 1000 = 10111$
   (e) $(48)(32) = 1294$ (f) $(604)(35) = 31406$

*8. Perform the indicated multiplications and divisions.
   (a) $(T2_{\text{twelve}})(58_{\text{twelve}})$ (b) $(234_{\text{six}}) \div (4_{\text{six}})$
   (c) $(10110_{\text{two}}) \div (10_{\text{two}})$ (d) $(58E_{\text{twelve}}) \div (E_{\text{twelve}})$
   (e) $(455_{\text{six}})(24_{\text{six}})$ (f) $(110_{\text{two}})(101_{\text{two}})$

9. Perform the following operations in base eight, base ten, and base twelve. Change all three answers to base ten.
   (a) $664 - 565$ (b) $462 + 726$
   (c) $337 - 23$ (d) $1111 \cdot 654$
   (e) $4246 + 35712$ (f) $6660 \cdot 32$

Suggested Reading

History of Numeration Systems: Bouwsma, Corle, Clemson, pp. 17–25. Byrne, pp. 68–75. Campbell, pp. 63–65. Garner, pp. 107–112. Graham, pp. 61–67, 70–72. Hutton, pp. 34–37. Meserve, Sobel, pp. 31–36. Nichols, Swain, pp. 54–68. Peterson, Hashisaki, pp. 1–10, 15–21. Podraza, Blevins, Hanson, Prall, pp. 21–23. Scandura, pp. 209–217. Smith, pp. 41–47. Willerding, pp. 70–75. Wren, pp. 21–33. Zwier, Nyhoff, pp. 71–76.

Decimal Numeration System: Armstrong, pp. 118–122, 245–249. Byrne, p. 80. Campbell, pp. 64–65. Garner, pp. 113–117. Graham, pp. 67–75. Hutton, pp. 39–52. Smith, pp. 48–59. Spector, pp. 147–150. Willerding, pp. 73–77. Wren, pp. 29–33. Zwier, Nyhoff, pp. 71–73.

Algorithms for Computation: Bouwsma, Corle, Clemson, pp. 115–121. Garner, pp. 132–137. Graham, pp. 125–135. Hutton, pp. 121–141. Peterson, Hashisaki, pp. 96–99. Scandura, pp. 218–228. Smith, pp. 78–79, 86–88, 109–110. Spector, pp. 176–199. Willerding, pp. 94–101. Wren, pp. 47–48, 61–65, 92–93.

Addition and Subtraction in Different Bases: Armstrong, pp. 293–298. Campbell, pp. 69–76. Hutton, pp. 124–127, 140–141. Meserve, Sobel, pp. 55–66. Nichols, Swain, pp. 77–80, 87–88, 90–92. Peterson, Hashisaki, pp. 113–124. Scandura, pp. 218–226. Smith, pp. 267–270, 280–287. Spector, pp. 158–167. Willerding, pp. 96–98. Wren, pp. 67–68. Zwier, Nyhoff, pp. 80–84.

Multiplication and Division in Various Bases: Armstrong, pp. 293–298. Campbell, pp. 76–82. Garner, pp. 142–143. Hutton, pp. 135–137, 146–147. Meserve, Sobel, pp. 57–59, 64–65. Nichols, Swain, pp. 80–83, 87–88, 90–92. Peterson, Hashisaki, pp. 113–121. Scandura, pp. 226–230. Smith, pp. 270–275, 280–287. Spector, pp. 161–167. Willerding, pp. 101–107. Wren, pp. 66–69. Zwier, Nyhoff, pp. 80–84.

5

The System of Integers

1

Introduction

From your previous studies, you are aware of numbers other than whole numbers. You undoubtedly remember such numbers as $^-3$, $^-5$, $^-7$, called *negative numbers*, and numbers denoted by 2/3, 4/7, 5/9, called *rational numbers*. How can we extend our system of whole numbers to include these new concepts? This question will be answered in this chapter and in Chapter 7.

Since we will be extending the system of whole numbers to include both negative numbers and rational numbers, a choice must be made concerning which to consider first. From the standpoint of an intuitive approach, it seems preferable to consider first the negative numbers and then the rational numbers. Thus, in this book the system of whole numbers will be extended to include negative numbers; then the resulting system will be enlarged to include rational numbers.

The need for negative numbers, even in everyday language, is evident. Such numbers provide the most convenient method for distinguishing between 10° above zero and 10° below zero, a gain of 6 yards on the football field or a loss of 6 yards, and "I owe $50" or "I have $50."

The concept of a negative number was called "absurd" by the famous mathematician Diophantus of Alexandria. In fact, in 1225 the mathematician Fibonacci stated that the equation $a + x = b$, with $a > b$, had no solution unless it expressed that a man was in debt. As late as the seventeenth century, negative numbers were referred to as "false" or "fictitious." Finally, as a result of Descartes' *La Geometrie*, published in 1637, negative numbers were given concrete meaning.

In this chapter we introduce negative integers and develop their properties. In Chapter 7, we discuss positive and negative rational numbers. Later, the rational numbers will be extended to the system of real numbers.

2

The Integers

In Chapter 3, the difference $a - b$ was defined in terms of the equality $a = b + x$. But, in the system of whole numbers, this equality does not always have a solution; that is, if a and b are whole numbers, $a - b$ is not necessarily a whole number. For example, $5 = 8 + x$ has no solution in the system of whole numbers; and thus $5 - 8$ is a meaningless expression in the system of whole numbers. Hence, the question arises, "Can we enlarge the system of whole numbers so that subtraction will always have meaning in the new system?"

In an attempt to make this extension, we consider the problem of subtraction from 0, namely, the problem $0 - a = ?$; or, what is x such that $a + x = 0$? Can we, for example, find numbers for x, y, and z such that $1 + x = 0$, $2 + y = 0$, or $5 + z = 0$? If by numbers we mean the set of whole numbers 0, 1, 2, ..., the answer is *no*. However, let us see if we cannot *create* some new numbers that will satisfy such requirements. If you earn \$10 and then spend the \$10, your net holding is \$0. If the money spent is symbolized as $^-$\$10, then we can say that

$$\$10 + (^-\$10) = \$0.$$

Similarly, if you begin a walk to the drugstore, travel three blocks, and then remember that your money is at home and return the three blocks, the distance you are from home amounts to zero. $3 + (^-3) = 0$. Let $^-1$ be a number such that $1 + {}^-1 = 0$; $^-2$ a number such that $2 + {}^-2 = 0$; and $^-5$ a number such that $5 + {}^-5 = 0$. In general, ^-n will be a number such that $n + {}^-n = 0$.

Definition 5-1: If n is a natural number, then ^-n is defined to be a unique number such that $n + {}^-n = {}^-n + n = 0$.

Thus, $^-1$ is the only number which when added to 1 gives 0; $^-20$ is the only number which when added to 20 gives 0; and in general, ^-n is the only number which when added to n gives 0, where n is a natural number. The new number ^-n is called "the *additive inverse* of n," "the negative of n," "minus n," or "the opposite of n." In like manner, if ^-n is the additive inverse of n, then n is the additive inverse of ^-n. For example, 6 is the additive inverse of $^-6$, because $6 + {}^-6 = 0$.

What is the additive inverse of 0? To be in agreement with Definition 5-1, this will be a number $^-0$ such that $0 + {}^-0 = 0$. But 0 itself has the property that $0 + 0 = 0$. In order that the additive inverse be unique, we thus insist that $^-0$ be the same as 0.

When n is a natural number, its additive inverse ^-n is called a *negative integer*. By forming the union of the negative integers and the whole numbers, a new classification of numbers is obtained.

Definition 5-2: The set of integers is the union of the set of whole numbers and the set $\{^-n\}$ such that for each natural number n, there is a number ^-n, where $n + {}^-n = {}^-n + n = 0$.

The natural numbers, as part of the integers, are often called *positive integers*. The set of positive integers is sometimes written as $\{+1, +2, +3, \ldots\}$ or as $\{^+1, {}^+2, {}^+3, \ldots\}$, but in this book the plus sign will be omitted. The positive integers and 0 are called the *nonnegative integers*.

Thus, the integers consist of the positive integers, 0, and the negative integers. The set of integers can be tabulated as

$$I = \{\ldots, {}^-3, {}^-2, {}^-1, 0, 1, 2, 3, \ldots\}.$$

A second approach to the development of the system of integers involves the idea of equivalence classes of ordered pairs of whole numbers. Note that the answer 4 can be obtained from an infinite number of subtractions, $4 - 0$, $5 - 1$, $6 - 2$, $7 - 3$, ... Thus 4 is identified with the class of ordered pairs of whole numbers, $\{(4 - 0), (5 - 1), (6 - 2), (7 - 3), \ldots\}$. In a similar manner, $^-3$ can be identified with the class of ordered pairs of whole numbers, $\{(0 - 3), (1 - 4), (2 - 5), (3 - 6), \ldots\}$. That is, $^-3$ is defined to be the answer obtained from any one of the subtractions within this class.

With this approach, the set of integers may be constructed as equivalence classes of subtractions of whole numbers. Each integer may be represented by any one set of the subtractions in the class.

0 identifies with $\{(0 - 0), (1 - 1), (2 - 2), \ldots\}$.
1 identifies with $\{(1 - 0), (2 - 1), (3 - 2), \ldots\}$.
$^-1$ identifies with $\{(0 - 1), (1 - 2), (2 - 3), \ldots\}$.
2 identifies with $\{(2 - 0), (3 - 1), (4 - 2), \ldots\}$.
$^-2$ identifies with $\{(0 - 2), (1 - 3), (2 - 4), \ldots\}$.

Thus, any integer may be represented as a difference $a - b$ of two whole numbers. Some books use an ordered pair (a, b) to represent the integer. If $a > b$, the integer is positive; if $a < b$, the integer is negative; and if $a = b$, the integer is zero.

Both the additive-inverse approach and the equivalence-class-of-subtractions approach are effective in extending the system of whole numbers to the system of integers. The additive-inverse approach will be used in the text narrative, and derivations using equivalence classes will be considered in the exercise sets.

Just as the natural numbers in Chapter 3 were considered as directed distances to the right on a number line, we can give special meaning to the integers. Consider a number line as extending endlessly in both directions. Mark an arbitrarily chosen point as 0, called the origin, and then measure equal segments to the *right* to determine points labeled 1, 2, 3, 4, ... In a like manner, measure segments of the same size to the *left* for points labeled $^-1$, $^-2$, $^-3$, $^-4$, ..., as illustrated in Figure 5-1. By this process, we are setting up a one-to-one correspondence between a subset of points on the line and the integers.

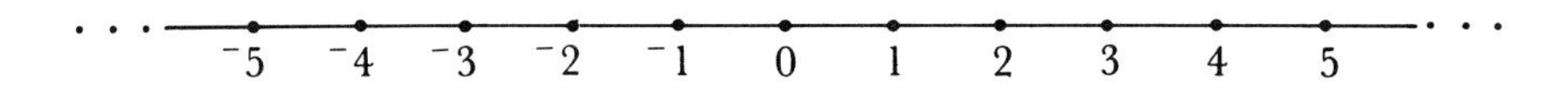

Figure 5-1

We sometimes use the term *distance* to indicate the length between any two points on the number line. We consider this distance to be the whole number of intervals between the points. That is, the distance between the point labeled 2 and the point labeled 5 is 3. The distance from $^-1$ to 2 is 3. The distance between $^-1$ and $^-5$ is 4. We use this concept of distance to define absolute value.

Definition 5-3: The distance of an integer from 0 is called the *absolute value* of the integer. The absolute value of an integer a is written $|a|$. Thus, for n a natural number,

$$
\begin{aligned}
|a| &= n \quad \text{if} \quad a = n,\\
|a| &= 0 \quad \text{if} \quad a = 0,\\
|a| &= n \quad \text{if} \quad a = {}^-n.
\end{aligned}
$$

Examples: $|{}^-6| = 6$; $|3| = 3$; $|{}^-4| = 4$

We now associate with each integer an arrow diagram in a manner similar to that used for natural numbers. To represent an integer, draw an arrow diagram from 0 to the point representing the integer. The integer 3 may be represented by an arrow diagram directed toward the right and of the same length as the line segment from 0 to 3 on the number line; it could also be represented by the line segment from 2 to 5 on the number line. Thus, an arrow diagram may be *translated* or moved along the number line, as shown in Figure 5-2. In the same figure, integer $^-4$ is represented by two arrow diagrams of length 4; note that both of these diagrams are directed toward the left.

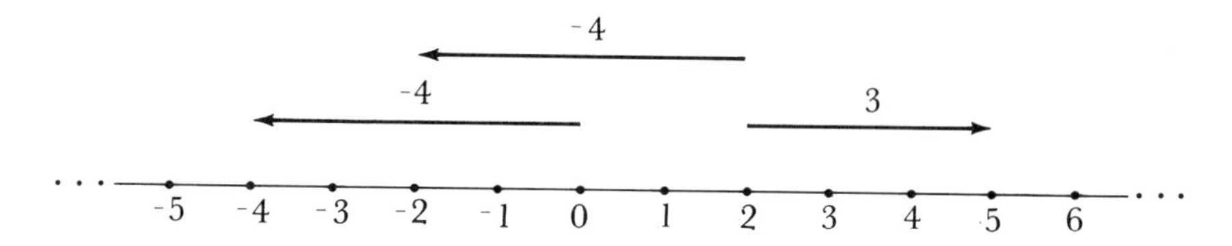

Figure 5-2

Figures 5-1 and 5-2 make it apparent that the positive integers may be thought of as extending indefinitely to the right of zero and the negative integers as extending indefinitely to the left of zero. Pairs of additive inverses such as $^-5$ and 5 are represented by points at equal distances on each side of zero.

Exercise Set 5-1

1. What are the additive inverses of the following integers?
   (a) 5 (b) $^-3$ (c) 0 (d) $^-8$
   (e) a (f) ^-a (g) $^-(a+b)$ (h) ^-b
2. Draw arrow diagrams on a number line to represent the following integers. Start each arrow diagram at 0.
   (a) $^-2$ (b) $^-4$ (c) $^-3$
   (d) 8 (e) 6 (f) $^-5$
3. Draw arrow diagrams starting at $^-3$ for each integer in Exercise 2.
4. Draw arrow diagrams to represent the additive inverses of each number in Exercise 2.
5. From the following list, select the pairs which are additive inverses.
   (a) $^-2$, 2 (b) $^-3$, $^-(^-3)$

(c) $7, {}^-8$ (d) $(2 \cdot 3), {}^-(3+3)$
(e) $(2+3), {}^-(2+3)$ (f) $0, 0$
(g) $2, 2$ (h) $1, {}^-1$

6. What integers will make the following true?
(a) ${}^-(\ \ )+5=0$ (b) $(5+3)+{}^-(2 \cdot 4)=(\ \ )$
(c) $(\ \ )+4=0$ (d) ${}^-(\ \ )+{}^-4=0$
(e) $(4+2)+{}^-(4+2)=(\ \ )$ (f) $4+{}^-(3+1)=(\ \ )$
(g) ${}^-2+(\ \ )=0$ (h) $3+{}^-(\ \ )=0$
7. What integer is represented by $|x|$ and ${}^-x$ if:
(a) $x=3$? (b) $x={}^-5$? (c) $x=0$? (d) $x={}^-({}^-1)$?
8. (a) Is there a smallest negative integer?
(b) What is the smallest positive integer?
(c) What is the smallest nonnegative integer?
(d) What is the greatest negative integer?
9. Let C be the set of counting numbers, W be the set of whole numbers, and I be the set of integers. The set of integers can be separated into ${}^+I$, the set of positive integers, and ${}^-I$, the set of negative integers, and zero. Determine if the following are true or false. If false, state why.
(a) $C \cup I = I$ (b) $C \cup {}^+I = W$
(c) $C \cap {}^-I = \{0\}$ (d) $W \subset {}^+I$
(e) ${}^+I \subseteq W$ (f) $W \cap {}^-I = \{0\}$
(g) $(W \cup {}^+I) \cap C = \varnothing$ (h) ${}^-I \cup {}^-I = I$
(i) ${}^-I \cup W = I$ (j) $C \cap W = C$
10. Find the absolute value and the additive inverse of each of the following.
(a) additive inverse of 3 (b) additive inverse of ${}^-2$
(c) ${}^-({}^-3)$ (d) ${}^-|{}^-5|$

*11. Let the difference $a-b$ of two natural numbers represent an integer.
(a) Find some ordered pairs of natural numbers to represent ${}^-4$, 3, and ${}^-7$.
(b) Is it ever the case that $a-b$ and $b-a$ are the same?
(c) Define the equality of two integers represented by $a-b$ and $c-d$ or by (a, b) and (c, d).

3

The System of Integers

In Section 2, we considered the set of integers as an extension or enlargement which contains as a subset the set of whole numbers. We now define what we mean by the *system of integers*. Recall that the system of whole numbers consists of a set W: $\{0, 1, 2, 3, \ldots\}$ and two operations $+$ and $\cdot$ that satisfy several properties—closure, associative, commutative, and distributive properties, and existence of additive and multiplicative identities. The extension of the whole numbers to the system of integers is ideally accomplished in such a way that the two operations $+$ and $\cdot$ still have these properties and one additional property postulating the existence of a unique additive inverse of each element. The operations of $+$ and $\cdot$ will be defined so that the operations involving whole numbers within the system of integers will be the same as within the system of whole numbers itself.

Definition 5-4: The system of integers consists of the set I of integers $\{\ldots, {}^-4, {}^-3, {}^-2, {}^-1, 0, 1, 2, 3, 4, \ldots\}$ and two binary operations, addition, $+$, and multiplication, $\cdot$, with the following properties for any integers a, b, and c.

Closure Properties

1. $a + b$ is a unique integer.
2. $a \cdot b$ is a unique integer.

Commutative Properties

3. $a + b = b + a$
4. $a \cdot b = b \cdot a$

Associative Properties

5. $(a + b) + c = a + (b + c)$
6. $(a \cdot b) \cdot c = a \cdot (b \cdot c)$

Distributive Property of Multiplication over Addition

7. $a \cdot (b + c) = a \cdot b + a \cdot c$
8. $(b + c) \cdot a = b \cdot a + c \cdot a$

Identity Elements

9. There is a unique integer, namely 0, such that

$$0 + a = a + 0 = a.$$

10. There is a unique integer, namely 1, such that

$$1 \cdot a = a \cdot 1 = a.$$

Additive Inverses

11. For each a in I, there exists a unique additive inverse ${}^-a$ such that $a + {}^-a = {}^-a + a = 0$.

The additive identity, 0, has the following multiplicative property. If x is any integer, then

$$0 \cdot x = x \cdot 0 = 0.$$

Examples:

$$8 \cdot 0 = 0$$
$$0 \cdot 4 = 0$$
$$^{-}6 \cdot 0 = 0$$

The following examples illustrate some of the reasoning involved in the formulation of the definition of the operation of addition of integers.

Example:

$(^{-}5 + {}^{-}3) + (5 + 3) = (^{-}5 + {}^{-}3) + (3 + 5)$	Commutative property of addition
$= {}^{-}5 + [^{-}3 + (3 + 5)]$	Associative property of addition
$= {}^{-}5 + [(^{-}3 + 3) + 5]$	Associative property of addition
$= {}^{-}5 + [0 + 5]$	Additive inverse
$= {}^{-}5 + 5$	Additive identity
$= 0$	Additive inverse

Therefore,

$(^{-}5 + {}^{-}3) + (5 + 3) = 0.$	Transitive property of equalities

Thus, the number $^{-}5 + {}^{-}3$ is the additive inverse of $5 + 3$. But by definition, the unique additive inverse of $5 + 3$ is $^{-}(5 + 3) = {}^{-}8$. Hence, $^{-}5 + {}^{-}3 = {}^{-}8$.

The reasonableness of this addition is further illustrated by considering the addition of $^{-}5 + {}^{-}3$ as indicated on the number line (Figure 5-3). Remember that in a demonstration of addition on a

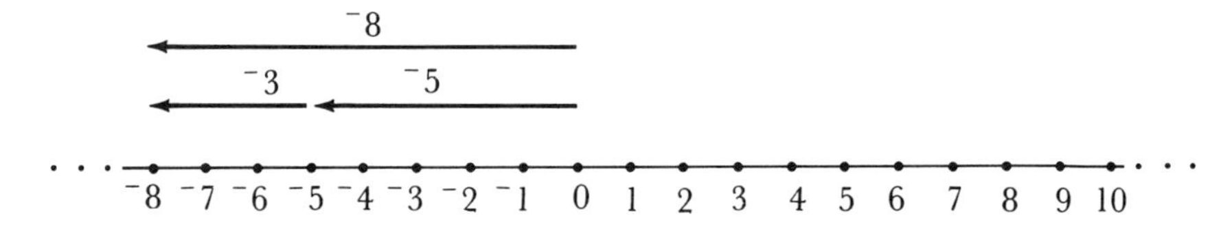

Figure 5-3

number line we place the starting end of the arrow diagram representing the second number at the arrow point of the arrow diagram of the first number. The resulting combined arrow diagram gives the sum of the two numbers. That is, the arrow on the second arrow diagram terminates at the answer. Note that $^-5 + {}^-3$ terminates at the previously obtained answer of $^-8$.

In general,

$$\begin{aligned} ({}^-a + {}^-b) + (a + b) &= ({}^-a + {}^-b) + (b + a) && \text{Why?} \\ &= {}^-a + [{}^-b + (b + a)] && \text{Why?} \\ &= {}^-a + [({}^-b + b) + a] && \text{Why?} \\ &= {}^-a + [0 + a] && \text{Why?} \\ &= {}^-a + a && \text{Why?} \\ &= 0 && \text{Why?} \end{aligned}$$

Therefore, $({}^-a + {}^-b) + (a + b) = 0.$ Why?

Thus, the number $^-a + {}^-b$ is the additive inverse of $a + b$. However, the unique additive inverse of $a + b$ is $^-(a + b)$. Thus, $^-a + {}^-b = {}^-(a + b)$.

Example: How should the sum $6 + {}^-2$ be defined to satisfy the properties of addition?

$$\begin{aligned} 6 + {}^-2 &= (4 + 2) + {}^-2 && \text{Renaming 6} \\ &= 4 + (2 + {}^-2) && \text{Associative property of addition} \\ &= 4 + 0 && \text{Additive inverse} \\ &= 4 && \text{Additive identity} \end{aligned}$$

Thus, $6 + {}^-2 = 4$. We recall from Chapter 3 that $6 - 2 = 4$. Hence, $6 + {}^-2 = 6 - 2$.

In general, consider $a + {}^-b$ where $a > b$. Since $a > b$, we can write $a = b + c$ where c is a natural number. Then

$$\begin{aligned} a + {}^-b &= (b + c) + {}^-b && \text{Renaming } a \\ &= (c + b) + {}^-b && \text{Commutative property of addition} \\ &= c + (b + {}^-b) && \text{Associative property of addition} \\ &= c + 0 && \text{Additive inverse} \\ &= c && \text{Additive identity} \\ &= a - b && c = a - b \text{ since } a = b + c \end{aligned}$$

Therefore,

$a + {}^-b = a - b$ when $a > b$.

Again the reasonableness of this answer can be emphasized by illustrating the addition of $6 + {}^-2$ on the number line (Figure 5-4). Note that $6 + {}^-2 = 4$.

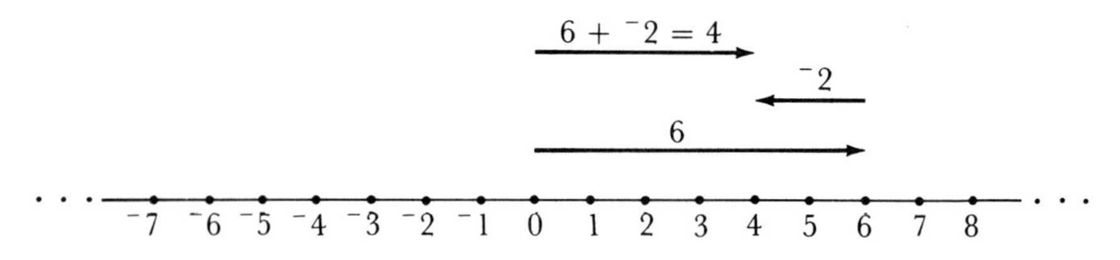

Figure 5-4

If the addition of integers is to be commutative, $a + {}^-b$ must equal ${}^-b + a$. However, is this result consistent with the other properties desired? Let us examine this question by considering the sum of ${}^-2 + 6$. Again, 6 can be written as $2 + 4$.

${}^-2 + 6 = {}^-2 + (2 + 4)$	Renaming 6
$= ({}^-2 + 2) + 4$	Associative property of addition
$= 0 + 4$	Additive inverse
$= 4$	Additive identity

Thus, ${}^-2 + 6 = 4 = 6 + {}^-2$.

Example: Consider now the sum ${}^-8 + 5$ so that our enumerated properties in Definition 5-4 are retained. From a previous example ${}^-8 = {}^-5 + {}^-3$. Thus,

${}^-8 + 5 = ({}^-5 + {}^-3) + 5$	Renaming ${}^-8$
$= ({}^-3 + {}^-5) + 5$	Commutative property of addition
$= {}^-3 + ({}^-5 + 5)$	Associative property of addition
$= {}^-3 + 0$	Additive inverse
$= {}^-3$	Additive identity

Therefore,

${}^-8 + 5 = {}^-3$.	Transitive property of equalities

In general, consider ${}^-a + b$ where $a > b$. Since $a > b$, write $a - b = c$ or $a = b + c$. Then

${}^-a + b = {}^-(b + c) + b$	Renaming a
$= ({}^-b + {}^-c) + b$	Previous example in this section
$= ({}^-c + {}^-b) + b$	Commutative property of addition
$= {}^-c + ({}^-b + b)$	Associative property of addition
$= {}^-c + 0$	Additive inverse
$= {}^-c$	Additive identity
$= {}^-(a - b)$	Renaming c

Therefore,

$^-a + b = {}^-(a - b)$ when $a > b$.

The addition of $^-8 + 5$ on the number line is illustrated in Figure 5-5. Note again that $^-8 + 5 = {}^-3$.

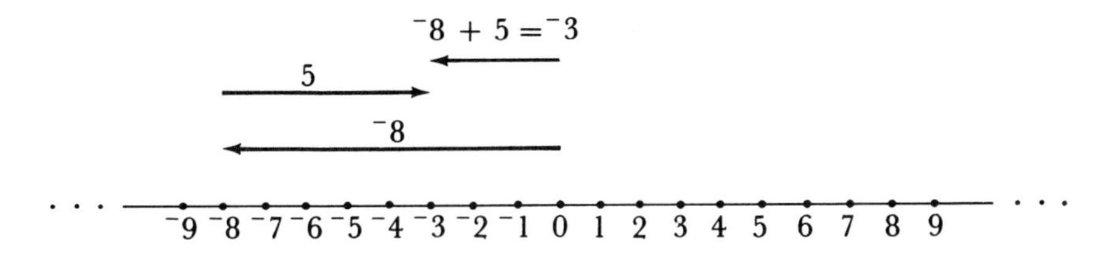

Figure 5-5

To satisfy the conditions imposed by Definition 5-4 and illustrated by the preceding examples, the definition of addition of integers is formulated as follows.

Definition 5-5: The sum of two integers (denoted by the operator $+$) is defined by the following cases, where a and b are whole numbers.

(a) $a + b = n(A \cup B)$, where $a = n(A)$, $b = n(B)$, and $A \cap B = \varnothing$ for sets A and B.
(b) $^-a + {}^-b = {}^-(a + b)$.
(c) $a + {}^-b = {}^-b + a = (a - b)$ if $a > b$.
(d) $a + {}^-b = {}^-b + a = 0$ if $a = b$.
(e) $a + {}^-b = {}^-b + a = {}^-(b - a)$ if $a < b$.

The following set of examples illustrates addition of integers.

Examples:

$$2 + ({}^-5) = {}^-(5 - 2) = {}^-3$$
$$7 + ({}^-4) = (7 - 4) = 3$$
$$^-3 + ({}^-5) = {}^-(3 + 5) = {}^-8$$
$$^-7 + 9 = 9 + ({}^-7) = (9 - 7) = 2$$
$$^-9 + 5 = 5 + ({}^-9) = {}^-(9 - 5) = {}^-4$$
$$^-16 + ({}^-17) = {}^-(16 + 17) = {}^-33$$

We note from Definition 5-5 that the sum of any two positive integers is a positive integer; likewise, the sum of any two negative integers is a negative integer. Thus, we can say that the set of positive integers is closed under the operation of addition; likewise, the set of negative integers is closed under addition.

The definition of subtraction for whole numbers was introduced in Chapter 3, where $a - b$ was defined to be a whole number c which when added to b gives an answer a. The set of whole numbers is not closed under the operation of subtraction. This is one of the reasons for extending the whole numbers to the set of integers. We now define subtraction for integers in a way which is consistent with subtraction for whole numbers, and we shall see that the integers are closed under the operation of subtraction.

Definition 5-6: If x and y are integers, the difference in the subtraction of y from x (denoted by $x - y$) is the integer z if and only if $x = y + z$.

Examples:

$$
\begin{array}{rcl}
6 - 4 = 2 & \text{because} & 6 = 4 + 2 \\
{}^{-}2 - 5 = {}^{-}7 & \text{because} & {}^{-}2 = 5 + {}^{-}7 \\
8 - {}^{-}3 = 11 & \text{because} & 8 = {}^{-}3 + 11 \\
{}^{-}5 - {}^{-}8 = 3 & \text{because} & {}^{-}5 = {}^{-}8 + 3
\end{array}
$$

Example: $4 - 3 = 1$ because $4 = 3 + 1$; since $4 = 3 + 1 = 1 + 3$,

$$4 + {}^{-}3 = (1 + 3) + {}^{-}3 = 1 + (3 + {}^{-}3) = 1 + 0 = 1,$$

so $4 + {}^{-}3 = 1$. Since $4 - 3$ and $4 + {}^{-}3$ both name 1,

$$4 - 3 = 4 + {}^{-}3.$$

Notice that the definition of subtraction does not indicate that for any pair of integers, x and y, an integer z always exists such that $x = y + z$. Neither does the definition state that the answer is unique. Both are true, however, and are verified by the following theorem.

Theorem 5-1: If x and y are integers, then there always exists a unique integer z such that $x - y = z$. z can be written as $z = x + {}^{-}y$.

Proof: To prove the existence of an answer for $x - y = ?$, we must find an integer z for any x and any y such that $x = y + z$. Such an integer is $z = x + {}^{-}y$ where ${}^{-}y$ is the unique additive inverse of y. To verify this fact substitute $z = x + {}^{-}y$ in $x = y + z$.

$x = y + (x + {}^{-}y)$	Substitution
$= y + ({}^{-}y + x)$	Commutative property of addition
$= (y + {}^{-}y) + x$	Associative property of addition
$= 0 + x$	Additive inverse
$= x$	Additive identity

Thus $z = x + {}^{-}y$ does satisfy the requirement that $x = y + z$. Hence, $x - y = x + {}^{-}y$.

To prove that z is unique assume that there are integers z and v such that $x - y = z$ and $x - y = v$. Then $x = y + z$ and $x = y + v$, which by the transitive property of equality means that $y + z = y + v$. But by the cancellation property of addition (stated and proved in Section 5 of this chapter), $z = v$, so the subtraction of integers gives a unique result. Thus, we know that the integers are closed under the operation of subtraction.

The preceding theorem states that to subtract, change the sign of the number being subtracted and then add. It enables one to reduce subtractions involving positive and negative integers to calculations involving addition only.

Examples:

$$5 - 8 = 5 + ({}^{-}8) = {}^{-}3$$
$$4 - ({}^{-}10) = 4 + 10 = 14$$
$$0 - ({}^{-}6) = 0 + 6 = 6$$
$${}^{-}2 - (4) = {}^{-}2 + ({}^{-}4) = {}^{-}6$$

Since the operation of subtracting 5 is equivalent to adding ${}^{-}5$, many books use the same symbol for the sign of a number and the operation of subtraction. That is, ${}^{-}5$ is indicated by -5, the same symbol as is used in subtracting 5 from 8, $8 - 5$.

Subtraction has a simple interpretation on the number line. Since subtraction may be considered as the addition of the additive inverse of the number to be subtracted, the following examples may be drawn on the number line in the same manner as addition.

Examples:

(a) $4 - 5 = {}^-1$ (b) ${}^-3 - 4 = {}^-7$ (c) ${}^-3 - ({}^-4) = 1$

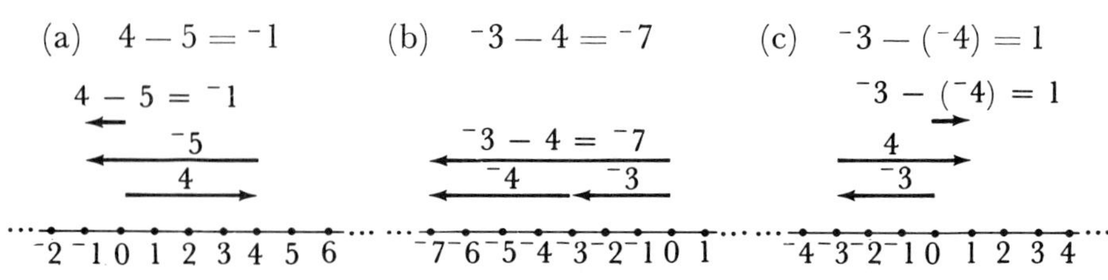

Figure 5-6

Exercise Set 5-2

1. Perform the indicated operations.
   (a) $2 - 5$ (b) $3 - {}^-1$ (c) ${}^-6 - 2$
   (d) $4 - 13$ (e) ${}^-10 + {}^-3$ (f) $6 - {}^-2$
   (g) ${}^-8 + {}^-3$ (h) $13 - {}^-6$ (i) ${}^-9 - 7$
   (j) $500 - {}^-5$ (k) $6 - 4 - {}^-2$ (l) ${}^-7 - (6 - 4)$
   (m) $({}^-4 + 8) + {}^-6$ (n) $(6 + {}^-6) + {}^-2$
   (o) $({}^-3 + {}^-2) + 5$ (p) $({}^-8 + {}^-7) + 8$
   (q) $(a + b) + {}^-b$ (r) $(x + y) + {}^-(x + y)$
2. Compute the following by using the number line.
   (a) ${}^-3 - 6$ (b) ${}^-3 - {}^-4$ (c) ${}^-6 + {}^-5$
   (d) $4 - {}^-3$ (e) $7 - {}^-1$ (f) $6 + {}^-3$
3. What important property does the system of integers have that the system of whole numbers does not have?
4. Demonstrate on the number line each of the following properties for integers a, b, and c.
   (a) Commutative property of addition
   (b) Additive inverse
   (c) Associative property for addition
5. For each of the following, verify by evaluating each side separately.
   (a) $({}^-2) + 3 + [2 + 3] = 3 + 3 + [({}^-2) + 2]$
   (b) $4 + [3 + ({}^-5)] + ({}^-7) = 4 + 3 + [({}^-7) + ({}^-5)]$
   (c) $[1 + 2] + 3 + ({}^-4) = 2 + [(1 + 3) + ({}^-4)]$
   (d) $[15 + ({}^-16)] + ({}^-18) = 15 + [({}^-18) + ({}^-16)]$
6. Suppose you have eighty-nine dollars in the bank. If you write one check for ninety-nine dollars and then deposit sixty-nine dollars, how much money will you have in the bank?
7. For each of the following properties, state whether or not the property holds for subtraction. If a property does not hold, then disprove it by a counterexample. If a property is valid, illustrate it with an example.

(a) Associative
(b) Commutative
(c) Identity

Exercises 8 through 11 are to be solved without Definition 5-5.

*8. Add $10 + {}^-4$, showing and justifying each step.
*9. Show that ${}^-7 + {}^-9 = {}^-(7 + 9)$. Justify each step.
*10. Add ${}^-5 + 3$ and justify each step. (Recall that ${}^-(a + b) = {}^-a + {}^-b$.)
*11. Show that $12 + {}^-17$ is another name for ${}^-5$ if it is known that ${}^-(a + b) = {}^-a + {}^-b$. Justify each step.
*12. Using the number-pair definition of an integer, we define addition to be $(a - b) + (c - d) = [(a + c) - (b + d)]$.
(a) Prove that addition is commutative.
(b) Prove that addition is associative.
(c) Find an additive identity.
(d) Find an additive inverse for $a - b$.

4

Multiplication and Division of Integers

We wish to define multiplication for the set of integers such that the closure property for multiplication, the commutative, associative, and identity properties of multiplication, the multiplicative property of 0, and the distributive property for multiplication over addition, as previously discussed for the set of whole numbers, will be valid for the set of integers.

Example: How should one define ${}^-6 \cdot 4$ in order for these multiplicative properties to hold? Consider

$({}^-6 \cdot 4) + (6 \cdot 4) = ({}^-6 + 6)4$ The distributive property of multiplication over addition
$= 0 \cdot 4$ Additive inverse
$= 0$ Multiplicative property of 0

Thus, ${}^-6 \cdot 4$ is the additive inverse of $6 \cdot 4$. However, by definition, the unique additive inverse of $6 \cdot 4$ is ${}^-(6 \cdot 4)$. Therefore,

$$^-6 \cdot 4 = {}^-(6 \cdot 4) = {}^-24.$$

In a like manner, $4 \cdot {}^-6 = {}^-(4 \cdot 6)$, because

$$(4 \cdot {}^-6) + (4 \cdot 6) = 4({}^-6 + 6) = 4 \cdot 0 = 0. \text{ (Why?)}$$

Thus, $4 \cdot {}^-6$ is the additive inverse of $4 \cdot 6$, so $4 \cdot {}^-6 = {}^-(4 \cdot 6) = {}^-24$.

We have shown that $({}^-6)(4) = (4)({}^-6)$, which is consistent with the requirement that multiplication be commutative.

The fact that $4 \cdot {}^-6 = {}^-24$ is consistent with the interpretation that multiplication by a counting number is simply repeated addition.

$$4 \cdot {}^-6 = {}^-6 + {}^-6 + {}^-6 + {}^-6 = {}^-24.$$

That is, if the stock market drops six points a day for 4 days, it will have dropped 24 points, which is represented by $^-24$.

Using multiplication as repeated addition, we note that $4({}^-6) = {}^-24$ on the number line in Figure 5-7.

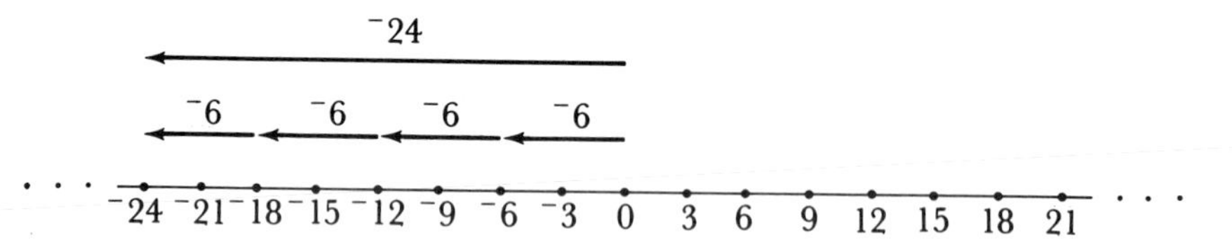

Figure 5-7

Now let us show, in general, that $({}^-a)(b) = {}^-(ab)$.

$({}^-a)(b) + ab = ({}^-a + a)b$	Distributive property of multiplication over addition
$= 0 \cdot b$	Additive inverse
$= 0$	Multiplicative property of zero

Thus $({}^-a)(b)$ is an additive inverse of ab. But the unique additive inverse of ab is $^-(ab)$. Thus, $({}^-a)(b) = {}^-(ab)$. Since multiplication is commutative, $({}^-a)(b) = (b)({}^-a) = {}^-(ab)$.

Example: How should we define $^-3 \cdot {}^-4$ if the properties of multiplication of whole numbers are to hold for integers? Consider

$(3 \cdot {}^{-}4) + ({}^{-}3 \cdot {}^{-}4) = (3 + {}^{-}3) \cdot {}^{-}4$	Distributive property of multiplication over addition
$= 0 \cdot {}^{-}4$	Additive inverse
$= 0$	Multiplicative property of zero
$3 \cdot {}^{-}4 = {}^{-}12$	Renaming $3 \cdot {}^{-}4$ from a previous example, i.e., ${}^{-}a \cdot b = {}^{-}ab$

Thus, ${}^{-}3 \cdot {}^{-}4$ is the additive inverse of ${}^{-}12$. But the unique additive inverse of ${}^{-}12$ is 12. Thus, ${}^{-}3 \cdot {}^{-}4 = 12$.

In general,

$(a)({}^{-}b) + ({}^{-}a)({}^{-}b) = (a + {}^{-}a)({}^{-}b)$	Distributive property of multiplication over addition
$= 0({}^{-}b)$	Additive inverse
$= 0$	Multiplicative property of zero

Thus, $({}^{-}a)({}^{-}b)$ is the additive inverse of $(a)({}^{-}b)$. Since $a({}^{-}b) = {}^{-}(ab)$ and since the unique additive inverse of ${}^{-}(ab)$ is ab, then $({}^{-}a)({}^{-}b) = ab$.

To satisfy the conditions imposed by Definition 5-4 and illustrated by the examples of this section, the multiplication of integers is defined as follows.

Definition 5-7: The product of two integers is defined by the following cases, where a and b are whole numbers.

(a) $a \cdot b = n(A \times B)$, where $a = n(A)$ and $b = n(B)$ for sets A and B.
(b) ${}^{-}a \cdot {}^{-}b = ab$.
(c) ${}^{-}a \cdot b = b \cdot {}^{-}a = {}^{-}(ab)$.

Examples:

$$3({}^{-}4) = {}^{-}12$$
$$5({}^{-}3) = {}^{-}15$$
$$({}^{-}7)(6) = {}^{-}42$$
$$({}^{-}4)({}^{-}3) = 12$$
$$({}^{-}6)({}^{-}17) = 102$$
$$\begin{aligned}({}^{-}2)({}^{-}5)({}^{-}3) &= ({}^{-}2)[({}^{-}5)({}^{-}3)] \\ &= ({}^{-}2)(15) \\ &= {}^{-}30\end{aligned}$$

Example: Associative property of multiplication.

$$^-7(4 \cdot {}^-2) = ({}^-7 \cdot 4) \cdot {}^-2$$

because

$$^-7(4 \cdot {}^-2) = {}^-7 \cdot {}^-8 = 56$$

and also

$$({}^-7 \cdot 4) \cdot {}^-2 = {}^-28 \cdot {}^-2 = 56.$$

Examples: Distributive property of multiplication over addition.

$$^-3(4 + {}^-2) = ({}^-3 \cdot 4) + ({}^-3 \cdot {}^-2)$$

because

$$^-3(4 + {}^-2) = {}^-3 \cdot 2 = {}^-6$$

and also

$$({}^-3 \cdot 4) + ({}^-3 \cdot {}^-2) = {}^-12 + 6 = {}^-6.$$

$$({}^-4 + 2) \cdot 5 = ({}^-4 \cdot 5) + (2 \cdot 5)$$

because

$$({}^-4 + 2) \cdot 5 = {}^-2 \cdot 5 = {}^-10$$

and also

$$({}^-4 \cdot 5) + (2 \cdot 5) = {}^-20 + 10 = {}^-10.$$

Example: Identify the properties used in showing that

$$({}^-4 + 5) \cdot {}^-7 = ({}^-7 \cdot 5) + ({}^-4 \cdot {}^-7).$$

$({}^-4 + 5) \cdot {}^-7 = ({}^-4 \cdot {}^-7) + (5 \cdot {}^-7)$	Distributive property of multiplication over addition
$= ({}^-4 \cdot {}^-7) + ({}^-7 \cdot 5)$	Commutative property of multiplication
$= ({}^-7 \cdot 5) + ({}^-4 \cdot {}^-7)$	Commutative property of addition

The discussion of division for whole numbers introduced in Chapter 3 defined $a \div b (b \neq 0)$ to be a whole number whose product with b

was equal to a. We will now define division between a pair of integers in the same way.

Definition 5-8: If x and y are integers with $y \neq 0$, then the division of x by y (denoted by $x \div y$) is equal to an integer z if and only if $x = y \cdot z$.

In order to discuss whether the answer is positive or negative, we refer to the definition of multiplication. Since $x \div y = z$ if and only if $x = y \cdot z$, then the sign of z will be fixed in order that the multiplication equals x. If $x = {}^-a$ and $y = {}^-b$, where a and b are natural numbers, then z will need to be a positive integer, by Definition 5-7, in order for the product with y to equal x (a negative integer). If $x = a$ and $y = {}^-b$, then z will have to be a negative integer. Also, if $x = {}^-a$ and $y = b$, z will have to be a negative integer. Of course, if x and y are both positive integers, then z will be a positive integer. We may summarize these results as follows: "The quotient, if it exists, of two positive or two negative integers is a positive integer; the quotient, if it exists, of a negative and a positive integer, in either order, is a negative integer."

$$
\begin{array}{lll}
{}^-12 \div 3 = {}^-4 & \text{because} & 3({}^-4) = {}^-12 \\
45 \div {}^-5 = {}^-9 & \text{because} & {}^-5({}^-9) = 45 \\
{}^-24 \div {}^-6 = 4 & \text{because} & {}^-6(4) = {}^-24 \\
{}^-16 \div 5 = \,? & \text{because} & 5(?) = {}^-16
\end{array}
$$

Obviously 5 does not divide ${}^-16$ because there is no integer that when multiplied by 5 gives a result of ${}^-16$. Thus, our extension to the integers gives us a set of numbers not closed under the operation of division. This continues to be a deficiency in our system of numbers, and the next extension of our number system will be made to alleviate this difficulty.

Once again, since 0 plays such a special role in division, we should issue a word of caution. *Division of an integer by* 0 *is not defined* in the system of integers. This is true because $z \div 0 (z \neq 0)$ cannot equal an integer y. If this were true, then $z = 0 \cdot y = 0$, which is a contradiction since $z \neq 0$. Expressions such as $6 \div 0$, $10 \div 0$, and $5 \div 0$ do not yield numbers.

Exercise Set 5-3

1. Write the following products and quotients as integers.
   (a) $5({}^-9)$ (b) ${}^-7({}^-4)$
   (c) $0 \div {}^-9$ (d) ${}^-24 \div {}^-3$

(e) ${}^{-}7(46)$ (f) ${}^{-}5(42)$
(g) ${}^{-}x({}^{-}y)$ (h) ${}^{-}h(g)$
(i) $(3 \cdot {}^{-}2) \div {}^{-}3$ (j) $8({}^{-}3) \div {}^{-}4$
(k) ${}^{-}9({}^{-}5 \cdot {}^{-}3)$ (l) ${}^{-}3({}^{-}2 \cdot 4)$

2. Perform the following computations.
(a) ${}^{-}3({}^{-}4 \cdot 6)$ (b) $({}^{-}2 \cdot {}^{-}5) \cdot 7$
(c) $({}^{-}6 \cdot {}^{-}4) \cdot (8 \cdot {}^{-}2)$ (d) $({}^{-}6 \cdot {}^{-}4) \cdot (0 \cdot 3)$
(e) $(23 - 5) \div {}^{-}6$ (f) $({}^{-}6 + {}^{-}18) \div {}^{-}4$
(g) $({}^{-}27 + {}^{-}5) \div {}^{-}4$ (h) ${}^{-}3(5 + {}^{-}1) \div 6$
(i) $({}^{-}4 + {}^{-}8) \cdot (5 + {}^{-}7)$ (j) $({}^{-}6 + {}^{-}1)({}^{-}3 + {}^{-}2)$
(k) $({}^{-}x \cdot {}^{-}y) \cdot {}^{-}z$ (l) $({}^{-}4 + 4)(3 + {}^{-}6)$

3. Compute each of the following in two different ways.
(a) ${}^{-}1(3 + {}^{-}5)$ (b) $0(3 + {}^{-}2)$
(c) $6(4 \cdot {}^{-}5)$ (d) ${}^{-}7(3 \cdot {}^{-}5)$
(e) $(3 + {}^{-}2)5$ (f) $(6 + {}^{-}1) \cdot {}^{-}7$

4. Working from left to right, identify the property or properties that make the following true.
(a) $(8 + 4) + {}^{-}2 = (8 + {}^{-}2) + 4$
(b) $(6 + {}^{-}6) + 0 = (0 + 6) + {}^{-}6$
(c) $({}^{-}4 + 6) + {}^{-}2 = ({}^{-}2 + 6) + {}^{-}4$
(d) $({}^{-}6 + {}^{-}4) + 2 = ({}^{-}6 + 2) + {}^{-}4$
(e) $(8 \cdot 4) \cdot {}^{-}2 = (8 \cdot {}^{-}2) \cdot 4$
(f) $({}^{-}4 \cdot 6) \cdot {}^{-}2 = {}^{-}2 \cdot (6 \cdot {}^{-}4)$
(g) ${}^{-}6(4 + {}^{-}3) = ({}^{-}6 \cdot 4) + ({}^{-}6 \cdot {}^{-}3)$
(h) $7({}^{-}2 + {}^{-}5) = (7 \cdot {}^{-}2) + (7 \cdot {}^{-}5)$
(i) $2({}^{-}6 \cdot {}^{-}4) = ({}^{-}6 \cdot 2) \cdot {}^{-}4$
(j) ${}^{-}2({}^{-}3 + {}^{-}7) = ({}^{-}7 \cdot {}^{-}2) + ({}^{-}2 \cdot {}^{-}3)$

5. Generalize the distributive law for integers and expand each of the following into the sum of three terms.
(a) ${}^{-}5(a + b + c)$ (b) $3(2x + y + z)$
(c) ${}^{-}y(3 + {}^{-}2x + 4z)$ (d) ${}^{-}x(2 + 3z + y)$

6. Label each of the following as either true or false.
(a) $(48 \div {}^{-}12) \div 2 = 48 \div ({}^{-}12 \div 2)$
(b) $18 \div (3 + 3) = (18 \div 3) + (18 \div 3)$
(c) $(12 + {}^{-}6) \div 3 = (12 \div 3) + ({}^{-}6 \div 3)$
(d) $3 + (9 \cdot 4) = 3 \cdot (9 + 4)$

7. Determine the numerical value of the following:
(a) $({}^{-}2)^2$ (b) $({}^{-}3)^2 \cdot ({}^{-}2)^3$ (c) $({}^{-}4)^3 \cdot ({}^{-}4)^2$
(d) $({}^{-}1)^n$ if n is an even counting number.
(e) $({}^{-}1)^n$ if n is an odd counting number.

8. (a) For what integral values of a and b does $a \div b = b \div a$?
(b) What is the value of ${}^{-}24 \div (4 \div {}^{-}2)$? $({}^{-}24 \div 4) \div {}^{-}2$?
(c) What can you conclude from (b) about the associativity of division for integers?

*9. Signs are placed at regular intervals along a mountain road to indicate changes in altitude. At Greenbriar Lodge, the sign indicates an elevation of 2,876 feet. Seventeen miles down the road another sign indicates that the elevation is 2,667 feet. What is the average elevation change per mile traveled from the lodge?

Work Exercises 10 through 14 without the use of Definition 5-7.

*10. Show that $4(^-5) = {}^-20$ and justify each step.
*11. Multiply $(^-6)(3)$ and justify each step.
*12. Show that $^-1(5) = {}^-5$.
*13. Show that $^-1(a) = {}^-a$.
*14. Show that $(^-3)(^-5) = 15$ and justify each step.

*15. For each integer below, give the additive inverse, additive identity, and multiplicative identity.
(a) $^-(6 \cdot {}^-3)$ (b) $^-(a + {}^-b)$ (c) $^-(^-a \cdot {}^-c)$
*16. Find the digits represented by △, □, Θ, ▭, and ◇ in the following multiplication and division:

(a)

```
        △  □  Θ
           □  □
   ─────────────
     ▭  Θ  □  □
  ▭  Θ  □  □
  ─────────────
  △  ◇  △  Θ  □
```

(b)

```
              Θ  Θ  Θ
      ┌────────────────
 Θ  Θ │ 4  3  9  5  Θ
        △  □  Θ
        ───────
           4  △  5
           △  □  Θ
           ───────
              △  □  Θ
              △  □  Θ
              ───────
```

*17. Prove that $x = 0$ or $y = 0$ if and only if $x \cdot y = 0$.
*18. Using the number-pair definition of an integer, we define multiplication to be $(a - b) \cdot (c - d) = [(ac + bd) - (ad + bc)]$.
(a) Prove that multiplication is commutative.
(b) Prove that multiplication is associative.
(c) Find a multiplicative identity.
*19. For a, b, and c integers, prove the following:
(a) $c(a - b) = ca - cb$
(b) $(a + b) - c = a + (b - c)$

5

Using Properties of Integers to Solve Equations and Inequalities

Now that we have a basic understanding of sets and have discussed the system of integers involving the operations of addition, subtraction, multiplication, and division, we are prepared to use these concepts in a study of some topics that are algebraic in nature. First of all, we extend the ideas of order, as discussed for whole numbers in Chapter 3, to the set of integers. In Section 2 of this chapter, a correspondence was set up between the integers and the points on a number line. You will notice in Figures 5-1 and 5-2 that the points representing the integers present themselves as *ordered* from left to right. Hence we may think of one number as being *less than* another number if its representation on the number line lies to the *left* of that number on the number line. For example, $^{-}2 < 3$, $^{-}4 < 2$, and $^{-}13 < {}^{-}4$. In like manner, $6 > 4$, $4 > {}^{-}1$, and $^{-}1 > {}^{-}5$. Thus, the integers are ordered as

$$\ldots {}^{-}5 < {}^{-}4 < {}^{-}3 < {}^{-}2 < {}^{-}1 < 0 < 1 < 2 < 3 < 4 < 5 \ldots.$$

We shall note later that this idea of order agrees with the formal definition of "less than" and with the previous notion of order for the whole numbers.

In the set of whole numbers, the relations $<$ and $>$ were defined in terms of the operation of addition of natural numbers. Recall that $a < b$ for whole numbers if and only if there exists a natural number c such that $a + c = b$. Since we have considered the system of integers as an extension of the system of whole numbers, the definition of $<$ for integers must hold for whole numbers considered as integers at all times that $<$ is true for whole numbers. The following definition satisfies this requirement.

Definition 5-9: If a and b are integers, then a is said to be less than b (written $a < b$) if and only if there exists a positive integer c such that $a + c = b$.

a is *greater than* b (written $a > b$) if and only if b is less than a or $b + c = a$ where c is a positive integer. Definition 5-9 is in agreement with the idea that a is less than b if the point labeled a is to the left of the point labeled b on a number line. If point a is to the left of point b, the addition of an arrow diagram representing a positive integer is required to get from a to b.

Examples:

$$6 < 8 \text{ because } 6 + 2 = 8 \text{ (2 is a positive integer)}$$
$$^-5 < {}^-2 \text{ because } {}^-5 + 3 = {}^-2 \text{ (3 is a positive integer)}$$
$$^-9 < {}^-5 \text{ because } {}^-9 + 4 = {}^-5 \text{ (4 is a positive integer)}$$

Sometimes we use the symbols $\leq$ and $\geq$. $\leq$ means less than or equal to, while $\geq$ means greater than or equal to.

Since ${}^-7 + 3 = {}^-2 + {}^-2$, verify by performing the operations that

$$({}^-7 + 3) + 5 = ({}^-2 + {}^-2) + 5$$

and that

$$({}^-7 + 3) + {}^-10 = ({}^-2 + {}^-2) + {}^-10.$$

Since

$${}^-7 + 3 = {}^-2 + {}^-2,$$

then, by Definition 5-9, ${}^-7 < {}^-2 + {}^-2$. By regrouping and using the associative and commutative properties, we find

$({}^-7 + 3) + 5 = ({}^-2 + {}^-2) + 5$ Verified in the preceding paragraph
${}^-7 + (3 + 5) = ({}^-2 + {}^-2) + 5$ Associative property of addition
${}^-7 + (5 + 3) = ({}^-2 + {}^-2) + 5$ Commutative property of addition
$({}^-7 + 5) + 3 = ({}^-2 + {}^-2) + 5$ Associative property of addition

Therefore,

${}^-7 + 5 < ({}^-2 + {}^-2) + 5$ Definition 5-9

This discussion leads us to the following generalizations where a, b, and c are integers.

Theorem 5-2: Let a, b, and c be integers.

(a) If $a = b$, then $a + c = b + c$.
(b) If $a < b$, then $a + c < b + c$.

Proof: We will prove (b) and leave the proof of (a) for an exercise. If $a < b$, then $a + d = b$ where d is a positive integer, by Definition 5-9.

$$\begin{array}{rl} (a+d)+c=b+c & \text{Property (a), Theorem 5-2} \\ c+(a+d)=b+c & \text{Commutative property of addition} \\ (c+a)+d=b+c & \text{Associative property of addition} \\ (a+c)+d=b+c & \text{Commutative property of addition} \\ a+c<b+c & \text{Definition 5-9} \end{array}$$

Beginning with the equality $^-7+3={}^-2+{}^-2$, verify the truth of the following by performing the operations.

$$(^-7+3)\cdot 6=(^-2+{}^-2)\cdot 6$$

and

$$(^-7+3)\cdot {}^-5=(^-2+{}^-2)\cdot {}^-5.$$

Likewise, starting with $^-7<{}^-2+{}^-2$ verify that

$$^-7\cdot 6<(^-2+{}^-2)\cdot 6$$

and

$$^-7\cdot 10<(^-2+{}^-2)\cdot 10.$$

However, note that

$$^-7\cdot {}^-3>(^-2+{}^-2)\cdot {}^-3$$

and

$$^-7\cdot {}^-10>(^-2+{}^-2)\cdot {}^-10.$$

We are now ready to make the following generalization:

Theorem 5-3: Let a, b, c, and d be integers.

(a) If $a=b$, then $ac=bc$; and if $a=b$ and $c=d$, then $ac=bd$.
(b) If $a<b$, then $ac<bc$ if $c>0$, and $ac>bc$ if $c<0$.

The proof of this theorem will be assigned as problems in Exercise Set 5-4.

Theorem 5-4: *Cancellation Properties.* Let a, b, and c be integers.

(a) If $a+c=b+c$, then $a=b$.
(b) If $ac=bc$, $c\neq 0$, then $a=b$.

Theorem 5-4(a) is easy to prove using Theorem 5-2(a).

$$\begin{aligned}(a+c)+{}^{-}c &= (b+c)+{}^{-}c\\ a+(c+{}^{-}c) &= b+(c+{}^{-}c)\\ a+0 &= b+0\\ a &= b\end{aligned}$$

To formulate methods for the solving of equations and inequalities, we introduce the following terminology:

Definition 5-10: A *variable* is a symbol that represents any member of the set of elements. This set of elements is called the domain of the variable.

Generally, the symbol selected to represent a variable is one of the letters of the alphabet, such as a, b, c, x, y, or z. Expressions such as $x+4=9$ and $x<5$ are called *open sentences*. They become statements when we substitute an integer in place of x. If we replace x by 2 in both expressions we get the statements

$$2+4=9 \quad \text{and} \quad 2<5.$$

The first statement is false and the second statement is true. $x+4=9$ is also called an *equation* and $x<5$ is called an *inequality*.

Open sentences in elementary arithmetic books make use of geometric figures to represent variables.

Example: $x+4=7$ and $8x=16$ are sometimes written as

$$\square+4=7 \quad \text{and} \quad 8\cdot\triangle=16.$$

The preceding examples of open sentences are examples of *simple* sentences. $2<x<7$ and $x\leq 3$ are called *compound* sentences. $2<x<7$ means x is less than 7 and x is greater than 2. The two simple sentences are combined by the connective *and.* $x\leq 3$ is a compound sentence consisting of $x<3$ or $x=3$, making use of the connective *or.*

Examples: $1<x-2<5$. $4x+7\geq 2$.

When a number is substituted for a variable in an equation or inequality, it is called the value of the variable. The set of all substitutions or replacements that make an open sentence true is called the *solution set* of the sentence.

Example: $x + 5 = 7$. Find the solution set if the domain of the variable is the set of integers. The solution set is $\{2\}$ and consists of one element. No other integer will satisfy the equation or make the sentence true.

Examples: Find the solution set for the following open sentences if the domain of the variable is $\{0, 1, 2, 3, 4, 5\}$.

(a) $x + 3$ is in the domain Answer: $\{0, 1, 2\}$
(b) $x + 3 = 5$ Answer: $\{2\}$
(c) $x < 4$ Answer: $\{0, 1, 2, 3\}$
(d) $x + x = 2x$ Answer: $\{0, 1, 2, 3, 4, 5\}$
(e) $x > 3$ and $x < 5$ Answer: $\{4\}$
(f) $x + 7$ is in the domain Answer: $\varnothing$

An equation or inequality in one variable is called an *identity* if its solution set includes all the elements of the domain. The following examples are identities where the domain is taken to be the set of integers.

(a) $(x - 2)(x + 2) = x^2 - 4$
(b) $x(x + 1) = x^2 + x$
(c) $(x + 2) + (x + 1) = 2x + 3$
(d) $x < x + 2$

A statement is called a *conditional equation* or *conditional inequality* if the solution set is a *nonempty proper* subset of the domain of the variable. For example, for the following conditional equalities and inequalities, where the domain of the variable is $\{8, 9, 10, 11, 12\}$, the solutions are proper subsets of the domain.

(a) $x - 5 = 7$ Solution set is $\{12\}$
(b) $x + 1 < 11$ Solution set is $\{8, 9\}$
(c) $x + 1 > 8$ and $x - 1 < 11$ Solution set is $\{8, 9, 10, 11\}$

In some cases, no values in the domain of the variable will satisfy an equation or inequality. Then we say that the solution set is the empty set or that it is impossible to solve the equation. For example,

(a) $x + 4 = x$ and
(b) $x + 2 < x$ are both impossible.

The theorems given in this chapter may be used for solving equations. To solve an equation, one may perform mathematical operations on each side of the equation until the variable is isolated on one side.

Example: Solve $x + 4 = 17$ when the domain is the set of integers.

$x + 4 = 17$	Given
$(x + 4) + {}^-4 = 17 + {}^-4$	Theorem 5-2
$(x + 4) + {}^-4 = 13$	Addition
$x + (4 + {}^-4) = 13$	Associative property of addition
$x + 0 = 13$	Additive inverses
$x = 13$	Additive identity

Thus, 13 is a solution, or root, of the equation. We say the solution set is $\{13\}$. A solution can be checked by substituting in the original equation: $(13) + 4 = 17$.

Example: Solve $x + {}^-3 = 5$ if the domain is the set of integers.

$x + {}^-3 = 5$	
$(x + {}^-3) + 3 = 5 + 3$	Why?
$(x + {}^-3) + 3 = 8$	Why?
$x + ({}^-3 + 3) = 8$	Why?
$x + 0 = 8$	Why?
$x = 8$	Why?

The solution set is $\{8\}$. As a check $(8) + {}^-3 = 5$.

The steps for solving inequalities are very similar to the steps for solving equalities. The properties of inequalities as discussed in this chapter will be used to solve inequalities.

Example: Solve $x + 3 < {}^-5$ on the integers.

$x + 3 < {}^-5$	
$(x + 3) + {}^-3 < {}^-5 + {}^-3$	Theorem 5-2
$(x + 3) + {}^-3 < {}^-8$	Addition
$x + (3 + {}^-3) < {}^-8$	Associative property of addition
$x + 0 < {}^-8$	Additive inverses
$x < {}^-8$	Additive identity

The solution set is $\{x \mid x \text{ is an integer less than } {}^-8\}$.

Examples: (a) $x + {}^{-}2 < 11$. Add 2 to both sides of this inequality to obtain

$$(x + {}^{-}2) + 2 < 11 + 2;$$

so $x < 13$. The solution set is $\{x \mid x \text{ is an integer less than } 13\}$.

(b) ${}^{-}x + {}^{-}3 < 1$. Add 3 to both sides of the inequality to obtain

$$({}^{-}x + {}^{-}3) + 3 < 1 + 3;$$

so ${}^{-}x < 4$. Then multiply both sides of the inequality by ${}^{-}1$ to obtain $x > {}^{-}4$. Why?

The solution set is $\{x \mid x \text{ is an integer greater than } {}^{-}4\}$ or $\{{}^{-}3, {}^{-}2, {}^{-}1, 0, 1, 2, 3, \ldots\}$.

Exercise Set 5-4

1. Write the following sets of numbers in increasing order of magnitude.
   (a) ${}^{-}7, 9, 0, {}^{-}4, 3$ (b) ${}^{-}4, {}^{-}9, 7, {}^{-}10, 1$
   (c) $0, {}^{-}2, {}^{-}100, {}^{-}5, 4$ (d) ${}^{-}8, 6, {}^{-}4, 9, {}^{-}10$
2. Show that the following are true by using Definition 5-9.
   (a) ${}^{-}5 < {}^{-}2$ (b) ${}^{-}7 < 0$
   (c) ${}^{-}10 < {}^{-}8$ (d) ${}^{-}9 < 4$
3. Find the solution set of the following if the domain is the set of whole numbers less than 10.
   (a) $t < 3$ (b) $t < 4$
   (c) x is divisible by 5 (d) $e + 4$ is not in the domain
   (e) $z + z = 2z$ (f) $y + 4 = 6$
   (g) $m < 4$ and $m > 4$ (h) $a + 5$ is in the domain and a is divisible by 2
   (i) $a < 7 - a$ (j) $k \cdot 1 = k$
   (k) $b + 0 = b$ (l) $x < 7$ or $x > 7$
   (m) $s + 0 = s$ and $s + 5 = 7$ (n) $p = 6$ or $p = 4$
   (o) $n + 8$ is not in the domain and $n = 5$ (p) $g < 2$, $g > 2$, or $g = 2$
4. Classify the following as identities, conditional equalities, or impossible. Explain your answers.
   (a) $x + 7 = 10$ (b) $3 + z = z + 3$
   (c) $x - 4 = 4 - x$ (d) $x = x + 1$
   (e) $x + x = 2x$ (f) $x + 3 = (x + 1) + 2$

(g) $x - 1 < x$ (h) $x + 3 = 3 - x$
(i) $5z - 3z = 2z$ (j) $6(h + 3) = 6h + 18$

5. Find the solution set for the following where the domain for the variable is the set of integers.
   (a) $y - 3 = 11$ (b) $y + 4 < {}^{-}6$
   (c) $z + {}^{-}5 \geq 7$ (d) $x + {}^{-}2 = 20$
   (e) $5x = 4x + 4$ (f) $2m + 3 = m + 7$
   (g) $2x + 5 \geq 7 + x$ (h) ${}^{-}2m + 1 > {}^{-}m + 2$
   (i) $2x + 3 = {}^{-}3 + x$ (j) $2x + 3 < x - 1$
   (k) $3x - 7 = 4x + {}^{-}2$ (l) $7y - 6 = 6y + {}^{-}2$
6. List the elements of the following sets.
   (a) $A = \{x | x$ is a whole number $< 6\}$
   (b) $B = \{x | x$ is an integer such that ${}^{-}6 \leq x < 1\}$
   (c) $C = \{x | x$ is a natural number such that $x > 10$ and $x \leq 15\}$
   (d) $D = \{x | x$ is a natural number less than or equal to $5\}$
   (e) $E = \{x | x$ is an integer $< 5\}$
7. State the reason or reasons for the listed steps in the proof that if $a < b$, then $ca < cb$ if $c > 0$. [Assume Theorem 5-3(a).]
   (a) $a + d = b$, where d is a positive integer
   (b) $c(a + d) = cb$
   (c) $ca + cd = cb$
   (d) cd is a positive integer
   (e) $ca < cb$
8. State the reason or reasons for each step in the proof that if $a < b$, then $ca > cb$ when $c < 0$. [Assume Theorem 5-3(a).]
   (a) $a + d = b$ where d is a positive integer
   (b) $c(a + d) = cb$
   (c) $ca + cd = cb$
   (d) $(ca + cd) + {}^{-}cd = cb + {}^{-}cd$
   (e) $ca + (cd + {}^{-}cd) = cb + {}^{-}cd$
   (f) $ca + 0 = cb + {}^{-}cd$
   (g) $ca = cb + {}^{-}cd$
   (h) ${}^{-}cd$ is a positive integer
   (i) $ca > cb$

*9. If $a = b$, prove $a + c = b + c$ where a, b, and c are integers.
*10. If $a = b$ and $c = d$, prove $a + c = b + d$ where a, b, c, and d are integers.
*11. If $a = b$, prove $ac = bc$ where a, b, and c are integers.
*12. If $a = b$ and $c = d$, prove $ac = bd$ where a, b, c, and d are integers.
*13. Prove that if $a < b$ and $b < c$, then $a < c$.
*14. Prove Theorem 5-4(b).

6

Summary

Our purpose in this chapter was to extend the system of whole numbers to gain new properties without losing any old ones. New mathematics is often created in just such a manner. It is obvious that the system of integers has an advantage over the system of whole numbers, for each integer has an inverse with respect to addition, thus making subtraction always possible. However, the operation of division is not closed in the system of integers; in fact, division is possible only in special cases. It is for this reason that Chapter 7 will be devoted to fractions and rational numbers.

Review Exercise Set 5-5

1. Perform the following calculations.
   (a) $36 - {}^-7 - 18$ (b) ${}^-12 - 15 + {}^-7 - {}^-26$
   (c) $50 - 64 - 13 + 10$ (d) $0 - (2 - {}^-3)$
   (e) ${}^-6 - ({}^-6)$ (f) ${}^-1 - {}^-3 - {}^-4 - 17$
   (g) ${}^-6(3 - {}^-5) - {}^-7$ (h) ${}^-x + (2x + {}^-3x)$
   (i) ${}^-x({}^-z + y)$ (j) ${}^-x(w + {}^-y) + xw$
   (k) $({}^-30 \div 6) \cdot {}^-8$ (l) ${}^-4 + ({}^-16 \div {}^-8)$
2. Each even integer is divisible by 2. The odd integers are those which are not even.
   (a) Write the sum of the first 5 positive odd integers.
   (b) Write the sum of the first 10 negative even integers.
   (c) Write in tabular form $\{x \mid {}^-8 < (x - 2) < {}^-4;\ x \text{ an odd integer}\}$.
   (d) What is the smallest nonnegative odd integer?
   (e) Write in tabular form $\{x \mid x \text{ is an odd integer and } {}^-x = x\}$.
   (f) Write in tabular form $\{x \mid x \text{ is an integer and } {}^-x = x\}$.
   (g) What odd integers divide ${}^-18$?
   (h) Is 0 even or odd?
   (i) Is the sum of two even integers even or odd? What about the sum of two odd integers?
3. Complete this crossword puzzle by placing digits in the blanks.

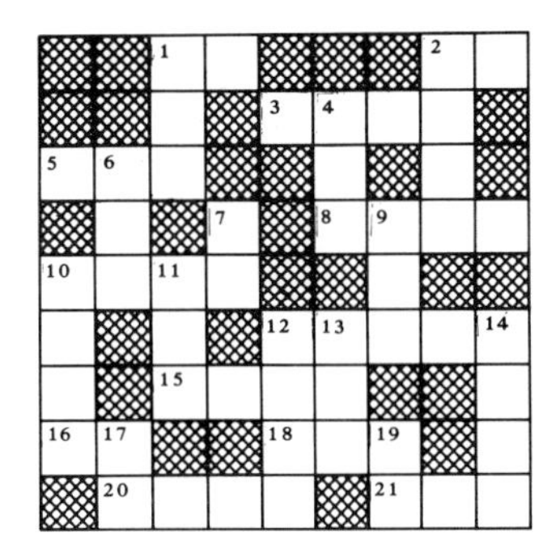

ACROSS:

1. Additive inverse of $^{-}11$
2. $^{-}6(^{-}30 \div 6) + [(2 + ^{-}6) + (8 \div 2)]$
3. Additive inverse of $51 \cdot ^{-}4 \cdot ^{-}5 \cdot ^{-}5$
5. $652 <$ ______ < 654
8. $(366 \cdot 6) - (5 - 2)$
10. If $35 \cdot ^{-}1665 = 7 \cdot ^{-}5 \cdot x$, then $x =$ ______.
12. Ordered solution set of $\{x \mid x$ is a natural number $< 6\}$
15. Ordered solution set of $\{x \mid x$ is an integer where $3 < x < 8\}$
16. $6(^{-}33 + 4) \div (3 \cdot ^{-}2)$
18. $13x - 366 + ^{-}4x + 101 = 5459$. $x =$ ______.
20. $^{-}(^{-}10{,}000 + 115) + (0 \div 16)$
21. 1st digit $= 3 \cdot 2 + 1$
    2nd digit = multiplicative identity of first digit
    3rd digit = first even integer

DOWN:

1. $^{-}(^{-}2829) \div 23$
2. $3(^{-}4 \cdot ^{-}73 + 721)$
4. $1{,}000{,}181 + ^{-}999{,}999$
6. The digits of # 5 across scrambled
7. $(^{-}3 \cdot 3) \cdot (25 \div ^{-}5)$
9. The difference between # 1 across and 114
10. Additive inverse of the number formed by the ordered solution set of $\{x \mid x$ is an integer and $^{-}1 \leq x < 3\}$
11. If $6x + ^{-}24 = 3780$, $x =$ ______.
12. The same as # 10 across
13. $0(^{-}6) + 273$
14. Reverse the digits of # 12 across and omit the new last digit.
17. Additive inverse of $(^{-}33 \cdot 3)$
19. $[2(52 + ^{-}11) \div 2] + (78 \div 3)$

4. Show that each of the following holds. Justify each step.
   (a) $4(^-3) + (^-4)(2) = 4(^-3 + ^-2)$
   (b) $^-[4 \cdot ^-3] = 3 \cdot 4$
   (c) $^-4(^-8 \cdot 2) = (8 \cdot ^-2) \cdot ^-4$
5. State the reason or reasons for the listed steps in the proof that $^-(x + y) = ^-x + ^-y$, where x and y are any integers.
   (a) $^-(x + y) + (x + y) = 0$
   (b) $[^-(x + y) + (x + y)] + ^-y = 0 + ^-y$
   (c) $^-(x + y) + [(x + y) + ^-y] = 0 + ^-y$
   (d) $^-(x + y) + [x + (y + ^-y)] = 0 + ^-y$
   (e) $^-(x + y) + (x + 0) = 0 + ^-y$
   (f) $^-(x + y) + x = ^-y$
   (g) $[^-(x + y) + x] + ^-x = ^-y + ^-x$
   (h) $^-(x + y) + (x + ^-x) = ^-y + ^-x$
   (i) $^-(x + y) + 0 = ^-y + ^-x$
   (j) $^-(x + y) = ^-y + ^-x$
   (k) $^-(x + y) = ^-x + ^-y$

*6. If symbols such as $5x$ mean 5 multiplied by x, prove $2x + 7x = 9x$, where x is any integer.

*7. Prove $(x + 3)(2x + 1) = 2x^2 + 7x + 3$.

*8. Explain why $x^2 - 6x + 13 = (x - 3)^2 + 4$.

*9. Prove (a) $^-(x - y) = y - x$.
   (b) $(x - y)(x + y) = x^2 - y^2$.

Suggested Reading

Addition: Allendoerfer, pp. 397–400. Armstrong, pp. 187–199. Bouwsma, Corle, Clemson, pp. 61–65, 67–68, 71–72. Campbell, pp. 88–92, 99–103. Garner, pp. 165–172. Graham, pp. 154–159. Hutton, pp. 191–196. Meserve, Sobel, pp. 149–151. Nichols, Swain, pp. 238–241. Ohmer, Aucoin, Cortez, pp. 154–159. Peterson, Hashisaki, pp. 130–135. Podraza, Blevins, Hanson, Prall, pp. 85–89. Scandura, pp. 284–289, 305–313. Smith, pp. 147–153, 164–165. Spector, pp. 254–257, 260. Willerding, pp. 230–233. Wren, pp. 81–89. Zwier, Nyhoff, pp. 116–121.

Multiplication: Allendoerfer, pp. 423–440. Armstrong, pp. 204–210. Bouwsma, Corle, Clemson, pp. 92–98, 102. Campbell, pp. 88–92, 99–103. Garner, pp. 181–182. Graham, pp. 162–165. Hutton, pp. 201–202. Nichols, Swain, pp. 246–249. Ohmer, Aucoin, Cortez, pp. 164–171. Peterson, Hashisaki, pp. 136–137. Podraza, Blevins, Hanson, Prall, pp. 94–98.

Scandura, pp. 294–299. Smith, pp. 157–161. Willerding, pp. 235–237. Zwier, Nyhoff, pp. 124–125.

Order Relations: Armstrong, pp. 186–187, 198–199. Campbell, pp. 104–107. Garner, p. 167. Graham, pp. 167–170. Hutton, pp. 191–192. Peterson, Hashisaki, pp. 152–155. Willerding, pp. 228–229. Wren, pp. 96–97.

Subtraction and Division: Armstrong, pp. 200–203, 211–213. Bouwsma, Corle, Clemson, pp. 66–68. Campbell, pp. 89–92, 99–103. Garner, pp. 172–180. Graham, pp. 160–161, 165–166. Hutton, pp. 197–200, 205–207. Nichols, Swain, pp. 242–245. Scandura, pp. 290–293, 300–304. Smith, pp. 153–156, 162–163. Willerding, pp. 233–235, 238–239. Wren, pp. 89–93. Zwier, Nyhoff, pp. 122–123.

Equations and Inequalities: Bouwsma, Corle, Clemson, pp. 41–43, 69–71. Nichols, Swain, pp. 279–282, 286–288. Podraza, Blevins, Hanson, Prall, pp. 102–107. Smith, pp. 18–19. Wren, p. 331.

6

Elementary Number Theory

1

Divisibility

In Chapter 5, we extended the system of whole numbers to the integers, and now we wish to consider some interesting subsets of the integers. Actually, some of the material we will be studying in this chapter was of much interest to Greek mathematicians more than twenty centuries ago. Within the last four hundred years, three men contributed significantly to the development of elementary number theory as we know it today. Pierre Fermat (1601–1665) and Leonhard Euler (1707–1783) developed important theories and relationships, several of which bear their names. The great mathematician Carl Friedrich Gauss (1777–1855) is reported to have said "Mathematics is the Queen of the sciences, and arithmetic is the Queen of mathematics." By *arithmetic*, Gauss was referring to the subject of this chapter—elementary number theory.

Definition 6-1: If a and b are any integers ($a \neq 0$), then a is said to divide b (denoted by $a \mid b$) if and only if there exists an integer c such that $b = ac$.

If a divides b, then a is said to be a *divisor* or *factor* of b, and b is said to be a *multiple* of a.

Example: $2 \mid 16$ because $2 \cdot 8 = 16$. Thus, 2 is a factor of 16, and 16 is a multiple of 2.

You should be very careful to note the distinction (made implicitly in Definition 6-1 and in the examples) between $a|b$ (a divides b) and the number a/b (a divided by b). $3|6$ is a true statement, and $3|7$ is false; these are statements and not numbers.

Example: $5|{}^{-}30$ because $5({}^{-}6) = {}^{-}30$. Again, 5 is a factor of ${}^{-}30$, and ${}^{-}30$ is a multiple of 5.

Example: $1|x$ because $1 \cdot x = x$, and $x|x$ because $x \cdot 1 = x$.

Example: Notice that $x|0$ for every nonzero integer x since $0 = x \cdot 0$.

We will use the notation $a \nmid b$ to indicate that a does not divide b; that is, $3 \nmid 8$ because there is no x such that $8 = 3x$, where x is an integer.

Example: Note that $4|20$, since $20 = 4 \cdot 5$. Furthermore $20|460$, since $460 = 20 \cdot 23$. Replacing 20 by $4 \cdot 5$ gives

$$460 = (4 \cdot 5) \cdot 23 = 4(5 \cdot 23).$$

Thus, $460 = 4 \cdot 115$. Hence, $4|460$. In general, if $x|y$ and $y|z$, then $x|z$. See Theorem 6-1.

Example: $4|24$ and $4|36$, since $24 = 4 \cdot 6$ and $36 = 4 \cdot 9$. Therefore,

$$24 + 36 = (4 \cdot 6) + (4 \cdot 9) = 4(6 + 9) = 4 \cdot 15.$$

So $4|(24 + 36)$. In general, if $x|y$ and $x|z$, then $x|(y + z)$. See Theorem 6-1.

This discussion should suggest some of the following properties of division.

Theorem 6-1: If x, y, and z are integers ($x \neq 0$ and $y \neq 0$), and

(a) if $x|y$ and $y|z$, then $x|z$.
(b) if $x|y$ and $x|z$, then $x|(y + z)$.
(c) if $x|y$ and $x|z$, then $x|(y - z)$.
(d) if $x|y$ and $x|(y + z)$ or $x|(y - z)$, then $x|z$.
(e) if $x|y$ or $x|z$, then $x|yz$.

Proof of Theorem 6-1(a): Let x, y, and z be any integers ($x \neq 0, y \neq 0$). Given that $x | y$ and $y | z$, $y = xk$ and $z = yl$, where k and l are integers, by Definition 6-1.

$yl = (xk)l$	Theorem 5-3
$z = (xk)l$	Transitive property of equality
$z = x(kl)$	Associative property of multiplication
kl is an integer	Closure property of multiplication of integers
$x \| z$	Definition 6-1

Theorem 6-1(b) will be proved in Exercise 8, (c) in Exercise 10, (d) in Exercise 9, and (e) in Exercise 15 of Exercise Set 6-1.

Example: $3 | 6$ and $6 | 24$; therefore, $3 | 24$. This illustrates the transitive property of the division relationship given in Theorem 6-1(a).

Example: $7 | 49$ and $7 | 84$; hence, $7 | 133$ because $133 = 49 + 84$. This illustrates the distributive property of the division relationship in Theorem 6-1(b).

Example: Note that although 7 does not divide 23 and 7 does not divide 26, $7 | (23 + 26)$; hence, a may divide $b + c$, although it does not divide either b or c.

Example: $6 | 30$ and $6 | 42$, so $6 | {}^{-}12$, since ${}^{-}12 = 30 - 42$ (Theorem 6-1(c)).

Example: Now, $7 | 21$ and $7 | 84$; thus, by Theorem 6-1(d), $7 | 63$, since $84 = 63 + 21$.

Example: Since $3 | 18$, then $3 | 18(20)$, although 3 does not divide 20 (Theorem 6-1(e)).

Theorem 6-1(b) may be generalized to include the sum of any finite number n of integers. For example, if

$$a | b_1, a | b_2, a | b_3, \ldots, a | b_n,$$

then

$$a \mid (b_1 + b_2 + b_3 + \cdots + b_n),$$

where the subscript n denotes that n integers are involved.

In a like manner, Theorem 6-1(d) may be generalized to include the sum of any finite number of integers. For example, if

$$a \mid b_1,\ a \mid b_2,\ \ldots,\ a \mid b_{n-1},$$

and if

$$a \mid (b_1 + b_2 + \cdots + b_{n-1} + b_n),$$

then $a \mid b_n$. This generalization states that if an integer divides the sum of n integers and also divides each one of $n - 1$ of these integers, then the integer divides the remaining integer.

The process of determining whether or not a number is divisible by another number often reduces to a matter of inspection. The following properties indicate some tests that may be applied to investigate divisibility by several small positive integers.

A number is divisible by 2 *if and only if the last digit of its base ten numeral is even (divisible by* 2). To understand divisibility by 2, we investigate whether or not $2 \mid 236$. Recall that 236 can be written as

$$236 = 2(10)^2 + 3(10) + 6.$$

Since $2 \mid 10^2$ and $2 \mid 10$, then by Theorem 6-1(b), $2 \mid 236$, since $2 \mid 6$. In general, any positive integer N can be written in the form

$$N = a_k\, 10^k + a_{k-1} 10^{k-1} + \cdots + a_2\, 10^2 + a_1 10 + a_0,$$

where each of the a's is a digit. Now $2 \mid 10$, $2 \mid 100$, $2 \mid 1000$, and, in general, $2 \mid 10^n$ for any natural number n; thus, by Theorem 6-1(b),

$$2 \mid (a_k\, 10^k + \cdots + a_2\, 10^2 + a_1 10 + a_0),$$

if $2 \mid a_0$.

One cannot tell whether or not a number is divisible by 3 by just looking at the last digit of its numeral. However, there is an interesting test for divisibility by 3.

A number is divisible by 3 *if and only if the sum of the digits of its base ten numeral is divisible by* 3. 756 is divisible by 3 because $3 \mid (7 + 5 + 6)$ or $3 \mid 18$.

In general, any positive integer N can be written in the form

$$N = a_k\, 10^k + \cdots + a_3\, 10^3 + a_2\, 10^2 + a_1 10 + a_0.$$

Rewrite as

$$N = a_k(10^k - 1 + 1) + \cdots + a_3(10^3 - 1 + 1) + a_2(10^2 - 1 + 1) + a_1(10 - 1 + 1) + a_0.$$

Rearrange as

$$a_k(10^k - 1) + \cdots + a_3(10^3 - 1) + a_2(10^2 - 1) + a_1(10 - 1) + (a_k + \cdots + a_3 + a_2 + a_1 + a_0).$$

Now $3 | (10 - 1), 3 | (10^2 - 1), 3 | (10^3 - 1)$, and, in general, $3 | (10^k - 1)$, where k is any natural number. Thus $3 | N$ if and only if

$$3 | (a_k + \cdots + a_3 + a_2 + a_1 + a_0).$$

You easily recognize this expression to be the sum of the digits of the number. Thus, a number is divisible by 3 if and only if the sum of the digits of its base ten numeral is divisible by 3.

A number is divisible by 4 *if and only if the last two digits of its base ten numeral represent a number divisible by* 4. One can verify that $4 | 10^n$ when $n \geq 2$. Thus, when writing a number in terms of powers of 10, 4 will divide all terms involving 10^k for $k \geq 2$. Therefore, if 4 divides the sum of the last two terms, $a_1(10) + a_0$, then 4 divides the number. For example, $4 | 536$ because $4 | 36$.

A number is divisible by 5 *if and only if the last digit of its base ten numeral is* 0 *or* 5. Since all integral powers of 10 are divisible by 5, all terms (except the units digit) of any number written in terms of the base ten will be divisible by 5. If the units digit is 0 or 5, then all terms are divisible by 5; and so the number is divisible by 5. Thus, the divisibility of a number by 5 depends only on whether or not the last digit of the numeral representing the number is either 0 or 5. For example, 678,324,570 is divisible by 5, as is 417,235.

A number is divisible by 6 *if and only if it is divisible both by* 2 *and by* 3. Since 6 is the product of 2 and 3, for a number to be divisible by 6, it must satisfy the requirements of divisibility both by 2 and by 3. For example, 6294 is divisible by 6, because 4 is divisible by 2 and $6 + 2 + 9 + 4 = 21$ is divisible by 3.

A number is divisible by 7 *if and only if it satisfies the following property. Double the right-hand digit; subtract the double of the right-hand digit from the number represented by the remaining digits. If the difference is divisible by* 7, *then the original number is divisible by* 7. The operation may be repeated until the difference is small enough so that divisibility by 7 is obvious. For example, $7 | 203$ because

$$20 - 2(3) = 14,$$

which is obviously divisible by 7. $7 \mid 6055$ because

$$605 - 2(5) = 595,$$

and because

$$59 - 2(5) = 49$$

and 7 obviously divides 49.

Write any number

$$N = a_k 10^k + \cdots + a_3 10^3 + a_2 10^2 + a_1 10 + a_0 .$$

Let

$$7 \mid (a_k 10^{k-1} + \cdots + a_3 10^2 + a_2 10 + a_1 - 2a_0).$$

Then

$$a_k 10^{k-1} + \cdots + a_3 10^2 + a_2 10 + a_1 - 2a_0 = 7t,$$

or

$$a_k 10^{k-1} + \cdots + a_3 10^2 + a_2 10 + a_1 = 2a_0 + 7t.$$

Using Theorem 5-3, multiply both sides by 10 to obtain

$$a_k 10^k + \cdots + a_3 10^3 + a_2 10^2 + a_1 10 = 20a_0 + 70t.$$

Add a_0 to both sides of the equation:

$$a_k 10^k + \cdots + a_3 10^3 + a_2 10^2 + a_1 10 + a_0 = 21a_0 + 70t = 7(3a_0 + 10t).$$

$$N = 7(3a_0 + 10t).$$

Since $3a_0 + 10t$ is an integer, $7 \mid N$.

A number is divisible by 8 *if and only if the last three digits of its base ten numeral represent a number divisible by* 8. The truth of this statement depends on the fact that $8 \mid 10^n$ if $n \geq 3$. 362,789,576 is divisible by 8 because $8 \mid 576$.

A number is divisible by 9 *if and only if the sum of the digits of the base ten numeral representing the number is divisible by* 9. For example, 216 is divisible by 9, because $2 + 1 + 6 = 9$ is divisible by 9.

A number is divisible by 10 *if and only if the last digit of its base ten numeral is* 0.

A number is divisible by 11 *if the sum of the odd-numbered digits of its base ten numeral (counting from. right to left) minus the sum of the even-numbered digits is divisible by* 11.

Is 91,839 divisible by 11? The answer is *yes*, because

$$11 \mid [(9+8+9)-(3+1)];$$

that is, $11 \mid 22$.

Exercise Set 6-1

1. List three distinct positive integers that are divisors of each of the following numbers.
   (a) 35 (b) 216 (c) 1001 (d) 299
2. Without actually dividing, test each of the following numbers for divisibility by 5, by 6, by 7, by 8, by 9, and by 11.
   (a) 6944 (b) 81,432 (c) 1,941,070
   (d) 50,177 (e) 1,254,866 (f) 162,122
   (g) 62,553 (h) 214,890 (i) 482,144
   (j) 150,024 (k) 22,814 (l) 63,372
3. (a) Is there any need to test for divisibility by 9 if a number is not divisible by 3?
   (b) Is there any need to test divisibility by 2 if a number is not divisible by 8?
4. Find the missing digits in order for divisibility to hold.
   (a) $9 \mid 24\_$ (b) $11 \mid 18\_2$
   (c) $36 \mid \_11\_$ (d) $24 \mid \_23\_$
5. Determine the truth values of each of the following statements.
   (a) $15 \mid 345$ (b) $45 \mid 3780$ (c) $16 \mid 1432$
   (d) $17 \mid 1581$ (e) $44 \mid 2772$ (f) $40 \mid 2640$
6. Discuss why each of the following is true or false. Illustrate each answer with an example for those that are true and produce a counterexample for those that are false.
   (a) If $4 \mid x$ and $3 \mid x$, then $12 \mid x$. (b) If $12 \nmid x$, then $3 \nmid x$.
   (c) If $6 \mid x$ and $2 \mid x$, then $12 \mid x$. (d) If $10 \mid x$, then $2 \mid x$.
   (e) If $10 \nmid x$, then $2 \nmid x$. (f) If $11 \nmid x$, then $121 \nmid x$.
7. (a) Write a five-digit number divisible by 9 and by 5.
   (b) Write a seven-digit number divisible by 11 and by 6.
   (c) Write a six-digit number divisible by 7 and by 8.
   (d) Write an eight-digit number divisible by 4 and by 3.
8. State the reason or reasons for each step in the proof that if $x \mid y$ and $x \mid z$, then $x \mid (y+z)$ (Theorem 6-1(b)).
   (a) $y = xk$ and $z = xl$, where k and l are integers.
   (b) $y + z = xk + z$

(c) $y + z = xk + xl$
(d) $y + z = x(k + l)$
(e) $k + l$ is an integer
(f) $x \mid (y + z)$

9. State the reason or reasons for each step in the proof that if $x \mid y$ and $x \mid (y + z)$, then $x \mid z$ (Theorem 6-1(d)).
   (a) Since $x \mid (y + z)$ and $x \mid y$, then $x \mid [(y + z) - y]$
   (b) $x \mid [(y + z) + {}^{-}y]$
   (c) $x \mid [(z + y) + {}^{-}y]$
   (d) $x \mid [z + (y + {}^{-}y)]$
   (e) $x \mid z$
10. Since, if $x \mid y$ and $x \mid z$, then $x \mid (y + z)$, write $y - z$ as $y + {}^{-}z$, and verify that $x \mid (y - z)$ (Theorem 6-1(c)).

*11. Formulate tests for divisibility by 15 and by 16. Test your conjectures with examples.

*12. (a) Make up an example to demonstrate that the test for divisibility by 3 in base eight is the same as the test for divisibility by 11 in base ten. Show this in general.
   (b) Show that the tests for divisibility by 2, 4, and 8 hold for base twelve numerals.
   (c) Does a test for divisibility by 2 in a base ten numeration system work in base four? In base three? Explain.
   (d) Make a conjecture for divisibility by 7 in base eight. Test your conjecture with an example.

*13. Verify the rule for divisibility by 9 for any positive integer.

*14. Verify the rule for divisibility by 11 for any positive integer.

*15. Prove Theorem 6-1(e).

2

Primes and Composite Numbers

About 200 B.C., a Greek mathematician named Eratosthenes devised a procedure for classifying numbers. This procedure today is called "the sieve of Eratosthenes." Let us apply "the sieve of Eratosthenes" to an array of the first one hundred positive integers arranged in columns of 10 each; see Table 6-1. Cross out the 1. Then cross out all the numerals representing numbers divisible by 2 (every other numeral). Do not cross out the 2. Next, cross out all numerals representing numbers divisible by 3, but do not cross out 3. There will be no need to cross out numerals that have already been marked. Notice that the numeral 4, representing the next consecutive number, has already been crossed out. Continue this process with 5 by crossing out all numerals representing numbers divisible by 5 except 5 itself. Continue this process for every positive integer up to 100 whose numeral has not been crossed out. After you have completed the task, your array will appear as shown in Table 6-1.

~~1~~	2	3	~~4~~	5	~~6~~	7	~~8~~	~~9~~	~~10~~
11	~~12~~	13	~~14~~	~~15~~	~~16~~	17	~~18~~	19	~~20~~
~~21~~	~~22~~	23	~~24~~	~~25~~	~~26~~	~~27~~	~~28~~	29	~~30~~
31	~~32~~	~~33~~	~~34~~	~~35~~	~~36~~	37	~~38~~	~~39~~	~~40~~
41	~~42~~	43	~~44~~	~~45~~	~~46~~	47	~~48~~	~~49~~	~~50~~
~~51~~	~~52~~	53	~~54~~	~~55~~	~~56~~	~~57~~	~~58~~	59	~~60~~
61	~~62~~	~~63~~	~~64~~	~~65~~	~~66~~	67	~~68~~	~~69~~	~~70~~
71	~~72~~	73	~~74~~	~~75~~	~~76~~	~~77~~	~~78~~	79	~~80~~
~~81~~	~~82~~	83	~~84~~	~~85~~	~~86~~	~~87~~	~~88~~	89	~~90~~
~~91~~	~~92~~	~~93~~	~~94~~	~~95~~	~~96~~	97	~~98~~	~~99~~	~~100~~

Table 6-1

The numerals not crossed out in this array will be the numerals that represent *prime numbers* less than or equal to 100.

Definition 6-2: A positive integer, p, greater than 1, is called a *prime* if and only if the only positive integral divisors of p are 1 and p.

Notice that the smallest prime is 2; also notice the fact that 2 is the only prime that is an even number.

The first ten primes in the order of increasing size are

$$2, 3, 5, 7, 11, 13, 17, 19, 23, 29, \ldots.$$

Of course, the dots at the end indicate that there are still more primes. How many more primes are there? Is there a largest prime? Can you find a prime greater than 100? Can you find a prime greater than 100,000?

Definition 6-3: A positive integer is said to be *composite* if it has a positive integral divisor other than itself and 1.

The first ten composite positive integers are

$$4, 6, 8, 9, 10, 12, 14, 15, 16, 18, \ldots.$$

Once again the dots indicate that there are many more composite numbers. Note that Definitions 6-2 and 6-3 exclude 1 as being either prime or composite. 1 is called a *unit*. Likewise, the factors of 0 form an infinite set, but 0 is neither prime nor composite.

To test whether a number N is prime, one may divide this number by all primes less than N. For example, to determine whether or not 91 is prime, one could test the divisibility by 2, 3, 5, 7, . . ., 89. However, it is not *necessary* to test all primes less than 91. It is sufficient to test all possible prime divisors less than a number whose square is greater than 91. If there were a divisor of 91 greater than 10 ($10^2 > 91$), then the quotient obtained by dividing 91 by this divisor would be less than 10. *In general, for any prime number p, each number not prime less than p^2 has a prime divisor less than p.*

Every positive integer greater than 1 may be classified as either a prime number or a composite number. This classification leads to the partition of the set of positive integers into three mutually exclusive and exhaustive subsets—the set {1}, the set of prime numbers, and the set of composite numbers. Thus, every positive integer is an element of one and only one of these subsets.

Another partition of the positive integers into subsets involves even and odd numbers. An *even* positive integer is one that is divisible by 2, and an *odd* positive integer is one that is not even. Thus, the set of positive integers can also be partitioned into two mutually exclusive, exhaustive subsets, the even positive integers and the odd positive integers.

Composite numbers may be written as products of prime factors. When a composite number is written as a product of all its prime divisors, we say we have a complete *factorization.*

Two methods are commonly employed to find all the prime factors of a composite number. The first method consists of repeated division starting with the smallest prime, 2, and continuing until all prime factors have been obtained. For example, what are the prime factors of 72?

$$\begin{aligned} 72 &= 2 \cdot 36 \\ 36 &= 2 \cdot 18 \\ 18 &= 2 \cdot 9 \\ 9 &= 3 \cdot 3 \\ 72 &= 2 \cdot 2 \cdot 2 \cdot 3 \cdot 3 \end{aligned}$$

The second method involves factoring the number into any two easily recognized factors, then factoring the factors.

$$72 = (12)(6) = (4 \cdot 3)(3 \cdot 2) = (2 \cdot 2 \cdot 3) \cdot (3 \cdot 2) = 2 \cdot 2 \cdot 2 \cdot 3 \cdot 3$$

Some elementary textbooks employ *factor trees* to illustrate factoring into prime factors.

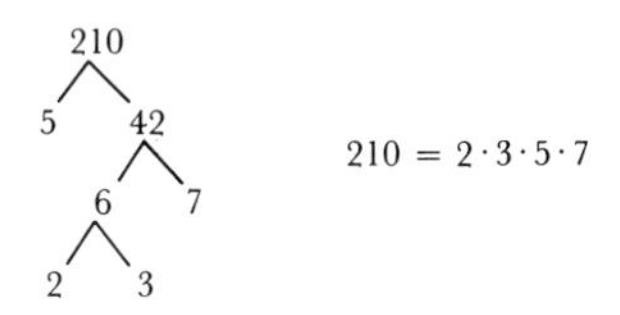

Figure 6-1

It is of interest to note that the factorization of a composite number into prime factors gives a unique answer that can be expressed with the factors arranged in different orders.

Theorem 6-2: *The Fundamental Theorem of Arithmetic.* Every composite positive integer can be factored uniquely into a product of primes.

The proof of this theorem will be given in Section 6 of this chapter.

Examples:

(a) $30 = 2 \cdot 3 \cdot 5$
(b) $72 = 2^3 \cdot 3^2$
(c) $900 = 2^2 \cdot 3^2 \cdot 5^2$
(d) $18{,}900 = 2^2 \cdot 3^3 \cdot 5^2 \cdot 7$

To illustrate the uniqueness of the factors in the prime factorization of a composite number, we will factor 360 in two different ways.

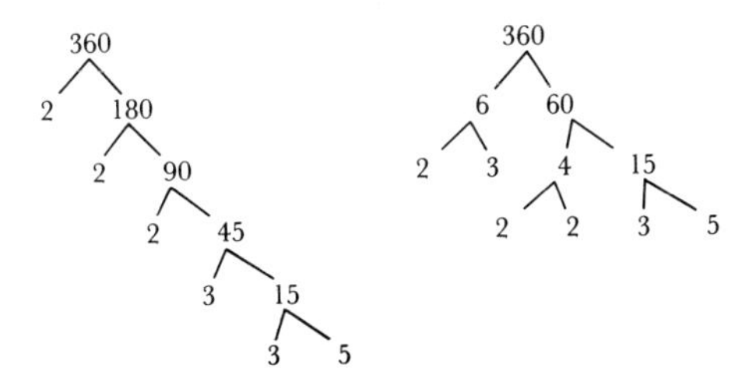

Figure 6-2

In both cases, $360 = 2 \cdot 2 \cdot 2 \cdot 3 \cdot 3 \cdot 5$.

About 300 B.C., Euclid proved that there are infinitely many primes. His theorem could have been stated in the following manner: "Given any list of prime numbers, it is always possible to find a new prime number not on the list; hence, the number of primes is infinite." His proof, by contradiction, is still a model of beauty and simplicity. The proof that we will give for this theorem is essentially the same as that given by Euclid.

Theorem 6-3: There are infinitely many primes.

Proof: Assume that there are but n primes, where n is a natural number. That is, assume there are only a finite number of primes. We may list all these primes as $2, 3, 5, 7, \ldots, p$. Now form a number R by multiplying together all these n primes; then add 1 to the result. $R = (2 \cdot 3 \cdot 5 \cdot 7 \cdots p) + 1$. R is either a prime number or a composite number, for it is clearly greater than 1. If R is a prime number, we will show that it is not one of the primes $2, 3, 5, 7, \ldots, p$. If R were equal to some p_i, then $p_i = (2 \cdot 3 \cdot 5 \cdots p_i \cdots p) + 1$. Since $p_i | p_i$ and since $p_i | (2 \cdot 3 \cdot 5 \cdots p_i \cdots p)$, then, by Theorem 6-1(d), $p_i | 1$. But then p_i would not be a prime. Thus R cannot be one of the primes listed. But if R is not one of the primes listed, we have a contradiction, for we assumed that we had listed all the primes.

The other possibility is that R is a composite number. If R is a composite number, some prime p_i will divide R. Now

$$R = (2 \cdot 3 \cdot 5 \cdots p_i \cdots p) + 1.$$

Since $p_i | R$ and since $p_i | (2 \cdot 3 \cdot 5 \cdots p_i \cdots p)$, then, by Theorem 6-1(d), $p_i | 1$. If this were true, p_i would again not be a prime, which, of course, means that the prime p_i that divides R is not one of the primes listed; and this contradicts our assumption that we have listed all the primes.

We assumed that $2, 3, 5, \ldots, p$ were all the primes. We then formed a new number R and considered two cases. If R were prime, we verified that it was different from $2, 3, 5, \ldots, p$. If R were composite, we verified that it was divisible by a prime different from $2, 3, 5, \ldots, p$. Consequently, no matter whether R is prime or composite (and these are the only possibilities), we obtain a contradiction to our assumption that there are only a finite number of primes. Thus, there are infinitely many primes. The digital computer has enabled mathematicians to discover large primes. For example, in 1963, $2^{11213} - 1$ was discovered to be a prime. In 1971 Dr. Bryant Tuckerman of I.B.M. discovered a much larger prime.

Other interesting and fascinating problems having to do with primes will be considered in other parts of this chapter.

Exercise Set 6-2

1. Test whether or not the following numbers are primes.
   (a) 149 (b) 89 (c) 87 (d) 43
   (e) 737 (f) 411 (g) 501 (h) 1003
2. List all prime divisors of the following.
   (a) 54 (b) 36 (c) 120
   (d) 51 (e) 141 (f) 76
3. Write the following numbers as products of prime factors.
   (a) 144 (b) 178 (c) 256
   (d) 2000 (e) 108 (f) 299
4. Apply the "sieve of Eratosthenes" to get all primes less than 150.
5. List the first four prime numbers in the sets determined by each of the following.
   (a) $\{2x+1 | x=0, 1, 2, \ldots\}$
   (b) $\{3x+2 | x=0, 1, 2, \ldots\}$
   (c) $\{4x+3 | x=0, 1, 2, \ldots\}$
6. Classify as odd or even.
   (a) $4 \cdot 5$ (b) $128-35$ (c) $3(4+6)$ (d) $7+9$
   (e) $3 \cdot 5 \cdot 9$ (f) 11_{three} (g) 13_{five} (h) $TE4_{\text{twelve}}$
7. What is the intersection of the set of prime numbers and the set of odd positive integers less than 25? What is the union?
8. There are how many prime numbers less than 50? Less than 100? There are how many composite numbers less than 50? Less than 100?
9. What is the smallest positive integer greater than zero that is divisible by every number from 1 to 10 inclusive?
10. Argue that $(7 \cdot 6 \cdot 5 \cdot 4 \cdot 3 \cdot 2)+1$ has a prime factor larger than 7.

*11. In general, argue that $n!+1$ has a prime factor larger than n.

*12. Pairs of primes that differ by 2 are called *twin primes*. List all the twin primes less than 50.

*13. One mathematical conjecture states that numbers that are the squares of integers may be written as the sum of two primes. For example, $4=2+2$ and $9=7+2$. Find two primes whose sum is the square of the following numbers.
   (a) 8 (b) 5 (c) 10 (d) 9

*14. An unproven conjecture, Goldbach's Conjecture, says that every even number greater than 4 is the sum of two odd primes. Show each of the following as the sum of two odd primes.
(a) 104 (b) 66 (c) 144 (d) 90

3

Greatest Common Divisor

If a positive integer d is a divisor of two integers b and c, then d is called a *common divisor* of b and c. Consider the set A made up of the positive integral divisors of 18: $A = \{1, 2, 3, 6, 9, 18\}$. The set of positive integral divisors of 42 may be expressed as $B = \{1, 2, 3, 6, 7, 14, 21, 42\}$. Then $A \cap B = \{1, 2, 3, 6\}$ is the set of common divisors of 18 and 42. Among the common divisors of any two numbers is a largest number, which we call the *greatest common divisor* (or *highest common factor*). The greatest common divisor of 18 and 42 is 6.

Definition 6-4: The *greatest common divisor* (abbreviated g.c.d.) of two nonzero integers, a and b, is the largest positive integer d such that $d | a$ and $d | b$.

Thus, the greatest common divisor of two integers is the largest positive integer that is a divisor of both numbers. It is usually denoted by

$$d = \text{g.c.d.}\ (a, b).$$

Since every negative integer such as ^-a can be written as $^-1 \cdot a$, by Theorem 6-1(e), every positive integral divisor of a is a divisor of ^-a. Let d be the largest positive integral divisor of ^-a. Then $^-a = d \cdot k$, where k is an integer.

$$(^-1)(^-a) = {}^-1(dk) \quad \text{and} \quad a = {}^-1(dk) = (dk)(^-1) = d(k \cdot {}^-1) = d(^-k).$$

Hence, $d | a$. Therefore, since the largest positive integral divisor of ^-a divides a, all of the examples we consider will involve positive integers. Actually the largest positive divisors of ^-a and a are the same. Can you prove this fact?

The problem of finding the greatest common divisor of two numbers is quite simple when the numbers are small. An easy way to find the g.c.d. of small numbers is to write the numbers as products of primes. We then note the primes that are common to both numbers. These are the primes that are factors of the g.c.d. We illustrate this method by the following example.

Example: Find the g.c.d. of 24 and 36.

$$24 = 2^3 \cdot 3 \quad \text{and} \quad 36 = 2^2 \cdot 3^2.$$

Since the g.c.d. divides both 24 and 36, the g.c.d. cannot contain the factor 2 more than twice or the factor 3 more than once. Hence the g.c.d. of 24 and 36 is $2^2 \cdot 3 = 12$.

The results of our reasoning in this example may be summarized by the following statements:

The g.c.d. of two positive integers is the product of any common factors that occur in the factoring of the numbers into prime factors.

The highest power of each factor in the g.c.d. will be the smallest power of the factor that occurs in any one of the factorizations of the numbers into prime factors.

Examples: (a) Find the g.c.d. of 70 and 90.

$$70 = 2 \cdot 5 \cdot 7 \quad \text{and} \quad 90 = 2 \cdot 3 \cdot 3 \cdot 5.$$

We note that 2 and 5 are divisors of both 70 and 90. Thus, the g.c.d. of 70 and 90 is $2 \cdot 5 = 10$.

(b) Find the g.c.d. of 144 and 180.

$$144 = 2 \cdot 2 \cdot 2 \cdot 2 \cdot 3 \cdot 3 = 2^4 \cdot 3^2,$$

and

$$180 = 2 \cdot 2 \cdot 3 \cdot 3 \cdot 5 = 2^2 \cdot 3^2 \cdot 5.$$

We note that two 2's and two 3's are divisors of both 144 and 180. Thus, g.c.d. $(144, 180) = 2^2 \cdot 3^2 = 36$.

The g.c.d. of three numbers a, b, and c can be found by pairing. First, find the g.c.d. $(a, b) = e$ and g.c.d. $(b, c) = f$. Then g.c.d. (a, b, c) $=$ g.c.d. (e, f).

Example: g.c.d. $(72, 84) = 12$; g.c.d. $(54, 72) = 18$; g.c.d. $(12, 18)$ $= 6$. Therefore, the g.c.d. $(54, 72, 84) = 6$.

The process of writing the numbers as products of primes and then using the highest powers of the primes common to the numbers applies for three or more numbers as well as for two numbers.

Example: Find the greatest common divisor of 84, 72, and 108.

$$84 = 2 \cdot 2 \cdot 3 \cdot 7, \quad 72 = 2 \cdot 2 \cdot 2 \cdot 3 \cdot 3, \quad \text{and} \quad 108 = 2 \cdot 2 \cdot 3 \cdot 3 \cdot 3.$$

Since both 2^2 and 3 are factors of all three numbers,

$$\text{g.c.d. } (84, 72, 108) = 2^2 \cdot 3 = 12.$$

Definition 6-5: If the greatest common divisor of two positive integers a and b is 1, we say that a and b are *relatively prime.*

Examples: (a) 5 and 7 are relatively prime, because g.c.d. $(5, 7) = 1$.
(b) 23 and 124 are relatively prime, because g.c.d. $(23, 124) = 1$.
(c) 111 and 6 are not relatively prime, because g.c.d. $(111, 6) = 3$.

Note that all the numbers less than a prime number are relatively prime to that number. For example, each of the numbers, 2, 3, 4, 5, and 6 are relatively prime to the prime number 7.

The writing of a positive integer as a product of prime factors is quite cumbersome when the numbers are large. Hence, we need a more practical method to use as we attempt to find a g.c.d. for large numbers. The method we consider at this time is based on repeated use of the Division Algorithm introduced in Chapter 4.

Suppose we wish to find the g.c.d. of 28 and 16. The Division Algorithm allows us to express

$$28 = 1(16) + 12, \quad \text{where } 0 \leq 12 < 16.$$

Note that by Theorem 6-1(d) any number that divides 28 and 16 must also divide 12. Thus, g.c.d. $(28, 16) = d$ must divide 12. This implies that d is a common divisor of 16 and 12. In addition, d is the g.c.d. of 16 and 12, because any larger divisor of 16 and 12 by Theorem 6-1(b) would divide 28, and then d would not be the g.c.d. of 28 and 16. We have shown that

$$\text{g.c.d. } (28, 16) = \text{g.c.d. } (16, 12).$$

The problem can be further reduced by applying the Division Algorithm again to obtain

$$16 = 1(12) + 4, \quad \text{where} \quad 0 \leq 4 < 12.$$

Using the same argument,

$$\text{g.c.d. } (16, 12) = \text{g.c.d. } (12, 4).$$

Thus,

$$\text{g.c.d. } (28, 16) = \text{g.c.d. } (12, 4).$$

Finally, $12 = 3 \cdot 4$. This last statement identifies g.c.d. (12, 4) to be 4. Thus g.c.d. (28, 16) = 4.

Example: Find the g.c.d. of 144 and 104.

$$\begin{aligned}
144 &= 1(104) + 40 & \text{g.c.d. } (144, 104) &= \text{g.c.d. } (104, 40) \\
104 &= 2(40) + 24 & \text{g.c.d. } (104, 40) &= \text{g.c.d. } (40, 24) \\
40 &= 1(24) + 16 & \text{g.c.d. } (40, 24) &= \text{g.c.d. } (24, 16) \\
24 &= 1(16) + 8 & \text{g.c.d. } (24, 16) &= \text{g.c.d. } (16, 8) \\
16 &= 2(8) & \text{g.c.d. } (16, 8) &= 8
\end{aligned}$$

So we see that 8 is the g.c.d. of 104 and 144.

This set of equations, obtained by successive applications of the Division Algorithm, is known as the *Euclidean Algorithm.*

Exercise Set 6-3

1. Find the greatest common divisor for each pair of numbers.
   (a) 105 and 30 (b) 72 and 54
   (c) 66 and 90 (d) 84 and 92
   (e) 46 and 38 (f) 57 and 90
   (g) 12 and 18 (h) 10 and 9
   (i) 188 and 72 (j) 84 and 288
2. Three numbers, a, b, and c, may be factored into primes in the manner below. Find the g.c.d. of the three numbers expressed as a product of primes.

$$\begin{aligned}
a &= 37^2 \cdot 7^3 \cdot 3 \\
b &= 31 \cdot 3^7 \cdot 7^2 \\
c &= 7^6 \cdot 31^3 \cdot 37^4 \cdot 3^4
\end{aligned}$$

3. (a) What is the g.c.d. of a and b if a and b are distinct primes?
   (b) If a is a prime number and b is a natural number such that $a|b$, what is g.c.d. (a, b)?

4. Find the greatest common divisor of the following.
   (a) 24, 30, and 42 (b) 144, 72, and 36
   (c) 36, 45, and 24 (d) 28, 63, and 42
   (e) 65, 42, and 66 (f) 42, 96, 104, and 18
5. Using the Euclidean Algorithm, find the greatest common divisor.
   (a) 1122 and 105 (b) 4652 and 232
   (c) 2244 and 418 (d) 735 and 850
   (e) 220 and 315 (f) 486 and 522
   (g) 912 and 19,656 (h) 7286 and 1684
6. Answer yes or no, and indicate why.
   (a) Is $2^3 \cdot 3^2 \cdot 5^7$ a factor of $2^4 \cdot 3^9 \cdot 5^7$?
   (b) Is $3^7 \cdot 5 \cdot 17^4$ a factor of $3^7 \cdot 5^2 \cdot 17^3$?
   (c) Does $7^2 \cdot 3^5$ divide $7^5 \cdot 3^6$?
   (d) Is $7^4 \cdot 11^5 \cdot 5^4$ a multiple of $7^5 \cdot 11^3 \cdot 5$?
7. List the natural numbers that are less than the x given below and that are relatively prime to that x.
   (a) $x = 7$ (b) $x = 15$ (c) $x = 23$ (d) $x = 20$

*8. (a) Must two relatively prime numbers be distinct prime numbers? Explain.
   (b) Can two composite numbers be relatively prime? Explain.

*9. If D_a and D_b represent the set of all divisors of a and b, respectively, prove the following where $a \neq b$.
   (a) $D_5 \subset D_{20}$
   (b) $D_3 \subset D_{15}$
   (c) $D_a \subset D_b$ if $a \mid b$ and $a \neq b$
   (d) $D_a \cap D_b = D_d$, where $d =$ g.c.d. (a, b)

*10. If d is the g.c.d. (a, b), prove that $a \div d$ and $b \div d$ are relatively prime.

4

Least Common Multiple

In the last section, we studied the greatest common divisor of two or more numbers. In this section, we are interested in multiples of numbers. Remember that if $b \mid c$, then c is said to be a *multiple* of b. The set of positive integers that are multiples of 5 may be written as $\{5, 10, 15, 20, 25, \ldots\}$. Observe that each element in this set is of the form $5k$, where k is a positive integer. In general, we consider the *multiples of a positive integer c to be kc, where* $k = 1, 2, 3, \ldots$. For example, the positive multiples of 5 may be written as

$$\{5, 10, 15, 20, \ldots, 5k, \ldots\}.$$

Similarly, the set of positive multiples of 2 is

$$\{2, 4, 6, 8, 10, \ldots, 2k, \ldots\}.$$

What is the set of numbers common to these two sets? From our knowledge of set theory, we know the answer to be the intersection of the two sets, namely, {10, 20, 30, ...}. Thus, a number m is a *common multiple* of two positive integers b and c if it is a multiple of b and a multiple of c. How many common positive multiples can two positive integers have? Do you agree that the answer is infinitely many? Now suppose we examine the set of common positive multiples of 6 and 9. The set of positive multiples of 6 is written as {6, 12, 18, 24, 30, ...}, and the set of positive multiples of 9 is {9, 18, 27, 36, ...}. The intersection of these two sets, or the set of common multiples, is {18, 36, 54, ...}. We wish to select now the smallest of the common positive multiples of 6 and 9. Obviously, 18 is the smallest common multiple, or the *least common multiple.*

Definition 6-6: A positive integer m is the *least common multiple* (abbreviated by l.c.m.) of two nonzero integers b and c if both b and c divide m and if m is the least positive integer divisible by both b and c.

Thus, the least common multiple of two integers is the smallest positive integer that is divisible by both of the integers. The l.c.m. of 3 and 4 is 12; the l.c.m. of 4 and 5 is 20; the l.c.m. of 6 and 15 is 30.

Since the positive integers that are multiples of $^{-}5$ are exactly the same as the multiples of 5, the examples will involve positive integers.

The procedure for finding the least common multiple by finding the set of common multiples and then selecting the smallest number is rather cumbersome, and additional methods for finding the l.c.m. must be considered. The first procedure considered will be a method involving prime factorization.

Examples: (a) Find l.c.m. (8, 12). Now, $8 = 2^3$ and $12 = 2^2 \cdot 3$. If each of these prime factors is to divide the l.c.m., then the l.c.m. must contain 2 as a factor three times and the factor 3 once. Thus, the l.c.m. of 8 and 12 is $2^3 \cdot 3 = 24$.

The results of our reasoning in this example may be summarized in the following statement:

The l.c.m. of two positive integers is the product of the highest powers of all the different prime factors that occur in factoring the numbers into prime factors.

(b) Find the l.c.m. of 120 and 180. Now, $120 = 2^3 \cdot 3 \cdot 5$, and $180 = 2^2 \cdot 3^2 \cdot 5$. Thus, l.c.m. $(120, 180) = 2^3 \cdot 3^2 \cdot 5 = 360$.

Another method for finding the l.c.m. of a pair of positive integers is to first find the g.c.d. of the two numbers and then divide the product of the two numbers by the g.c.d.

Theorem 6-4: If d and m are g.c.d. and l.c.m., respectively, of a and b, then $dm = ab$.

Example: The greatest common divisor d of 120 and 180 is 60. The least common multiple m is 360.

$$dm = ab, \quad \text{since } 60 \cdot 360 = 120 \cdot 180 = 21{,}600.$$

The proof of Theorem 6-4 will be delayed until the next section since Theorem 6-7 is used in the proof.

Theorem 6-4 provides a second method for finding the l.c.m. of a pair of numbers; that is,

$$\text{l.c.m. } (a, b) = \frac{ab}{\text{g.c.d. } (a, b)}.$$

In many situations, this formula provides the easiest method for finding the l.c.m., especially when the numbers are large.

Example: Find the l.c.m. of 144 and 180.

$$\text{l.c.m. } (144, 180) = \frac{144 \cdot 180}{\text{g.c.d. } (144, 180)} = \frac{144 \cdot 180}{36} = 720$$

The l.c.m. of three or more positive integers can be found by grouping the numbers by pairs. For example, to find the l.c.m. of a, b, c, and d, find the least common multiple, m_1, of a and b and the least common multiple, m_2, of c and d. Then the l.c.m. of a, b, c, and d is the l.c.m. of m_1 and m_2. This idea is illustrated by the following example.

Example: Find the l.c.m. of 8, 12, 18, and 36. The l.c.m. of 8 and 12 is 24. The l.c.m. of 18 and 36 is 36. Therefore, the l.c.m. of 8, 12, 18, and 36 is the l.c.m. of 24 and 36, which is 72.

What would happen if we grouped by pairs in another way? The l.c.m. of 12 and 18 is 36. The l.c.m. of 8 and 36 is 72. Then the l.c.m. of 8, 12, 18, and 36 is the l.c.m. of 72 and 36, which is 72. This is the same answer that was obtained by the first grouping.

If the l.c.m. of several positive integers is to be found, a procedure involving division by primes sometimes provides a quicker process of computation. Let us find the l.c.m. of 12 and 30 by this process. Divide both 12 and 30 by 2, getting 6 and 15; divide 6 and 15 by 3, getting 2 and 5. Since 2 and 5 are primes, the process is complete. Note that in order to obtain the l.c.m. by this procedure, the division process must be continued until the row of answers (after division) consists of ones and relatively prime numbers.

$$\begin{array}{r|rr} 2 & 12 & 30 \\ \hline 3 & 6 & 15 \\ \hline & 2 & 5 \end{array} \qquad \text{l.c.m. } (12, 30) = 2 \cdot 3 \cdot 2 \cdot 5 = 60$$

Now suppose we change the order of the division and divide by 2, 2, and then 3.

$$\begin{array}{r|rr} 2 & 12 & 30 \\ \hline 2 & 6 & 15 \\ \hline 3 & 3 & 15 \\ \hline & 1 & 5 \end{array} \qquad \text{l.c.m. } (12, 30) = 2 \cdot 2 \cdot 3 \cdot 1 \cdot 5 = 60$$

In the second line of this procedure, notice that the 2 does not divide 15; thus, we simply bring down the 15.

Example: Find l.c.m. (24, 15, 20, 6).

$$\begin{array}{r|rrrr} 2 & 24 & 15 & 20 & 6 \\ \hline 3 & 12 & 15 & 10 & 3 \\ \hline 5 & 4 & 5 & 10 & 1 \\ \hline 2 & 4 & 1 & 2 & 1 \\ \hline & 2 & 1 & 1 & 1 \end{array}$$

$$\text{l.c.m. } (24, 15, 20, 6) = 2 \cdot 3 \cdot 5 \cdot 2 \cdot 2 \cdot 1 \cdot 1 \cdot 1 = 120$$

Exercise Set 6-4

1. Find the least common multiple of the following.
   (a) 8 and 20 (b) 15 and 21
   (c) 56 and 21 (d) 60 and 108
   (e) 42 and 66 (f) 16 and 42
   (g) 22 and 90 (h) 252 and 96
2. What is the least common multiple of 1 and 4, 1 and 101, 1 and a?
3. Find the l.c.m. of each of the following.
   (a) 20, 22, 12 (b) 26, 36, 39
   (c) 4600, 224, 228 (d) 561, 27, 30
   (e) 18, 26, 12, 39 (f) 15, 39, 30, 21, 70
   (g) 324, 306, 180, 144 (h) 205, 250, 306, 200
4. Find the least common multiple of the following by *three* different processes.
   (a) 44 and 92 (b) 45 and 72 (c) 146 and 124
   (d) 252 and 74 (e) 840 and 1800 (f) 8125 and 1980
5. (a) Express the l.c.m. (a, b) as a product of primes.

$$a = 3^3 \cdot 4^2 \cdot 2$$
$$b = 3 \cdot 4 \cdot 2^3$$

   (b) Express the l.c.m. (a, b) as a product of primes where r, s, and t are primes.

$$a = r \cdot s^3 \cdot t^2$$
$$b = r^4 \cdot s^2 \cdot t^5$$

6. If $a|b$, compute both g.c.d. (a, b) and l.c.m. (a, b).
7. Compute both g.c.d. (a, a) and l.c.m. (a, a) where a is any positive integer.
8. (a) If a and b are prime, $a \neq b$, what is l.c.m. (a, b)?
   (b) If a and b are composite, under what conditions will l.c.m. $(a, b) = ab$?
   (c) What is the relationship between a and b if l.c.m. $(a, b) = a$?

*9. If M_a and M_b represent the set of all multiples of a and b, respectively, prove each of the following.
   (a) $M_{15} \subset M_5$
   (b) $M_8 \subset M_4$
   (c) $M_b \subset M_a$ if $a|b$ and $a \neq b$
   (d) $M_a \cap M_b = M_m$, where $m =$ l.c.m. (a, b)

*10. Prove that the l.c.m. of a and b equals ab if and only if g.c.d. $(a, b) = 1$.

*11. Prove that if k is any natural number, the l.c.m. of ka and kb is k times the l.c.m. of a and b.

*5

The Division Algorithm

In the preceding sections, we made use of the Division Algorithm in finding the greatest common divisor of two numbers. We now develop the theory associated with this procedure. For emphasis, we repeat the statement of the algorithm.

Theorem 6-5: *The Division Algorithm.* If a and b are any two positive integers, then there exist unique nonnegative integers q and r such that $a = bq + r$, where $0 \leq r < b$.

Proof: Consider the set of multiples of b, $\{0b, 1b, 2b, 3b, \ldots\}$. Now, $0(b) = 0 < a$, since a is a positive integer. This verifies that at least one element of the set is less than a. From the Archimedian Property, we know that there exists a natural number k such that $a \leq kb$. Thus, there are elements of the set greater than a. Consequently, some numbers of the set are less than a and some are greater than a. Let q be the largest positive integer such that $qb \leq a$. Then take $a - qb = r \geq 0$.

To prove $r < b$, we show that $r \geq b$ is impossible. If $r \geq b$, consider the following: $a - (q + 1)b = a - qb - b = r - b$ since $a - qb = r$. But since we assumed $r \geq b$, $r - b \geq 0$; so $a - (q + 1)b \geq 0$ or $a \geq (q + 1)b$. This contradicts our assumption that qb was the largest multiple of b less than or equal to a. Thus, by the trichotomy principle, $r < b$.

To prove that q and r are unique for given a and b, we suppose $a = q_1b + r_1$ and $a = q_2b + r_2$, where $0 \leq r_1 < b$ and $0 \leq r_2 < b$. Now, by the trichotomy principle, $q_1 > q_2$, or $q_1 < q_2$, or $q_1 = q_2$. Consider $q_1 > q_2$. Then $q_1 \geq q_2 + 1$. So $q_1b + r_1 \geq (q_2 + 1)b + r_1$. Why? Thus $a \geq q_2b + b + r_1$. Since $b > r_2$, $a > q_2b + r_2 + r_1$, so that $a > q_2b + r_2$. But $q_2b + r_2 = a$; so $a > a$. This is, of course, a contradiction, and hence q_1 is not greater than q_2. If we assume $q_1 < q_2$, a similar proof is given by interchanging q_1 and q_2. Thus q_1 is not less than q_2. Hence, by the trichotomy principle, $q_1 = q_2$. Now $q_1b + r_1 = q_2b + r_2$; and since $q_1 = q_2$, $q_1b + r_1 = q_1b + r_2$. Therefore $r_1 = r_2$ by the cancellation property of addition. Consequently, if a and b are positive integers, then there exist unique whole numbers q and r such that $a = bq + r$, where $0 \leq r < b$.

The repeated use of the Division Algorithm may be listed as follows.

$$
\begin{array}{rlr}
a &= q_1 b + r_1 & 0 \leq r_1 < b \\
b &= q_2 r_1 + r_2 & 0 \leq r_2 < r_1 \\
r_1 &= q_3 r_2 + r_3 & 0 \leq r_3 < r_2 \\
&\vdots & \vdots \\
r_{n-3} &= q_{n-1} r_{n-2} + r_{n-1} & 0 \leq r_{n-1} < r_{n-2} \\
r_{n-2} &= q_n r_{n-1} + r_n & 0 \leq r_n < r_{n-1} \\
r_{n-1} &= q_{n+1} r_n + 0 &
\end{array}
$$

The inequalities associated with the foregoing equations require that $b > r_1 > r_2 > \cdots > r_n$; that is, the remainders are decreasing. Since they are natural numbers and decreasing, after a finite number of steps the remainder must be 0. Let r_n be the last remainder not zero. Now $r_n | r_{n-1}$ from the last equation; then $r_n | r_{n-2}$ from the next to last equation; from this result $r_n | r_{n-3}$ in the third from last equation; in turn, r_n divides each remainder and eventually both b and a; thus r_n is a common divisor of b and a.

Now if x is any common divisor of a and b, then from the first equation x divides r_1; in the second equation $x | r_2$, and from the third equation $x | r_3$. In turn, x divides each remainder, so that finally $x | r_n$. Thus r_n is the greatest common divisor of a and b.

One of the most useful and interesting properties of the g.c.d. is that there exist integers x and y such that the g.c.d. can be written as $xa + yb$. This result is a consequence of the Euclidean Algorithm.

Theorem 6-6: If $d =$ g.c.d. (a, b), then there exist integers x and y such that $d = xa + yb$. In particular, if a and b are relatively prime, there exist integers x and y such that $xa + yb = 1$.

Example: Write the g.c.d. of 72 and 86 as $xa + yb$, where x and y are integers. To get g.c.d. (72, 86), we use the Euclidean Algorithm. $86 = 1(72) + 14$; $72 = 5(14) + 2$; and $14 = 7(2)$. Thus

$$d = \text{g.c.d. } (72, 86) = 2.$$

Now, using the next to the last equation of the Euclidean Algorithm, solve for 2. $2 = 1(72) + {}^{-}5(14)$. In the first equation solve for 14. $14 = 1(86) + {}^{-}1(72)$. Replace 14 in

$$2 = 1(72) + {}^{-}5(14) \quad \text{by} \quad 14 = 1(86) + {}^{-}1(72),$$

getting

$2 = 1(72) + {}^-5[1(86) + {}^-1(72)]$
$= 1(72) + {}^-5(86) + 5(72) = 6(72) + {}^-5(86).$

In this example, $x = 6$ and $y = {}^-5$.

Example: Express g.c.d. (147, 130) in the form $x \cdot 147 + y \cdot 130 = d$. Now $147 = 1(130) + 17$; $130 = 7(17) + 11$; $17 = 1(11) + 6$; $11 = 1(6) + 5$; $6 = 1(5) + 1$; $5 = 5(1)$. Thus the g.c.d. of 147 and 130 is 1. In each of the preceding equations, solve for the remainder in terms of the other parts of the equation. $17 = 1(147) + {}^-1(130)$; $11 = 1(130) + {}^-7(17)$; $6 = 1(17) + {}^-1(11)$; $5 = 1(11) + {}^-1(6)$; and, finally, $1 = 1(6) + {}^-1(5)$. If we replace 5 in the last equation by the expression for 5 in the next to the last equation, we get

$$1 = 1(6) + {}^-1[1(11) + {}^-1(6)],$$

so

$$1 = 2(6) + {}^-1(11).$$

Now replace the 6 to obtain $1 = 2[1(17) + {}^-1(11)] + {}^-1(11)$ or $1 = 2(17) + {}^-3(11)$. By replacing 11, one gets

$$1 = 2(17) + {}^-3[1(130) + {}^-7(17)] = 23(17) + {}^-3(130).$$

Finally, replacing 17,

$$1 = 23[1(147) + {}^-1(130)] + {}^-3(130) = 23(147) + {}^-26(130).$$

Thus, 1 is expressed as $x(147) + y(130)$, where $x = 23$ and $y = {}^-26$.

Proof: Suppose we examine the last equations in the application of the Euclidean Algorithm as given in the beginning of this section. Three of the last four equations may be written as

$$r_{n-4} = q_{n-2} r_{n-3} + r_{n-2},$$
$$r_{n-3} = q_{n-1} r_{n-2} + r_{n-1},$$
$$r_{n-2} = q_n r_{n-1} + r_n.$$

In each of these, solve for the remainder: $r_{n-2} = r_{n-4} + {}^-q_{n-2} r_{n-3}$; $r_{n-1} = r_{n-3} + {}^-q_{n-1} r_{n-2}$; and $r_n = r_{n-2} + {}^-q_n r_{n-1}$. Replacing r_{n-1} in the last equation by the expression for r_{n-1} in the next to the last equation yields

$$r_n = r_{n-2} + {}^-q_n(r_{n-3} + {}^-q_{n-1} r_{n-2}) = (1 + q_n q_{n-1}) r_{n-2} + {}^-q_n r_{n-3}.$$

Now substitute for r_{n-2}; this produces

$$r_n = (1 + q_n q_{n-1})(r_{n-4} + {}^-q_{n-2} r_{n-3}) + {}^-q_n r_{n-3}$$
$$= (1 + q_n q_{n-1}) r_{n-4} + ({}^-q_{n-2} + {}^-q_n q_{n-1} q_{n-2} + {}^-q_n) r_{n-3}.$$

Note that the coefficient of r_{n-4} is an integer. Why? Since there are a finite number of steps from the first equation in the Euclidean Algorithm to the last equation in the algorithm, this procedure can be continued until r_n is finally expressed in terms of a and b as

$$r_n = xa + yb,$$

where x and y are integers.

Theorem 6-7: If $a | bc$ and g.c.d. $(a, b) = 1$, then $a | c$.

Example: The g.c.d. $(15, 19) = 1$. Since $15 | (19 \cdot 30)$, then $15 | 30$.

This theorem is rather important in that it is used in the proof of the Fundamental Theorem of Arithmetic, which we shall consider in the next section of this chapter. After a bit of thought, this theorem does not seem too remarkable, particularly if we consider a, b, and c to be factored into primes. However, remember that we have not proved the uniqueness of factorization into primes. Even without the uniqueness of factorization into primes, the proof of this theorem is rather simple.

Proof: Since g.c.d. $(a, b) = 1$, then, by Theorem 6-6, $1 = ax + by$ for integers x and y. Multiply both sides of the equation by c to obtain $c = cax + cby$. Now, $a | cby$ because $a | bc$ by hypothesis. Also, $a | cax$. Then, by Theorem 6-1(b), $a | c$, which completes the proof.

A result that is easily derived from the preceding theorem may be stated as follows:

Theorem 6-8: Let p be a prime such that $p | bc$ and $p \nmid b$ (recall that $p \nmid b$ means p does not divide b). Then $p | c$.

Examples: (a) $3 | 24$. Consider 24 as $4 \cdot 6$. $3 \nmid 4$; therefore $3 | 6$.
(b) $13 | 1300$. Write $1300 = 50(26)$. $13 \nmid 50$; therefore $13 | 26$.

Proof: If $p \nmid b$, then g.c.d. $(p, b) = 1$, since p is a prime and hence has no divisors other than 1 and itself. Thus, p and b are numbers that satisfy the hypothesis of Theorem 6-7. Hence $p | c$.

Theorem 6-4 of the preceding section stated that if d and m are g.c.d. and l.c.m., respectively, of a and b, then $dm = ab$. This theorem may be proved by using Theorem 6-7. Let d be the g.c.d. of a and b. Then $a = dx$ and $b = dy$, where g.c.d. $(x, y) = 1$, because if there were some number that would divide both x and y, then d would not be the g.c.d. of a and b. Since $ya = ydx = xdy$ and $xb = xdy$, xdy is a multiple of both a and b. We now show that xdy is the l.c.m. of a and b. Any common multiple of a and b can be written as ka. Thus $b \mid ka$ since ka is a common multiple. Replace b with yd and a with xd. Then $yd \mid kxd$, so $y \mid kx$. Why? Now $y \mid kx$ implies y must divide k since g.c.d. $(x, y) = 1$. Therefore, xdy divides any common multiple of a and b such as ka because $xd = a$ and $y \mid k$. Hence, xdy is the least common multiple of m of a and b. Now, $ab = (dx)(dy) = d(xdy)$. Replace xdy by m to obtain $ab = dm$.

Theorem 6-9: If p is a prime and $p \mid ab$, then either $p \mid a$ or $p \mid b$ (or both).

Examples:

3 is a prime; $3 \mid (6 \cdot 5)$; $3 \mid 6$.
7 is a prime; $7 \mid (3 \cdot 14)$; $7 \mid 14$.
11 is a prime; $11 \mid (33 \cdot 55)$; $11 \mid 33$ and $11 \mid 55$.

Proof: If $p \mid a$, the proof is complete. If p does not divide a, then since p is a prime, the g.c.d. of p and a is 1. Therefore, by Theorem 6-8, $p \mid b$. Hence p divides either a or b (or maybe both).

Theorem 6-10: If a and b are relatively prime and $a \mid c$ and $b \mid c$, then $ab \mid c$.

Example: $3 \mid 72$ and $8 \mid 72$. Since 3 and 8 are relatively prime, then $(3 \cdot 8) \mid 72$.

Proof: $a \mid c$ implies $c = ak$, where k is a positive integer. Since $c = ak$, $b \mid c$ means $b \mid ak$. But g.c.d. $(a, b) = 1$ by hypothesis. Therefore, by Theorem 6-7, $b \mid k$. Hence $k = bm$, where m is some natural number. Replacing k in $c = ak$ by bm yields $c = a(bm) = (ab)m$. Since $ab \mid (ab)m$, then $ab \mid c$.

This theorem is useful in testing the divisibility of numbers without performing the division.

Example: Without performing the actual division, show that 1008 is divisible by 12. Now $12 = 4 \cdot 3$ and 4 and 3 are relatively prime. Since $4 | 1008$ and $3 | 1008$, then $12 | 1008$.

Theorem 6-11: Let p and $p_1, p_2, \ldots, p_n$ be primes. If

$$p | p_1 p_2 p_3 \cdots p_n,$$

then p is equal to one of the primes p_i.

Example: Consider 210 written as a product of primes. $7 | 210$ so 7 must equal one of the primes when 210 is written as a product of primes; $210 = 2 \cdot 3 \cdot 5 \cdot 7$. This, of course, does include a 7. Consider 429 as a product of primes. $13 | 429$, so 13 must equal one of the primes in the product. The product is $429 = 3 \cdot 11 \cdot 13$.

Proof: Group $p_1 p_2 p_3 \cdots p_n = p_1(p_2 p_3 \cdots p_n)$. Now

$$p | p_1(p_2 p_3 \cdots p_n);$$

hence, by Theorem 6-9, $p | p_1$ or $p | (p_2 p_3 \cdots p_n)$. If $p | p_1$ then $p = p_1$, for both p and p_1 are primes, and so the proof is complete. If $p \neq p_1$, then $p \nmid p_1$ so $p | (p_2 p_3 \cdots p_n)$. Now write $p_2 p_3 \cdots p_n = p_2(p_3 p_4 \cdots p_n)$ and proceed with the same argument. The process will terminate as soon as $p | p_i$. If p does not divide any of $p_1, p_2, \ldots, p_{n-2}$, then $p | p_{n-1} p_n$. This implies $p | p_{n-1}$ or $p | p_n$ by Theorem 6-9. Thus, p is equal either to p_{n-1} or to p_n, and the proof is complete.

Exercise Set 6-5

1. Without using actual division, determine the truth values of each of the following statements.
   (a) $15 | 345$ (b) $45 | 3780$ (c) $44 | 2772$ (d) $40 | 2640$
   (e) $30 | 2910$ (f) $150 | 2700$ (g) $45 | 1170$ (h) $165 | 2805$
2. If it is true that $9 | n$ and $10 | n$, does it follow that $90 | n$? Why?
3. In each of the following, find the greatest common divisor of a and b, and then find x and y such that g.c.d. $(a, b) = xa + yb$.
   (a) $a = 48, b = 130$ (b) $a = 63, b = 230$
   (c) $a = 62, b = 154$ (d) $a = 66, b = 390$
   (e) $a = 37, b = 53$ (f) $a = 23, b = 41$

4. Prove that if $d = \text{g.c.d.}\ (a, b)$ and $a = rd$ and $b = sd$, then $\text{g.c.d.}\ (r, s) = 1$.
5. Prove that if the product of two relatively prime integers is a square, each integer is a square.
6. Prove that if $a \mid b$, where a is a prime, then:
   (a) $\text{g.c.d.}\ (a, b) = a$ (b) $\text{l.c.m.}\ (a, b) = b$
7. Prove that if $a \nmid b$, where a is a prime, then:
   (a) $\text{g.c.d.}\ (a, b) = 1$ (b) $\text{l.c.m.}\ (a, b) = ab$
8. Prove that if $d \mid x$ and $d \nmid y$, then $d \nmid (x + y)$ and $d \nmid (x - y)$.
9. (a) Prove, using the divisibility theorem, that the sum of an odd number a and an even number b is an odd number c.
   (b) Prove that the product of two odd numbers a and b is an odd number c.

*10. If k is any natural number, prove that if $r = \text{g.c.d}\ (ka, kb)$ and $s = \text{g.c.d.}\ (a, b)$, then $r = ks$.

*6

The Fundamental Theorem of Arithmetic

Any composite number can be written as a product of primes in one and only one way, except for the order in which the multiplication of the primes occurs. This result was stated as the Fundamental Theorem of Arithmetic (Theorem 6-2).

This theorem states that if x is any composite positive integer, then x can be written as $x = p_1 p_2 \cdots p_n$, where each p_i is a prime. Moreover, if $x = p_1 p_2 \cdots p_k$ and $x = q_1 q_2 \cdots q_n$, where each q_j is a prime, then the $p_1, p_2, \ldots, p_k$ are the same primes as $q_1, q_2, \ldots, q_n$ in some order.

You may feel that the Fundamental Theorem of Arithmetic is obvious and needs no proof. If so, then perhaps you need to understand that there are number systems in which we can define integers, divisibility, and primes so that when an integer is factored into primes, the factorization is not necessarily unique. Such an example will be given at the end of this section.

Proof: We will divide the proof of this theorem into two parts. In the first part, we will show that every composite positive integer can be factored into primes. In the second part, it will be verified that the factorization is unique.

(a) If N is a composite positive integer, we will show that it can be written as a product of prime factors. Since N is a composite number, then $N = x_1 \cdot x_2$, where $1 < x_1 < N$ and $1 < x_2 < N$. If x_1 and x_2 are both primes, we have a prime factorization and the proof is complete. If x_1 is not a prime, factor it into two factors $x_1 = x_3 \cdot x_4$ such that

$$1 < x_3 < x_1 \quad \text{and} \quad 1 < x_4 < x_1.$$

Do the same for x_2 if it is not a prime. $x_2 = x_5 \cdot x_6$, where $1 < x_5 < x_2$ and $1 < x_6 < x_2$. Then $N = x_3 \cdot x_4 \cdot x_5 \cdot x_6$. If x_3, x_4, x_5, and x_6 are all primes, the proof is complete. Otherwise, repeat the same process. You should notice in this process that the factors get smaller and smaller. Since the factors are positive integers, they will eventually decrease until they reach a smallest prime. Thus, N has a factorization into primes.

(b) We will now verify that this factorization is unique. To prove that the factorization is unique, suppose there are two prime factorizations of N, $p_1 p_2 p_3 \cdots p_k$ and $q_1 q_2 q_3 \cdots q_m$. Since both factorizations are equal to N, then they are equal; $p_1 p_2 p_3 \cdots p_k = q_1 q_2 q_3 \cdots q_m$. Now since $p_1 | p_1 p_2 p_3 \cdots p_k$, it must divide $q_1 q_2 q_3 \cdots q_m$. By Theorem 6-11, $p_1 = q_i$ for some i. Let q_1 be the symbol for the q_i equal to p_1. Then divide the equation by p_1 (which is equal to q_1), getting $p_2 p_3 \cdots p_k = q_2 q_3 \cdots q_m$. Repeat the same process and let $q_2 = p_2$, $q_3 = p_3$, etc. If $k < m$, the repetition of this process yields $1 = q_{k+1} q_{k+2} \cdots q_m$. Since it is impossible for a product of primes to equal 1, we have reached a contradiction for $k < m$. If $k > m$, the above reasoning would yield $p_{m+1} p_{m+2} \cdots p_k = 1$; this again is a contradiction. Thus for $k > m$ and $k < m$, a contradiction has been reached; consequently, for these two cases the prime factorization is unique. If $k = m$, by the preceding reasoning, the p's and the q's are equal by pairs and the factors are actually the same. Thus, the factorization of a positive integer into primes is unique.

At this point, we wish to present a set of numbers that is closed, associative, and commutative with respect to multiplication; at the same time, these numbers will not satisfy many of the theorems in this chapter and definitely will not satisfy the Fundamental Theorem of Arithmetic. Consider a set of numbers of the form $\{3x + 1 | x = 0, 1, 2, 3, \ldots\}$. This set of numbers can be written as $\{1, 4, 7, 10, 13, 16, \ldots\}$. Notice that all members of this set can be described as "one added to the product of 3 and a whole number." To show that this set is closed under the operation of multiplication would require that a product be expressible as "3 times a whole number added to 1." Consider two numbers in the set, $3x + 1$ and $3y + 1$, where x and y are whole numbers.

$$(3x+1)(3y+1) = 3x(3y+1) + 1(3y+1)$$
$$= 9xy + 3x + 3y + 1 = 3(3xy + x + y) + 1.$$

Now $3xy + x + y$ is a whole number; hence $3(3xy + x + y) + 1$ is "3 times a whole number plus 1," and so the set is closed under the operation of multiplication. In like manner, multiplication can be shown to be commutative and associative.

Let us define a prime in this set to be a number greater than 1 divisible only by itself and 1. If a number is divisible by a prime other than itself, it is called a composite number. Some primes of this set are $\{4, 7, 10, 13, 19, \ldots\}$. Theorem 6-9 asserts that if $p \mid ab$ and p is a prime, then $p \mid a$ or $p \mid b$. Does this theorem hold for this new set of numbers? Consider the statement $10 \mid 100$. Now both 100 and 10 are in this set because $100 = 3(33) + 1$ and $10 = 3(3) + 1$. Since neither 4 nor 7 divides 10, it must be a prime. Now write 100 as $4(25)$, and the problem becomes $10 \mid 4 \cdot 25$. But $10 \nmid 4$ and $10 \nmid 25$. Thus, Theorem 6-9 is not true for this set of numbers.

Next, notice the two different factorizations of 100. $100 = 10 \cdot 10$ and $100 = 4 \cdot 25$. Note that 4, 10, and 25 are all primes in the system under discussion. Thus, we are able to factor 100 into a product of primes in two different ways. Hence, the Fundamental Theorem of Arithmetic is not true for this system of numbers.

Exercise Set 6-6

1. Represent the following numbers as products of primes.
   (a) 48 (b) 184 (c) 4802
   (d) 259,308 (e) 518,616 (f) 345,740
2. Although there is only one unique set of prime factors for any given number, a number can be expressed as a product in other ways not involving all primes. Express each of the following as (1) the product of two equal numbers, (2) the product of a composite number and a prime number, and (3) the product of two composite numbers.
   (a) 3^6 (b) 64 (c) 100 (d) 144
3. Consider the subset of natural numbers defined by

$$\{2x + 1 \mid x = 0, 1, 2, \ldots\}.$$

   (a) Show that this set is closed under multiplication.
   (b) Test for commutativity and associativity.
   (c) Define a prime on this set of numbers; list the first ten primes.
   (d) Does the Fundamental Theorem of Arithmetic hold?

4. Do all parts of Exercise 3 where the set is defined as

$$\{4x + 1 \mid x = 0, 1, 2, 3, \ldots\}.$$

5. Do all parts of Exercise 3 where the set is defined as

$$\{5x + 1 \mid x = 0, 1, 2, 3, \ldots\}.$$

*6. Consider the subset of natural numbers defined by $\{n^2 \mid n = 1, 2, 3, 4, \ldots\}$.
   (a) Is this set closed under multiplication?
   (b) Test for commutativity and associativity.
   (c) Define a prime on this set of numbers; list the first four primes.
   (d) Does the Fundamental Theorem of Arithmetic hold?

*7. Consider the system 3, 4, 5, 6, ... of natural numbers greater than 2. Is this system closed under multiplication? If so, write down the first ten primes. Does every number of the system factor uniquely as a product of primes?

*7

Some Solved and Unsolved Problems in Number Theory

Many people believe that all mathematical problems have already been solved, that mathematics today is simply a matter of learning what has already been done. However, exactly the opposite is true. The more mathematics that is developed, the more problems seem to arise. Someone has said that more mathematics has been developed since 1900 than in all the rest of history. However, this is simply a conjecture; and we would not want to even guess its validity. We do know that each day there are many new problems to be solved. However, we are not interested at this time in new problems. We want to look at some famous problems that seem to have defied solution and yet have not been shown to be unsolvable.

We consider first some unsolved problems concerning *perfect numbers*. Perfect numbers were highly esteemed by the Pythagoreans, the members of a school of philosophy and mathematics founded by the Greek mathematician Pythagoras. The Pythagoreans felt that numbers were the basis of all substance and that numbers had mystical and sacred powers. For example, it was said that God chose to create all things in six days because of His realization of the perfect qualities of the number 6. 6 also represented marriage, health, and beauty for the Pythagoreans.

Definition 6-7: A number is said to be *perfect* if it is equal to the sum of its proper divisors. (The proper divisors of a number do not include the number.)

Is 10 a perfect number? The divisors of 10 are 1, 2, 5; since $1+2+5=8\neq 10$, then 10 is not a perfect number according to our definition.

Suppose we try 18. The divisors of 18 are 1, 2, 3, 6, and 9.

$$1+2+3+6+9=21\neq 18.$$

Thus, 18 is not a perfect number.

Suppose you select some other numbers by chance and test to see if they are perfect numbers. You will find that you obtain many more numbers that are not perfect than numbers that are perfect. Thus, you soon acquire the idea that perfect numbers are rather scarce.

Now let us take a look at some perfect numbers. 6 is the smallest perfect number, since $1+2+3=6$, and none of the numbers 2, 3, 4, or 5 is perfect. The next perfect number is 28: $1+2+4+7+14=28$. The third perfect number is 496, the fourth 8128, and the fifth 33,550,336. You can guess that finding perfect numbers is a rather difficult task. We might ask the question "How many perfect numbers are there?" The answer is "No one knows." Perhaps there are an infinitude of perfect numbers, but can you prove that this is true? Can you answer the question "Are there any odd perfect numbers?" All the perfect numbers that have been found at this time are even numbers.

Theorem 6-12: If 2^n-1 is a prime, then $N=2^{n-1}(2^n-1)$ is perfect, and every even perfect number is of this form.

Example: $2^2-1=3$ is a prime. Then $N=2(3)=6$ is perfect.

Example: $2^3-1=7$ is a prime. Then $N=2^2(2^3-1)=28$ is an even perfect number.

Example: $2^5-1=31$ is a prime. Then $N=2^4(2^5-1)=496$ is an even perfect number because

$$1 + 2 + 4 + 8 + 16 + 31 + 62 + 124 + 248 = 496.$$

Some additional unsolved problems occur in the study of amicable numbers. From Greek to Medieval times, these amicable numbers were believed to have a divine or mystic relationship to human friendship.

Definition 6-8: Two numbers are said to be *amicable* if each one is the sum of the proper divisors of the other.

Some examples of amicable number pairs are 220 and 284; 17,296 and 18,416; and 1184 and 1210. This last pair of amicable numbers was discovered by a 16-year-old boy. Perhaps you can discover a pair. Of course, a mathematical question is "How many amicable numbers are there?" Are there infinitely many? If you think there are infinitely many amicable numbers, can you prove this conjecture?

Another famous unsolved problem deals with twin primes. As you learned in Exercise Set 6-2, *twin primes* are primes that differ by 2. There are many examples of twin primes, such as 3 and 5, 5 and 7, 17 and 19, and 29 and 31. Yet, an interesting question is "Are there infinitely many such pairs?" No one knows, although many people have worked on the problem.

The last unsolved problems we will consider are called *Goldbach's Conjectures*, discussed briefly in Exercise Set 6-2. This eighteenth-century mathematician advanced two conjectures:

(a) Every even number greater than 4 is the sum of two odd primes.

Examples:

$$\begin{aligned} 6 &= 3 + 3 \\ 8 &= 3 + 5 \\ 10 &= 5 + 5 = 3 + 7 \\ 12 &= 5 + 7 \\ 14 &= 7 + 7 = 3 + 11 \\ 16 &= 5 + 11 = 3 + 13 \end{aligned}$$

(b) Every odd number greater than or equal to 9 is the sum of three odd primes.

Examples:

$$\begin{aligned} 9 &= 3 + 3 + 3 \\ 11 &= 3 + 3 + 5 \\ 13 &= 5 + 5 + 3 = 3 + 3 + 7 \\ 15 &= 5 + 5 + 5 = 5 + 3 + 7 \\ 17 &= 3 + 3 + 11 = 5 + 5 + 7 = 3 + 7 + 7 \end{aligned}$$

No one knows how to prove these conjectures. No one has been able to prove them incorrect. All one has to do to prove them incorrect is to find one example that does not work. However, no such example has ever been found.

We trust that in this section you have been convinced that there are many unsolved problems relating to the theory of integers. The same is true in all branches of mathematics.

Exercise Set 6-7

1. The great mathematician Fermat (1608–1665) proved the following theorem: "If p is a prime and a is any integer, the integer given by $a^p - a$ is divisible by p." Verify this theorem when
   (a) $a = 6, p = 5$ (b) $a = 4, p = 7$ (c) $a = 3, p = 11$
2. An eighteenth-century English mathematician by the name of Wilson proved the following theorem: "The number $N = [1 \cdot 2 \cdot 3 \cdots (p - 1)] + 1$, where p is a prime, is divisible by p." Verify that this is true when $p = 5$, when $p = 11$, when $p = 17$.
3. Show that 220 and 284 are amicable numbers; do the same for 1184 and 1210.
4. Verify Theorem 6-12 with one example different from the three examples explained in this section.
5. Prime numbers of the form 3, 5, 7 are called *prime triplets*. These three numbers are consecutive primes that differ by 2. Can you find another set of prime triplets? Make a conjecture about prime triplets. Prove your conjecture.
6. For each of the following, give an example or explain why you cannot do so.
   (a) A composite number of form $4n + 1$
   (b) A composite number of form $3^p + 1$, where p is a prime
   (c) A prime number of the form $n^2 + n$, when n is a natural number

(d) A prime number of the form $4n + 1$, when n is a natural number
(e) A prime number of the form $8n + 1$, where n is a natural number

7. Part (b) of Goldbach's Conjectures (stated in the preceding section) is that every odd number greater than or equal to 9 can be expressed as the sum of three (not necessarily different) odd primes. Write the following as such a sum.
(a) 9 (b) 25 (c) 77 (d) 43

*8. A *deficient* number is defined as a number with the property that the sum of its proper divisors is less than the number. An *abundant* number is a number with the property that the sum of its proper divisors is greater than the number.
(a) Give an example of a number that is deficient and an example of a number that is abundant.
(b) Classify the integers x such that $5 < x < 20$ as perfect, abundant, or deficient.

8

Summary

In this chapter, we have studied some interesting properties of subsets of integers. Do you remember the meaning of and do you understand *multiple, least common multiple, divisor, factor, greatest common divisor, perfect number, amicable number, Euclidean Algorithm,* and the *Fundamental Theorem of Arithmetic*? If not, it is possible that further study of this chapter is needed.

Review Exercise Set 6-8

1. What is the greatest common divisor of 5734 and 12,862? What is the least common multiple?
2. Show without actual division that 12 divides 936. Give reasons for your answer.
3. Is it possible to have exactly six consecutive composite numbers between two primes? Justify your answer by finding an example.
4. Consider an operation on the set of natural numbers as follows: For natural numbers x and y, $x \otimes y = \text{l.c.m.}\ (x, y)$.
(a) Is the set of natural numbers closed with respect to this operation?

(b) Is this operation commutative?
(c) Is this operation associative?

5. If 1 is the greatest common divisor of two numbers, what can you say about their least common multiple?
6. Suppose the least common multiple of two numbers is the same as their greatest common divisor. What can you say about the numbers?
7. Let us define an operation on the set of integers as follows: For integers a and b, $a \odot b =$ g.c.d. (a, b).
   (a) Is the set of integers closed with respect to this operation?
   (b) Is this operation commutative?
   (c) Is this operation associative?
8. Show that 8732_{eleven} is divisible by T_{eleven}. Why?
9. In base seven, write the first ten multiples of 6.
10. Prove that a number of the form *abcabc*, such as 243243, is divisible by 7, 11, and 13.

Suggested Reading

Divisibility: Allendoerfer, pp. 360–361. Armstrong, pp. 211–212. Campbell, p. 109. Copeland, pp. 149–150. Graham, pp. 206–210. Hutton, pp. 156–163. Nichols, Swain, pp. 188–189. Ohmer, Aucoin, pp. 137–141. Ohmer, Aucoin, Cortez, pp. 179–184. Podraza, Blevins, Hanson, Prall, pp. 113–114. Smith, pp. 126–128. Spector, pp. 397–401. Webber, Brown, pp. 109–112. Willerding, pp. 168–170. Wren, pp. 182–187. Zwier, Nyhoff, pp. 93–94.

Primes and Composite Numbers: Allendoerfer, pp. 453–454. Armstrong, pp. 123–131. Bouwsma, Corle, Clemson, pp. 163–172. Byrne, pp. 142–145. Campbell, pp. 109–110. Copeland, pp. 146–154. Garner, pp. 93–95. Garstens, Jackson, pp. 53–58. Graham, pp. 177–180. Meserve, Sobel, pp. 134–139. Nichols, Swain, pp. 173–174. Ohmer, Aucoin, pp. 146–152. Ohmer, Aucoin, Cortez, pp. 190–194. Peterson, Hashisaki, pp. 140–141. Podraza, Blevins, Hanson, Prall, pp. 116–117. Scandura, pp. 200–201. Smith, pp. 129–132. Spector, pp. 403–410, 419–421. Webber, Brown, pp. 113–114. Willerding, pp. 161–164. Wren, pp. 98–100. Zwier, Nyhoff, pp. 93–95.

Least Common Multiple and Greatest Common Divisor: Allendoerfer, pp. 453–454. Armstrong, pp. 132–136. Bouwsma, Corle, Clemson, pp. 176–186. Byrne, pp. 146–151. Campbell, pp. 111–120. Copeland, pp. 150–153. Garner, pp. 98–102. Graham, pp. 186–191. Keedy, pp. 165–167.

Meserve, Sobel, pp. 141–144. Nichols, Swain, pp. 179–185. Ohmer, Aucoin, Cortez, pp. 191–204. Peterson, Hashisaki, pp. 145–151. Podraza, Blevins, Hanson, Prall, pp. 121–122. Scandura, pp. 202–204. Smith, pp. 135–138. Spector, pp. 413–415. Webber, Brown, pp. 115–120. Willerding, pp. 173–178. Wren, pp. 120–134. Zwier, Nyhoff, pp. 95–97.

Division Algorithm, Euclid's Algorithm, Fundamental Theorem of Arithmetic: Bouwsma, Corle, Clemson, p. 183. Byrne, pp. 150–152. Ohmer, Aucoin, Cortez, p. 191. Webber, Brown, p. 117. Wren, pp. 120–121.

7

Fractions and Rational Numbers

1

Introduction

In Chapter 5, the system of whole numbers was extended to the system of integers to provide for closure for subtraction, an operation which is necessary for solving certain problems. In a similar manner, we extend the system of integers to the system of rational numbers to provide closure for division (excluding division by zero). This extension will allow us to solve a wider class of problems, and it will be made so that the relations and operations for integers hold for rational numbers.

We now introduce rational numbers through a discussion of fractions. You are well aware of numerals of the form 2/3, 4/7, 15/9, called *fractions*. In fact, you were introduced to numbers represented this way very early in your school work. The symbol 2/3 involves the pair of integers 2 and 3. 2 is the *numerator*, and 3 is the *denominator*. Because of the positions they occupy, we refer to this numeral as an ordered pair of integers. Ordinarily, we write ordered pairs as (a, b), but we prefer in this case to use the notation a/b.

Actually, these number-pair symbols take on three different meanings or interpretations. These interpretations may be summarized as:

(a) a "partition" or "part of" interpretation,
(b) a "division" interpretation,
(c) an "element of a mathematical system" interpretation.

Thus, we use the number pair 2/3 to represent:

(a) 2/3 of something, or two parts out of three,
(b) the division of 2 by 3,
(c) the rational number, 2/3.

We will attempt in this chapter to relate these three interpretations. But before discussing these concepts, let us consider the historical development of these number-pair symbols.

Historical records verify that the Egyptians worked with fractions. At first, they used only fractions with numerators of 1, such as 1/2, 1/3, and 1/4. Later, other fractions were added to their set of numerals. The Babylonians used fractions with denominators of 60 or powers of 60. Since our units of time were borrowed from the Babylonians, one minute is 1/60 of an hour and one second is 1/60 of a minute or 1/3600 of an hour. The Romans made little use of fractions; most of their fractions had denominators of 12 and were associated with measurements. Many different notations have been used for fractions. The notation illustrated by the fraction $\frac{2}{3}$, or 2/3, came into general use about the sixteenth century. Fractions can be made to correspond to parts of figures, to parts of sets, to points on a number line, and to various other items or ideas.

The need for fractions in everyday language is evident. Many situations exist in which objects are actually cut into pieces. For example, a pie may be cut into six equal parts called *sixths*. A length of string may be divided into two equal parts called *halves* or into three equal parts called *thirds*. A young child has an intuitive notion of what it means to cut an apple into four equal parts, each of which is called one-fourth of the whole.

The concept of fractions can be seen by dividing a plane into regions having the same size and shape. Let us consider a set of "like" regions, such as the nine squares in (a), (b), and (c) of Figure 7-1. Now, let us choose certain squares in each set and indicate them by shading. If we associate the number of shaded squares with the total number of squares and write the result in the form 1/9 (Figure 7-1(a)), 2/9 (Figure 7-1(b)), and 3/9 (Figure 7-1(c)), we begin to get an intuitive idea of the type of

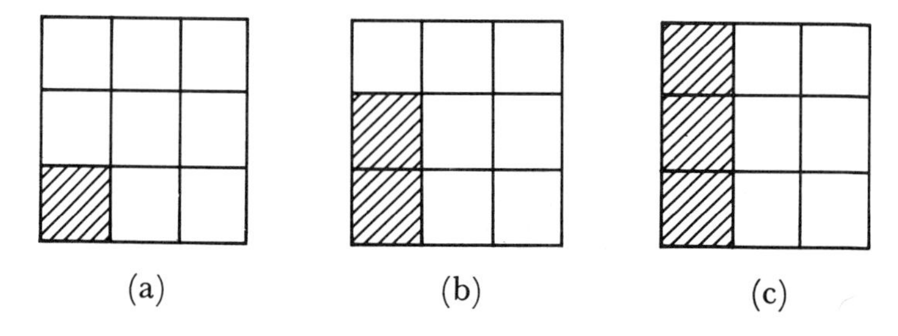

Figure 7-1

number that fractions represent. Notice that the three squares out of nine squares could be considered as three regions out of nine and denoted by 3/9, or as one region out of three and denoted by 1/3.

Now let us concentrate our attention on "like" line segments. By "like" line segments, we mean line segments of equal length. In Chapter 10 we will call such line segments *congruent* segments. Let us consider a unit length on a number line as being subdivided into "like" segments. First of all, in Figure 7-2, we divide the line segment representing one unit into two "like" segments. Each segment represents one part of two, denoted by 1/2. We then divide the same segment into three and four equal parts as indicated in Figure 7-2.

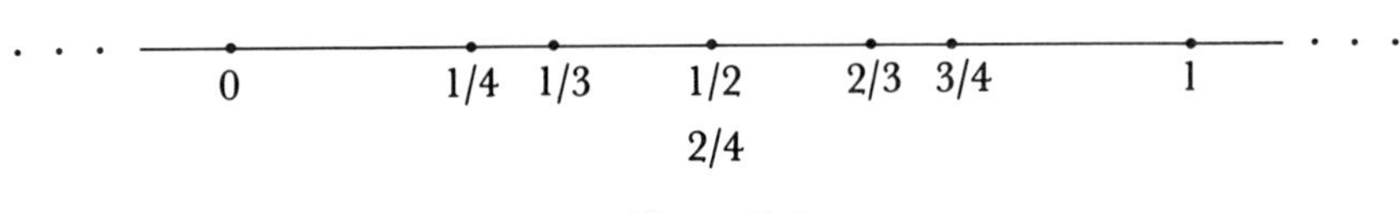

Figure 7-2

It is the *partition* or *part-of* interpretation of the number pair a/b that suggests the name that we use for a/b, namely a *fraction*. "Fraction" means "part of." The name *denominator* of the fraction, which is here called b, designates the number of parts into which the whole is divided. Naturally, $b \neq 0$. "Denominator" comes from the Latin word *denominatus*, to call by name. "Numerator" comes from the Latin word *numeratus*, to want. The numerator of a fraction, which we denote here by a, counts the parts under consideration. Now that we have named the number pairs $a/b (b \neq 0)$ as fractions based on their physical structure, we will proceed to define numbers denoted by $\frac{a}{b}$ or a/b in terms of division.

Division, $a \div b$, was defined in Chapters 3 and 5 by means of the equation $a = bx$. Of course, in the system of whole numbers and in the system of integers, this equation does not always have a solution. For example, $5 = 8x$ has no solution in the system of integers. Thus, the question arises "Can we enlarge the system of integers so that division will always have a meaning in the new system except when the divisor is zero?"

In the process of making this extension, we consider the problem of solving $1 \div a = x$, or of determining what x exists such that $a \cdot x = 1$. Can we find numbers for x, y, and z such that $1 = 2x$, $1 = 3y$, and

$1 = 4z$? If by a number we mean an integer, the answer is *no*. However, let us create some new numbers that will satisfy our requirement. Denote by $1/2$ a number such that $2(1/2) = 1$, by $1/3$ a number such that $3(1/3) = 1$, and by $1/4$ a number such that $4(1/4) = 1$. In general, let $1/a$ represent a number such that $a(1/a) = 1$. This new number $1/a$ may be called "the reciprocal of a," or "the multiplicative inverse of a." There is only one integer that we cannot define to have a multiplicative inverse, and this integer is zero. Make sure you understand why.

There is additional need for other numbers if one considers $5 = 8x$. How do we get a solution for this equation? Suppose we again invent a new number, symbolized by $5/8$, such that $8(5/8) = 5$. In general, $bx = a$ would have a solution, a/b (where, of course, $b \neq 0$).

Note that we have not given an explicit definition for a fraction. We have merely written symbols like $\frac{a}{b}$ and a/b, where a and b are integers with $b \neq 0$, and then we have interpreted these to be points on a number line. Although the traditional notation for fractions has been used, observe that a/b denotes simply an ordered pair of integers and could just as well have been written (a, b). By extending the number-pair concept to the solutions of equations so that $a = bx$ $(b \neq 0)$ has a solution a/b, we define

$$b \cdot (a/b) = a.$$

2

Rational Numbers

Note from our defined solution of the equation $a = bx$ that the solution is not unique. For example, $3x = 12$ has the solution $12/3$. However, 4 is also a solution. Therefore, 4 and $12/3$ must be names for the same number if the solution is to be unique. That is, 4 and $12/3$ should be represented by the same point on a number line. This leads to the question as to what relationship exists between two numerals representing the same number.

Definition 7-1: Two fractions a/b and c/d are said to be *equivalent* (denoted by $a/b \simeq c/d$) if and only if they represent the same point on the number line.

This definition leads to the following theorem, which presents a method for testing whether or not two fractions are equivalent.

Theorem 7-1: The fractions a/b and c/d are equivalent if and only if $ad = bc$.

Proof: Now suppose that a/b and c/d are equivalent. a/b is a point representing the solution of $bx = a$, or $b(a/b) = a$. Likewise, since c/d represents the same point as a/b and since we want to define the solution of an equation to be unique, $b(c/d) = a$. Using Theorem 5-3 for integers,

$$db(c/d) = da,$$

or

$$bd(c/d) = ad,$$

or

$$b[d(c/d)] = ad.$$

However, $d(c/d) = c$ by the definition of c/d. Thus, if $a/b \simeq c/d$, then $bc = ad$ or $ad = bc$.

Now let $ad = bc$. Consider $bd(a/b) = db(a/b) = d[b(a/b)] = da = bc$. Then, by Theorem 5-4 (b), $d(a/b) = c$. But $d(c/d) = c$. Thus, $a/b \simeq c/d$.

Examples:

$$\frac{2}{5} \simeq \frac{6}{15} \quad \text{because} \quad 2(15) = 5(6) = 30.$$

$$\frac{3}{8} \simeq \frac{^-9}{^-24} \quad \text{because} \quad 3(^-24) = 8(^-9) = {^-72}.$$

In Chapter 2, we studied three properties of relations: the reflexive property, the symmetric property, and the transitive property. A relation that has these three properties is called an *equivalence relation.* We note that the relation $\simeq$ for the set of fractions is an equivalence relation.

Theorem 7-2: The relation $\simeq$ for the set of fractions is an equivalence relation.

Proof: To verify that this relation is an equivalence relation, we must test to see if it is reflexive, symmetric, and transitive. If a/b is any fraction from the set of all fractions, we wish to prove that $a/b \simeq a/b$. By Theorem 7-1, $a/b \simeq a/b$ if and only if $ab = ba$. Since a and b are integers, then by the commutative property of the multiplication of integers, $ab = ba$. Thus, $\simeq$ is reflexive.

The second property to be verified is the symmetric property of equivalences. If a/b and c/d are any fractions from the set of all fractions, and if $a/b \simeq c/d$, we will prove $c/d \simeq a/b$.

$a/b \simeq c/d$	Given
$ad = bc$	Theorem 7-1
$bc = ad$	Symmetric property of equality for integers
$cb = da$	Commutative property of multiplication of integers
$c/d \simeq a/b$	Theorem 7-1

Finally, we must prove the transitive property of this relation. If $a/b \simeq c/d$ and $c/d \simeq e/f$, where a/b, c/d, and e/f are any fractions from the set of all fractions, then $a/b \simeq e/f$ because

$a/b \simeq c/d$, and $c/d \simeq e/f$	Given
$ad = bc,\ cf = de$	Theorem 7-1
$(ad)f = (bc)f$	Theorem 5-3(a)
$(af)d = b(cf)$	Commutative and associative properties of multiplication of integers
$(af)d = b(de)$	Substitution, since $cf = de$
$(af)d = d(be)$	Commutative and associative properties of multiplication of integers
$[(af)d]/d = be$	Definition 5-8, $d \neq 0$
$[(af)d]/d = af$	Definition 5-8 applied to $(af)d = d(af)$
$af = be$	Transitive property of equalities
$a/b \simeq e/f$	Theorem 7-1

Thus, the relation $\simeq$ for the set of fractions is an equivalence relation.

The preceding theorem suggests that the set of all fractions can be considered as being partitioned into equivalence classes. The class of equivalent fractions to which 1/3 belongs may be written as

$$\left\{\ldots, \frac{^-4}{^-12}, \frac{^-3}{^-9}, \frac{^-2}{^-6}, \frac{^-1}{^-3}, \frac{1}{3}, \frac{2}{6}, \frac{3}{9}, \frac{4}{12}, \ldots\right\}.$$

Similarly, 0/1 belongs to the class

$$\left\{\ldots, \frac{0}{^-5}, \frac{0}{^-4}, \frac{0}{^-3}, \frac{0}{^-2}, \frac{0}{^-1}, \frac{0}{1}, \frac{0}{2}, \frac{0}{3}, \frac{0}{4}, \frac{0}{5}, \ldots\right\}.$$

We now have an infinite set of equivalent fractions, each of which is a numeral for the solution of an equation or a point on the number line. To avoid these complications we introduce a new set of numbers called the *rational numbers.* Intuitively, a *rational number* is a number represented by any one of the fractions in an equivalence class of fractions. Any member of the set

$$\left\{\ldots, \frac{^-15}{^-9}, \frac{^-10}{^-6}, \frac{^-5}{^-3}, \frac{5}{3}, \frac{10}{6}, \frac{15}{9}, \ldots\right\}$$

represents the rational number solution of $3x = 5$, so any of the fractions in the set, such as 10/6, could represent the rational number solution of $3x = 5$. The formal definition of a rational number is as follows:

Definition 7-2: A *rational number* is a number represented by any one of the fractions in an equivalence class of fractions.

Thus, when we discuss operations with rational numbers, we will symbolize a rational number by some fraction such as a/b that is a member of the class. For example, the fraction 2/5 denotes the rational number associated with the class of fractions equivalent to 2/5. For this reason, we will use the expression "the rational number 2/5."

Therefore, a rational number is an abstract mathematical concept just as are the counting numbers and the integers. The numeral we use to represent the rational number will be any fraction from a class of equivalent fractions.

Figure 7-3 illustrates the fact that fractions in an equivalence class are

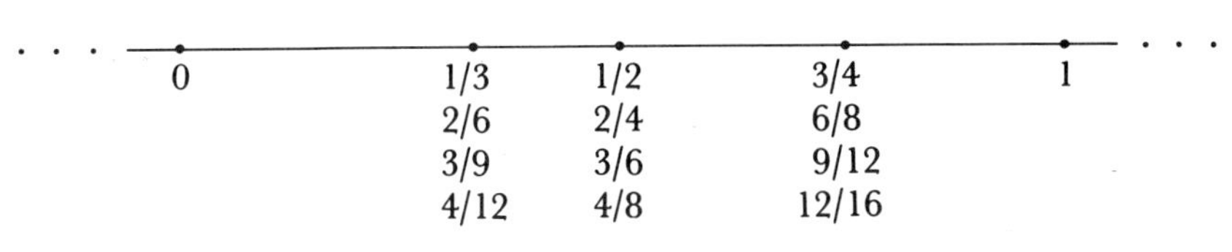

Figure 7-3

represented by one point on the number line, so we can refer to "the point corresponding to a given rational number."

Two rational numbers are equal if they are represented by fractions from the same equivalence class. Thus, if a/b represents rational number r, and c/d represents rational number s, then $r = s$ if and only if $a/b \simeq c/d$.

Example: The rational numbers represented by

$$\frac{2}{3} \quad \text{and} \quad \frac{^-4}{^-6}$$

are equal because

$$\frac{2}{3} \simeq \frac{^-4}{^-6},$$

since $2(^-6) = 3(^-4)$.

Example: Consider the fractions

$$\frac{2}{3} \quad \text{and} \quad \frac{2 \cdot 6}{3 \cdot 6}.$$

Since $2 \cdot (3 \cdot 6) = 3 \cdot (2 \cdot 6)$,

$$\frac{2}{3} \simeq \frac{2 \cdot 6}{3 \cdot 6}.$$

In a like manner,

$$\frac{2}{3} \simeq \frac{2x}{3x}, \qquad x \neq 0,$$

since $2(3x) = 3(2x)$. In general,

$$\frac{a}{b} \simeq \frac{ac}{bc}, \qquad c \neq 0,$$

since $a(bc) = b(ac)$. These results are stated in the following property of fractions.

Theorem 7-3: *Fundamental Law of Fractions.* If a, b, and c are integers, with $b \neq 0$ and $c \neq 0$, then $a/b \simeq ac/bc$.

Proof: The proof of this theorem makes use of Theorem 7-1, which states the requirement for two fractions to be equivalent. Hence this theorem is true if $a(bc) = b(ac)$. Since a, b, and c are integers, by the commutative and associative properties for the multiplication of integers, $a(bc) = b(ac)$ can be derived from $abc = abc$. The steps of the proof are left as Exercise 17 of Exercise Set 7-1.

Example:

$$\frac{2}{3} \simeq \frac{2 \cdot 5}{3 \cdot 5} = \frac{10}{15}, \quad \text{for} \quad 2(15) = 3(10).$$

Example:

$$\frac{9}{15} \quad \text{can be written as} \quad \frac{3 \cdot 3}{5 \cdot 3},$$

for $9 = 3 \cdot 3$ and $15 = 5 \cdot 3$. Thus, by Theorem 7-3, $9/15 \simeq 3/5$. This is easily verified, for $9 \cdot 5 = 15 \cdot 3$.

Example:

$$\frac{r^2}{r^5} \simeq \frac{1}{r^3}$$

because $r^2 \cdot r^3 = r^5 \cdot 1$.

Therefore, if a/b is a fraction in an equivalence class, then $(ac)/(bc)$, $c \neq 0$, is in the same class. Consequently, we can say that the rational numbers represented by a/b and ac/bc are equal because the fractions a/b and $(ac)/(bc)$ are equivalent.

Theorem 7-4: The rational numbers represented by a/b and ac/bc, $c \neq 0$, are equal.

The preceding discussion establishes the fact that if the numerator and denominator of a fraction are each multiplied by a nonzero integer, the resulting fraction names the same rational number as does the given fraction. Similarly, if there exists some nonzero integer c such that c divides the numerator and c divides the denominator, then the fraction obtained by dividing the numerator and denominator by c names the same rational number as does the original fraction.

When one sees a fraction with a large numerator and a large denominator, one should investigate to see if there is a fraction with a smaller numerator and denominator that is equivalent to the given fraction. The process of replacing a fraction by a fraction whose numerator and denominator are smaller is called *reducing the fraction.* The simplest form of a fraction is obtained when the denominator is positive and the only positive integer that will divide both numerator and denominator is 1. Notice that 2/3 is in reduced form, that 5/7 is in reduced form, and that 14/18 is not in reduced form. We may use Theorem 7-4 to reduce 14/18 to 7/9.

Definition 7-3: If a rational number is represented by a fraction a/b, $b > 0$, where the g.c.d. of the numerator and denominator is 1, then the rational number is represented by a *reduced fraction,* or a fraction in reduced form.

In this chapter, we have developed the set of rational numbers by extending or enlarging the set of integers. It was our purpose to make this extension so that the set of integers would be a subset of the set of rational numbers. Let us consider the following one-to-one correspondence between a subset of rational numbers and the integers.

$$\begin{array}{cccccccc} \ldots, & {}^-2, & {}^-1, & 0, & 1, & 2, & 3, & \ldots \\ & \updownarrow & \updownarrow & \updownarrow & \updownarrow & \updownarrow & \updownarrow & \\ \ldots, & \frac{{}^-2}{1}, & \frac{{}^-1}{1}, & \frac{0}{1}, & \frac{1}{1}, & \frac{2}{1}, & \frac{3}{1}, & \ldots \end{array}$$

This seems to be a natural way to associate integers with a subset of the rational numbers. We can strengthen the reason for choosing this particular correspondence by examining the points that represent $a/1$ and a (where a is any integer) on a number line.

$$\begin{array}{cccccccc} \ldots & \frac{{}^-1}{1} & \frac{0}{1} & \frac{1}{1} & \frac{2}{1} & \frac{3}{1} & \frac{4}{1} & \ldots \\ & \text{or} & \text{or} & \text{or} & \text{or} & \text{or} & \text{or} & \\ & {}^-1 & 0 & 1 & 2 & 3 & 4 & \end{array}$$

Figure 7-4

We often use the integer notation 4 for the rational number 4/1 or the notation ${}^-3$ for the rational number denoted by ${}^-12/4$ or ${}^-3/1$. Likewise we sometimes use the notation $^{-}(12/4)$ to indicate that the fraction is negative.

Exercise Set 7-1

1. Find the value of x that will make the following fractions equivalent.

   (a) $\frac{1}{6} \simeq \frac{x}{18}$ (b) $\frac{8}{5} \simeq \frac{^{-}16}{x}$ (c) $\frac{0}{5} \simeq \frac{x}{4}$

   (d) $\frac{x}{^{-}7} \simeq \frac{3}{21}$ (e) $\frac{6}{^{-}9} \simeq \frac{3x}{27}$ (f) $\frac{^{-}2}{5} \simeq \frac{6}{x}$

2. Write five fractions equivalent to the following.

   (a) $\frac{3}{4}$ (b) $\frac{2}{^{-}2}$ (c) $\frac{0}{8}$

3. Which of the following are equivalent fractions?

   (a) $\frac{4}{1}$ and $\frac{8}{2}$ (b) $\frac{6}{1}$ and $\frac{18}{2}$

   (c) $\frac{0}{^{-}1}$ and $\frac{0}{2}$ (d) $\frac{^{-}7}{20}$ and $\frac{21}{^{-}60}$

   (e) $\frac{3a+b}{3c}$ and $\frac{a+b}{c}$, where, a, b, and c are any integers, with $c \neq 0$

   (f) $\frac{7a}{7b+c}$ and $\frac{a}{b+c}$, where $7b+c \neq 0$ and $b+c \neq 0$

4. Write integers that correspond to the following fractions.

   (a) $\frac{^{-}15}{5}$ (b) $\frac{6}{^{-}2}$ (c) $\frac{^{-}27}{9}$

5. Find the reduced fraction for each of the following.

   (a) $\frac{162}{88}$ (b) $\frac{252}{210}$ (c) $\frac{308}{418}$ (d) $\frac{14}{261}$

   (e) $\frac{x^2y}{yz^2}$ (f) $\frac{r^2s^3t^4}{r^2s^4t}$ (g) $\frac{x^3yz}{x^2y^2z^4}$ (h) $\frac{xy+xz}{x^2y+x^2z}$

6. Find the rational number with specified numerator or denominator that equals the given rational number a/b. Note that in some problems it is impossible.

(a) $\frac{7}{3}$, $b = 21$ (b) $\frac{2}{3}$, $b = 27$ (c) $\frac{4}{9}$, $b = 14$

(d) $\frac{5}{3}$, $b = 10$ (e) $\frac{3}{4}$, $a = 30$ (f) $\frac{9}{5}$, $a = 27$

(g) $\frac{5}{9}$, $a = 8$ (h) $\frac{7}{16}$, $a = 32$ (i) $\frac{^-5}{12}$, $a = 10$

(j) $\frac{^-7}{3}$, $a = 14$ (k) $\frac{3}{^-5}$, $b = 10$ (l) $\frac{7}{^-3}$, $b = 12$

7. Use diagrams to interpret the meaning of the following fractions.

(a) $\frac{2}{5}$ (b) $\frac{7}{12}$ (c) $\frac{5}{8}$ (d) $\frac{1}{20}$

8. Write five members of the set described, where a and b are integers.

(a) $\{x \mid x = a/b \text{ and } a + b = 5\}$ (b) $\{x \mid x = a/b \text{ and } a + b < 7\}$
(c) $\{x \mid x = a/b \text{ and } a - b < 2\}$ (d) $\{x \mid x = a/b \text{ and } a \cdot b = a + b\}$
(e) $\{x \mid x = a/b \text{ and } a - b < 2,\ b < 0\}$
(f) $\{x \mid x = a/b \text{ and } a = b\}$

9. If W is the set of whole numbers, I is the set of integers, and R is the set of rational numbers, indicate which of the following are true and which are false.
(a) $W \subset R$ (b) $W \subset (R \cup I)$
(c) $W \subseteq I$ (d) $I \cap R = W$
(e) $W \cap R = W \cap I$ (f) $W \cup R = I \cup W$

10. Prove that if the numerator of some fraction in an equivalence class is 0, then the numerator of every fraction in that class is 0.

11. (a) If $a = c$, is a/d equal to c/d? Why?
(b) If $b = d$, is a/b equal to c/d? Why?
(c) If $a/b = c/d$ and $b = d$, what relationship exists between a and c?

12. (a) What do we mean when we write $0/11 = 0/6$?
(b) What do we mean when we write $4/4 = {}^-3/{}^-3$?
(c) What do we mean when we write ${}^-3/4 = 3/{}^-4$?

13. Consider the set of all fractions a/b such that $0 < a \leq b \leq 8$. Partition this set into subsets of equivalent fractions using the definition of this section. How many equivalence classes can you find?

14. Explain the difference between a rational number and a fraction.

*15. (a) Suppose $a/b \not\doteq c/d$ (not equivalent to) and $c/d \not\doteq e/f$. Is it true that $a/b \not\doteq e/f$? Explain your answer.
(b) Suppose $a/b = c/d$ and $c/d \neq e/f$. Is it then true that $a/b \neq e/f$? Explain your answer.
(c) If $K(a/b) \neq a/b$ in the case where $a/b \neq 0$, then K is not equal to what?
(d) Show that m/n and mn/n^2 belong to the same equivalence class.

*16. Use the Euclidean Algorithm to find the reduced form of each fraction below.
(a) 506/759 (b) 357/3099
(c) 2261/3059 (d) 6851/7905

*17. Complete the proof of Theorem 7-3.

*18. Define a rational number as an ordered pair (a, b), where a and b are integers. Formulate a requirement that two ordered pairs (a, b) and (c, d) represent the same rational number.

3

Addition of Rational Numbers

We say that the integers are "embedded in" or "contained as a subset of" the set of rational numbers. Consequently, the definitions that we formulate for the operations in the system of rational numbers must be consistent with the definitions of these operations on integers.

Recall that the system of integers consists of the set $\{\ldots, {}^-4, {}^-3, {}^-2, {}^-1, 0, 1, 2, 3, 4, \ldots\}$ and two operations, $+$ and $\cdot$, which satisfy several properties—closure, associative, commutative, and distributive properties, the existence of additive and multiplicative identities and additive inverses, as well as other properties developed from these. The extension of the integers to the rational numbers should be accomplished in such a way that the operations will still have these properties. Furthermore, the operations of $+$ and $\cdot$ must be so defined that when they involve integers in the system of rational numbers, the results will be the same as within the system of integers. We use these requirements to illustrate some of the reasoning for formulating our definitions for the operations on the set of rational numbers.

Let us consider addition on the number line to see if we can get ideas for defining the sum of two rational numbers. We already know that in the addition of positive integers, $3 + 4 = 7$ or, using the corresponding rational numbers, $(3/1) + (4/1) = 7/1$. In general, we should

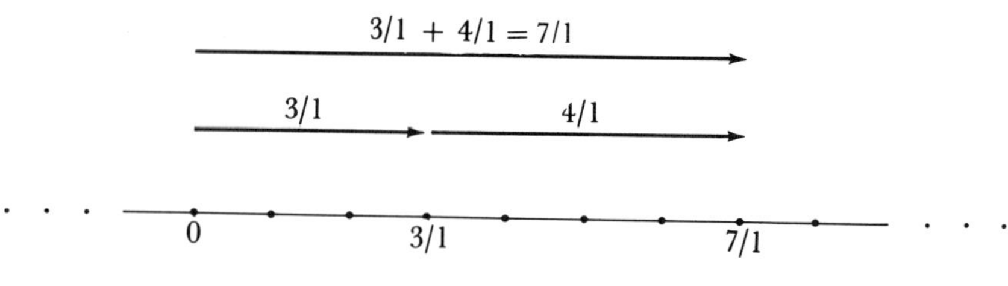

Figure 7-5

define the addition of rational numbers with denominators of 1 to be such that

$$(a/1) + (b/1) = (a + b)/1.$$

We now consider the case where both denominators are equal but not necessarily 1. On the number line, we observe the sum 3/5 and 4/5. Thus, it seems we should define $(3/5) + (4/5) = (3 + 4)/5 = 7/5$, and in general, $(a/c) + (b/c) = (a + b)/c$.

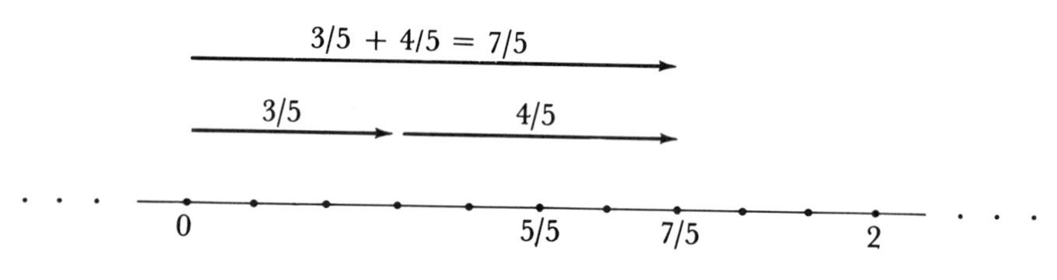

Figure 7-6

Finally, we consider the case in which the denominators are unequal. Add 2/3 and 4/5. Now by Theorem 7-4,

$$\frac{2}{3} = \frac{2 \cdot 5}{3 \cdot 5} \quad \text{and} \quad \frac{4}{5} = \frac{3 \cdot 4}{3 \cdot 5}.$$

The denominators are now the same, and we add according to the preceding discussion. Thus

$$\frac{2}{3} + \frac{4}{5} = \frac{(2 \cdot 5) + (3 \cdot 4)}{3 \cdot 5}.$$

In general, let a/b and c/d represent rational numbers. The denominators can be made identical by the following procedure.

$$\frac{a}{b}+\frac{c}{d}=\frac{ad}{bd}+\frac{bc}{bd}=\frac{ad+bc}{bd}$$

Because of these considerations, we define the operation $+$ for addition of rational numbers in the following manner.

Definition 7-4: If a/b and c/d are any two rational numbers, then $(a/b)+(c/d)=(ad+bc)/bd$.

Examples:

(a) $\frac{2}{3}+\frac{4}{7}=\frac{2(7)+3(4)}{3\cdot 7}=\frac{14+12}{21}=\frac{26}{21}$

(b) $\frac{5}{7}+\frac{^-2}{4}=\frac{5(4)+7(^-2)}{7\cdot 4}=\frac{20+{}^-14}{28}=\frac{6}{28}=\frac{3}{14}$

We now establish the properties of addition for rational numbers by appealing to our knowledge of the properties of integers.

(a) *Closure Property of Addition.* The set of rational numbers is closed under addition. That is, the sum of any two rational numbers a/b and c/d exists and is unique. Since

$$\frac{a}{b}+\frac{c}{d}=\frac{ad+bc}{bd}$$

and since $ad+bc$ is an integer and bd is an integer, the sum of two rational numbers is a rational number. Now we demonstrate that the sum is unique.

Theorem 7-5: The sum of rational numbers is unique.

Proof: Let r_1, a rational number, be represented by a/b and by c/d; and let r_2 be represented by e/f and g/h. If we consider r_1+r_2 as $a/b+e/f$ and also as $c/d+g/h$, we would have answers

$$\frac{af+be}{bf} \quad \text{and} \quad \frac{ch+dg}{dh}.$$

By Theorem 7-1,

$$\frac{af+be}{bf}=\frac{ch+dg}{dh} \quad \text{if and only if} \quad (af+be)dh=bf(ch+dg).$$

Hence we shall prove that $(af+be)dh=bf(ch+dg)$. Since $a/b \simeq c/d$ and $e/f \simeq g/h$, it follows from Theorem 7-1 that $ad=bc$ and $eh=fg$. Thus $fh(ad)=fh(bc)$ and $bd\,(eh)=bd(fg)$. Hence

$$fhad+bdeh=fhbc+bdfg \quad \text{or} \quad dh(af+be)=bf(ch+dg).$$

Hence

$$\frac{af+be}{bf}=\frac{ch+dg}{dh}.$$

Therefore, r_1+r_2 is unique.

(b) *Commutative Property of Addition.*

$$\frac{a}{b}+\frac{c}{d}=\frac{c}{d}+\frac{a}{b}.$$

(For proof, see Exercise 6, Exercise Set 7-2.)

Example:

$$\frac{2}{3}+\frac{4}{5}=\frac{2(5)+3(4)}{3\cdot 5}=\frac{22}{15}$$

$$\frac{4}{5}+\frac{2}{3}=\frac{4(3)+5(2)}{3\cdot 5}=\frac{22}{15}$$

Thus

$$\frac{2}{3}+\frac{4}{5}=\frac{4}{5}+\frac{2}{3}.$$

(c) *Associative Property of Addition.*

$$\frac{a}{b}+\left(\frac{c}{d}+\frac{e}{f}\right)=\left(\frac{a}{b}+\frac{c}{d}\right)+\frac{e}{f}.$$

(For proof, see Problem 7, Exercise Set 7-2.)

Example:

$$\frac{2}{3}+\left(\frac{1}{4}+\frac{1}{2}\right)=\frac{2}{3}+\left(\frac{2+4}{8}\right)=\frac{2}{3}+\frac{6}{8}=\frac{16+18}{24}=\frac{34}{24}$$

Similarly,

$$\left(\frac{2}{3}+\frac{1}{4}\right)+\frac{1}{2}=\left(\frac{8+3}{12}\right)+\frac{1}{2}=\frac{11}{12}+\frac{1}{2}=\frac{22+12}{24}=\frac{34}{24}.$$

Thus,

$$\left(\frac{2}{3}+\frac{1}{4}\right)+\frac{1}{2}=\frac{2}{3}+\left(\frac{1}{4}+\frac{1}{2}\right).$$

(d) *Identity Element for Addition.* There is a unique rational number 0/1 such that for any a/b,

$$\frac{a}{b}+\frac{0}{1}=\frac{0}{1}+\frac{a}{b}=\frac{a}{b}.$$

Note that

$$\frac{a}{b}+\frac{0}{d}=\frac{ad+b\cdot 0}{bd}=\frac{ad}{bd}=\frac{a}{b}.$$

Since $0/d=0/1$, 0/1 or the corresponding integer 0 is usually used to represent the additive identity.

Example:

$$\frac{2}{3}+\frac{0}{4}=\frac{2(4)+3(0)}{12}=\frac{8}{12}=\frac{2}{3}$$

(e) *Additive Inverse.* For each rational number a/b, define $^{-}(a/b)$ to be the additive inverse such that $a/b+{}^{-}(a/b)=0/1$, where 0/1 is the additive identity. By Definition 7-4,

$$\frac{a}{b}+\frac{^{-}a}{b}=\frac{(ab)+(b\cdot {}^{-}a)}{b^2}=\frac{ab+{}^{-}ab}{b^2}=\frac{0}{b^2}=\frac{0}{1}.$$

Thus, $\frac{^{-}a}{b}$ is also an additive inverse of $\frac{a}{b}$.

In a like manner

$$\frac{a}{b} + \frac{a}{{}^-b} = \frac{{}^-ab + ba}{{}^-b^2} = \frac{0}{{}^-b^2} = \frac{0}{1},$$

and $\frac{a}{{}^-b}$ is an additive inverse of $\frac{a}{b}$.

Suppose a rational number r has two additive inverses r_1 and r_2. Then $r + r_1 = 0$ and $r + r_2 = 0$, which implies $r + r_1 = r + r_2$. Add an additive inverse, ${}^-r$, to both members to obtain

$$\begin{array}{rcll} {}^-r + (r + r_1) &=& {}^-r + (r + r_2) & \text{See Theorem 7-8(a).} \\ ({}^-r + r) + r_1 &=& ({}^-r + r) + r_2 & \text{Why?} \\ 0 + r_1 &=& 0 + r_2 & \text{Why?} \\ r_1 &=& r_2 & \text{Why?} \end{array}$$

Thus, if r_1 and r_2 are any two additive inverses of the rational number r, then $r_1 = r_2$. This states that the additive inverse of a rational number is unique. Hence, we have proved the following theorem.

Theorem 7-6: The additive inverse of a rational number is unique. If the rational number is denoted by a/b, the additive inverse may be written as

$$^-(a/b) \quad \text{or} \quad {}^-a/b \quad \text{or} \quad a/{}^-b.$$

The additive inverse of 2/3 is $^-2/3$ because

$$\frac{2}{3} + \frac{^-2}{3} = \frac{6 + {}^-6}{9} = \frac{0}{9} = \frac{0}{1},$$

or the additive inverse of 2/3 is $^-(2/3)$ because

$$(2/3) + {}^-(2/3) = 0/1.$$

The procedure for adding rational numbers used in the foregoing discussion is perhaps not exactly the same procedure you are accustomed to using. You are probably familiar with a method called "finding the least common denominator." The two methods involve essentially the same amount of work when the denominators have no common divisors except 1. If the denominators have common factors, then the "least common denominator method" involves less manipulation. The following theorem leads to this process for addition.

Theorem 7-7: For rational numbers a/c and b/c $(c \neq 0)$, $(a/c) + (b/c) = (a+b)/c$.

Proof: By Definition 7-4,

$$\frac{a}{c} + \frac{b}{c} = \frac{ac + cb}{c \cdot c}.$$

By the commutative and distributive properties for integers,

$$\frac{a}{c} + \frac{b}{c} = \frac{ac + bc}{c \cdot c} = \frac{(a+b)c}{c \cdot c}.$$

Then by Theorem 7-4,

$$\frac{a}{c} + \frac{b}{c} = \frac{a+b}{c}.$$

Thus the theorem is true.

Examples:

$$\frac{2}{3} + \frac{5}{3} = \frac{7}{3} \quad \text{and} \quad \frac{5}{9} + \frac{2}{9} = \frac{7}{9}$$

Now if the denominators are not the same, the Fundamental Law of Fractions may be applied to each fraction to make the denominators identical. In order to make the manipulation as simple as possible, we look for what is called "the least common denominator." The least common denominator will be the l.c.m. of the denominators of the fractions to be added. Theorem 7-4 can be applied to each fraction to obtain a fraction with the l.c.m. as a denominator.

Examples:

$$\frac{5}{6} + \frac{7}{15} = \frac{5 \cdot 5}{6 \cdot 5} + \frac{7 \cdot 2}{15 \cdot 2} = \frac{25}{30} + \frac{14}{30} = \frac{39}{30}$$

$$\frac{4}{9} + \frac{^-5}{6} = \frac{4 \cdot 2}{9 \cdot 2} + \frac{^-5 \cdot 3}{6 \cdot 3} = \frac{8}{18} + \frac{^-15}{18} = \frac{^-7}{18}$$

$$\frac{5}{24} + \frac{11}{36} = \frac{5}{2^3 \cdot 3} + \frac{11}{2^2 \cdot 3^2} = \frac{5 \cdot 3}{2^3 \cdot 3^2} + \frac{11 \cdot 2}{2^3 \cdot 3^2} = \frac{15 + 22}{2^3 \cdot 3^2} = \frac{37}{72}$$

Sometimes we wish to indicate the addition of a fraction that corresponds to an integer and a fraction that does not correspond to an integer without performing the operation of addition. For example, consider $4/1 + 2/3$. The addition $4 + 2/3$ is often denoted by the numeral $4\frac{2}{3}$, which is called a *mixed fraction* because it involves a notation for an integer and a notation for a fraction. Thus, when you see $7\frac{3}{8}$ you will recognize it to be $7 + 3/8$, or $7/1 + 3/8 = 59/8$. Fractions like $59/8$ are often called *improper* fractions. That is, an improper fraction is one in which the integral numerator is larger in absolute value than the integral denominator. A fraction with an integral numerator and denominator where the numerator is smaller in absolute value than the denominator is called a *proper* fraction.

Examples: (a) Change $5\frac{3}{4}$ to an improper fraction.

$$5\frac{3}{4} = 5 + \frac{3}{4} = \frac{5}{1} + \frac{3}{4} = \frac{20 + 3}{1 \cdot 4} = \frac{23}{4}$$

(b) Change $\frac{17}{3}$ to a mixed fraction.

$$\frac{17}{3} = \frac{15 + 2}{3} = \frac{15}{3} + \frac{2}{3} = \frac{5}{1} + \frac{2}{3} = 5 + \frac{2}{3} = 5\frac{2}{3}$$

(c) Add $4\frac{1}{3}$ and $6\frac{1}{2}$.

$$4\frac{1}{3} + 6\frac{1}{2} = \left(4 + \frac{1}{3}\right) + \left(6 + \frac{1}{2}\right) = (4 + 6) + \left(\frac{1}{3} + \frac{1}{2}\right) = 10 + \frac{5}{6} = 10\frac{5}{6}$$

(d) Change $^{-}3\frac{5}{8}$ to an improper fraction.

$$^{-}3\frac{5}{8} = {}^{-}\left(3 + \frac{5}{8}\right) = {}^{-}3 + \frac{^{-}5}{8} = \frac{^{-}24}{8} + \frac{^{-}5}{8} = \frac{^{-}29}{8}$$

The following properties are easy to validate using the theory of this chapter.

Theorem 7-8: Let r, s, t, and u be rational numbers.

(a) If $r = s$, then $r + t = s + t$.
(b) If $r = s$ and $t = u$, then $r + t = s + u$.

Example: An equation such as $x + 3/5 = 7/5$ can be easily solved using the additive inverse.

$$\begin{aligned} x + 3/5 &= 7/5 \\ (x + 3/5) + {}^{-}3/5 &= 7/5 + {}^{-}3/5 \\ x + (3/5 + {}^{-}3/5) &= 7/5 + {}^{-}3/5 \\ x + 0 &= 4/5 \\ x &= 4/5 \end{aligned}$$

Theorem 7-9: *The Cancellation Property of Addition.* Let r, s, and t be rational numbers. If $r + t = s + t$, then $r = s$.

The proofs of Theorems 7-8 and 7-9 will be a part of the next exercise set.

Exercise Set 7-2

1. Add the following, using Definition 7-4, and reduce answers if possible. Leave all answers with positive denominators.

   (a) $\frac{3}{4} + 0$ (b) $\frac{2}{3} + \frac{7}{8}$ (c) $\frac{5}{9} + \frac{-1}{4}$

   (d) $\frac{2}{3} + \frac{7}{4} + \frac{-3}{5}$ (e) $\frac{4}{x} + \frac{5}{-2x}$ (f) $\frac{7}{xy} + \frac{4}{y} + \frac{5}{x}$

2. Find the least common denominator of the following fractions and then add the fractions.

   (a) $\frac{5}{8}, \frac{3}{2}, \frac{1}{4}$ (b) $\frac{5}{32}, \frac{5}{4}, \frac{7}{8}$

   (c) $\frac{7}{54}, \frac{4}{27}, \frac{5}{6}$ (d) $\frac{3}{4}, \frac{5}{3}, \frac{7}{5}$

   (e) $\frac{5}{11}, \frac{11}{5}, \frac{3}{7}$ (f) $\frac{3}{3}, \frac{4}{3}, \frac{7}{9}$

3. The sum of two fractions is given below. Determine the two fractions, reducing each to lowest terms. For example,

$$\frac{15 + 24}{36} = \frac{15}{36} + \frac{24}{36} = \frac{5}{12} + \frac{2}{3}.$$

   (a) $\frac{64 + 96}{512}$ (b) $\frac{4(5) + 12(9)}{9(5)}$

(c) $\dfrac{15 + 19}{4}$ (d) $\dfrac{4(4) + 3(5)}{5(4)}$

4. Perform each computation in two ways, and then show that the answers are equal.

(a) $\dfrac{5}{18} + \dfrac{7}{24}$ (b) $\dfrac{10}{28} + \dfrac{9}{77}$

(c) $\dfrac{7}{50} + \dfrac{41}{210} + \dfrac{^-6}{20}$ (d) $\dfrac{7}{30} + \dfrac{^-6}{35} + \dfrac{8}{40}$

(e) $4 + \dfrac{^-2}{3} + 7$ (f) $18 + \dfrac{^-3}{7} + \dfrac{15}{2}$

(g) $\dfrac{3}{ab} + \dfrac{5}{a}$ (h) $\dfrac{d}{a^3bc^2} + \dfrac{c}{ab^3c}$

5. Carry out the indicated operations.

(a) $5\dfrac{1}{2} + 4\dfrac{3}{4}$ (b) $\left(4\dfrac{1}{2}\right) + \left(2\dfrac{3}{4}\right)$

(c) $\left(3\dfrac{5}{9}\right) + \left(2\dfrac{2}{3}\right)$ (d) $2\dfrac{5}{9} + 4\dfrac{2}{3}$

(e) $3\dfrac{5}{12} + {}^-7\dfrac{2}{5}$ (f) $\left({}^-17\dfrac{7}{12}\right) + \left(7\dfrac{2}{5}\right)$

(g) ${}^-11\dfrac{8}{21} + {}^-3\dfrac{5}{9}$ (h) ${}^-11\dfrac{8}{21} + 3\dfrac{5}{9}$

6. List the reason or reasons in showing that $\dfrac{a}{b} + \dfrac{c}{d} = \dfrac{c}{d} + \dfrac{a}{b}$ (commutative property of addition).

(a) $\dfrac{a}{b} + \dfrac{c}{d} = \dfrac{ad + bc}{bd}$ (b) $\dfrac{a}{b} + \dfrac{c}{d} = \dfrac{da + cb}{db}$

(c) $\dfrac{a}{b} + \dfrac{c}{d} = \dfrac{cb + da}{db}$ (d) $\dfrac{a}{b} + \dfrac{c}{d} = \dfrac{c}{d} + \dfrac{a}{b}$

7. List the reason or reasons that $\dfrac{a}{b} + \left(\dfrac{c}{d} + \dfrac{e}{f}\right) = \left(\dfrac{a}{b} + \dfrac{c}{d}\right) + \dfrac{e}{f}$ (associative property of addition).

(a) $\frac{a}{b}+\left(\frac{c}{d}+\frac{e}{f}\right)=\frac{a}{b}+\frac{cf+de}{df}$

(b) $=\frac{a(df)+b(cf+de)}{b(df)}$

(c) $=\frac{a(df)+[b(cf)+b(de)]}{b(df)}$

(d) $=\frac{[a(df)+b(cf)]+b(de)}{b(df)}$

(e) $=\frac{[(ad)f+(bc)f]+(bd)e}{(bd)f}$

(f) $=\frac{(ad+bc)f+(bd)e}{(bd)f}$

(g) $=\frac{(ad+bc)f}{(bd)f}+\frac{(bd)e}{(bd)f}$

(h) $=\frac{ad+bc}{bd}+\frac{e}{f}$

(i) $=\left(\frac{a}{b}+\frac{c}{d}\right)+\frac{e}{f}$

8. Let the integer m correspond to $m/1$ and the integer n correspond to $n/1$. Show that addition of rational numbers corresponds to the addition of integers.

*9. Prove that for no positive integers a and b is it true that

$$1/a+1/b=1/(a+b).$$

*10. Can you give some examples where $\frac{a}{b}+\frac{c}{d}=\frac{a+c}{b+d}$? Now find several examples where $\frac{a}{b}+\frac{c}{d}\neq\frac{a+c}{b+d}$.

*11. (a) Prove Theorem 7-8(a).
(b) Prove Theorem 7-8(b).
(c) Prove Theorem 7-9.

*12. Using the notation of rational numbers in Exercise 18, Exercise Set 7-1, define addition by $(a, b)+(c, d)=(ad+bc, bd)$. Prove:
(a) Addition is commutative.
(b) Addition is associative.

(c) Demonstrate an additive identity.
(d) Demonstrate an additive inverse.

4

Multiplication of Rational Numbers

In the preceding section, we extended the system of integers to new numbers, called rational numbers, in such a manner that the addition of integers remained the same after the extension was made. We wish now to obtain a similar result for the multiplication of rational numbers. Once again, we assume that all the properties of multiplication for integers hold for these new numbers; under this assumption, we investigate the form of the definition for multiplication. Of course, once we have the form that we wish to use for the multiplication of rational numbers, we will then state the result as a definition.

First of all we investigate fractions representing rational numbers that correspond to integers. For example, consider $2/1 \leftrightarrow 2$, $3/1 \leftrightarrow 3$, and $6/1 \leftrightarrow 6$. Note that we have made the product involving fractions correspond to the product for integers.

$$2 \cdot 3 = 6 \quad \text{or} \quad \frac{2}{1} \cdot \frac{3}{1} = \frac{6}{1}$$

The answer we have given to $(2/1) \cdot (3/1)$ can be obtained by multiplying numerators and denominators.

Now interpret the multiplication of $(1/3) \cdot (2/5)$ to be $1/3$ of $2/5$. Geometrically, this interpretation may be accomplished by dividing a square into fifteen like regions with three columns and five rows.

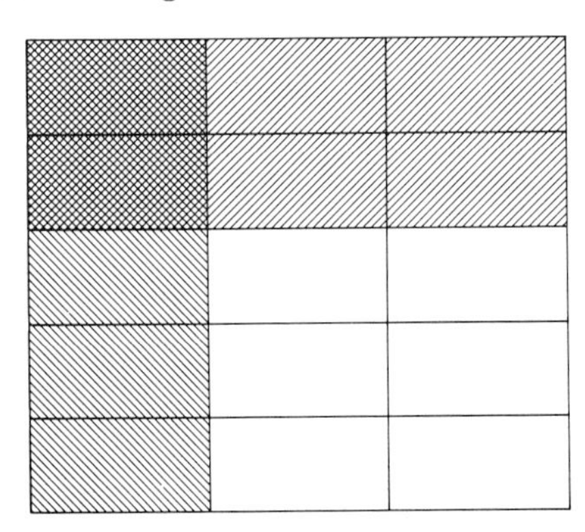

Figure 7-7

To show the result of $(1/3) \cdot (2/5)$, shade any one of the three columns and then shade any two of the five rows. By observation, we see that two parts have been shaded twice, so $(1/3) \cdot (2/5)$ can be interpreted to be $2/15$. Again, this result could have been obtained by multiplying the numerators and denominators of the two fractions.

Consider $x = a/b$ and $y = c/d$, where a, b, c, and d are integers.

$$a = bx \quad \text{and} \quad c = dy.$$

Using Theorem 5-3(a),

$$ac = (bx)(dy),$$
$$ac = (bd)(xy).$$

Therefore, $\dfrac{ac}{bd} = xy = \dfrac{a}{b} \cdot \dfrac{c}{d}$. The concepts demonstrate the reasonableness of the following definition.

Definition 7-5: If a/b and c/d are any rational numbers, then $(a/b) \cdot (c/d) = ac/bd$.

Examples:

(a) $\dfrac{4}{5} \cdot \dfrac{1}{2} = \dfrac{4}{10}$

(b) $\dfrac{2}{3} \cdot \dfrac{^-7}{11} = \dfrac{^-14}{33}$

Often we use Definition 7-5 to multiply two rational numbers when one is represented by a numeral for an integer. For example, multiply $2(1/8)$. Now 2 may be written as $(2/1)$, so the multiplication becomes $(2/1)(1/8) = (2/8)$. Note that this result is the same as multiplying the integer times the numerator of the fraction.

Example:

$$3 \cdot \frac{7}{40} = \frac{3}{1} \cdot \frac{7}{40} = \frac{21}{40}$$

We now use the properties of the multiplication of integers to develop corresponding properties of rational numbers where a/b, c/d, and e/f are any rational numbers.

(a) *Closure Property of Multiplication.* The rational numbers are closed under multiplication. That is, the product ac/bd for any rational numbers a/b and c/d must exist and be unique. From the closure property of integers, ac and bd exist as integers. Also $bd \neq 0$ since $b \neq 0$ and $d \neq 0$. Hence ac/bd exists.

Let us use the fraction 3/4 to represent a rational number and the fraction 12/16 to represent the same rational number. Do we get the same answer when we multiply by 5/8? Now

$$(5/8) \cdot (3/4) = (5 \cdot 3)/(8 \cdot 4) = 15/32 \quad \text{and} \quad (5/8) \cdot (12/16) = 60/128.$$

The fractions 15/32 and 60/128 are equivalent, since

$$15(128) = 32(60) = 1920.$$

Thus, the fractions 15/32 and 60/128 represent the same rational number. This work suggests the following theorem.

Theorem 7-10: The multiplication of rational numbers is unique.

Proof: Let r_1, a rational number, be represented by a/b and by c/d and let r_2 be represented by e/f and g/h. If we consider $r_1 \cdot r_2$ as $(a/b) \cdot (e/f)$ and also as $(c/d) \cdot (g/h)$, do we get the same answer? Now $(a/b) \cdot (e/f) = ae/bf$, and $(c/d) \cdot (g/h) = cg/dh$. Since $a/b \simeq c/d$ and $e/f \simeq g/h$, $ad = bc$ and $eh = fg$. Since ad, bc, eh, and fg are integers, $(ad)(eh) = (bc)(fg)$. This can be rearranged as $(ae)(dh) = (bf)(cg)$ by the associative and commutative properties of the multiplication of integers. Therefore, $ae/bf \simeq cg/dh$, and the two fractions represent the same rational answer.

(b) *Commutative Property of Multiplication.*

$$\frac{a}{b} \cdot \frac{c}{d} = \frac{c}{d} \cdot \frac{a}{b}$$

By definition, $(a/b) \cdot (c/d) = ac/bd$ and $(c/d) \cdot (a/b) = ca/db$. These two rational numbers are equal if and only if $(ac)(db) = (bd)(ca)$. Now $ac = ca$ and $bd = db$, by the commutative property of integers. Therefore, $(ac)(db) = (ca)(bd) = (bd)(ca)$. Why? Hence,

$$(a/b) \cdot (c/d) = (c/d) \cdot (a/b).$$

(c) *Associative Property of Multiplication.*

$$\left(\frac{a}{b}\cdot\frac{c}{d}\right)\cdot\frac{e}{f}=\frac{a}{b}\cdot\left(\frac{c}{d}\cdot\frac{e}{f}\right)$$

(d) *Distributive Property of Multiplication over Addition.*

$$\frac{a}{b}\left(\frac{c}{d}+\frac{e}{f}\right)=\frac{a}{b}\cdot\frac{c}{d}+\frac{a}{b}\cdot\frac{e}{f}$$

(e) *The Identity Element for Multiplication.* $1/1$ represents the multiplicative identity. It is easy to verify this fact since

$$\frac{c}{d}\cdot\frac{1}{1}=\frac{c\cdot 1}{d\cdot 1}=\frac{c}{d}.$$

Also, a/a, where a is any integer unequal to zero, represents the multiplicative identity, since $1/1=a/a$. In fact, the rational number that serves as the multiplicative identity may be represented by any fraction in the class of fractions $\{\ldots, {}^-3/{}^-3, {}^-2/{}^-2, {}^-1/{}^-1, 1/1, 2/2, 3/3, \ldots\}$. This rational number thus corresponds to the integer 1.

(f) *Multiplication by Zero.* The rational number that corresponds to the integer zero is represented by any fraction from the set $\{0/x\}$, where x is any integer except zero. Since $0/x=0/1$, we use $0/1$ as zero. By definition, $(a/b)\cdot(0/1)=0/b=0/1$.

(g) *Multiplicative Inverse.* If $x\neq 0$ and $y\neq 0$, then x/y is a multiplicative inverse of y/x because $(x/y)\cdot(y/x)=xy/yx$. But $xy=yx$ since x and y are integers. Thus $(x/y)\cdot(y/x)=xy/xy=1/1$, where $1/1$ is the multiplicative identity; consequently, x/y and y/x are multiplicative inverses.

It should be noted that the multiplicative inverse for x/y where $x=0$ does not exist. That is, the rational number represented by fractions whose numerators are zero does not have an inverse. $0/a$ multiplied by any rational number r/s gives $(0/a)\cdot(r/s)=0/as=0/1$. Thus when $0/a$ is multiplied by any rational number, the result is the rational number $0/1$ and not the multiplicative identity.

Theorem 7-11: If $r/s=u/v$ and $x/y=w/z$, then

$$(r/s)\cdot(x/y)=(u/v)\cdot(w/z).$$

Proof: $(r/s) \cdot (x/y) = rx/sy$ and $(u/v) \cdot (w/z) = uw/vz$ by Definition 7-5. $rx/sy = uw/vz$ if and only if $rx(vz) = sy(uw)$ by Theorem 7-1. However, $r/s = u/v$ and $x/y = w/z$ by hypothesis. Thus by Theorem 7-1, $rv = su$ and $xz = yw$. Then $rv(xz) = su(yw)$, for r, v, x, z, s, u, y, and w are integers. Then by the properties of multiplication of integers, $rv(xz) = su(yw)$ can be rearranged so that $rx(vz) = sy(uw)$. Thus, $(r/s) \cdot (x/y) = (u/v) \cdot (w/z)$.

As a special case of this theorem, we have the following when $x/y = e/f = w/z$.

Theorem 7-12: Let r/s, u/v, and e/f be rational numbers. If $r/s = u/v$, then $(r/s) \cdot (e/f) = (u/v) \cdot (e/f)$.

The preceding theorem may be used to solve equations.

Example: Solve $(2/3)x = 6$. To solve this equation, use Theorem 7-12 to multiply both sides of the equation by 3/2.

$$(3/2)[(2/3)x] = (3/2)6$$
$$x = 9$$

Theorem 7-13: *Cancellation Property of Multiplication.* If

$$\frac{a}{b} \cdot \frac{c}{d} = \frac{a}{b} \cdot \frac{e}{f}$$

where $\frac{a}{b} \neq \frac{0}{1}$, then $\frac{c}{d} = \frac{e}{f}$.

Example: $(2/3) \cdot x = (2/3) \cdot (6)$, thus $x = 6$.

Multiplication involving mixed numbers may be performed in two ways. The mixed numbers may be changed to improper fractions in order to use Definition 7-5, or the distributive property of multiplication over addition may be utilized. Both methods are illustrated in the following example.

Example:

Multiply $2\frac{3}{4} \cdot 7\frac{1}{8}$.

$$2\frac{3}{4} \cdot 7\frac{1}{8} = \frac{11}{4} \cdot \frac{57}{8} = \frac{627}{32} = 19\frac{19}{32}$$

Likewise,

$$\begin{aligned} 2\frac{3}{4} \cdot 7\frac{1}{8} &= \left(2 + \frac{3}{4}\right)\left(7 + \frac{1}{8}\right) \\ &= 2\left(7 + \frac{1}{8}\right) + \frac{3}{4}\left(7 + \frac{1}{8}\right) \\ &= 2(7) + 2\left(\frac{1}{8}\right) + \left(\frac{3}{4}\right)(7) + \left(\frac{3}{4}\right)\left(\frac{1}{8}\right) \\ &= 14 + \frac{1}{4} + \frac{21}{4} + \frac{3}{32} \\ &= 14 + \frac{1}{4} + \left(5 + \frac{1}{4}\right) + \frac{3}{32} \\ &= 19 + \left(\frac{1}{4} + \frac{1}{4} + \frac{3}{32}\right) \\ &= 19\frac{19}{32} \end{aligned}$$

As you recall from Section 2 of this chapter, the set of integers and the set of rational numbers with denominators of 1 may be put into one-to-one correspondence.

$$\begin{array}{ccccccccc} \cdots & {}^-3 & {}^-2 & {}^-1 & 0 & 1 & 2 & 3 & \cdots \\ & \updownarrow & \updownarrow & \updownarrow & \updownarrow & \updownarrow & \updownarrow & \updownarrow & \\ \cdots & \frac{{}^-3}{1} & \frac{{}^-2}{1} & \frac{{}^-1}{1} & \frac{0}{1} & \frac{1}{1} & \frac{2}{1} & \frac{3}{1} & \cdots \end{array}$$

It can be demonstrated that the sum of any two integers corresponds to the rational number sum of the corresponding rational numbers. For example,

$$^-2+3=1 \quad \text{and} \quad \frac{^-2}{1}+\frac{3}{1}=\frac{1}{1}.$$

$$1 \leftrightarrow \frac{1}{1}.$$

Likewise, it can be demonstrated that the product of any two integers corresponds to the product of the corresponding rational numbers. For example,

$$^-2 \cdot 3 = {^-6} \quad \text{and} \quad \frac{^-2}{1}\cdot\frac{3}{1}=\frac{^-6}{1}.$$

$$^-6 \leftrightarrow \frac{^-6}{1}.$$

In general,

$$a+b \leftrightarrow \frac{a}{1}+\frac{b}{1}=\frac{a+b}{1} \quad \text{and} \quad a\cdot b \leftrightarrow \frac{a}{1}\cdot\frac{b}{1}=\frac{ab}{1}.$$

Since

$$a+b \leftrightarrow \frac{a+b}{1} \quad \text{and} \quad a\cdot b \leftrightarrow \frac{a\cdot b}{1},$$

the integers are said to be *isomorphic* to a subset of the rational numbers. They may be thought of as a subsystem of the rationals or as embedded within the rationals.

Exercise Set 7-3

1. Write the multiplicative inverse of each of the following, where possible, and perform the operations necessary to verify your answers.

(a) $\frac{7}{1}$ (b) $\frac{^-8}{3}$ (c) $\frac{^-3}{^-5}$

(d) $\frac{0}{8}$ (e) $\frac{6}{^-11}$ (f) $\frac{0}{^-4}$

(g) $\frac{x}{y}$ (h) $\frac{w}{^-z}$ (i) $\frac{^-z}{w}$

2. Perform the following multiplications and write your answer as a reduced fraction.

   (a) $\frac{3}{4}\left(\frac{1}{2}\cdot\frac{2}{3}\right)$ (b) $\frac{4}{3}\left(\frac{1}{3}\cdot\frac{^-1}{2}\right)$

   (c) $\frac{4}{7}\left(\frac{5}{8}\cdot\frac{2}{3}\right)$ (d) $\left(\frac{2}{3}\cdot\frac{0}{1}\right)\left(\frac{13}{2}\right)$

3. Although the concepts used in dealing with variables are the same as the concepts used for numbers, some students have difficulty working with statements involving variables. Simplify each of the following statements, using the same concepts you use when simplifying with figures.

   (a) $\frac{a}{b}\cdot\frac{abc}{ac}$ (b) $\frac{x+y}{x}\cdot\frac{x}{y}$

   (c) $\frac{ab}{c}\left(\frac{a}{e}+\frac{b}{d}\right)$ (d) $2x\cdot(3/xy)\cdot(y/z)$

   (e) $\frac{y^2}{x}\cdot\frac{x^3}{y^4}$ (f) $\frac{r^2s^3}{t}\cdot\frac{t^2}{rs^2}$

4. Perform in two different ways the following multiplications involving mixed numbers.

   (a) $7\frac{1}{8}\cdot 6\frac{1}{4}$ (b) $^-8\frac{1}{7}\cdot 3\frac{1}{2}$

   (c) $2\frac{1}{10}\cdot {}^-8\frac{1}{3}$ (d) $16\frac{2}{3}\cdot {}^-4\frac{1}{4}$

5. Solve each of the following equations.
   (a) $3x = 1/6 + 3/4$ (b) $(2/3)x = {}^-5 + 3/4$
   (c) $(1/6)x = {}^-3 + (5/6)$ (d) $(1/3) - (1/4)x = 2$
6. Consider the multiplication of $(3/4)\cdot(7/11)$. Select fractions equivalent to 3/4 and 7/11 and show that multiplication of equivalent fractions gives the same rational number answer.
7. (a) If $r_1 \cdot r_2 = 0/a$, what can you say about r_1 or r_2?
   (b) If $r_1 \cdot r_2 = a/a$, what can you say about r_1 and r_2?
8. (a) John has 3/4 of a pound of candy. He gives his friend Joel 1/3 of the candy by weight. What fraction of a pound of the candy did he give Joel?

(b) Catherine has eaten 5/6 of a box of candy. She gives Mary 1/2 of the remainder. How much does she have left?

9. If x is the multiplicative inverse of y, what is the multiplicative inverse of x?

10. List the reason or reasons for the following steps in the verification of the distributive property of multiplication over addition.

(a) $$\frac{a}{b}\left(\frac{c}{d}+\frac{e}{f}\right)=\frac{a}{b}\left(\frac{cf+de}{df}\right)$$

(b) $$=\frac{a(cf+de)}{b(df)}$$

(c) $$=\frac{acf+ade}{bdf}$$

(d) $$\frac{a}{b}\left(\frac{c}{d}+\frac{e}{f}\right)=\frac{acf+ade}{bdf}$$

(e) $$\left(\frac{a}{b}\cdot\frac{c}{d}\right)+\left(\frac{a}{b}\cdot\frac{e}{f}\right)=\frac{ac}{bd}+\frac{ae}{bf}$$

(f) $$=\frac{acf}{bdf}+\frac{aed}{bfd}$$

(g) $$=\frac{acf}{bdf}+\frac{ade}{bdf}$$

(h) $$=\frac{acf+ade}{bdf}$$

(i) $$\left(\frac{a}{b}\cdot\frac{c}{d}\right)+\left(\frac{a}{b}\cdot\frac{e}{f}\right)=\frac{acf+ade}{bdf}$$

(j) $$\frac{a}{b}\left(\frac{c}{d}+\frac{e}{f}\right)=\left(\frac{a}{b}\cdot\frac{c}{d}\right)+\left(\frac{a}{b}\cdot\frac{e}{f}\right)$$

11. List the reason or reasons for the following steps in the verification of the associative property of multiplication.

(a) $\left(\frac{a}{b}\cdot\frac{c}{d}\right)\cdot\frac{e}{f}=\left(\frac{ac}{bd}\right)\frac{e}{f}$ and $\frac{a}{b}\cdot\left(\frac{c}{d}\cdot\frac{e}{f}\right)=\frac{a}{b}\left(\frac{ce}{df}\right)$

(b) $\left(\frac{ac}{bd}\right)\cdot\frac{e}{f}=\frac{ace}{bdf}$ and $\frac{a}{b}\cdot\left(\frac{ce}{df}\right)=\frac{ace}{bdf}$

(c) $\left(\frac{a}{b} \cdot \frac{c}{d}\right) \cdot \frac{e}{f} = \frac{a}{b} \cdot \left(\frac{c}{d} \cdot \frac{e}{f}\right)$

*12. Prove each of the following statements.

(a) $\frac{x}{y} \cdot {}^{-}\left(\frac{a}{b}\right) = {}^{-}\left(\frac{x}{y}\right) \cdot \frac{a}{b}$

(b) $\frac{^{-}x}{y} \cdot \frac{^{-}a}{b} = \frac{x}{y} \cdot \frac{a}{b}$

(c) $\frac{x}{^{-}y} \cdot \frac{a}{b} = {}^{-}\left(\frac{x}{y} \cdot \frac{a}{b}\right)$

*13. The multiplicative inverse can be considered as a relation $\odot$ such that $a \odot b$ if $a \cdot b = 1$; that is, $a \odot b$ exists if a and b are multiplicative inverses.
(a) Determine if this relation is reflexive.
(b) Determine if this relation if symmetric.
(c) Determine if this relation is transitive.
(d) Is $\odot$ an equivalence relation?

*14. Prove Theorem 7-13.

*15. If a rational number is defined as in Exercise 18, Exercise Set 7-1, define

$$(a, b) \cdot (c, d) = (ac, bd).$$

(a) Prove multiplication is commutative.
(b) Prove multiplication is associative.
(c) Develop a distributive property of multiplication over addition.
(d) Find a multiplicative inverse.
(e) Find a multiplicative identity.

5

Subtraction and Division of Rational Numbers

Before introducing a discussion of the operation of subtraction, we would like to review the results previously obtained. The identity for addition in the system of rational numbers is $0/a$ or, more simply, 0. We have also shown that any rational number x/y has an additive inverse $^-x/y$ such that $x/y + {}^-x/y = 0/y$. These ideas will be used in the discussion of subtraction for rational numbers. Subtraction will now be defined as the inverse operation of addition in a manner similar to the definition of subtraction for whole numbers and integers.

Definition 7-6: *Subtraction of Rational Numbers.* If a/d and b/e represent rational numbers, then $(a/d) - (b/e) = c/f$, a rational number, if and only if $(a/d) = (b/e) + (c/f)$.

Examples:

(a) $\frac{7}{3} - \frac{5}{3} = \frac{2}{3}$ because $\frac{7}{3} = \frac{5}{3} + \frac{2}{3}$

(b) $\frac{5}{4} - \frac{^-1}{6} = \frac{17}{12}$ because $\frac{5}{4} = \frac{^-1}{6} + \frac{17}{12}$

Now

$$\frac{^-7}{9} - \frac{4}{9} = \frac{^-11}{9} \quad \text{because} \quad \frac{^-7}{9} = \frac{4}{9} + \frac{^-11}{9} \quad \text{or} \quad \frac{^-7}{9} = \frac{^-11}{9} + \frac{4}{9}.$$

So

$$\frac{^-7}{9} + \frac{^-4}{9} = \left(\frac{^-11}{9} + \frac{4}{9}\right) + \frac{^-4}{9} = \frac{^-11}{9} + \left(\frac{4}{9} + \frac{^-4}{9}\right) = \frac{^-11}{9}.$$

Therefore,

$$\frac{^-7}{9} - \frac{4}{9} = \frac{^-7}{9} + \frac{^-4}{9}.$$

In general, if

$$\frac{a}{b} - \frac{c}{d} = \frac{e}{f},$$

then

$$\frac{a}{b} = \frac{e}{f} + \frac{c}{d} \quad \text{and} \quad \frac{a}{b} + \frac{^-c}{d} = \left(\frac{e}{f} + \frac{c}{d}\right) + \frac{^-c}{d} = \frac{e}{f} + \left(\frac{c}{d} + \frac{^-c}{d}\right) = \frac{e}{f}.$$

Thus,

$$\frac{a}{b} - \frac{c}{d} = \frac{a}{b} + \frac{^-c}{d}.$$

Theorem 7-14: If a/b and c/d are rational numbers, then

$$\frac{a}{b} - \frac{c}{d} = \frac{a}{b} + \frac{^-c}{d}.$$

This theorem actually changes subtraction problems to problems involving addition. Note that since every rational number has an additive inverse, this theorem shows that subtraction is always possible.

Examples:

(a) $\frac{7}{3} - \frac{5}{3} = \frac{7}{3} + \frac{^-5}{3} = \frac{2}{3}$

(b) $\frac{^-4}{5} - \frac{2}{3} = \frac{^-4}{5} + \frac{^-2}{3} = \frac{^-12 + {^-10}}{15} = \frac{^-22}{15}$

(c) $\frac{5}{4} - \frac{^-1}{6} = \frac{5}{4} + \frac{1}{6} = \frac{30 + 4}{24} = \frac{34}{24} = \frac{17}{12}$

Since $a/b - c/d = a/b + {^-c}/d$, we have shown that an answer for subtraction always exists for rational numbers. Now we investigate whether or not this answer is unique. Suppose one answer is e/f and a second answer is g/h. Then $a/b = c/d + e/f$ and $a/b = c/d + g/h$. Thus, $c/d + e/f = c/d + g/h$. By Theorem 7-9, $(e/f) = (g/h)$, so the answer is unique. We have shown that the rational numbers are closed relative to subtraction.

Since subtraction has been reduced to the addition of the inverse of the number to be subtracted, two additional properties of inverses for rational numbers should be considered. Both of these can be derived directly from the corresponding properties of integers.

Theorem 7-15: For rational numbers a/b and c/d,

(a) $^-\left(\frac{a}{b} + \frac{c}{d}\right) = \frac{^-a}{b} + \frac{^-c}{d}$ and (b) $^-\left(\frac{^-a}{b}\right) = \frac{a}{b}.$

Examples:

(a) $^-\left(\frac{^-4}{5}\right) = \frac{4}{5}$

(b) $^-\left(\frac{5}{7} + \frac{3}{4}\right) = \frac{^-5}{7} + \frac{^-3}{4}$

(c) $^{-}\left(\frac{^{-}2}{3}+\frac{4}{5}\right)={}^{-}\left(\frac{^{-}2}{3}\right)+\frac{^{-}4}{5}=\frac{2}{3}+\frac{^{-}4}{5}$

Theorem 7-16: *Distributive Property of Multiplication over Subtraction.* For rational numbers x, y, z,

$$x(y-z)=xy-xz.$$

Proof:

$y-z=y+{}^{-}z$	Theorem 7-14
$x(y-z)=x(y+{}^{-}z)$	Theorem 7-12
$=xy+x({}^{-}z)$	Distributive property of multiplication over addition
$=xy+{}^{-}xz$	Definition of multiplication
$=xy-xz$	Theorem 7-14

Let x and y be rational numbers. Since $x-y=x+{}^{-}y$, most books in advanced mathematics use the notation $-x$ instead of ${}^{-}x$. Thus, in advanced books you might see -2 instead of ${}^{-}2$ and $-\frac{2}{3}$ instead of ${}^{-}\left(\frac{2}{3}\right)$ or $\frac{^{-}2}{3}$. We will use this standard notation from advanced mathematics books for a negative number after we complete this chapter.

Definition 7-7: If x and y are rational numbers with $y \neq 0$, then x divided by y (denoted by $x \div y$) is equal to a rational number z if and only if $x=yz$.

This definition corresponds to the definition for division as given in whole numbers and in integers. It actually defines division as the inverse of multiplication.

Examples:

$$\frac{2}{3}\div\frac{4}{5}=\frac{5}{6} \quad \text{because} \quad \frac{4}{5}\cdot\frac{5}{6}=\frac{20}{30}=\frac{2}{3}$$

$$\frac{^{-}5}{4}\div\frac{^{-}3}{2}=\frac{5}{6} \quad \text{because} \quad \frac{^{-}3}{2}\cdot\frac{5}{6}=\frac{^{-}15}{12}=\frac{^{-}5}{4}$$

Since

$$\frac{5}{8} \div \frac{6}{3} = \frac{15}{48},$$

then

$$\frac{5}{8} = \frac{6}{3} \cdot \frac{15}{48} = \frac{15}{48} \cdot \frac{6}{3}.$$

Now

$$\frac{5}{8} \cdot \frac{3}{6} = \left(\frac{15}{48} \cdot \frac{6}{3}\right) \cdot \frac{3}{6} = \frac{15}{48}.$$

Thus,

$$\frac{5}{8} \div \frac{6}{3} = \frac{5}{8} \cdot \frac{3}{6}.$$

Generally,

$$\frac{p}{q} \div \frac{r}{s} = \frac{e}{f} \quad \text{because} \quad \frac{p}{q} = \frac{r}{s} \cdot \frac{e}{f} = \frac{e}{f} \cdot \frac{r}{s}.$$

Then

$$\frac{p}{q} \cdot \frac{s}{r} = \left(\frac{e}{f} \cdot \frac{r}{s}\right) \cdot \frac{s}{r} = \frac{e}{f} \cdot \left(\frac{r}{s} \cdot \frac{s}{r}\right) = \frac{e}{f}.$$

Thus,

$$\frac{p}{q} \div \frac{r}{s} = \frac{p}{q} \cdot \frac{s}{r}.$$

Theorem 7-17: If p/q and r/s are rational numbers, with $r/s \neq 0$, then $p/q \div r/s = (p/q)(s/r)$, where s/r is the multiplicative inverse of r/s.

Examples:

(a) $\frac{2}{3} \div \frac{4}{5} = \frac{2}{3} \cdot \frac{5}{4} = \frac{10}{12} = \frac{5}{6}$ (b) $\frac{^-4}{7} \div \frac{3}{8} = \frac{^-4}{7} \cdot \frac{8}{3} = \frac{^-32}{21}$

(c) $\frac{^-1}{4} \div \frac{^-1}{3} = \frac{^-1}{4} \cdot \frac{3}{^-1} = \frac{^-3}{^-4} = \frac{3}{4}$

Note that since every rational number except 0 has a multiplicative inverse, the preceding theorem shows that division is always possible in the system of rational numbers except when the divisor is 0.

We shall now show that the quotient for the division of rational numbers is unique and, in so doing, we shall prove that the set of rational numbers (excluding division by 0) is closed with respect to division.

Suppose $a/b \div c/d = e/f$ and $a/b \div c/d = g/h$. Then $ad/bc = e/f$ and $ad/bc = g/h$. Thus, $e/f = g/h$.

Consider the quotient

$$x \div y = \frac{x}{1} \div \frac{y}{1} = \frac{x}{1} \cdot \frac{1}{y} = \frac{x}{y}.$$

Thus, the symbol $\frac{x}{y}$ can be used to represent the division, $x \div y$. For example, $\frac{3}{4}$ can be considered as $3 \div 4$. Sometimes we expand this notation to involve the quotient of two rational numbers. $a/b \div c/d$ can be written as $\frac{a/b}{c/d}$. Thus, $\frac{a/b}{c/d} = \frac{a}{b} \cdot \frac{d}{c}$.

Example:

$$\frac{\frac{3}{4} + 2}{1 - \frac{{}^{-}1}{4}} = \frac{\frac{3}{4} + \frac{2}{1}}{\frac{1}{1} - \frac{{}^{-}1}{4}} = \frac{\frac{3+8}{4}}{\frac{4+1}{4}}$$

$$= \frac{\frac{11}{4}}{\frac{5}{4}} = \frac{11}{4} \cdot \frac{4}{5} = \frac{44}{20} = \frac{11}{5}$$

Example: On the set of rational numbers, find the solution set of $3x/2 - 1/2 = {}^{-}5$.

$$\frac{3}{2} \cdot x - \frac{1}{2} = {}^{-}5 \qquad \text{Why?}$$

$$\frac{3}{2} \cdot x + \frac{{}^{-}1}{2} = {}^{-}5 \qquad \text{Why?}$$

$$\left(\frac{3}{2} \cdot x + \frac{^-1}{2}\right) + \frac{1}{2} = {^-5} + \frac{1}{2} \quad \text{Why?}$$

$$\frac{3}{2} \cdot x + \left(\frac{^-1}{2} + \frac{1}{2}\right) = {^-5} + \frac{1}{2} \quad \text{Why?}$$

$$\frac{3}{2} \cdot x = \frac{^-9}{2} \quad \text{Why?}$$

$$\frac{2}{3}\left(\frac{3}{2} \cdot x\right) = \frac{2}{3} \cdot \frac{^-9}{2} \quad \text{Why?}$$

$$x = \frac{^-9}{3} = {^-3} \quad \text{Why?}$$

The solution set is $\{x \mid x = {^-3}\}$.

Exercise Set 7-4

1. Perform the computations indicated in the problems below.

(a) $\frac{^-7}{2} - \frac{2}{3}$

(b) $\frac{9}{7} - \frac{^-1}{3}$

(c) $\left(\frac{7}{12} - \frac{^-4}{12}\right) - \frac{2}{12}$

(d) $\left(\frac{2}{9} - \frac{^-5}{9}\right) - \frac{3}{9}$

(e) $\left(\frac{2}{5} + \frac{3}{8}\right) - \left(\frac{2}{5} + \frac{^-6}{9}\right)$

(f) $\left(\frac{5}{8} - \frac{^-4}{9}\right) + \frac{^-3}{5}$

(g) $\frac{^-41}{30} - \left(\frac{2}{5} - \frac{1}{3}\right)$

(h) $\left(\frac{^-12}{35} - \frac{2}{3}\right) - 1$

2. Perform the following subtractions, when fractions are considered as mixed numbers.

(a) $5\frac{7}{12} - \left({^-3}\frac{1}{4}\right)$

(b) ${^-2}\frac{9}{10} - 4\frac{1}{2}$

(c) ${^-7}\frac{1}{8} - \left({^-3}\frac{1}{3}\right)$

(d) ${^-8}\frac{1}{4} - \left(3\frac{1}{3}\right)$

(e) ${^-10}\frac{1}{4} - 3\frac{1}{5}$

(f) $7\frac{1}{2} - \left({^-1}\frac{1}{4}\right)$

3. Perform the following divisions, leaving your answer in simplest form.

(a) $\frac{2}{3} \div \frac{{}^{-}7}{8}$ (b) $\frac{{}^{-}5}{8} \div \frac{4}{{}^{-}5}$ (c) $\frac{2}{3} \div 4$

(d) $\frac{{}^{-}7}{8} \div 6$ (e) ${}^{-}4 \div \frac{1}{9}$ (f) ${}^{-}8 \div \frac{1}{4}$

(g) $\left(\frac{2}{3} + \frac{{}^{-}1}{5}\right) \div \frac{{}^{-}7}{10}$ (h) $\left(8 \div \frac{{}^{-}5}{11}\right) \cdot \left(\frac{2}{3} \div {}^{-}9\right)$

(i) $\left(\frac{4}{3} \div \frac{{}^{-}7}{8}\right) \cdot \left(\frac{{}^{-}1}{2} \div \frac{{}^{-}3}{4}\right)$ (j) $\left(\frac{{}^{-}3}{2} + \frac{7}{12}\right) \div \left(\frac{{}^{-}2}{5} + \frac{3}{8}\right)$

4. Express each of the following in the form a/b.

(a) ${}^{-}\left(\frac{3}{5} + \frac{7}{8}\right)$ (b) ${}^{-}\left(\frac{{}^{-}7}{5} + \frac{{}^{-}3}{7}\right)$

(c) ${}^{-}\left(\frac{6}{5} + \frac{{}^{-}7}{9}\right)$ (d) ${}^{-}\left(\frac{7}{3} - \frac{6}{5}\right)$

5. Perform the indicated computations.

(a) ${}^{-}7\frac{1}{2} \div 4\frac{1}{3}$ (b) $\frac{16}{3} \div 2\frac{1}{6}$

(c) $4\frac{7}{8} \div 1\frac{3}{4}$ (d) $6\frac{1}{4} \div \frac{4}{11}$

6. Subtract and simplify if possible.

(a) $\frac{r}{xy} - \frac{s}{yz}$ (b) $\frac{w}{x^2z} - \frac{t}{xz^2}$

(c) $\frac{2}{a^3bc^2} - \frac{3}{ab^3c}$ (d) $\frac{2}{x-y} - \frac{3}{x+y}$

7. Find a solution set of rational numbers for each of the following.

(a) $\frac{x}{3} = \frac{{}^{-}18}{54}$ (b) $2x + {}^{-}4 - 16$

(c) ${}^{-}3x - 7 = 41$ (d) $4x - 3 = 5$

(e) $\frac{x}{100} = \frac{480}{54}$ (f) ${}^{-}3x + 7 = {}^{-}5$

(g) $\frac{{}^{-}3x}{2} + 6 = \frac{1}{2}x - \frac{1}{3}$ (h) $\frac{{}^{-}8}{5} - \frac{3x}{10} = \frac{{}^{-}3}{5}x + \frac{7}{10}$

(i) $\frac{3}{4} \cdot x - \frac{2}{3} = \frac{{}^{-}1}{2}$ (j) $\frac{4}{5}x - \left(\frac{{}^{-}2}{3}\right) = \frac{{}^{-}3}{7}$

(k) $\left(3\frac{1}{2}\right)x - 4\frac{2}{3} = 5\frac{1}{4}$ (l) $\left({}^{-}4\frac{2}{3}\right)x + 7\frac{1}{2} = 1\frac{5}{8}$

8. If $x = 2$, determine the value of each of the following rational numbers.

(a) $\frac{6}{x^2 + 4}$ (b) $\frac{a - 1}{2a - x}$ (c) $\frac{a + 1}{3(x + 2)}$

(d) $\frac{6x}{3 + (x - 2)}$ (e) $\frac{2x^2 - 3x - 2}{3 - x^2}$

9. List the reasons for each step in the proof of Theorem 7-15(a).

(a) $${}^{-}\left(\frac{a}{b} + \frac{c}{d}\right) + \left(\frac{a}{b} + \frac{c}{d}\right) = \frac{0}{1}$$

(b) $${}^{-}\left(\frac{a}{b}\right) + {}^{-}\left(\frac{c}{d}\right) + \left(\frac{a}{b} + \frac{c}{d}\right) = {}^{-}\left(\frac{a}{b}\right) + \left[{}^{-}\left(\frac{c}{d}\right) + \left(\frac{a}{b} + \frac{c}{d}\right)\right]$$

(c) $$= {}^{-}\left(\frac{a}{b}\right) + \left[{}^{-}\left(\frac{c}{d}\right) + \frac{a}{b}\right] + \frac{c}{d}$$

(d) $$= {}^{-}\left(\frac{a}{b}\right) + \left[\frac{a}{b} + {}^{-}\left(\frac{c}{d}\right)\right] + \frac{c}{d}$$

(e) $$= \left[\frac{{}^{-}a}{b} + \frac{a}{b}\right] + \left[{}^{-}\left(\frac{c}{d}\right) + \frac{c}{d}\right]$$

(f) $$= \frac{0}{1}$$

(g) $${}^{-}\left(\frac{a}{b} + \frac{c}{d}\right) = {}^{-}\left(\frac{a}{b}\right) + {}^{-}\left(\frac{c}{d}\right)$$

10. (a) Make up two examples to show that the operation of subtraction of rational numbers is neither commutative nor associative.
    (b) Make up two examples to show that division is neither commutative nor associative.
    (c) Do rational numbers have a division identity? Give reasons to support your answer.

11. Express each of the following as a fraction in simplest form.

(a) $\dfrac{3/4 + 2/7}{3/4 - 1/2}$ (b) $\dfrac{2/3 - 1/2}{2/3}$ (c) $\dfrac{1/3 - 2/5}{5/6 - 3/5}$

(d) $\dfrac{1/3 + 1/4}{7/12}$ (e) $\dfrac{2/3 - 6/7}{{}^{-}13/27}$ (f) $\dfrac{1}{1/2 + 3/4}$

*12. Martha's employer told her that her new salary would be 3/4 of her salary now. If Martha's weekly pay was $96.00, what will her new salary be? What would be her salary if her employer had told her he was going to reduce her salary by 3/4?

*13. Does the distributive property of division over addition hold true? Over subtraction? Prove your answers to these questions.

*14. Prove Theorem 7-15(b).

6

Ordering the Rational Numbers

In preceding chapters, we defined the relation *less than* ($<$) for whole numbers and for integers and used it as an order relation. Now let us consider the question "Can we define an order relation on the rationals which is consistent with our definition for ordering the integers and the whole numbers?"

Such a definition would be stated in the following manner. $a/b < c/d$ if and only if there exists a positive rational number e/f such that $a/b + e/f = c/d$. Possibly the easiest approach to the definition of an appropriate order relationship between two rational numbers is to refer to the definition of equality. Recall that $p/q = r/s$ if and only if $ps = qr$. This fact suggests a way in which order can be defined for rational numbers.

Definition 7-8: If p/q and r/s are rational numbers expressed with positive denominators, then p/q is less than r/s, denoted by $p/q < r/s$, if and only if $ps < qr$. If $p/q < r/s$, then $r/s > p/q$, and we say that, in this case, r/s is greater than p/q.

The equivalence of these two definitions will be discussed in the exercise set.

Examples:

$$\frac{^-4}{5} < \frac{1}{4} \quad \text{because} \quad ^-4 \cdot 4 < 5 \cdot 1.$$

$$\frac{^-7}{3} < \frac{^-2}{5} \quad \text{because} \quad ^-7 \cdot 5 < 3 \cdot {^-2}.$$

The following properties of "less than" follow from the corresponding properties of integers, where p/q, r/s, and t/v are any rational numbers.

Theorem 7-18: *Trichotomy Property for Rationals.* If p/q and r/s are any two rational numbers with positive denominators, then exactly one of the following must be true.

(a) $\frac{p}{q} = \frac{r}{s}$, (b) $\frac{p}{q} < \frac{r}{s}$, (c) $\frac{p}{q} > \frac{r}{s}$.

Proof: $p/q = r/s$, $p/q < r/s$, $p/q > r/s$ are true if and only if $ps = qr$, $ps < qr$, $ps > qr$, respectively. But ps and qr are integers; so exactly one of the cases $ps = qr$, $ps < qr$, or $ps > qr$ must be true by the trichotomy property of integers. Thus, exactly one of the cases $p/q = r/s$, $p/q < r/s$, and $p/q > r/s$ must be true for rational numbers.

Theorem 7-19: Let p/q, r/s, t/v be rational numbers such that $p/q < r/s$. Then

(a) $\frac{p}{q} + \frac{t}{v} < \frac{r}{s} + \frac{t}{v}$,

(b) $\frac{p}{q} \cdot \frac{t}{v} < \frac{r}{s} \cdot \frac{t}{v}$ if $\frac{t}{v} > 0$,

(c) $\frac{p}{q} \cdot \frac{t}{v} > \frac{r}{s} \cdot \frac{t}{v}$ if $\frac{t}{v} < 0$.

Examples: Consider $2/3 < 7/8$; then

(a) $\frac{2}{3} + \frac{1}{4} < \frac{7}{8} + \frac{1}{4}$ or $\frac{11}{12} < \frac{36}{32}$

(b) $\frac{2}{3} \cdot \frac{1}{4} < \frac{7}{8} \cdot \frac{1}{4}$ or $\frac{2}{12} < \frac{7}{32}$

(c) $\frac{2}{3} \cdot \frac{^-1}{4} > \frac{7}{8} \cdot \frac{^-1}{4}$ or $\frac{^-2}{12} > \frac{^-7}{32}$

Theorem 7-20: *Transitive Property of Order for the Rational Numbers.* If $a/b < c/d$ and $c/d < e/f$, where $b > 0$, $d > 0$, and $f > 0$, then $a/b < e/f$.

Proof:

$b > 0$, $d > 0$, and $f > 0$	Given
$\frac{a}{b} < \frac{c}{d}$ and $\frac{c}{d} < \frac{e}{f}$	Given
$ad < bc$ and $cf < de$	Definition 7-8
$adf < bcf$ and $bcf < bde$	Theorem 5-3(b)
$adf < bde$	Transitive property of less than for integers
$(af)d < (be)d$	Associative and commutative properties of multiplication of integers
$af < be$	Theorem 7-19(b), multiplication by $1/d$
$a/b < e/f$	Definition 7-8

The preceding theorems can be used to solve simple inequalities. Find the solution set for the following examples.

Examples: (a) $3x + (^-4)/3 < 14/3$. By Theorem 7-19, add $4/3$ to both sides of this inequality to obtain $3x < 6$. Since $1/3$ is the multiplicative inverse of 3, $(1/3)(3x) < (1/3)(6)$ or $x < 2$. Thus the solution set is $\{x \mid x < 2$ and a rational number$\}$.

(b) $x/5 + 4/9 < 5/9$. Add $^-4/9$ to both sides of the inequality to obtain $x/5 < 1/9$. Then multiply by 5, thus getting an answer, $x < 5/9$. So the solution set is $\{x \mid x < 5/9$ and a rational number$\}$.

As before, the number line serves as a good visual aid for interpreting the idea of order for numbers. In the preceding chapter, we considered "less than" for integers to mean "to the left of" on the number line and "greater than" to mean "to the right of." We will now demonstrate that the same idea may be applied to rational numbers. Take any rational number such as 3/4. Plot this number on a number line. Take any number to the right of 3/4 on a number line, such as 4/5. You will notice that $3/4 < 4/5$ by definition.

Now investigate $^-3/4$ and $^-4/5$. $^-4/5 < {}^-3/4$ and $^-4/5$ is to the left of $^-3/4$ on a number line.

A very interesting property of rational numbers may be stated as follows: If a and b are any two unequal rational numbers, it is always possible to find another rational number between a and b. If a and b are rational numbers with $a < b$, then we can always find a rational number c such that $a < c < b$. This property is called *denseness*. When the statement is made that the rational numbers are *dense*, it means that between any two rational numbers one can always find another rational number. This fact indicates that there are infinitely many rational numbers between any two rational numbers on a number line, so there is no "next" rational number after a given rational number in the sense that 2 is next after 1 in the positive integers.

For example, we will seek a rational number between 3/5 and 2/3. Such a rational number is

$$\left(\frac{1}{2}\right)\left(\frac{3}{5}+\frac{2}{3}\right)=\left(\frac{1}{2}\right)\left(\frac{19}{15}\right)=\frac{19}{30}.$$

Now $3/5 < 19/30$ because $90 < 95$, and $19/30 < 2/3$ because $57 < 60$. Thus $3/5 < 19/30 < 2/3$.

This number is the *average* or *arithmetic mean* of the two numbers. Consider two rational numbers a/b and c/d with $a/b < c/d$, where $b \neq 0$ and $d \neq 0$. Now the average of these two numbers is

$$\frac{a/b + c/d}{2} = \frac{ad + bc}{2bd}.$$

Let us verify that this number is between a/b and c/d. Now since $a/b < c/d$, $ad < bc$ and $(ad)d < (bc)d$. Add to each side of this inequality bcd to get

$$(ad)d + (bc)d < 2bcd \quad \text{or} \quad (ad + bc)d < 2bcd.$$

Rearranging $2bcd$ as $(ad + bc)d < (2bd)c$, we have by Definition 7-8

$$\frac{ad + bc}{2bd} < \frac{c}{d}.$$

Similarly,

$$\frac{a}{b} < \frac{ad + bc}{2bd}.$$

Therefore, between any two rational numbers, there is a third; then between the first and third there must be a fourth, and so on. This process can be continued indefinitely. It follows that between any two rational numbers there are infinitely many rational numbers.

Exercise Set 7-5

1. Use the definition of "less than" to prove or disprove the following.

   (a) $\frac{^-2}{3} < \frac{^-3}{4}$ (b) $\frac{0}{7} < \frac{^-5}{1}$ (c) $\frac{1}{2} < \frac{7}{3}$

   (d) $\frac{^-7}{3} < \frac{^-5}{3}$ (e) $\frac{a}{b} < \frac{a}{3b}$ (f) $\frac{x}{y} < \frac{0}{y}$

2. Arrange the following sets of rational numbers in increasing order, such as $^-1/2 < 2 < 7/2$.

   (a) $\frac{71}{100}, \frac{^-3}{2}, \frac{23}{30}$ (b) $\frac{^-9}{2}, {}^-4, \frac{^-165}{41}$

   (c) $\frac{2}{3}, \frac{11}{18}, \frac{16}{27}, \frac{67}{100}$ (d) $\frac{25}{28}, \frac{^-27}{20}, \frac{^-14}{16}, \frac{79}{90}$

   (e) $\frac{22}{7}, \frac{7}{2}, \frac{10}{3}, \frac{156}{50}$ (f) $\frac{^-51}{95}, \frac{^-19}{36}, \frac{^-17}{30}, \frac{^-14}{29}$

3. Find all rational numbers x in the solution set.

   (a) $x + \frac{^-1}{2} < \frac{5}{2}$ (b) $3x + \frac{3}{10} < \frac{^-9}{10}$

   (c) $2x + \frac{^-5}{6} < \frac{^-7}{4}$ (d) $2x + 5 < \frac{^-3}{10}$

   (e) $3x + {}^-5 < 7$ (f) $\frac{x}{4} + {}^-1 < {}^-7$

   (g) $\frac{x}{3} + ({}^-2) < {}^-5$ (h) $4x + ({}^-3) < {}^-7$

4. Insert three rational numbers between each pair of fractions, using the average of two numbers.

(a) $\frac{5}{13}$ and $\frac{7}{18}$ (b) $\frac{6}{11}$ and $\frac{5}{7}$

(c) $\frac{^-4}{3}$ and $\frac{2}{^-5}$ (d) $\frac{^-4}{7}$ and $\frac{3}{^-5}$

5. Is it true that if a/b and c/d are rational numbers and $a/b < c/d$, then $ad < bc$? Is it true that if $ad < bc$, then $a/b < c/d$? Give an example to illustrate each answer. (Use your work here to help you better understand Definition 7-8.)
6. Given the set $A = \{x \mid x \text{ is a rational number such that } ^-7 \leq x < 3\}$.
   (a) List the set of positive integers in A.
   (b) List at least ten members of the set of negative rational numbers defined in A.
   (c) How many negative integers are members of set A?
7. Given the first relationship in each item below, is the conclusion necessarily true?

   (a) If $\frac{6}{7} + \frac{1}{3} < \frac{29}{23}$, then $\frac{6}{7} < \frac{64}{69}$.

   (b) If $\frac{11}{12} > \frac{2}{5}$, then $\frac{^-11}{12} > \frac{^-2}{5}$.

   (c) If $\frac{^-63}{87} < \frac{1}{3}$ and $\frac{1}{3} \leq \frac{23}{69}$, then $\frac{^-63}{87} < \frac{23}{69}$.

   (d) If $\frac{a}{b} < 0$ and $\frac{c}{d} > 0$, then $\frac{a}{b} \cdot \frac{c}{d} > 0$.

   (e) If $\frac{a}{b} < \frac{c}{d}$ and $x < 0$, then $\left(\frac{a}{b}\right) x < \left(\frac{c}{d}\right) x$.

8. (a) How many integers are between $^-2$ and 3?
   (b) How many rational numbers are between $^-2$ and 3? Use set notation to identify them. Name three of them.

9. (a) What is the smallest positive integer?
   (b) Is there a smallest positive rational number?
   (c) What is the largest negative integer?
   (d) Is there a largest negative rational number?
10. If x and y are positive integers such that $x > y$, which one of the following is true: $1/x > 1/y$, $1/x = 1/y$, $1/x < 1/y$? Why?
11. Repeat Exercise 10 with x and y negative integers.

*12. Define $\frac{a}{b} < \frac{c}{d}$, if there exists an $\frac{e}{f} > 0$ such that $\frac{a}{b} + \frac{e}{f} = \frac{c}{d}$. List the reason or reasons for each step in proving that $\frac{a}{b} < \frac{c}{d}$ is equivalent to $ad < bc$ where $b > 0$, $d > 0$, and $f > 0$.

(a) $\frac{a}{b} + \frac{e}{f} = \frac{c}{d}$ (b) $\frac{af + be}{bf} = \frac{c}{d}$

(c) $(af + be)d = (bf)c$ (d) $(af)d + (be)d = (bf)c$

(e) $(be)d > 0$ (f) $(af)d < (bf)c$

(g) $f(ad) < f(bc)$ (h) $ad < bc$

*13. (a) Prove Theorem 7-19(a).
(b) Prove Theorem 7-19(b).
(c) Prove Theorem 7-19(c).

*14. Is it true that if a/b and c/d are distinct rational numbers, then $(a + c)/(b + d)$ is between them? Explain your answer.

7

Summary

In this chapter, we developed the set of rational numbers, defined the operations of addition, subtraction, multiplication, and division for any two rational numbers (except for division by 0), defined an order relation, and developed properties relative to operations on rational numbers. Some of the results of this chapter are summarized as follows.

The system of rational numbers consists of the set $R = \{x \mid x$ is represented by any fraction in an equivalence class of fractions$\}$, an equals relation $a/b = c/d$ if and only if $ad = bc$, an order relation so that for any two rational numbers a/b and c/d, either $a/b < c/d$ or $a/b = c/d$ or $a/b > c/d$, and operations of addition and multiplication defined by $a/b + c/d = (ad + bc)/bd$ and $(a/b)(c/d) = (a \cdot c)/(b \cdot d)$. The operations of subtraction and division were defined as inverse operations of addition and multiplication. It was demonstrated that the operations on rational numbers are extensions of similar properties for operations on integers.

To complete our discussion of rational numbers, we make the final statement that the rational numbers form a *mathematical field.* A *field, F,* has two binary operations $(+, \cdot)$ and satisfies eleven properties. Any set of elements or any system of numbers satisfying these requirements is called a *number field,* and the properties are called *field properties.* These eleven properties are listed on the left of the page and the characteristics of the rational numbers are listed on the right. These properties hold for all rational numbers.

1. $\frac{a}{d} + \frac{b}{e}$ is uniquely defined.	Rational numbers are closed with respect to addition.
2. $\frac{a}{d} \cdot \frac{b}{e}$ is uniquely defined.	Rational numbers are closed with respect to multiplication.
3. $\frac{a}{d} + \frac{b}{e} = \frac{b}{e} + \frac{a}{d}$.	Addition of rational numbers is commutative.
4. $\frac{a}{d} + \left(\frac{b}{e} + \frac{c}{f}\right) = \left(\frac{a}{d} + \frac{b}{e}\right) + \frac{c}{f}$.	Addition of rational numbers is associative.
5. $\frac{a}{d} \cdot \frac{b}{e} = \frac{b}{e} \cdot \frac{a}{d}$.	Multiplication of rational numbers is commutative.
6. $\frac{a}{d}\left(\frac{b}{e} \cdot \frac{c}{f}\right) = \left(\frac{a}{d} \cdot \frac{b}{e}\right)\frac{c}{f}$.	Multiplication of rational numbers is associative.
7. $\frac{a}{d}\left(\frac{b}{e} + \frac{c}{f}\right) = \left(\frac{a}{d} \cdot \frac{b}{e}\right) + \left(\frac{a}{d} \cdot \frac{c}{f}\right)$.	Multiplication of rational numbers is distributive over addition.
8. For any $\frac{a}{b}$ there is an element c such that $\frac{a}{b} + c = \frac{a}{b}$.	$c = \frac{0}{1}$ is the element.
9. For any $\frac{a}{b}$ there is an element c such that $\frac{a}{b} \cdot c = \frac{a}{b}$.	$c = \frac{1}{1}$ is the element.
10. For any $\frac{a}{b}$ there is an element c such that $\frac{a}{b} + c = \frac{0}{b}$.	$c = \frac{^{-}a}{b}$ is the element.
11. For any $\frac{a}{b}$ there is an element c such that $\frac{a}{b} \cdot c = \frac{1}{1}$.	$c = \frac{b}{a}$ is the element if $a \neq 0$.

A comparison of the operations on whole numbers, on integers, and on rational numbers is given in the following table.

Property	*Operation*	*Whole Numbers*	*Integers*	*Rational Numbers*
Closure	Addition	yes	yes	yes
Closure	Multiplication	yes	yes	yes
Closure	Subtraction	no	yes	yes
Closure	Division	no	no	yes
Commutativity	Addition	yes	yes	yes
Commutativity	Multiplication	yes	yes	yes
Commutativity	Subtraction	no	no	no
Commutativity	Division	no	no	no
Associativity	Addition	yes	yes	yes
Associativity	Multiplication	yes	yes	yes
Associativity	Subtraction	no	no	no
Associativity	Division	no	no	no
Distributivity	Over addition	yes	yes	yes
Distributivity	Over subtraction	yes	yes	yes
Identity	Addition	yes	yes	yes
Identity	Multiplication	yes	yes	yes
Inverse	Addition	no	yes	yes
Inverse	Multiplication	no	no	yes

Table 7-1

Review Exercise Set 7-6

1. What is the inverse of

   (a) $\frac{2}{3}$ under multiplication? (b) $\frac{^-1}{3}$ under addition?

   (c) $\frac{^-4}{3}$ under multiplication? (d) 0 under addition?

2. Add, multiply, subtract, and divide the following pairs of fractions.

   (a) $\frac{2}{^-7}$ and $\frac{^-2}{5}$ (b) $\frac{^-11}{9}$ and $\frac{4}{5}$

(c) $\frac{0}{10}$ and $\frac{^-3}{4}$ (d) $\frac{4}{3}$ and $\frac{^-5}{6}$

3. Replace x with an integer such that each statement is true.

(a) $\frac{0}{3}+\frac{2}{5}=\frac{2}{x}$ (b) $\frac{1}{8}+\frac{^-3}{5}=\frac{^-19}{x}$

(c) $\frac{3}{5}+\frac{x}{4}=\frac{7}{20}$ (d) $\frac{x}{17}-\frac{2}{3}=\frac{241}{51}$

(e) $\frac{4}{3}+\frac{^-x}{4}=\frac{13}{12}$ (f) $\frac{^-5}{8}+\frac{^-7}{3}=\frac{^-71}{x}$

4. Discuss the truth or falsity of each of the following.

(a) $\frac{a}{0}$ is a rational number.

(b) If a and b are rational numbers with $ab=0$ and $b=0$, then $a=0$.

(c) Division is always possible in the rational number system.

(d) $\frac{2}{3}$ is the largest rational number less than 3.

(e) 1 is the smallest positive integer as well as the smallest positive rational number.

(f) 8 is the largest integer less than 9.

(g) The set of rational numbers has more elements than the set of integers.

(h) Zero is a positive rational.

(i) The set of rational numbers between 2 and 3 is dense.

(j) If a is any rational number, $a^2>0$.

(k) $^-1$ is the largest negative rational number.

5. Perform the following operations.

(a) $\frac{^-72}{5}-\left(\frac{23}{16}-\frac{13}{12}\right)+\left(\frac{^-9}{8}-\frac{3}{20}\right)$

(b) $\left(\frac{^-7}{10}+\frac{3}{25}\right)\cdot\left(\frac{24}{5}-\frac{22}{15}\right)$

(c) $\left[\left(\frac{3}{8}\cdot\frac{4}{27}\right)+\frac{1}{3}\right]\cdot\left(\frac{17}{2}-4\right)$

(d) $\left(\frac{17}{10} - \frac{1}{2}\right) - \left(\frac{{}^{-}8}{9} - \frac{2}{5}\right)$

6. Simplify the following fractions.

(a) $\dfrac{1/8}{2/3}$ (b) $\dfrac{{}^{-}7/8}{{}^{-}1/4}$

(c) $\dfrac{3\frac{1}{5}}{{}^{-}2/10}$ (d) $\dfrac{12\frac{1}{2}}{{}^{-}100/4}$

(e) $\dfrac{a/b}{c/d}$ (f) $\dfrac{[x(a-b)]/c + y/d}{x/cd}$

7. For each of the following pairs of fractions (call the first a and the second b), find $a + b$, $a - b$, $a \cdot b$, and a/b.
(a) 1/16, 3/4 (b) $1/2x$, $3/x$ (c) $1/2x$, $x/3$
(d) 63/107, 3/13 (e) $13/3x$, 1/6 (f) 33/35, $2/(2x^2 + 2)$
8. Name at least one property satisfying each of the following qualifications.
(a) Satisfied on the rationals but not on the integers
(b) Satisfied on the rationals but not on the counting numbers
(c) Satisfied on the integers but not on the whole numbers
(d) Satisfied on the rationals but not on the whole numbers
(e) Satisfied on the counting numbers, whole numbers, integers, and rational numbers

*9. Indicate which of the following sets are fields. If the set is not a field, tell why it fails to meet all of the qualifications.
(a) Integers
(b) {0, 1} such that

+	0	1
0	0	1
1	1	0

and

·	0	1
0	0	0
1	0	1

(c) Positive rationals
(d) Counting numbers greater than 1
(e) Prime numbers

*10. Prove that if $c > 1$, then $1/c < 1$, where c is any rational number.

Suggested Reading

Fractions and Rational Numbers: Allendoerfer, pp. 467–470. Armstrong, pp. 215–219. Bouwsma, Corle, Clemson, pp. 221–230. Byrne, pp. 154–160. Campbell, pp. 127–129, 143–146. Garner, pp. 187–189, 196–197. Graham, pp. 213–214, 218–221, 234–235. Meserve, Sobel, pp. 152–157. Nichols, Swain, pp. 197–198. Ohmer, Aucoin, pp. 166–170. Ohmer, Aucoin, Cortez, pp. 215–220. Peterson, Hashisaki, pp. 170–175. Podraza, Blevins, Hanson, Prall, pp. 127–131. Scandura, pp. 232–241. Smith, pp. 167–171. Spector, pp. 218–223. Willerding, p. 241. Wren, pp. 104–109. Zwier, Nyhoff, pp. 225–230.

Addition and Multiplication of Rational Numbers: Allendoerfer, pp. 473–475. Armstrong, pp. 221–226. Bouwsma, Corle, Clemson, pp. 193–194, 250–252, 255–256. Byrne, pp. 171–175. Campbell, pp. 130–133. Garner, pp. 213–225. Graham, pp. 222–228. Hutton, pp. 235–237. Nichols, Swain, pp. 209–212. Ohmer, Aucoin, Cortez, pp. 226–230. Peterson, Hashisaki, pp. 177–179, 185–188. Podraza, Blevins, Hanson, Prall, pp. 131–134, 135–136. Scandura, pp. 247–251. Smith, pp. 180–186. Spector, pp. 226–231. Willerding, pp. 242–245. Wren, pp. 110–111. Zwier, Nyhoff, pp. 231–236.

Subtraction and Division of Rational Numbers: Allendoerfer, pp. 473, 478. Armstrong, pp. 226–228, 231. Bouwsma, Corle, Clemson, pp. 193–194, 199–202. Byrne, pp. 176–177, 182–184. Campbell, pp. 134–138. Garner, pp. 226–229, 241–247. Graham, pp. 231–233, 241–245. Hutton, pp. 230–231, 238–239. Nichols, Swain, pp. 202–204. Ohmer, Aucoin, Cortez, pp. 232–235, 241–246. Peterson, Hashisaki, p. 194. Podraza, Blevins, Hanson, Prall, pp. 133–136, 140–141. Scandura, pp. 252–255, 261–263. Smith, pp. 188–192, 212–220. Spector, 230. Willerding, p. 129. Wren, pp. 111–114.

Ordering: Armstrong, pp. 220, 234–241. Bouwsma, Corle, Clemson, pp. 262–263, 271–273. Byrne, pp. 166–170. Campbell, pp. 147–149, 163–167. Garner, p. 212. Garstens, Jackson, pp. 274–277. Graham, pp. 247–250. Hutton, pp. 243–246. Nichols, Swain, pp. 250–253. Ohmer, Aucoin, pp. 197–200. Ohmer, Aucoin, Cortez, pp. 248–253. Smith, pp. 173–175. Spector, pp. 239–244. Willerding, pp. 246–248. Wren, pp. 127–128.

8

Decimals and Real Numbers

1

Introduction

The extension from whole numbers to integers and then to rational numbers was carried out in the preceding chapters, and this development resulted in a number system in which addition, subtraction, multiplication, and division were possible, with the exception of division by 0. With respect to the four basic operations of arithmetic, the system of rational numbers is a finished product. It might appear that the system of rational numbers is adequate for all our needs, for it seems to serve adequately for counting, measurement, and other common applications. But certain weak points in the rational number system should be noted.

For example, it is easy to see in the rational number system that certain numbers have no square roots or cube roots. Another way of stating this inadequacy is to say that certain equations have no solutions in the system of rational numbers. Many *quadratic equations*, $ax^2 + bx + c = 0$, with $a \neq 0$, do not have solutions in this system of numbers. For instance, $x^2 - 2 = 0$ has no solution in the system of rationals, because there is no rational number which when multiplied by itself equals 2.

Thus, the need to extend the rational numbers to a new set of numbers (called the *real numbers*) is evident. In seeking an appropriate extension, we have selected decimal representation as the best means of developing the new concepts. The decimal representation of rational numbers will be discussed next, and then an extension will be made to real numbers.

2

Decimals

The extension of the positional numeration notation that we have used for writing integers, so that fractions can be written in a similar form, is actually a modern idea when compared with other mathematical concepts. A Dutchman named Simon Stevin published the first work on decimals in the sixteenth century. Stevin, a quartermaster in the Dutch army, recognized the serious need for developing a technique to shorten calculations with fractions.

The extension of fractions to decimal fractions was accomplished by placing a dot, called a *decimal point,* after the units digit and letting the digits to the right of the dot denote in turn so many tenths, so many hundredths, so many thousandths, and so forth.

Examples:

$$\frac{3}{10} = 0.3, \qquad \frac{19}{10} = 1.9, \qquad \frac{165}{10} = 16.5.$$

Note in the preceding examples that if the denominator is 10, there is one digit to the right of the decimal point. If the denominator is 100, we note in the next example that there are two digits to the right of the decimal point. If the denominator is 1000, there are three, and so on.

Examples:

$$\frac{31}{100} = 0.31, \qquad \frac{191}{100} = 1.91, \qquad \frac{1654}{100} = 16.54.$$

Example:

$$\frac{426}{1000} = 0.426.$$

This can also be written as

$$\frac{426}{1000} = \frac{4(10)^2 + 2(10) + 6}{10^3} = 4\left(\frac{1}{10}\right) + 2\left(\frac{1}{10}\right)^2 + 6\left(\frac{1}{10}\right)^3,$$

so

$$0.426 = 4\left(\frac{1}{10}\right) + 2\left(\frac{1}{10}\right)^2 + 6\left(\frac{1}{10}\right)^3.$$

If we denote 1/10 by 10^{-1}, $(1/10)^2$ by 10^{-2}, and $(1/10)^n$ by 10^{-n}, then

$$0.426 = 4(10)^{-1} + 2(10)^{-2} + 6(10)^{-3}.$$

Similarly,

$$6532.417 = 6 \cdot 10^3 + 5 \cdot 10^2 + 3 \cdot 10 + 2 \cdot 10^0 + 4 \cdot 10^{-1} + 1 \cdot 10^{-2} + 7 \cdot 10^{-3}.$$

Numbers expressed in the form

$$N = a_n \cdot 10^n + \cdots + a_1 \cdot 10 + a_0 \cdot 10^0 + b_1 \cdot 10^{-1} + b_2 \cdot 10^{-2} + \cdots + b_m \cdot 10^{-m} = a_n \cdots a_1 a_0 . b_1 b_2 \cdots b_m$$

are called *terminating decimal fractions*. Decimals that have only a finite number of digits to the right of the decimal point represent rational numbers although some infinite decimal expansions also represent rational numbers, as we shall see in Section 5.

It is good practice to write 0.57 instead of .57 since the 0 emphasizes the decimal point. If desired, additional zeros may be added to the right of a decimal expansion without changing the value. Thus, 2.3400 represents the same rational number as 2.34.

Examples:

(a) $$\frac{16{,}845}{1000} = 16.845 = 1(10)^1 + 6(10)^0 + 8(10)^{-1} + 4(10)^{-2} + 5(10)^{-3}$$

(b) The decimal fraction 3.64 represents

$3 + 6(10)^{-1} + 4(10)^{-2}$ or

$$3 + \frac{6}{10} + \frac{4}{(10)^2} = \frac{300 + 60 + 4}{100} = \frac{364}{100}.$$

In the following examples, logical steps are performed to change a fraction to a decimal fraction. At the same time, the steps of an algorithm are listed. This algorithm converts fractions to decimal fractions by dividing the denominator into the numerator, with zeros annexed following the decimal point so that division is exact.

Example:

$$\frac{3}{5} = \frac{3(10)}{5(10)}$$

$$= \frac{30}{5} \cdot \frac{1}{10}$$

```
   0.6
5 |3.0
   3.0
   ---
```

$$= 6 \cdot \frac{1}{10}$$

$$= 0.6$$

In this algorithm, since the divisor is a whole number, the dividend and quotient have the same number of decimal places. By placing the decimal point of the quotient directly above the decimal point of the dividend, the correct position of the decimal point of the quotient is automatically obtained.

Example:

$$\frac{5}{8} = \frac{5(1000)}{8(1000)}$$

$$= \frac{5000}{8} \cdot \frac{1}{1000}$$

$$= 625 \cdot \frac{1}{1000}$$

$$= 0.625$$

```
   0.625
8 |5.000
   48
   --
    20
    16
    --
     40
     40
     --
```

Example: Find a decimal representation of 13/40.

```
     0.325
40 |13.000
    12 0
    ----
     1 00
       80
     ----
      200
      200
      ---
```

Since the remainder is now 0, we can assert that $13/40 = 0.325$.

To justify this procedure we write

$$\frac{13}{40} = \frac{13}{40} \cdot \frac{1000}{1000}$$

$$= \frac{13{,}000}{40} \cdot \frac{1}{1000}$$

$$= 325 \cdot \frac{1}{1000}$$

$$= 0.325.$$

When can one change a rational number to a terminating decimal fraction? This question is answered by either of the following theorems.

Theorem 8-1: A positive rational number, N, can be written as a terminating decimal fraction if and only if there is a whole number, r, such that $10^r \cdot N$ is a whole number.

Theorem 8-2: A rational number a/b (in lowest terms) can be expressed as a terminating decimal fraction if and only if the prime factorization of b is of the form $b = 2^x 5^y$, where x and y are whole numbers.

Example: $2/15$ cannot be written as a terminating decimal fraction because $15 = 3 \cdot 5$; that is, the prime factorization of 15 contains a 3.

Example: $13/40$ can be written as a terminating decimal fraction because $40 = 2^3 \cdot 5$.

In Chapter 4 we studied how whole numbers written in other bases could be written in base ten. We now consider how to denote in other bases a fraction similar to a decimal fraction and how this fraction can be converted to a base ten expression.

Examples:

$$(10.1101)_{\text{two}}$$
$$= [1(10) + 0 + 1(10)^{-1} + 1(10)^{-2} + 0(10)^{-3} + 1(10)^{-4}]_{\text{two}}$$

$$= [1(2) + 0 + 1(2)^{-1} + 1(2)^{-2} + 0(2)^{-3} + 1(2)^{-4}]_{ten}$$
$$= [1(2) + 0 + 1(1/2) + 1(1/2)^{2} + 0(1/2)^{3} + 1(1/2)^{4}]_{ten}$$

$(TE.74)_{twelve}$

$$= [T(10) + E + 7(10)^{-1} + 4(10)^{-2}]_{twelve}$$
$$= [10(12) + 11 + 7(12)^{-1} + 4(12)^{-2}]_{ten}$$
$$= [10(12) + 11 + 7(1/12) + 4(1/12)^{2}]_{ten}$$

$(654.513)_{seven}$

$$= [6(10)^{2} + 5(10) + 4 + 5(10)^{-1} + 1(10)^{-2} + 3(10)^{-3}]_{seven}$$
$$= [6(7)^{2} + 5(7) + 4 + 5(7)^{-1} + 1(7)^{-2} + 3(7)^{-3}]_{ten}$$
$$= [6(7)^{2} + 5(7) + 4 + 5(1/7) + 1(1/7)^{2} + 3(1/7)^{3}]_{ten}$$

Exercise Set 8-1

1. Express the following decimal fractions as sums of powers of ten.
   (a) 6.71 (b) 14.001 (c) 86.6
   (d) 0.0005 (e) 0.001 (f) 6.01
2. Express each of the following decimal fractions as fractions.
   (a) 0.085 (b) 3.25 (c) 12.74
   (d) 0.84 (e) 2.75 (f) 0.00025
3. Express the following fractions as decimal fractions.
   (a) $6\frac{9}{100}$ (b) $\frac{316}{10}$ (c) $\frac{41}{1000}$
   (d) $\frac{602}{10{,}000{,}000}$ (e) $\frac{18}{15}$ (f) $\frac{63}{70}$
   (g) $\frac{651}{120}$ (h) $\frac{91}{140}$ (i) $\frac{126}{105}$
   (j) $\frac{5}{16}$ (k) $\frac{3}{8}$ (l) $18\frac{2}{5}$
4. Find the decimal representation of each fraction.
   (a) $\frac{1}{8}$ (b) $\frac{11}{40}$ (c) $\frac{18}{80}$
   (d) $\frac{3}{16}$ (e) $\frac{33}{64}$ (f) $\frac{13}{8}$
5. Which of the following can be expressed as terminating decimal fractions?

(a) 36/15 (b) 21/70 (c) 48/36
(d) 27/60 (e) 10/30 (f) 21/36

6. Arrange the following rational numbers in increasing order.
   (a) 0.365, 0.037, 0.307, 0.370, 0.360
   (b) 0.5086, 0.586, 0.508, 0.5065, 0.5

*7. Find the correct base for each of the following.
   (a) $3.4_b = 3.5_{ten}$ (b) $3.3_b = 3.75_{ten}$
   (c) $3.4_b = 3.8_{ten}$ (d) $3.3_b = 3.25_{ten}$

*8. Express each of the following as a base ten fraction.
   (a) $(0.7)_{twelve}$ (b) $(10.11)_{two}$ (c) $(12.012)_{three}$
   (d) $(3.T)_{twelve}$ (e) $(1T.0T)_{twelve}$ (f) $(0.664)_{seven}$
   (g) $(5.E)_{twelve}$ (h) $(T.EE)_{twelve}$ (i) $(2.13)_{four}$

*9. (a) Prove Theorem 8-1.
   (b) Prove Theorem 8-2.

3

Arithmetic of Decimals

We have already shown that a subset of the rational numbers has a decimal representation. Later in this chapter, we shall show that all rational numbers have decimal representations. Consequently, it is important that we learn the computation techniques for decimals.

Adding, subtracting, multiplying, and dividing rational numbers can be extended easily to decimal fractions. In this extension, one must keep in mind the meaning of the digits in the numeration scheme. The commutative, associative, and distributive properties enable one to perform arithmetic with decimals by using algorithms similar to those given in Chapter 4.

Example: Suppose we wish to add $0.24 + 0.37$.

$$0.24 + 0.37 = \frac{24}{100} + \frac{37}{100} = \frac{61}{100} = 0.61$$

Note in this example that if we ignore the decimal points and add 24 and 37, the answer is 61, which has the same digits as 0.61. This result suggests that there may be an algorithm that instructs one to add the decimals considered as whole numbers and then to insert the decimal point.

Example: To add $0.253 + 0.14$, express each fraction with a denominator of 1000.

$$\frac{253}{1000} + \frac{140}{1000} = \frac{393}{1000} = 0.393$$

You will notice that, in the first example, we could have performed the operation by adding the decimals as if they were whole numbers and obtained the answer by inserting the decimal point in the appropriate place. In the second example, we should affix a 0 on the end of the 0.14 to get 0.140. Then we can add the two numbers as whole numbers and insert the decimal in the answer. This process for adding decimals is given by the following algorithm.

Addition and Subtraction of Decimals: To add or subtract $a_r a_{r-1} \cdots a_1 . b_1 b_2 \cdots b_m$ and $c_s \cdots c_1 . d_1 d_2 \cdots d_n$, where a_i, b_i, c_i, and d_i are digits and $m < n$, insert $n - m$ zeros to the right of b_m and add or subtract as if one were operating with whole numbers. Insert a decimal point n positions from the right in your answer.

We shall now demonstrate the validity of this algorithm, using fractions.

Example: Add $3.452 + 5.73 + 24.2$.

$$\begin{array}{rcccl} 3.452 & = & 3452/1000 & = & 3452/1000 \\ 5.730 & = & 573/100 & = & 5730/1000 \\ \underline{24.200} & = & 242/10 & = & \underline{24{,}200/1000} \\ 33.382 & & & & 33{,}382/1000 = 33.382 \end{array}$$

Examples:

(a) Find $3.716 + 23.4$. Solution:
$$\begin{array}{r} 3.716 \\ \underline{23.400} \\ 27.116 \end{array}$$

(b) Find $16.14 - 0.237$. Solution:
$$\begin{array}{r} 16.140 \\ \underline{-0.237} \\ 15.903 \end{array}$$

This same algorithm may be used to add decimal expressions in other bases.

Examples:

$$\begin{array}{r} 1.201_{\text{four}} \\ +0.312_{\text{four}} \\ \hline 2.113_{\text{four}} \end{array} \qquad \begin{array}{r} 4.TE_{\text{twelve}} \\ +17.45_{\text{twelve}} \\ \hline 20.34_{\text{twelve}} \end{array}$$

In order to discuss easily the multiplication of two numbers involving decimals, we examine the multiplication of fractions expressed with denominators as powers of 10. Consider the following examples.

Examples: (a) Multiply 26.2 by 0.03. Now

$$26.2 = \frac{262}{10} \quad \text{and} \quad 0.03 = \frac{3}{(10)^2}.$$

Thus,

$$(26.2)(0.03) = \frac{262}{10} \cdot \frac{3}{(10)^2} = \frac{786}{(10)^3} = 0.786.$$

(b) Multiply 12.47 by 0.623.

$$(12.47)(0.623) = \frac{1247}{(10)^2} \cdot \frac{623}{(10)^3} = \frac{776,881}{(10)^5} = 7.76881.$$

Notice that this answer could have been obtained in the following manner. Multiply 12.47 by 0.623, considering each numeral as a whole number. To place the decimal point in the answer, add the number of decimal places in each of the two numbers to be multiplied. This sum is the number of decimal places that will appear in the answer. Simply count from right to left that many digits and place the decimal point after the last counted digit.

Example:

$$\begin{array}{rl} 12.47 & \text{(2 digits to the right of the decimal point)} \\ 0.623 & \text{(3 digits to the right of the decimal point)} \\ \hline 3741 & \\ 2494 & \\ 7\ 482 & \\ \hline 7.76881 & \text{(5 digits to the right of the decimal point)} \end{array}$$

Division involving decimal fractions can be easily changed to division involving whole numbers by performing the following operations. Consider the division problem as a fraction and multiply the numerator and denominator of the fraction by a power of 10 that will make the denominator of the fraction a whole number. Then the problem reduces to a division by a whole number.

Example:

$$106.08 \div 1.7 = \frac{106.08}{1.7} = \frac{106.08(10)}{1.7(10)} = \frac{1060.8}{17}$$

$$\begin{array}{r} 62.4 \\ 17\overline{)1060.8} \\ \underline{102} \\ 40 \\ \underline{34} \\ 6\,8 \\ \underline{6\,8} \end{array}$$

Another form for writing fractions is in terms of *percent*, denoted by %. A percent represents a fraction whose denominator is one hundred. That is,

$$37\% = \frac{37}{100} = 0.37.$$

Examples: Convert to decimals: 16%, 1.6%, and 247%.

$$16\% = \frac{16}{100} = 0.16$$

$$1.6\% = \frac{1.6}{100} = \frac{16}{1000} = 0.016$$

$$247\% = \frac{247}{100} = 2.47$$

Note in the preceding examples that changing a decimal fraction to a percent is merely a process of moving the decimal point two places to the right.

Examples: Convert to percents: 0.047, 0.769, 3.56, and 0.00071.

$$0.047 = 4.7\% \qquad 0.769 = 76.9\%$$
$$3.56 = 356\% \qquad 0.00071 = 0.071\%$$

To change a fraction to a percent involves two steps: change the fraction to a decimal; then change the decimal to a percent.

Problems involving percent can be classified into three categories. First, we consider examples in which the problem is to find a given percent of a number.

Examples: (a) Find 25% of 12,000. $0.25(12{,}000) = 3{,}000$.

(b) In a class of 48 people, 25% made A's. How many made A's? $0.25(48) = 12$ people making A's.

A second type of problem involves finding what percent one number is of another number.

Examples: (a) 92 is what percentage of 400?

$$\frac{92}{400} = 0.23 = 23\%, \text{ so 92 is 23\% of 400.}$$

(b) A solution contains 80 grams of substance A and 20 grams of substance B. What percent of the solution is substance B?

$$\frac{20}{80 + 20} = 0.20 = 20\%$$

The third type of problem involves finding a number when a certain percent of it is known.

Example: Eighteen percent of the freshmen at Swan University failed freshman English. If 396 freshmen failed English, how many freshmen are enrolled at Swan University?

Let n represent the number of freshmen at Swan University. Then

$$0.18n = 396,$$

$$n = \frac{396}{0.18} = 2200 \text{ freshmen.}$$

Exercise Set 8-2

1. Perform the operations indicated in the problems below.
   (a) $0.6 + 0.04 + 0.502$
   (b) $1 + 0.61 - 0.5805$
   (c) $0.8 + 0.603 - (2.251 + 0.006)$
   (d) $0.06 + 0.0083 + 0.30007 + 3.0068$
   (e) $30 + 2.9 + 0.0008 - 0.67905$
   (f) $37.4507 + 6.32 - 40.1$
   (g) $4.86 - 9.11 + 8.1 - (-2.347)$
   (h) $3.196 - 6.5 - (-0.8025)$
2. Perform each operation using fractions and then using decimal fractions. Check your answers.

   (a) $\frac{71}{100} - \frac{45}{50}$ (b) $\frac{1}{8} + \frac{3}{4}$ (c) $\frac{1}{2} + \frac{1}{4} + \frac{750}{2000}$

   (d) $\frac{1}{2} - \frac{3}{8}$ (e) $\frac{8}{2} - \frac{5}{4}$ (f) $\frac{255}{10} - \frac{30}{20}$

3. Perform the following operations using decimal fractions and justify the algorithm used by means of fractions.
   (a) $28.32 + 7.521$ (b) $0.56 + 0.006$
   (c) $0.3 + 5.00311$ (d) $2.04 + 4.1$
   (e) $354.51 - 38.64$ (f) $389.27 - 63.99$
   (g) $691.7 - 8.526$ (h) $84.5 - 9.72$
   (i) $(7.05)(0.006)$ (j) $(0.04)(6.011)$
   (k) $(2.04)(3.25)$ (l) $(21.7)(74.65)$
4. Perform the following divisions by first writing each division as a fraction having a denominator without a decimal.

   (a) $4.1\overline{)66.83}$ (b) $0.42\overline{)4.116}$

   (c) $0.31\overline{)5.301}$ (d) $2.6\overline{)445.64}$

5. Change the following fractions to decimal fractions.

   (a) $\frac{3}{8}$ (b) $\frac{3}{5}$ (c) $\frac{7}{8}$

   (d) $\frac{3}{20}$ (e) $\frac{27}{8}$ (f) $\frac{19}{4}$

6. Convert to decimals.
   (a) 5.5% (b) 0.012% (c) 0.31%
   (d) 426% (e) 43.6% (f) 18.4%
7. Convert to percents.
   (a) $\frac{18}{15}$ (b) $\frac{33}{150}$ (c) $\frac{6}{15}$
   (d) $\frac{18}{4}$ (e) $\frac{11}{40}$ (f) $\frac{144}{80}$
8. (a) What percent of 48 is 12?
   (b) 24 is what percent of 96?
   (c) Find 16% of 200.
   (d) Find 0.01% of 1.632.
   (e) What is 78% of 16?
   (f) What percent of 150 is 6?
   (g) What percent of 18 is 54?
   (h) What number is 130% of 96?
9. If your automobile payments are $105 a month, what percent of your $750 per month salary must be set aside to pay for your automobile?
10. In Exercise 9, you receive a 4% raise. What is your new monthly salary? What percent of your salary now must be set aside to pay for your automobile?
11. Of the 80,000 football seats in the stadium, 48,640 were filled. What percent of the stadium was filled?
12. If your income tax is 26% of $7200, what is your tax?
13. Dick wants to buy a sweater that costs $20. His mother gave him $2.55 and his father gave him $4.85. What percent of the cost remains for Dick to pay?
14. Mrs. Green spends $150 a month for groceries. If this is 16% of her monthly income, what is her income?
15. If a variety store sold 475 of 500 items, what percent of the stock was sold?
16. Ken and Bob have 43 coins. Bob has 11 more than Ken. What percent of the coins does Ken have?
17. A teacher has a salary of $10,200. This is a 6% increase over last year's salary. What was the salary last year?

*18. Compute the following additions.

(a)
$$\begin{array}{r} 2.213_{\text{four}} \\ 0.33_{\text{four}} \\ \hline \end{array}$$

(b)
$$\begin{array}{r} \text{T.E}6_{\text{twelve}} \\ 0.78_{\text{twelve}} \\ \hline \end{array}$$

(c) $\begin{array}{r} 1.1011_{\text{two}} \\ 0.101_{\text{two}} \\ \hline \end{array}$ (d) $\begin{array}{r} 7.657_{\text{eight}} \\ 0.56_{\text{eight}} \\ \hline \end{array}$

*19. Compute the following differences.

(a) $\begin{array}{r} 3.04_{\text{six}} \\ -0.15_{\text{six}} \\ \hline \end{array}$ (b) $\begin{array}{r} 13.72_{\text{twelve}} \\ -\text{E.TE}_{\text{twelve}} \\ \hline \end{array}$

(c) $\begin{array}{r} 1101.110_{\text{two}} \\ -10.101_{\text{two}} \\ \hline \end{array}$ (d) $\begin{array}{r} 132.45_{\text{nine}} \\ -7.68_{\text{nine}} \\ \hline \end{array}$

*4

Approximate Numbers

Numbers, like many other concepts, can be distinguished on the basis of certain characteristics. A number associated with a counting process may be classified as an *exact* number. In contrast, measurements of all types are usually considered as *approximate* numbers. For example, 54 students in the class and a bill for $14.63 represent exact numbers, whereas a chemical sample weighing 2.0347 grams involves an approximate number. (Can you explain why?)

With approximate numbers, it is often desirable to use only the first few digits of their decimal representation. When one number is used as an approximation for another number, we use the symbol $\approx$ to indicate this approximation.

$$0.42 \approx 0.4217; \qquad 0.3 \approx 1/3.$$

A decimal fraction obtained by omitting some of the ending digits of the original decimal representation is said to have been *rounded off*. The following rules are generally used for rounding off numbers.

(1) If the digit to the right of the last digit to be retained is less than 5, leave the last digit unchanged.
(2) If the digit to the right of the last digit to be retained is greater than 5, increase the last digit to be retained by 1.
(3) If the digit to the right of the last digit to be retained is equal to 5, then
   (a) Increase the last digit retained by 1 when there is at least one nonzero digit to the right of the 5.
   (b) When there is no nonzero digit to the right of the 5, increase the last digit retained by 1 if and only if the last digit retained is odd.

Examples: Round off the following numbers to two decimal places.

$$\begin{array}{ll} 3.1671 \approx 3.17 & 5.315 \approx 5.32 \\ 8.134 \approx 8.13 & 7.245 \approx 7.24 \\ 7.1451 \approx 7.15 & 3.1416 \approx 3.14 \end{array}$$

It is evident from the rules given above that if an approximate number, 5.84, is a rounded-off representation of an exact number, N, then

$$5.835 \leq N \leq 5.845.$$

We say the *precision* of the approximation is 0.01, which is the difference between 5.835 and 5.845. The *error* is 0.005 or one-half of the precision. For example, the approximation 2.1 has an error of 0.05, and the approximation 2.11 has an error of 0.005.

In order to determine the precision of a sum, consider an addition of the following approximate numbers.

$$\begin{array}{ll} 3.4 & 3.35 \leq N_1 \leq 3.45 \\ \underline{11.373} & \underline{11.3725 < N_2 < 11.3735} \\ & 14.7225 < N_1 + N_2 < 14.8235 \end{array}$$

Note that the precision of 3.4 is 0.1 and the precision of 11.373 is 0.001. The sum, $N_1 + N_2$, has a value between 14.7225 and 14.8235, or a precision of 0.101. This work suggests that the answer can be no more accurate than the least accurate number in a sum. This result may be stated as follows: *When approximate numbers are added (or subtracted), their sum (or difference) has at most the precision of the term with the greatest error.*

Example: Find the precision of $3.15 + 0.621$. The precision of 3.15 is 0.01, and the precision of 0.621 is 0.001. Thus, the precision of the sum is 0.01.

Often the accuracy of a number is indicated by writing the number in *scientific notation*. In scientific notation, a number N is written in the form

$$N = A \cdot 10^n,$$

where A is a number between 1 and 10 (such as 2.61) and n is an integral power of 10. The digits in A are called the *significant* digits of the number N.

The number of significant digits in a number N is called the *accuracy* of N.

Examples:

$1.40 \cdot (10)^5$ has three significant digits.
$1.4 \cdot (10)^2$ has two significant digits.
$1.041 \cdot (10)^7$ has four significant digits.

Example: Round off 57,148 to three significant digits and write the result in scientific notation.

$$57{,}148 \approx 5.71 \cdot (10)^4$$

Example: Express 0.00012 in scientific notation.

$$0.00012 = 1.2 \cdot (10)^{-4}$$

It should be emphasized that precision is sometimes inadequate in the comparison of errors. For example, 1,000,000.2 and 3.3 have the same precision and yet the error in the first number is not as large relative to its value as is the case with the second number. Thus, we need a new measure of accuracy and this measure is provided by the *relative error*. Relative error may be defined to be the error of a number N divided by the number N.

$$\text{Relative error of } N = \frac{\text{error of } N}{N}$$

Example: Find the relative errors of the approximate numbers 0.05 and 2.1.

$$\text{Relative error of } 0.05 = \frac{0.005}{0.05} = 0.1$$

$$\text{Relative error of } 2.1 = \frac{0.05}{2.1} = 0.02$$

When numbers are multiplied (or divided), their product (or quotient) is, at most, as accurate as the factor with the greatest relative error.

Example: Multiply the following approximate numbers: (0.271)(1.2).
1.2 has the greatest relative error and has two significant digits. 0.271 has three significant digits. By the preceding rule, the product can have only two significant digits. Thus,

$$(0.271)(1.2) \approx 0.33.$$

Exercise Set 8-3

1. Represent each of the following in scientific notation.
   (a) 0.000132 (b) 5,640,000 (c) 1.46
   (d) 0.00127 (e) 8,200,000 (f) 0.020003
2. Determine the precision, relative error, and accuracy of each of the following.
   (a) 23.41 (b) $2.7 \cdot (10)^{-3}$ (c) $0.301 \cdot (10)^4$
   (d) 300.4 (e) $168 \cdot (10)^{-3}$ (f) 0.0012
3. Express each part in scientific notation, and then perform the computation, rounding off the answer appropriately.
   (a) $0.00023 \cdot 526{,}000$ (b) $682{,}000 \div 0.00024$
   (c) $2.71 \cdot (10)^{-3} \div 0.0002$ (d) $2.71 \cdot (10)^{-3} \cdot 1.6$
4. (a) Find the quotient of $17.842 \div 6.3$ to the nearest thousandth; the nearest tenth.
   (b) Find a decimal approximation for 13/9 that has an error less than 0.001.
5. Perform the following additions and subtractions, rounding off answers to the appropriate number of decimal places.
   (a) $2.036 + 2.21 - 0.0072$
   (b) $40.672 - 0.05 + 31.6$
   (c) $17.6041 + 2.6 \cdot 10^{-2} - 0.014$

*6. Make up a rule corresponding to the base ten rule given in this section for rounding off in the following systems.
   (a) Base two (b) Base seven
   (c) Base twelve (d) Base five

5

Repeating Decimals

In this section, we shall consider the decimal representation of all positive rational numbers. Of course, the same results can be extended to negative rational numbers. Examine the following examples, which utilize the algorithm for division as discussed in the preceding section.

Examples: Change 3/8, 1/3, and 3/11 to decimals.

(a)
$$\begin{array}{r} 0.375 \\ 8\overline{)3.000} \\ \underline{2\,4} \\ 60 \\ \underline{56} \\ 40 \\ \underline{40} \\ 0 \end{array}$$

(b)
$$\begin{array}{r} 0.333 \\ 3\overline{)1.000} \\ \underline{9} \\ 10 \\ \underline{9} \\ 10 \\ \underline{9} \\ 1 \end{array}$$

(c)
$$\begin{array}{r} 0.2727 \\ 11\overline{)3.0000} \\ \underline{2\,2} \\ 80 \\ \underline{77} \\ 30 \\ \underline{22} \\ 80 \\ \underline{77} \\ 3 \end{array}$$

Notice in the first example that a remainder of 0 is obtained and thus the decimal is said to *terminate.* If division were continued in the first example, one would get 0.3750000 . . . , indicating that a terminating decimal may be written as an unending decimal.

In the other two examples, it should be clear that a remainder of zero will *never* be attained. These decimals are called *nonterminating.* However, these nonterminating decimals have interesting properties. In the second example, the 3 repeats, while in the last example, the pair of digits 27 repeats. Such decimals, called *repeating decimals,* are illustrated by the following examples.

Examples:

(a) $\frac{1}{3} = 0.3333\ldots = 0.\overline{3}$

(b) $\frac{1}{7} = 0.142857142857\ldots$

$= 0.\overline{142857}$

(c) $\frac{1}{9} = 0.11111\ldots = 0.\overline{1}$

(d) $\frac{3}{11} = 0.27272727\ldots = 0.\overline{27}$

(e) $\frac{19}{111} = 0.171171171\ldots = 0.\overline{171}$

(f) $\frac{143}{82} = 1.7\overline{43902}$

Notice that each of these decimal representations has repeating blocks of digits varying from one to six digits. The fact that a block of digits is repeated indefinitely is denoted by a bar above the digits being repeated. We will now show in general that if the division process does not yield a terminating decimal, it will yield a repeating decimal.

In the repeated application of the division algorithm,

$$a = bq + r \qquad (0 \leq r < b),$$

it is true that each remainder r is one of the elements of the set

$$\{0, 1, 2, \ldots, b - 1\},$$

a set with b distinct elements. Hence, if the division algorithm is applied again and again, not more than b applications can be made until a remainder will appear that has appeared previously. As soon as this happens, the digits will start repeating. Thus, each rational number is represented by either a terminating decimal or by a repeating decimal.

We have demonstrated in the preceding paragraphs that a rational number can be expressed as a repeating decimal or a terminating decimal. Is the converse of this statement true? If a number is represented by a repeating decimal, is it always a rational number? Consider part (d) of the preceding example.

Example: We shall try to find a rational number representation for $0.\overline{27}$. Let $N = 0.272727\ldots$. The block of repeating digits consists of "27." Since there are two digits in this repeating block, we multiply by 100. (If there were three, we would multiply by 1000, and so on.) Multiplying by 100 shifts the decimal point two places to the right, and we get

$$100N = 27.2727\ldots.$$

We subtract $\quad N = \;\; 0.2727\ldots$

to obtain $\quad 99N = 27 \quad \text{or} \quad N = \dfrac{27}{99} = \dfrac{3}{11}.$

Did you notice in the preceding discussion that we were performing operations that have not been defined, namely, subtracting unending decimals and multiplying unending decimals by powers of 10? Operations with such numbers will be given precise definitions later in this chapter. At this point, we will accept the operations as illustrated in the examples.

In the following examples, we will multiply by a power of ten that will place the decimal point at the beginning of the block of repeating decimals. Then we will work the problem as in the preceding example.

Example: Change 0.00133133133 ... to a fraction representing a rational number.

Let

$$N = 0.00\overline{133}.$$

Then

$$10^2N = 0.\overline{133}$$

and

$$10^3 \cdot 10^2N = 133.\overline{133}.$$

Subtract 10^2N from $10^3 \cdot 10^2N$:

$$10^3 \cdot 10^2N - 10^2N = 133.\overline{133} - 0.\overline{133}.$$

Thus,

$$(100{,}000 - 100)N = 133$$

and

$$N = \frac{133}{99900}.$$

We check this computation by changing 133/99900 back to a decimal fraction to see if we obtain the repeating decimal with which we started.

```
         0.00133133 ...
99900 ) 133.000000
         99 900
         33 1000
         29 9700
          3 13000
          2 99700
            13300
```

Thus, $0.00\overline{133}$ and 133/99900 represent the same rational number.

Example: Change $3.87\overline{6}$ to a fraction.

Let

$$N = 3.87\overline{6}.$$

Then

$$10^2 N = 387.\overline{6}$$

and

$$10^3 N = 3876.\overline{6}.$$

Subtract $10^2 N$ from $10^3 N$:

$$(10^3 - 10^2)N = 3489.$$

Thus,

$$N = \frac{3489}{900} = \frac{1163}{300}.$$

Once again, we check our computation by changing 1163/300 back to a decimal to see if we obtain the original repeating decimal.

```
         3.8766 ...
300 | 1163.0000
       900
       2630
       2400
        2300
        2100
      ┌→2000
      │  1800
      └→ 2000
```

Hence $3.87\overline{6} = 1163/300$.

The preceding examples are illustrations of the general fact that every decimal which eventually repeats represents a rational number. To see that this statement is always true, consider a repeating decimal in the form $N = b_0 b_1 \cdots b_n . b_{n+1} b_{n+2} \cdots b_{n+r} c_1 c_2 \cdots c_m c_1 c_2 \cdots c_m \cdots$. First we write

$$10^r N = 10^r [b_0 b_1 \cdots b_n . b_{n+1} b_{n+2} \cdots b_{n+r} c_1 c_2 \cdots c_m c_1 c_2 \cdots c_m \cdots]$$
$$10^r N = b_0 b_1 \cdots b_{n+r} . c_1 c_2 \cdots c_m c_1 c_2 \cdots c_m \cdots$$

in order to place the decimal point at the beginning of the block of repeating decimals. Since there are m digits in each block of this repeating decimal, multiply by 10^m.

$$
\begin{array}{lrl}
 & 10^m \cdot 10^r N = & b_0 b_1 \cdots b_{n+r} c_1 c_2 \cdots c_m . c_1 c_2 \cdots c_m c_1 c_2 \cdots c_m \cdots \\
\text{Subtract} & 10^r N = & \phantom{b_0 b_1 \cdots b_{n+r} c_1 c_2 \cdots c_m} b_0 b_1 \cdots b_{n+r} . c_1 c_2 \cdots c_m c_1 c_2 \cdots c_m \cdots \\
\hline
\text{to obtain} & (10^{m+r} - 10^r) N = & b_0 b_1 \cdots b_{n+r} c_1 c_2 \cdots c_m - b_0 b_1 \cdots b_{n+r}
\end{array}
$$

or

$$N = \frac{b_0 b_1 \cdots b_{n+r} c_1 c_2 \cdots c_m - b_0 b_1 \cdots b_{n+r}}{10^{m+r} - 10^r}.$$

Since $b_0 b_1 \cdots b_{n+r} c_1 c_2 \cdots c_m - b_0 b_1 \cdots b_{n+r}$ and $10^{m+r} - 10^r$ are both integers, N is a rational number.

The preceding discussion is the proof for the following theorem.

Theorem 8-3: Every rational number can be represented by a decimal fraction that either terminates or eventually repeats; conversely, every decimal that either terminates or eventually repeats represents a rational number.

Exercise Set 8-4

1. Obtain a decimal representation for each of the following.

(a) $\frac{11}{16}$ (b) $\frac{36}{7}$ (c) $\frac{7}{875}$ (d) $\frac{11}{400}$ (e) $\frac{5}{13}$

(f) $\frac{51}{8}$ (g) 16 (h) 14 (i) $12\frac{3}{5}$ (j) $2\frac{1}{5}$

2. Write each of the following decimal expressions as a fraction.

(a) 0.346 (b) $0.\overline{9}$

(c) $0.\overline{18}$ (d) $3.\overline{845}$

(e) 16.4 (f) 146.1

(g) $1.2\overline{54}$ (h) $0.01\overline{79}$

(i) $6.\overline{8}$ (j) $3.\overline{25}$

(k) $6.\overline{01}$ (l) $4.\overline{247}$

(m) $0.0\overline{09}$ (n) $0.0\overline{132}$

3. Arrange the following rational numbers in increasing order.
   (a) $0.\overline{8}$, 0.8, 0.84, 0.89, $0.\overline{89}$,
   (b) 0.32, $0.\overline{3}$, $0.\overline{32}$, 0.322, $0.\overline{322}$

*4. Consider a repeating decimal of the form $.\overline{xyz}$, where x, y, and z are integers from 0 to 9 inclusive. Give a fraction representation of this rational number.

*5. Note that

$$\frac{1}{9} = 0.111\ \ldots \text{ (one digit repeats);}$$

$$\frac{1}{27} = \frac{1}{3}\left(\frac{1}{9}\right) = \frac{1}{3}\,(0.111\ \ldots) = 0.037037037\ \ldots \text{ (three digits repeat);}$$

$$\frac{1}{81} = \frac{1}{3}\left(\frac{1}{27}\right) = \frac{1}{3}\,(0.037037037\ \ldots) = 0.012345679012345679\ \ldots$$

(nine digits repeat).

Can you guess how many digits repeat in 1/243?

*6. The rational number x as a decimal repeats in a block of m digits. The rational number y repeats in a block of n digits. What can you assert about $x + y$?

6

Irrational Numbers

In the preceding section, we established that every rational number can be represented by either a terminating decimal or a repeating decimal; conversely, every terminating decimal or repeating decimal represents a rational number (Theorem 8-3). This theorem implies that if a number in decimal form does not repeat, then it must represent a number other than a rational number. Such a decimal numeral represents what is called an *irrational number.*

If we can describe some process for writing an infinite or endless decimal so that there is no repeating cycle of digits, then we will have exhibited an irrational number. This can be done in many ways. Consider 0.13113111311113111113 . . . . This is an irrational number because it is a nonterminating, nonrepeating decimal. It is nonrepeating because each 3 has one more 1 preceding it than the preceding 3.

Perhaps the most familiar irrational number is the number π, the ratio of the circumference of a circle to the diameter of the circle. π is not $3\frac{1}{7}$ or 3.1416; π is an unending, nonrepeating decimal.

The solutions of $x^2 - 2 = 0$, $x^2 - 5 = 0$, and $x^2 - 2x - 5 = 0$ are all irrational numbers, expressed as $\pm\sqrt{2}$, $\pm\sqrt{5}$, and $1 \pm \sqrt{6}$, respectively.

Suppose we consider $\sqrt{2}$, one solution of $x^2 - 2 = 0$. The following discussion will give an approximate value for $\sqrt{2}$ and will demonstrate that $\sqrt{2}$ is irrational. Since $1^2 < 2$ and $2^2 = 4 > 2$, we agree that $\sqrt{2}$ lies between 1 and 2. Now divide the interval between 1 and 2 into ten equal parts. From the multiplication listed below, we see that $\sqrt{2}$ lies between the fourth and fifth of these intervals. Thus, $(1 + 4/10) < \sqrt{2} < (1 + 5/10)$. Subdivide the interval between 1.4 and 1.5 into ten equal parts and note in the computation following this paragraph that $\sqrt{2}$ lies in the second interval. Then $(1 + 4/10 + 1/10^2) < \sqrt{2} < (1 + 4/10 + 2/10^2)$. If we continue this process indefinitely, we obtain an infinite decimal that can be approximated by

$$a_0 + a_1(1/10) + a_2(1/10)^2 + \cdots + a_n(1/10)^n < \sqrt{2} < a_0 + a_1(1/10) + a_2(1/10)^2 + \cdots + (a_n + 1)(1/10)^n.$$

This approximation represents the number to the extent that, by stopping the process at some stage n, we obtain a rational approximation to the irrational number with an error as small as we wish. The following inequalities demonstrate the steps in this computation.

$$1^2 = 1 < 2 < 4 = (2)^2 \quad \text{so} \quad 1 < \sqrt{2} < 2$$
$$(1.4)^2 = 1.96 < 2 < 2.25 = (1.5)^2 \quad \text{so} \quad 1.4 < \sqrt{2} < 1.5$$
$$(1.41)^2 = 1.9881 < 2 < 2.0164 = (1.42)^2 \quad \text{so} \quad 1.41 < \sqrt{2} < 1.42$$
$$(1.414)^2 = 1.999396 < 2 < 2.002225 = (1.415)^2 \quad \text{so} \quad 1.414 < \sqrt{2} < 1.415$$
$$(1.4142)^2 = 1.99996164 < 2 < 2.00024449 = (1.4143)^2 \quad \text{so} \quad 1.4142 < \sqrt{2} < 1.4143$$

We could continue in this manner to obtain $\sqrt{2}$ to as many decimal places as desired, such as $\sqrt{2} = 1.414213 \ldots$. However, no matter how long we continue to find digits, we will never be able to find a repeating sequence of digits. This is true because $\sqrt{2}$ is an irrational number, which we shall prove at this time.

The indirect proof we will present is sometimes called Euclid's proof that $\sqrt{2}$ is an irrational number. Assume that $\sqrt{2}$ is a rational number. Then it can be written as a quotient of two integers a/b. By Theorem 7-4, we can reduce this fraction so that the numerator and denominator have no common divisors except 1; thus we assume that a/b is a reduced fraction. If $a/b = \sqrt{2}$, then $a = b\sqrt{2}$. Then $a^2 = 2b^2$. $2b^2$ is an even integer. Thus, a^2 is even. If a^2 is even, then a must be even. (Why?) Since a is even, write it as $2c$. Then a^2 is $4c^2$, so $4c^2 = 2b^2$ or $2c^2 = b^2$. Thus, b^2 is even; hence b is even. Therefore we have shown, under our assumption that $\sqrt{2}$ was the rational a/b, that a is even and b is even. Since both are even, they have a common factor of 2. But this is contrary to our assumption that a/b was in reduced form. Since we have reached an obvious contradiction, our hypothesis that $\sqrt{2}$ is a rational number must be false. Thus, $\sqrt{2}$ is irrational.

In Chapter 7, it was noted that the rational numbers are *dense*, which means that for any two points on the number line there is a rational number between them. Thus, it seems that the rational numbers may "fill up" the number line. But this is not correct. In fact, we shall demonstrate now that the irrational numbers can also be represented by points on the number line.

Place one side of a unit square on the number line as indicated in Figure 8-1. In Chapter 10, we will learn that for right triangles the

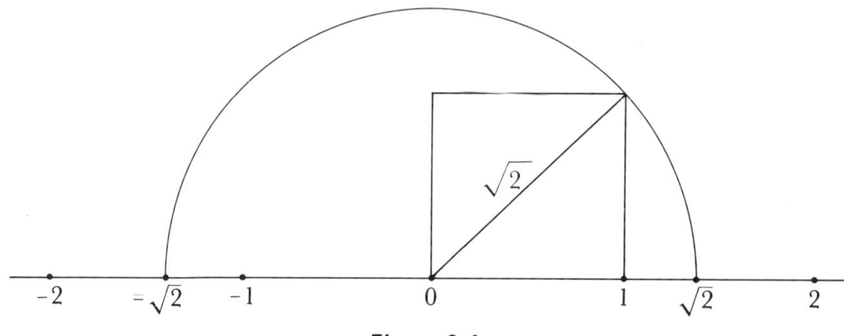

Figure 8-1

sum of the squares of two sides of the right triangle is equal to the square of the side opposite the right angle. Thus, the diagonal of the unit square is of length $\sqrt{1^2 + 1^2} = \sqrt{2}$. Consider the diagonal with one end at 0, and rotate this diagonal clockwise about the point 0 until it lies on the number line; the end will mark a point on the number line that is a distance of $\sqrt{2}$ from the origin, a point that represents the irrational number $\sqrt{2}$. In a like manner, if we rotate the diagonal counterclockwise about the origin, we can mark the point $-\sqrt{2}$.

Thus, there are points on the number line ($\sqrt{2}$ and $-\sqrt{2}$, for example) that cannot be represented by rational numbers. Hence, it is necessary for us to consider a further extension of our number system.

In the preceding discussion, we verified the existence of one irrational number, namely $\sqrt{2}$. Are there more? Is it reasonable to believe that there are more irrational numbers than rational numbers? How can you tell if a number is rational or irrational?

We can easily show that $\sqrt{3}$ is irrational by the following reasoning. Assume $\sqrt{3}$ equals a rational number a/b in reduced form; then $3b^2 = a^2$. If a ends in 0, 1, 2, 3, 4, 5, 6, 7, 8, or 9, then a^2 ends in 0, 1, 4, 9, 6, 5, 6, 9, 4, or 1, respectively. If b ends in 0, 1, 2, 3, 4, 5, 6, 7, 8, or 9, then $3b^2$ ends in 0, 3, 2, 7, 8, 5, 8, 7, 2, or 3. If $3b^2 = a^2$, $3b^2$ and a^2 must end with the same digit, and the only possibilities are 0 and 5. Thus, a and b both end with a 0 or with a 5. But then a/b is not in reduced form; a contradiction has been reached, and our assumption that $\sqrt{3}$ is a rational number is incorrect. Thus, $\sqrt{3}$ is an irrational number.

In Figure 8-2, consider the right triangle OAB. $\overline{OA}$ is of length one, as indicated on the number line. Using a compass (described in Chapter 10), measure AB to be of length $\sqrt{2}$ (obtained from Figure 8-1). Again using the fact that the sum of the squares of the sides of a right triangle is equal to the square of the side opposite the right angle, $\overline{OB}$ is of length $\sqrt{1^2 + (\sqrt{2})^2} = \sqrt{1+2} = \sqrt{3}$. Rotating OB clockwise about point O, we can mark on the number line a point that is a distance of $\sqrt{3}$ from O.

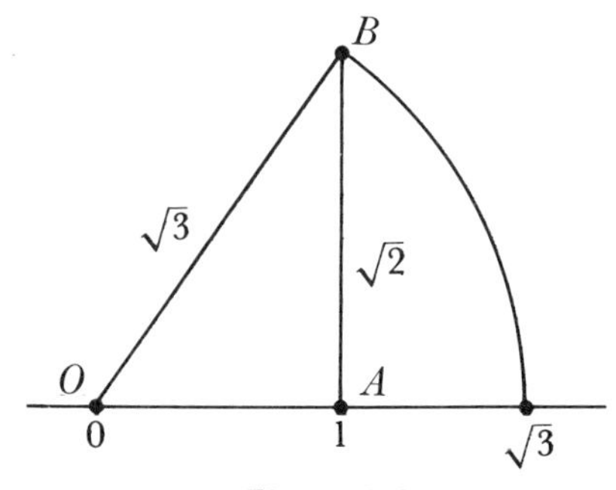

Figure 8-2

Now, returning to work with $\sqrt{2}$, let us translate our unit square so that its lower left corner is at 3/2 (Figure 8-3). Next, rotate the diagonal

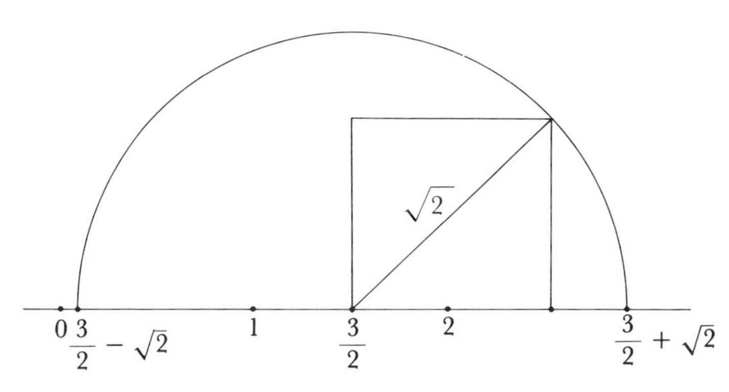

Figure 8-3

in both directions until it coincides with the number line. The points located at the ends of the diagonal correspond to $3/2 + \sqrt{2}$ and $3/2 - \sqrt{2}$. Again the question arises "Is $3/2 + \sqrt{2}$ a rational number or is it an irrational number?" If we assume it to be a rational number p/q, then $3/2 + \sqrt{2} = p/q$ or $\sqrt{2} = p/q - 3/2$. Since the rational numbers are closed under the operations of addition and subtraction, $p/q - 3/2$ is a rational number, which implies that $\sqrt{2}$ is a rational number. This, of course, is a contradiction; thus, $3/2 + \sqrt{2}$ is an irrational number.

This example illustrates that the sum of an irrational number and a rational number is an irrational number. In fact, the geometric process of placing the left corner of the unit square at a rational number a/b and then finding $a/b + \sqrt{2}$ and $a/b - \sqrt{2}$ always yields irrational numbers. Thus, it is seen that there are as many irrational numbers of the form $a/b + \sqrt{2}$ or $a/b - \sqrt{2}$ as there are rational numbers a/b. The same argument could be used to show that $a/b + \pi$, $a/b + \sqrt{3}$, $a/b + \sqrt{5}$, and so on are irrational numbers, and that there are as many in each class as there are rational numbers.

Since the square roots of many rational numbers are not rational but irrational numbers, we need an algorithm for determining a rational number approximation for square roots. There are many such algorithms. One of the easiest algorithms to learn is called the method of averaging. The steps in this algorithm are simple.

(a) Find an approximate value of the square root by estimating. It does not matter whether the estimate is too large or too small.

(b) Using the estimate as a divisor, find the quotient of the number and the estimated square root to as many decimal places as desired.

(c) Average (add and divide by 2) the quotient and the estimated square root. Use the average as an estimate. The process may be continued until the answer is as accurate as desired.

Example: Approximate $\sqrt{2}$.

Since $(1.4)^2 = 1.96$, we will start with 1.4 as our estimate.

$$2 \div 1.4 = 1.4286$$
$$(1.4 + 1.4286)/2 = 1.4143$$

Repeat the process.

$$2 \div 1.4143 = 1.4141$$
$$(1.4143 + 1.4141)/2 = 1.4142$$

Since a repetition of the algorithm changed only the fourth decimal place and this change was only one unit, a rational approximation to $\sqrt{2}$ can be given, accurate to three decimal places, as 1.414.

Example: Find $\sqrt{294.687}$.

Now, $(17)^2 = 289$, so 17 is a rough estimate or approximation for $\sqrt{294.687}$.

$$\frac{294.687}{17} = 17.3345$$

$$\frac{17.3345 + 17}{2} = 17.1673$$

$$\frac{294.687}{17.1673} = 17.165997$$

17.17 is an approximation for $\sqrt{294.687}$ accurate to two decimal places. If additional decimal places are desired we repeat the process.

$$\frac{17.165997 + 17.1673}{2} = 17.166648$$

$$\frac{294.687}{17.166648} = 17.16625$$

An answer of 17.166 is accurate to three decimal places.

The mere fact that a decimal does not repeat after a hundred, or two hundred, or even a thousand digits does not mean that the number is an irrational number. Generally it will be an irrational number, but there are exceptions. We could not tell that $\sqrt{2}$ was an irrational number by looking at any part of its decimal representation; we verified that it was an irrational number by a different approach. The same is true for other irrational numbers.

Would you be surprised to learn that irrational numbers are far more numerous than rational numbers? The square root of every integer that is not already a square is an irrational number. In addition, the cube root of every integer that is not the cube of an integer is irrational. Similar relations hold for other roots. This type of irrational number is called an *algebraic number*, described as a solution of an equation such as $x^2 = a$ or $ax^3 + bx^2 + cx + d = 0$. Other irrational numbers exist which are not algebraic. They are called *transcendental* numbers. The most familiar example of this type of number is π. The existence of transcendental numbers was discussed in 1814 by the French mathematician Joseph Liouville. Additional examples of transcendental numbers involve logarithms, trigonometric functions, and so forth.

Exercise Set 8-5

1. (a) Demonstrate with an example that the product of an irrational number and a rational number is irrational.
   (b) Prove this assertion.
2. (a) Demonstrate with an example that the sum of an irrational number and a rational number is irrational.
   (b) Prove this assertion.
3. $2\sqrt{3}$ and $5\sqrt{3}$ are both irrational numbers. (Why?)
   (a) Is the product of these two numbers irrational?
   (b) What about the quotient?
   (c) Is the sum of these two irrational numbers irrational?
4. Demonstrate that the product of two irrational numbers is not necessarily irrational. Do the same for the sum of two irrational numbers. Are the irrational numbers closed with respect to addition? With respect to multiplication?
5. Classify the following numbers as rational or irrational.

   (a) $\sqrt{3} - 1$ (b) $6 \cdot \sqrt{3}$ (c) 3π (d) $\dfrac{\sqrt{2}}{4\sqrt{2}}$

(e) $2 \div \sqrt{2}$ (f) $\sqrt{2}+7$ (g) $\dfrac{\sqrt{2}}{\sqrt{3}}$ (h) $\sqrt{2}-\sqrt{3}$

6. A certain number, approximated to four decimal places, is 2.3135. Can you tell whether the original number was rational or irrational?
7. Show both geometrically and algebraically that the following numbers are irrational numbers.

   (a) $-1+\sqrt{2}$ (b) $3+\sqrt{2}$

   (c) $5+\sqrt{2}$ (d) $\dfrac{-1}{4}+\sqrt{2}$

8. (a) Is there a largest integer less than 5?
   (b) Is there a largest rational number less than 5?
   (c) Is there a largest real number less than 5?
9. By constructing $\sqrt{2}$, then $\sqrt{3}$, and finally $\sqrt{4}$, show geometrically that $\sqrt{4}=2$.
10. Construct $\sqrt{5}$. Prove that $\sqrt{5}$ is irrational.
11. Find a rational approximation for the square roots of the following numbers correct to three decimal places.
   (a) 3 (b) 5 (c) 9
   (d) 15 (e) 21 (f) 49
   (g) 174.8 (h) 946.74 (i) 445.21
   (j) 0.0621 (k) 0.0071 (l) 0.00961

*12. Show that there is no rational number
   (a) whose square is 3/4. (b) whose square is 4/3.
   (c) whose cube is 4. (d) whose cube is 3.

7

Real Numbers

The problem of developing the real numbers as an extension of the rational numbers is undoubtedly the most difficult task we will explore in this book. Various procedures have been developed to accomplish this result.

Richard Dedekind (1831–1916) conceived the geometric idea that all rational points on a line could be separated into two equal classes such that every point in one class is on the right of every point in the other class and such that there exists one and only one point on a line that separates the line into these two sets. Thus Dedekind considered a real number as a point, a cut, or a separation on a line that partitions the rational numbers into two disjoint sets.

Another method of developing the real numbers uses the principle of nested intervals. A sequence of intervals $I_1, I_2, I_3, \ldots, I_n, \ldots$ can be considered as a sequence of nested intervals if $I_{n+1} \subset I_n$ and if I_n is as small as you please for n large enough. The sequence of nested intervals can be considered as "closing down on" or containing a unique point that is called a real number. And there are still other methods of defining real numbers, most of which are well beyond the scope of the material in this book.

A strict development of the idea of real numbers by either of the aforementioned methods would lead to difficulties much more formidable than any encountered in this book, and a precise mathematical development does not seem called for here. To define a real number we will use a decimal approach that is entirely adequate for our purpose.

Definition 8-1: The *real numbers* are defined to be the set of all decimal numbers.

The algorithms for performing addition and multiplication of rational numbers expressed as decimals can be used to *approximate* the sum and product of real numbers and to formulate intuitively a definition of addition and multiplication of real numbers.

Example: $16.41715\ldots + 3.54163\ldots$ is that number approached by the following sequence of rational approximations.

$$\begin{aligned} 16.4 + 3.5 &= 19.9 \\ 16.41 + 3.54 &= 19.95 \\ 16.417 + 3.541 &= 19.958 \\ 16.4171 + 3.5416 &= 19.9587 \\ 16.41715 + 3.54163 &= 19.95878 \\ \vdots \qquad \vdots \quad &\qquad \vdots \end{aligned}$$

We content ourselves to write the answer as $19.958\ldots$. We do not use the last two digits, 78, in our answer because we do not know whether or not they will remain as 7 and 8 or become something else in the addition of the next rational approximation.

Example: Add $6.33333\ldots$ and $2.55555\ldots$. The answer is the number approached by the following sequence.

$$
\begin{aligned}
6.3 + 2.5 &= 8.8 \\
6.33 + 2.55 &= 8.88 \\
6.333 + 2.555 &= 8.888 \\
6.3333 + 2.5555 &= 8.8888 \\
6.33333 + 2.55555 &= 8.88888 \\
\vdots \quad\quad \vdots \quad &\quad\quad \vdots
\end{aligned}
$$

The answer can be written as 8.88888 .... We call attention to the fact that 6.33333 ... equals 19/3, and 2.55555 ... equals 23/9.

$$19/3 + 23/9 = 80/9,$$

and 80/9 can be written as 8.88888 .... Hence, this example suggests that the rational numbers are embedded as part of the real numbers, where the addition of two rational numbers, when considered as real numbers, gives the same answer as when considered as rational numbers.

Example: (2.341752 ...) · (1.712561 ...) is equal to that number to which the following sequence gets closer and closer.

$$
\begin{aligned}
(2.3)(1.7) &= 3.91 \\
(2.34)(1.71) &= 4.0014 \\
(2.341)(1.712) &= 4.007792 \\
(2.3417)(1.7125) &= 4.01016125 \\
(2.34175)(1.71256) &= 4.01038738 \\
(2.341752)(1.712561) &= 4.010393146872 \\
\vdots \quad\quad \vdots \quad &\quad\quad \vdots
\end{aligned}
$$

We write the answer as 4.010 ..., since the other digits are not yet determined by the rational approximations considered.

This intuitive approach to multiplication is so constructed that it will hold for rational numbers considered as infinite decimals. The following example illustrates this fact.

Example: 4 · (7.666 ...) is the number approached by the following approximations.

$$
\begin{aligned}
(4.0)(7.6) &= 30.4 \\
(4.00)(7.66) &= 30.64 \\
(4.000)(7.666) &= 30.664 \\
(4.0000)(7.6666) &= 30.6664 \\
(4.00000)(7.66666) &= 30.66664 \\
(4.000000)(7.666666) &= 30.666664 \\
\vdots \quad\quad \vdots \quad &\quad\quad \vdots
\end{aligned}
$$

It is obvious that $4 \cdot (7.666\ldots) = 30.666\ldots$. Compare this result with the fact that $4 \cdot (7.6666\ldots) = 4 \cdot 7\frac{2}{3} = 30\frac{2}{3} = 30.666\ldots$.

The concepts of order, subtraction, and division for real numbers are extensions of the corresponding concepts for rational numbers: $x < y$ if and only if there exists a positive real number z such that $x + z = y$; $x - y = z$ if and only if $x = y + z$; and $x/y = z$ $(y \neq 0)$ if and only if $x = yz$.

Let us now consider the following example. In Figure 8.4(b), let the number line be divided into intervals so that there are ten segments between any pair of integers. Then mark 1.4 as a first approximation to $\sqrt{2}$. Verify by squaring 1.4. Now further divide the number line (Figure 8-4(c)) so that there are one hundred intervals between integers. 1.41 would approximate $\sqrt{2}$. Again, verify by squaring 1.41. Next, consider a thousand intervals (see Figure 8-4(d)) and mark 1.414, or jump to a million intervals and mark 1.414213. We see that this sequence of rational numbers 1.4, 1.41, 1.414, 1.4142, 1.41421, 1.414213, and so on, "sneaks up" on the place in the number line that represents $\sqrt{2}$.

Now, if the number line is divided into billionths, there would be one billion segments between any two consecutive integers. If one continues this process, it should suggest intuitively that the real numbers are *dense* on the number line. That is, between any two real numbers, no matter how close they are together, there are other real numbers.

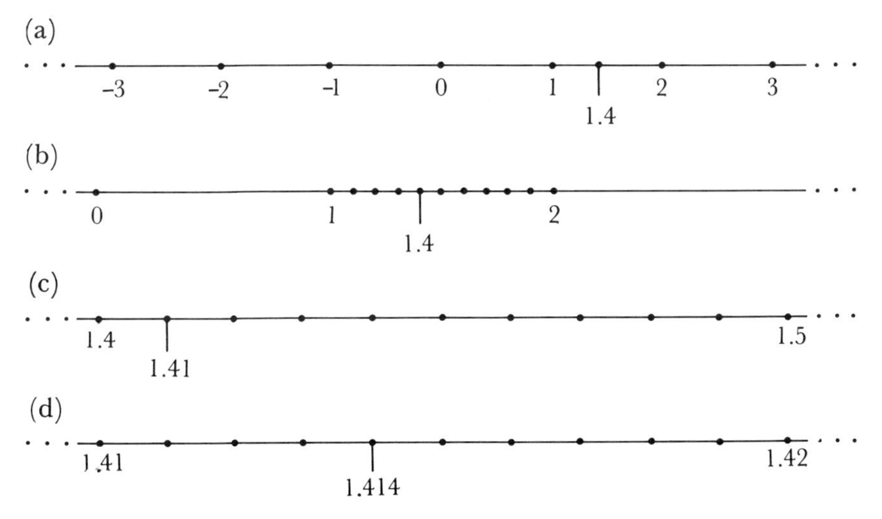

Figure 8-4

Now consider two different real numbers as infinite decimals. Since they are different numbers, let $x = a_k \cdots a_1 a_0 . b_1 b_2 \cdots$, and let $y = c_k \cdots c_1 c_0 . d_1 d_2 \cdots$. Assume all $a_j = c_j$ for all j and all $b_i = d_i$ for $i < p$, and let $b_p < d_p$. Take a rational number whose digits agree with x through b_p and whose next digit is greater than b_{p+1}. Let this rational number terminate with this last specified digit. Then this rational number lies between x and y. Therefore, since we can put another real number between any two real numbers, the real numbers are dense.

Intuitively, we have associated with every point on the number line a real number. As a result, we accept the fact *that there is a real number corresponding to each point on the number line and, conversely, every real number determines a unique point on the line.*

Exercise Set 8-6

1. For the given values of x and y, determine $x + y$, $x \cdot y$, and $x - y$, using the definitions of addition, subtraction, and multiplication for real numbers. If you cannot determine a pattern, round off your answer to four decimal places.
   (a) $x = 0.\overline{81}$ $\quad y = 0.\overline{8}$
   (b) $x = 6.666\ldots$ $\quad y = 2.2323\ldots$
   (c) $x = 3.\overline{412}$ $\quad y = 0.0\overline{132}$
   (d) $x = 0.\overline{027}$ $\quad y = 0.\overline{12}$
   (e) $x = 0.010101\ldots$ $\quad y = 99$
   (f) $x = 0.09796985\ldots$ $\quad y = 0.111\ldots$
   (g) $x = 0.999\ldots$ $\quad y = 0.247247\ldots$
   (h) $x = 0.182838\ldots$ $\quad y = 0.010101\ldots$
2. In Exercise 1, (a) through (e), change each numeral to a fraction and then perform the operations. Then change the answers in Exercise 1 to fractions. Do the answers as fractions agree?
3. Order the following numbers using a "less than" relationship.
   (a) $2\frac{1}{2}$, $2.\overline{3}$, 2.51, $\frac{6000}{29}$
   (b) 3.33, $\frac{577}{154}$, $3.\overline{12}$, $3\frac{5}{8}$
   (c) 22, $\frac{32}{11}$, $\frac{477}{154}$, $2.\overline{71}$

4. Find $y \div x$ for (e) and (g) in Exercise 1.

*5. $x^2 - y^2 = (x + y)(x - y)$ holds true when x and y are rational numbers. Is it true when x and y are real numbers? Why?

*6. Show that there is always an irrational number between any two rational numbers.

*7. Show that there is always a rational number between any two irrational numbers.

*8. Approximate π by rationals a/b when the decimal representation of a/b is

(a) $3.\overline{1}$. (b) $3.\overline{14}$. (c) $3.\overline{141}$.
(d) $3.\overline{1415}$. (e) $3.\overline{14159}$.

8

Summary

In the first eight chapters of this book, we have attempted to indicate the desirability of constructing number systems so as to maximize the number of properties that we are able to prove and to minimize the number of basic assumptions or postulates. We have made extensions from the whole numbers to the integers, then to the rational numbers, and, finally, to the real numbers. The most difficult link in this chain of development has been the extension from the rational numbers to the real numbers.

In our extension to the system of real numbers, we actually demonstrated that there were numbers that could not be represented as fractions. These numbers, called irrational numbers, obey the same rules for the operations of addition, subtraction, multiplication, and division as rational numbers because both were combined into the system of real numbers through the use of unending decimals. This presentation, along with geometric plausibility, illustrates the fact that the real number system satisfies the following properties.

(a) It is closed under addition.
(b) It is closed under subtraction.
(c) It is closed under multiplication.
(d) It is closed under division—division by zero being excluded.
(e) The elements of the set satisfy the commutative and associative properties of addition and multiplication.

(f) The elements of the set satisfy the distributive property of multiplication over addition.
(g) There exist additive and multiplicative identities and inverses.
(h) It is an ordered set.
(i) It is dense.

In fact, the real number system is an ordered field. An additional property of the real number system is called the *completeness property*. An upper bound of a set of numbers is a number that is greater than or equal to every element of the set. A lower bound is less than or equal to every element of the set. The completeness property states that any nonempty subset of real numbers that has an upper bound has a *least upper bound* and any nonempty subset that has a lower bound has a *greatest lower bound*.

The relationship between the systems of numbers that we have considered so far in this course is indicated by Figure 8-5.

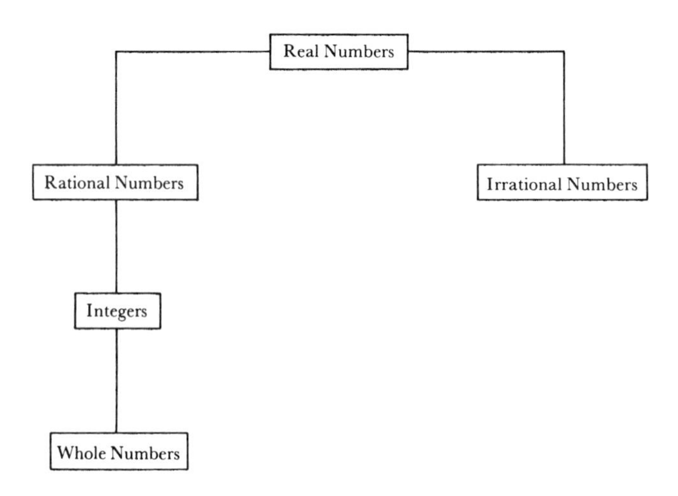

Figure 8-5

Notice that the whole numbers are included within the integers, within the rational numbers, and finally within the real numbers.

The following table summarizes the characteristics of real numbers as compared to rational, integral, and whole numbers. Y stands for *yes* and N for *no*. In all cases, division by zero is excluded.

	Whole Numbers				*Integers*			
	$+$	$\cdot$	$-$	$\div$	$+$	$\cdot$	$-$	$\div$
Closure	Y	Y	N	N	Y	Y	Y	N
Commutative Property	Y	Y	N	N	Y	Y	N	N
Associative Property	Y	Y	N	N	Y	Y	N	N

	Rational Numbers				*Real Numbers*			
	$+$	$\cdot$	$-$	$\div$	$+$	$\cdot$	$-$	$\div$
Closure	Y	Y	Y	Y	Y	Y	Y	Y
Commutative Property	Y	Y	N	N	Y	Y	N	N
Associative Property	Y	Y	N	N	Y	Y	N	N

Review Exercise Set 8-7

1. Determine the fraction a/b that corresponds to each of the following repeating decimals.
   (a) $14.2\overline{3}$ (b) $0.07\overline{201}$
   (c) $30.0\overline{12}$ (d) $0.002\overline{46}$
   (e) $-0.01\overline{79}$ (f) $3.3\overline{51}$
   (g) $6.5\overline{0}$ (h) $3.66\overline{0}$
2. Indicate whether each statement is true or false. If false, explain your answer.
   (a) 4 is an irrational number.
   (b) Every rational number is a real number.
   (c) $1.76 \cdot (10)^{-4} = 0.0000176$.
   (d) Every point on a number line represents a real number.
   (e) Every real number is a rational number.
   (f) The irrational numbers are closed under multiplication.
   (g) $231\% = 0.231$.
   (h) Every repeating decimal represents a real number.
   (i) Every irrational number can be represented by a repeating decimal.
   (j) $0.5\% = \frac{1}{200}$.
3. Let A be a set containing infinitely many real numbers such that each member of set A is less than 3.
   (a) Where does set A lie on the number line?

(b) Could there be a number in set A between 2 and 3?
(c) Could there be a number in set A between 2.99 and 3?
(d) Is it true that there is a real number in A such that each number in A is less than or equal to this number?
(e) A subset of A is defined by $\{x \mid \frac{1}{2} \leq x \leq 1$ and x is a real number$\}$. For this subset is there a real number such that every element of the subset is less than or equal to this number?

4. Order the following numbers, using a "less than" relationship.

(a) $3.14, 3\frac{1}{2}, \frac{677}{236}$ (b) $\sqrt{2}, \frac{16}{11}, \frac{477}{308}$

(c) $\sqrt{5}, 2\frac{1}{2}, \frac{23}{11}$ (d) $\sqrt{3}, \frac{3}{2}, \frac{577}{308}$

5. For the given values of x and y, determine $x+y$, $x \cdot y$, and $x-y$, using the definitions of addition, subtraction, and multiplication for real numbers. If you cannot determine a pattern, round off your answer to four decimal places.

(a) $x = 17.\overline{23}$ $y = 6.\overline{012}$
(b) $x = 8.333\ldots$ $y = 0.0024646\ldots$
(c) $x = 0.5000\ldots$ $y = 0.017979\ldots$
(d) $x = 1.2\overline{54}$ $y = 0.\overline{8}$

Suggested Reading

Terminating Decimals: Byrne, p. 244. Copeland, p. 207. Garner, pp. 276–279. Graham, pp. 258–259. Hutton, pp. 265–267. Keedy, p. 172. Willerding, p. 215.

Operations: Copeland, pp. 203–206. Garner, pp. 280–284. Graham, pp. 261–264. Keedy, pp. 174–175. Nichols, Swain, pp. 222–226.

Repeating Decimals: Campbell, pp. 176–177. Copeland, p. 207. Garner, pp. 290–295. Graham, pp. 265–267. Hutton, pp. 265–269. Keedy, p. 171. Wren, p. 139. Zwier, Nyhoff, pp. 254–261.

Irrationals: Campbell, pp. 195–196. Copeland, pp. 228–231. Garner, pp. 329–332. Graham, p. 257. Hutton, pp. 308–310. Nichols, Swain, pp. 257–261. Wren, pp. 157–161.

9

Informal Nonmetric Geometry

1

Introduction

For more than 4000 years, man has been studying geometry and its applications to the world in which he lives. The geometry of the Babylonian civilizations was concerned primarily with areas and volumes. The Egyptians used geometry in building their well-known pyramids. The Greeks made a great contribution when they introduced reason and logic into the presentation of geometric ideas and thus started the consideration of geometry as an abstract mathematical system. Euclid, a famous Greek mathematician, attempted to formulate basic geometric concepts in a precise and orderly manner in his book, *Elements*. The geometries studied today are deductive developments from sets of undefined terms and axioms, the truth of which is neither known nor questioned. As one changes the assumptions or axioms, one correspondingly changes the geometry.

No effort will be made in this book to develop the system of Euclidean geometry in the logical and formal manner often presented in high school. Instead, we study geometry informally. This intuitive approach involves a description of undefined terms and a discussion of definitions and axioms. Some theorems are stated and proved in order to emphasize selected ideas. We rely strongly on the intuition of the reader throughout this chapter.

The language of geometry used here involves the language of sets in the same manner in which we used sets of numbers or sets of symbols. In this chapter, our main concern is with sets of points satisfying certain geometric properties. This use of sets should facilitate your understanding of geometric ideas and illustrate for you the relationship between geometry and earlier topics in this book.

2

Points, Lines, Planes, and Space

Everyone has a notion of what is meant by a point. However, check your dictionary for a definition. You will find it expressed in words having less meaning to you than the word itself. Similarly, everyone has a notion or a definite idea of the meanings of "line," "plane," and "space," but attempting to give these terms precise mathematical definitions is often more confusing than helpful. Thus, we shall say that *point*, *line*, *plane*, *space*, and the idea of *betweenness* are undefined geometric ideas.

To picture a point, mark a dot on your paper. The smaller the dot, the better. In fact, if you can imagine the dot growing smaller and smaller until finally it cannot be seen, you have a good idea of the meaning of a geometric point. Although we use a dot to represent a point, no one can actually see a point, because it has no dimension. In this book a point, indicated in figures by ·, will be denoted by a capital letter such as A, B, P, or Q. An example of a point is the place where two lines intersect.

We shall think of space as being the set of all points. Thus, we visualize the physical universe as completely filled, or dense, with points. We may say there are an unlimited number of points in space. In space exist all the points that comprise the geometric figures we will study.

A line is a particular set of points in space. If you make a straight row of dots and keep adding more and more dots to the same row until finally you have a solid mark, you have a good picture representation of a line. However, no one can really see a line, for it is made up of points that we have already described as invisible. A line has no thickness, but it does extend indefinitely in two directions. We name a line by naming any pair of points on the line and by placing a double arrow over them. Thus, we may denote the line in Figure 9-1 by $\overleftrightarrow{AB}$, $\overleftrightarrow{AC}$, or $\overleftrightarrow{BC}$. Sometimes lines are denoted by single small letters such as a, b, l, m, r, and s. The set of points on $\overleftrightarrow{AB}$ is denoted by $\{x \mid x \subset \overleftrightarrow{AB}\}$.

When we visualize a plane, we usually think of a set of points lying on some flat surface, such as the top of a table. However, a plane has no thickness, even though it extends indefinitely in all other directions.

We sometimes say that a plane is two dimensional with length and width whereas a line is one dimensional with length. However, a flat surface is *not* a plane. A plane cannot be drawn or constructed physically. Thus, an extension of a flat surface, like the blackboard or the wall, is simply what we imagine a plane to be. We shall use Greek letters, α, β, and so on, to represent planes. The set of points on α is denoted by $\{x \mid x \in \alpha\}$.

If the members of a set of points are on a line, they are said to be *collinear*, and the line is said to *pass through* them or to *contain* them. If three points are collinear, then one of the points is *between* the other two. Which one of the points in Figure 9-1 is between the other two?

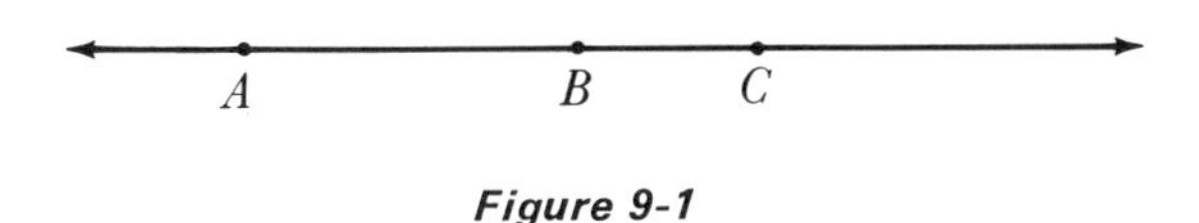

Figure 9-1

Consider a point in space. How many lines can contain (that is, pass through) this point? It should be easy for you to visualize that infinitely many lines can contain one single point. Now, how about two points? How many lines in space can contain two different points in space? Your answer to this question should agree with the following axiom concerning points and lines in space.

Axiom 9-1: There is exactly one line that contains two different points in space.

We may think of this as a "line of sight." Take any two points in a room. Between these two points stretch a string. Any other string stretched between these two points would occupy the same position as the first string. Thus, we see that two different points *uniquely determine* a straight line, or, *for two different points in space, there is exactly one line containing these two points.*

The operation of intersection for sets was defined and discussed in Chapter 2, and you are encouraged to review this material. This definition of intersection is applied now to particular pairs of point sets such as two lines.

Theorem 9-1: If two distinct lines intersect (contain a point in common), then the intersection is *exactly* one point.

Proof: An indirect method of proving this theorem involves the assumption that the intersection of two lines consists of more than one point. Assume that two distinct lines, a and b, are such that $a \cap b$ contains two or more distinct points, P and Q. By Axiom 9-1, only one line can contain both P and Q. Thus a and b must be the same line. This contradicts the fact that a and b are distinct lines. Thus, two distinct lines intersect in at most one point.

Any point on a line separates the line into three sets of points, each of which is a subset of the points forming the line; two of the sets are the infinite sets lying on each side of the point, while the third set is the point itself. The two infinite sets are called half-lines. Let point F in Figure 9-2 separate a line into two half-lines. If points A and C are on the same half-line, then A is between F and C or C is between A and F. If B and C are on different half-lines, then F is between B and C.

Definition 9-1: The union of the point that separates a line into half-lines and either of the half-lines is called a *ray*.

The first letter used to denote a ray names the endpoint, and the second letter names some point on the half-line. A is the endpoint of ray $\overrightarrow{AB}$. If D is a point on $\overrightarrow{AB}$ other than A, then $\overrightarrow{AB}$ and $\overrightarrow{AD}$ name the same rays. In Figure 9-2, A divides the line into two rays, $\overrightarrow{AB}$ and $\overrightarrow{AC}$, which are called opposite rays. Note that ray $\overrightarrow{AB}$ points toward the right and ray $\overrightarrow{AC}$ toward the left. Notice also that

$$\overrightarrow{AB} \cap \overrightarrow{AC} = A.$$

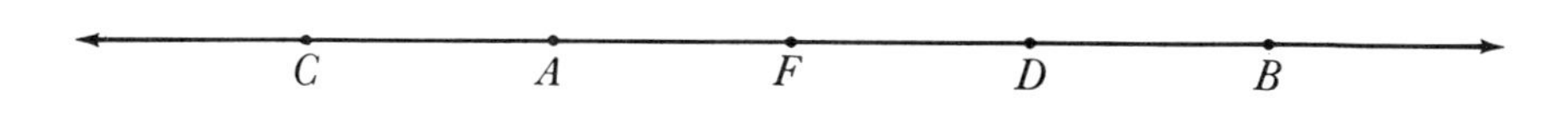

Figure 9-2

The idea of a ray is illustrated by a searchlight beam.

Definition 9-2: A *line segment* is a subset of a line consisting of two points and the set of points between them.

Consider the points A and B in Figure 9-2. These two endpoints and all the points between them will be called "segment AB" and will be denoted by either $\overline{AB}$ or $\overline{BA}$. Thus, $\overline{AB}$ consists of the points that $\overrightarrow{AB}$ and $\overrightarrow{BA}$ have in common, or $\overline{AB} = \overrightarrow{AB} \cap \overrightarrow{BA}$.

For two line segments to be equal, the endpoints must be identical. That is, if $\overline{AB} = \overline{CD}$, then the points A and B are the same as C and D or D and C; in fact, all points on $\overline{AB}$ are identical to all points on $\overline{CD}$.

Examples: (a) Find $\overline{CD} \cap \overline{AB}$ as given in Figure 9-2. The answer is $\overline{AD}$, because all the points in $\overline{AD}$ are common to both $\overline{CD}$ and $\overline{AB}$.

(b) Find $\overrightarrow{CA} \cap \overrightarrow{AC}$. The answer is $\overline{AC}$.

(c) Find $\overline{CA} \cap \overline{DB}$. The answer is $\varnothing$.

Just as a point separates a line into half-lines, a line separates a plane into half-planes. That is, any line on a plane separates the plane into three disjoint sets: two half-planes and the line. For example, in Figure 9-3 the shaded region represents one half-plane, while the remainder, excluding the line, represents the other half-plane. That portion of the plane α that contains point A is called the A side of line s, and the other half-plane is called the B side of line s. If A and B were on the same side of line s, then segment $\overline{AB}$ would lie entirely in one half-plane. Would $\overline{AB}$ in Figure 9-3 lie in one half-plane? Why or why not?

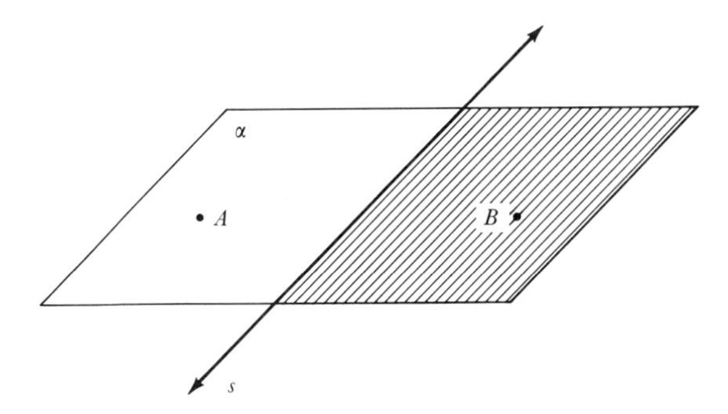

Figure 9-3

We now extend this concept of separation to space with the statement that a plane separates space into three infinite sets of points, each of which is a subset of the points in space. These three sets are sometimes described as the half-space on one side of the plane, the plane, and the half-space on the other side of the plane. In Figure 9-4, D lies in the half-space on the D side of plane α, and E lies in the

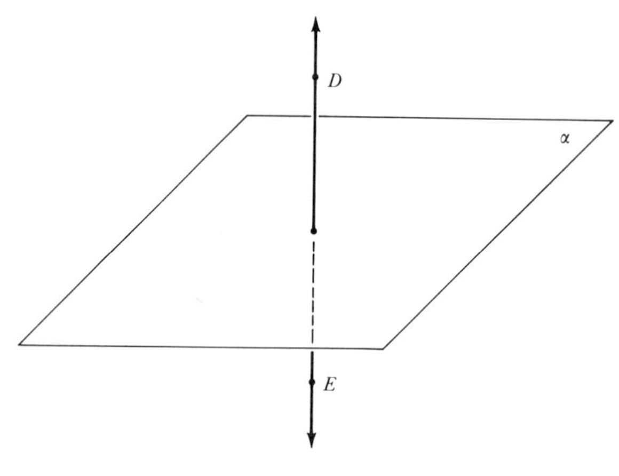

Figure 9-4

half-space on the E side of α. D and E lie in different half-spaces.

Note that the intersection of two half-lines separated by a point, two half-planes separated by a line, or two half-spaces separated by a plane is empty; that is, none of the above pairs intersect. The half-lines are separated by an endpoint, the half-planes by a line called the boundary, and the half-spaces by a plane.

Exercise Set 9-1

1. What is the greatest number of points of intersection determined by two distinct lines in the same plane? Three distinct lines in a plane? Four distinct lines in a plane? Three lines in space? Four lines in space?
2. How many different lines can you draw through one point? Two distinct points? Three noncollinear points?
3. Draw figures to show that the intersection of a line segment and a ray may be any of the following.
   (a) The given segment
   (b) The null set
   (c) A point
   (d) A line segment different from the given segment
4. Indicate whether each statement is true or false. Rewrite the false

statements as true statements.

(a) A line may have an infinite number of different line segments on it.

(b) Space is the set of all points.

(c) A point separates a line into two disjoint subsets.

(d) A line segment contains infinitely many points.

(e) $\overrightarrow{AB}$ is the same set of points as $\overrightarrow{BA}$.

(f) If A, B, and C are collinear with B between A and C, then $\overrightarrow{AB} \cap \overrightarrow{BC} = \overrightarrow{BC}$.

(g) In (f), $\overline{AB} \subset \overline{BC}$.

(h) In (f), $\overline{BC} \cap \overrightarrow{BA} = \overrightarrow{CA}$.

(i) In (f), $\overrightarrow{BA} \cup \overline{BC} = \overline{AB}$.

(j) In (f), $\overrightarrow{BA} \cap \overline{BC}$ is an infinite set of points.

(k) The endpoints of $\overleftrightarrow{BA}$ are B and A.

(l) $\overline{AB}$ is the same line segment as $\overline{BA}$.

(m) The two half-planes formed by a line include all the points of the plane.

(n) If A, B, and C are noncollinear points, then $\overrightarrow{AB} \cap C = \varnothing$.

(o) In (n), $\overleftrightarrow{AB} = \overleftrightarrow{AC}$.

(p) In (n), $\overrightarrow{BC} \cap \overrightarrow{AC} = C$.

(q) A line separates a plane into two disjoint sets.

(r) The symbol for a segment from A to B is $\overleftrightarrow{AB}$.

(s) A ray has two endpoints.

(t) Space is contained in a plane.

5. Using the number line given, determine each of the following.

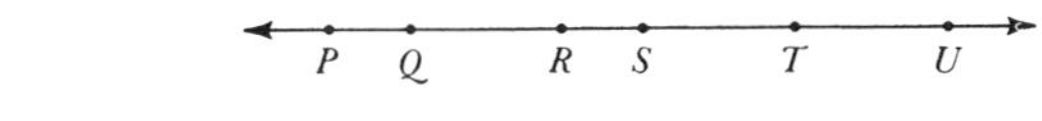

(a) $\overline{PR} \cap \overline{QT}$ (b) $\overline{RS} \cap \overline{ST}$

(c) $\overline{QS} \cap \overrightarrow{RU}$ (d) $\overline{PQ} \cap \overline{ST}$

(e) $\overline{RS} \cap \overrightarrow{QU}$ (f) $\overrightarrow{QS} \cap \overrightarrow{RQ}$

(g) $\overrightarrow{RP} \cap \overrightarrow{RU}$ (h) $\overrightarrow{RP} \cap \overrightarrow{RQ}$

6. In Exercise 5, which of the following are empty?

(a) $\overline{QR} \cap \overline{ST}$ (b) $\overrightarrow{QR} \cap \overline{ST}$ (c) $\overrightarrow{TS} \cap \overline{PQ}$

(d) $\overline{RS} \cap \overline{ST}$ (e) $\overline{PQ} \cap \overrightarrow{ST}$ (f) $\overrightarrow{RP} \cap \overrightarrow{UT}$

7. Draw two segments $\overline{UV}$ and $\overline{XW}$ for which $\overline{UV} \cap \overline{XW}$ is empty but for which $\overleftrightarrow{UV} \cup \overleftrightarrow{XW}$ consists of one line.

8. Draw two segments $\overline{AB}$ and $\overline{CD}$ for which $\overline{AB} \cap \overline{CD}$ is empty but for which $\overleftrightarrow{AB}$ is not the same as $\overleftrightarrow{CD}$.

9. Draw $\overline{AB}$ and $\overline{CD}$ on the same line so that
   (a) $\overline{AB} \cap \overline{CD} = \overline{CD}$.
   (b) $\overline{AB} \cap \overline{CD} = \overline{BC}$.
   (c) $\overline{AB} \cap \overline{CD} = \overline{AB}$.
   (d) $\overline{AC} \cap \overline{CD} = C$.
   (e) $\overline{AB} \cap \overline{CD} = \varnothing$.
   (f) $\overline{AB} \cap \overline{CD} = \overline{AD}$.

10. Consider the figure at the right. Find
   (a) $\overrightarrow{AC} \cap \overrightarrow{BA}$.
   (b) $\overline{AC} \cap (\overline{BC} \cap \overline{BA})$.
   (c) $\overrightarrow{AB} \cap \overrightarrow{BA}$.
   (d) $\overline{AC} \cap \overline{BA}$.

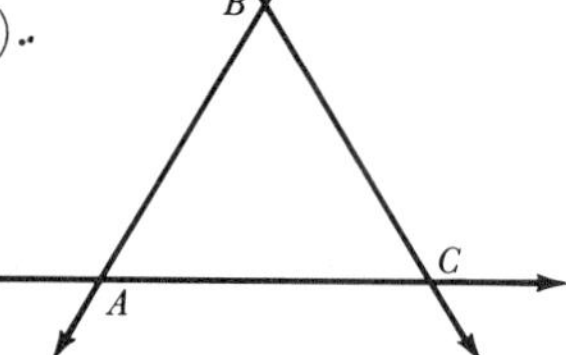

11. Line l contains the four points A, B, C, and D.

Name two rays on l with the following characteristics.
   (a) Their intersection is a ray.
   (b) Their intersection is a point.
   (c) Their intersection is a segment.
   (d) Their intersection is empty.

12. In Exercise 11, name two line segments such that:
   (a) their intersection is one of the segments.
   (b) their intersection is B.
   (c) their intersection is $\overline{BC}$.
   (d) their intersection is empty.

*13. Using terms such as segments, half-lines and so on, name the following sets where x represents points on a number line.
   (a) $\{x \mid x > 1\}$
   (b) $\{x \mid x \leq -2\}$
   (c) $\{x \mid x \geq 3\} \cap \{x \mid x \leq 5\}$
   (d) $\{x \mid x \leq 3\}$ and $\{x \mid x \geq 2\}$

*14. Consider the following models and determine whether or not they satisfy Axiom 9-1. Consider buildings as points and streets as lines.
   (a) Consider a new subdivision consisting of exactly three houses, not all on the same street, but with a street connecting any given pair.
   (b) Let a building on a given campus be a point and let a given pair of buildings on the campus be a line.

3

Properties of Lines and Planes

How many points are necessary to determine a plane? Is a plane fixed with two points? Fold several pieces of paper together and put a paper clip along the fold so that they will stay together. Let A and B be two points on the fold. Each half-page of paper is part of a plane containing points A and B (see Figure 9-5). Thus, there can be many planes through two distinct points, and therefore two distinct points do not determine a plane.

If you sit on a three-legged stool whose legs are not the same length, all three legs will contact the floor. However, try sitting in a four-legged chair having one leg shorter than the other three legs. Now how many legs will touch the floor? Did you get three? This suggests that *there is only one plane containing three distinct noncollinear points.*

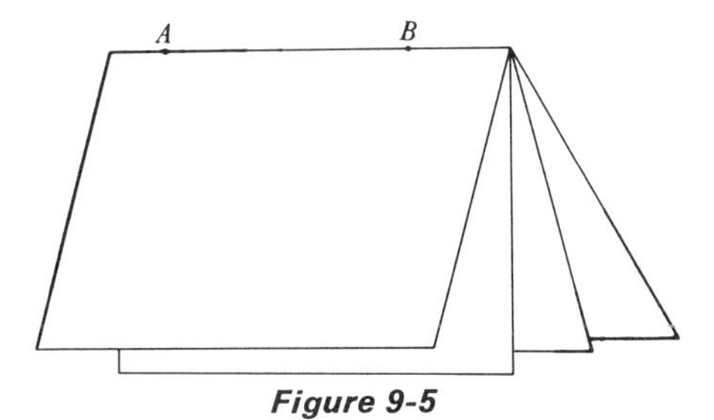

Figure 9-5

Axiom 9-2: There is one and only one plane containing three distinct noncollinear points.

Now examine the walls of a room. Let the walls represent parts of planes, since planes extend without bounds. The intersection of two walls (parts of planes) is clearly a line segment. We can generalize this idea with the following axiom.

Axiom 9-3: If the intersection of two distinct planes is not empty, then the intersection is a line.

Now mark two points on the blackboard, and stretch a rubber band between them. Notice how the stretched rubber band seems to cling to the blackboard. Intuitively, this should suggest the following theorem.

Theorem 9-2: If a line contains two different points of a plane, it lies in the plane.

Proof: Let A and B be two points of a plane on a given line. Consider a point C not on the given plane. By Axiom 9-2, point C and points A and B of the given plane determine a second plane. The intersection of the second plane and the given plane, by Axiom 9-3, is a line. This intersection also contains points A and B. By Axiom 9-1, there is only one line containing A and B. Thus, the given line and the line of intersection are the same. Consequently, the given line is in the plane.

Planes do not always intersect. Examine the opposite walls of a room. We suggest that these walls will not intersect no matter how far they are extended. Such planes are said to be *parallel.*

Definition 9-3: If the intersection of two distinct planes is empty, then the planes are said to be *parallel.*

A similar definition may be formulated for two lines in a plane where the intersection of the two lines is empty.

Definition 9-4: Let r and s be any two distinct lines contained in a plane. If $r \cap s$ is empty, then r and s are said to be *parallel* (Figure 9-6).

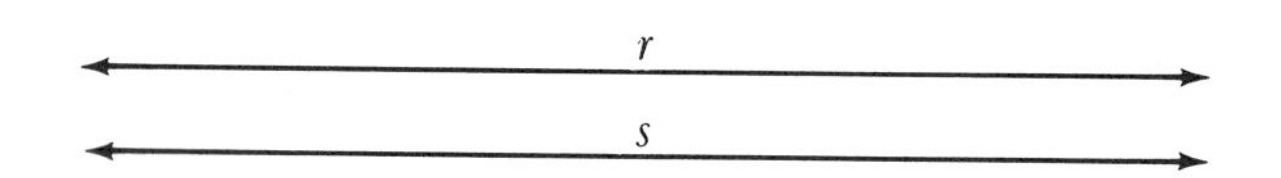

Figure 9-6

When we say that two rays or two line segments are parallel, we mean that lines containing these segments or rays are parallel.

Can you imagine two lines that do not meet anywhere but still are not parallel? This could happen if the two lines are not in the same plane. Notice in Figure 9-7 that r and s will not intersect no matter how far they are extended. Note also that they are not parallel. Such lines are called *skew lines.*

Definition 9-5: If lines r and s are not in the same plane and $r \cap s$ is empty, then r and s are said to be *skew lines* (Figure 9-7).

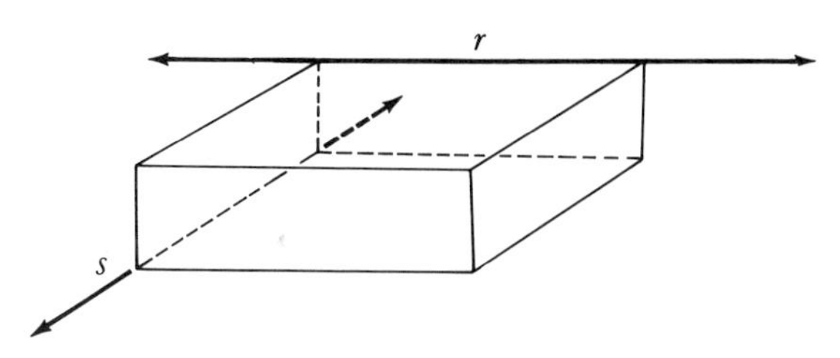

Figure 9-7

The intersection of a line and a plane may be summarized by the following discussion. Imagine a sheet of paper representing a plane punctured by a needle representing a line. This illustration suggests, as shown in Figure 9-8(a), that the intersection of a line and a plane is

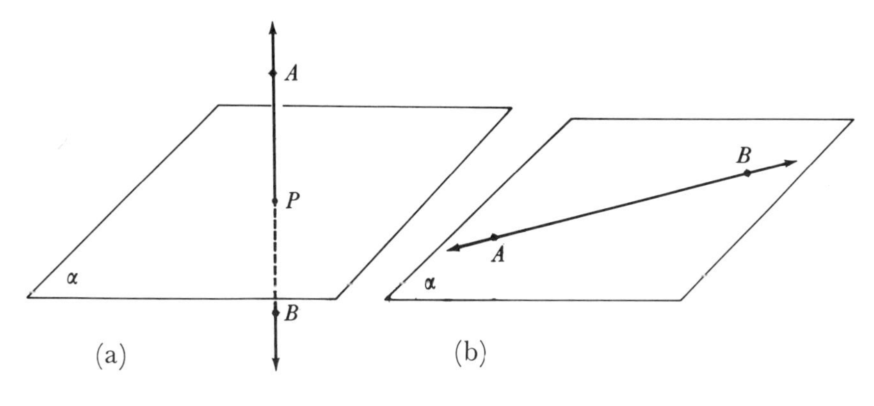

Figure 9-8

a point P; that is, $\alpha \cap \overleftrightarrow{AB} = P$. This discussion leads to the following theorem.

Theorem 9-3: If a line intersects a plane not containing it, then the intersection is a single point.

However, lay the needle flat on the paper. This suggests that the line may be contained entirely in the plane (Figure 9-8(b)).

Finally, hold the needle so that if it were extended indefinitely, it would never intersect the plane. The line and the plane are said to be *parallel*.

The intersection of a line and a plane may be summarized by the following statements. Let α represent a plane and s a line.

(a) If s is contained in α, then $\alpha \cap s = s$.
(b) If α and s are parallel, then $\alpha \cap s = \varnothing$.
(c) If s is not contained in α and if s and α do intersect, then $\alpha \cap s$ is exactly one point.

There are many ways to determine a plane. The most familiar ones are the four below.

(a) Three noncollinear points determine a plane (Axiom 9-2).
(b) A line and a point not on the line determine a plane.
(c) Two parallel lines determine a plane.
(d) Two intersecting lines determine a plane.

Can you prove, using Axiom 9-2, that each of the preceding determines a plane? (See Exercise 11, Exercise Set 9-2.)

Some interesting properties of lines and planes seem plausible on the basis of our experience with physical objects. We first state an axiom and then tabulate several of these properties.

Axiom 9-4: Let P and r be a point and a line in a plane with P not contained in r. Then, in the plane, there is one and only one line that contains P and is parallel to r.

Using Axioms 9-2 and 9-4 and the theory of this chapter, can you verify these properties? (See Exercise 12, Exercise Set 9-2.)

(a) If two parallel planes are cut by a third plane, the lines of intersection are parallel.
(b) If a plane contains one and only one of two parallel lines, it is parallel to the other line.
(c) If two lines are parallel to a third line, they are parallel to each other.
(d) Through a point not in a given plane, there is exactly one plane parallel to a given plane.
(e) If two intersecting lines are each parallel to a plane, then the plane of these lines is parallel to the given plane.

Exercise Set 9-2

1. Suppose we are given four points A, B, C, and D, no three of which

are collinear. If they are coplanar (lying in same plane), how many lines do they determine? If they are not coplanar?

2. Carefully fold a piece of paper in half. Hold the paper so that the fold is on the table top. Consider the three planes—the table top and the two parts of the folded paper. What is the intersection of these three planes? Now stand the folded paper on its end. What is the intersection of the three planes now?
3. Indicate whether each statement is true or false. If false, discuss why the statement is false.
   (a) Three points may be noncoplanar.
   (b) Two lines either intersect or are parallel.
   (c) Two planes either intersect or are parallel.
   (d) A line and a plane either intersect or are parallel.
   (e) If two different planes intersect, their intersection is a line.
   (f) Two distinct lines may intersect in more than one point.
   (g) If points A and B lie in plane α and $x \in \overleftrightarrow{AB}$, then $x \in \alpha$.
   (h) If line l intersects plane β, then $x \in l$ implies $x \in \beta$.
4. Given the figure below, let α and β represent the set of points on each of two intersecting planes. Find the following.

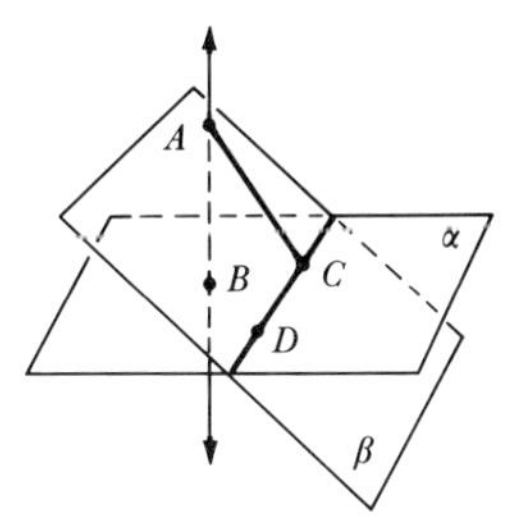

(a) Two skew lines
(b) $\overleftrightarrow{AC} \cap \beta$
(c) $\overleftrightarrow{AB} \cap \alpha$
(d) The intersection of a half-space and a ray
(e) $\overleftrightarrow{AB} \cap \beta$
(f) The intersection of a half-space and a plane
(g) $\overleftrightarrow{AB} \cap \overleftrightarrow{CD}$
(h) The intersection of a half-space and a line

5. Use the given figure with $\overleftrightarrow{RS}$ and points A, B, C, and D in plane α to answer the following questions.
   (a) $\overline{CD} \cap \overleftrightarrow{RS} = ?$
   (b) $\overline{AC} \cap \overleftrightarrow{RS} = ?$
   (c) Why do we say B and C are in different half-planes?
   (d) $\overrightarrow{XS} \cap \overline{CD} = ?$

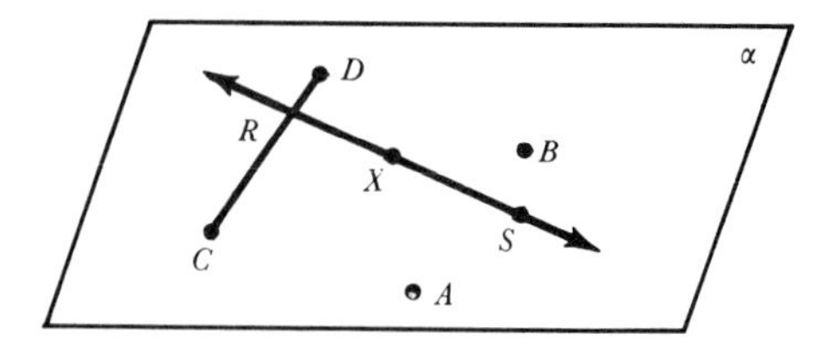

6. (a) Can skew lines be coplanar?
   (b) Do two distinct lines always determine a plane?
   (c) How many planes are determined by five points in space, no three of which are collinear and no four of which are coplanar?
7. (a) Planes α and β have points A, B, and C in common. Suppose A, B, and C are noncollinear. What conclusion can you draw about α and β?
   (b) If α and β are distinct planes having points A, B, and C in common, what conclusion can you make about points, A, B, and C?
8. If l is a line and A and B are points not on l, how many planes are there which contain the points A and B and at least one point on line l?
9. Suppose that two planes α and β intersect at line m to give the four half-planes α_1, α_2, β_1, and β_2, where α is composed of α_1 and α_2 and β is composed of β_1 and β_2.
   (a) If point $A \in \alpha_1$ and point $B \in \alpha_2$, describe the intersection of line $\overleftrightarrow{AB}$ with β.
   (b) If line l intersects α_1 and β_1, then does l intersect m? Explain your answer.

*10. Suppose that a geometric set is composed of exactly five points in space, A, B, C, D, and E, no three of which are collinear and all five of which are not in the same plane. The following three axioms operate within this geometry:
   (1) Two points determine a line.
   (2) Four points determine a plane.
   (3) The intersection of two planes is three points.

   Now, using the points and axioms defined for this geometry, answer the following questions.
   (a) How many lines can be formed in this set?
   (b) How many planes can be formed?
   (c) Using axioms (1) and (2), determine the number of lines that can be formed in a plane.
   (d) What is the intersection of planes $ABCD$ and $BCDE$? Of $ACDE$ and $ABDE$?

*11. Use Axiom 9-2 to verify the four ways to determine a plane as listed in this section.

*12. Use the axioms and theorems of this chapter to validate the five properties listed at the end of this section.

4

Angles

So far in our discussion, we have considered the intersections of sets of points. In like manner, the union of sets of points may be used to discuss geometric relationships among various sets.

In Figure 9-9(a), $\overline{AC} \cup \overline{AB} = \overline{AC}$, $\overline{AB} \cup \overline{BC} = \overline{AC}$, but $\overline{AB} \cup \overline{CD}$ consists of two disjoint sets, $\overline{AB}$ and $\overline{CD}$. Thus, if two segments have no common points, the union consists of disjoint sets. Note that the two segments $\overline{AB}$ and $\overline{BC}$ in Figure 9-9(b) do have a common endpoint. However, the union is not $\overline{AC}$. In this case, the union is called a *broken line*. *DABC* in Figure 9-9(b) is also a broken line.

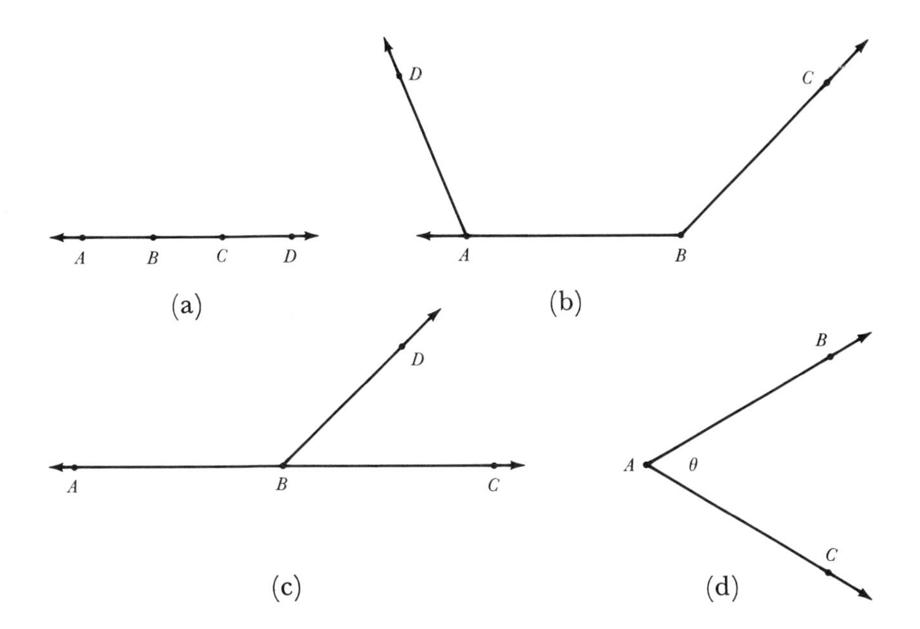

Figure 9-9

Figure 9-9(c) is used to discuss the union of rays. $\overrightarrow{BA} \cup \overrightarrow{BC}$ is $\overleftrightarrow{AC}$; $\overrightarrow{BD} \cup \overrightarrow{BA}$ is the broken line DBA; and $\overrightarrow{BD} \cup \overrightarrow{BC}$ is the broken line DBC. The union of the half-line on $\overleftrightarrow{BC}$ to the right of B and point B is $\overrightarrow{BC}$. The union of this half-line and $\overline{AB}$ is the ray $\overrightarrow{AC}$. The union of this half-line and $\overrightarrow{BD}$ is the broken line DBC. The union of the half-plane above $\overleftrightarrow{AC}$ and the half-plane below $\overleftrightarrow{AC}$ is all of the plane except $\overleftrightarrow{AC}$.

The preceding discussion of broken lines provides the terminology and background material necessary in order to introduce the concept of angles.

Definition 9-6: An *angle* is the union of two rays that have the same endpoint.

Each angle consists of two different rays. The two rays have the same endpoint, but in order to avoid ambiguity we agree here that the two rays forming an angle will not be on the same line. This restriction is introduced to avoid, for the present, considering a line as an angle.

In Figure 9-9(d) the rays $\overrightarrow{AB}$ and $\overrightarrow{AC}$ are called the *sides* of the angle, while the common endpoint A is called the *vertex*. Although an angle cannot actually be drawn, a figure such as the one in Figure 9-9(d) is a visual representation of an angle. An angle may be named in a number of ways:

(a) by using the symbol $\angle$ with the vertex, such as $\angle A$.

(b) by using the symbol $\angle$ and three points A, B, and C, such as $\angle BAC$ or $\angle CAB$. The vertex point is indicated by the middle letter.

(c) by a Greek letter such as θ or a letter of the alphabet such as x or y placed as indicated in Figure 9-9(d).

An angle, in a manner similar to a line separating a plane, separates a plane into disjoint sets of points which are called the *interior* and the *exterior* of the angle. The shaded portion in Figure 9-10(a) is the set of points on the C side of $\overleftrightarrow{AB}$; and in Figure 9-10(b), the shaded portion is the set of points on the B side of $\overleftrightarrow{AC}$. The intersection of these two sets is called the interior of $\angle BAC$ (double shading of Figure 9-10(c)).

Definition 9-7: A set of points is the *interior* of an angle ($\angle BAC$, Figure 9-10) if and only if it is the intersection of the half-plane on the B side of $\overleftrightarrow{AC}$ and the half-plane on the C side of $\overleftrightarrow{AB}$.

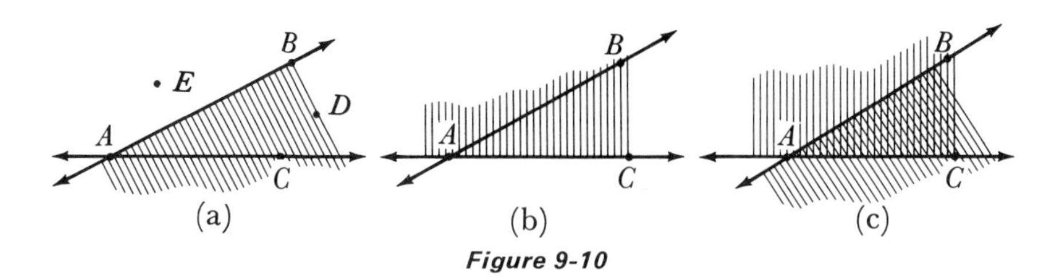

Figure 9-10

In Figure 9-10(a), D is said to be in the interior of $\angle BAC$. The *exterior* of an angle is the set of points of the plane not on the angle or in the interior of the angle. In Figure 9-10(a), E is in the exterior of the angle.

Angles in space may be defined in terms of intersecting half-planes.

Definition 9-8: A *dihedral angle* is the union of two distinct half-planes and their line of intersection (Figure 9-11).

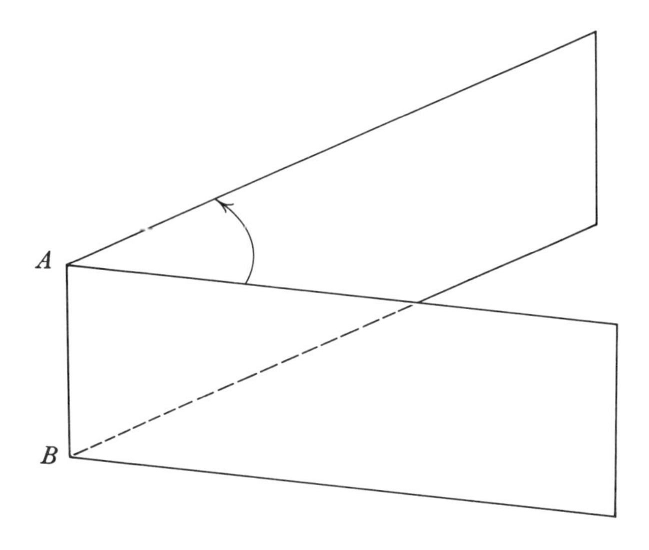

Figure 9-11

Note that the common edge of the dihedral angle, $\overleftrightarrow{AB}$ in Figure 9-11, does not belong to either half-plane. Thus, it was included in the definition so that it would be considered as part of a dihedral angle.

Figure 9-12 shows the intersection of two lines. Observe that four angles, $\angle AOB$, $\angle AOC$, $\angle COD$, and $\angle DOB$, are formed by the intersection of these two lines. Since angles AOB and AOC have a common ray $\overrightarrow{OA}$, we call them *adjacent angles.* Two angles that have a common vertex and a common side are called *adjacent angles* if the intersection of the interiors of the angles is empty. Angles COA and DOB

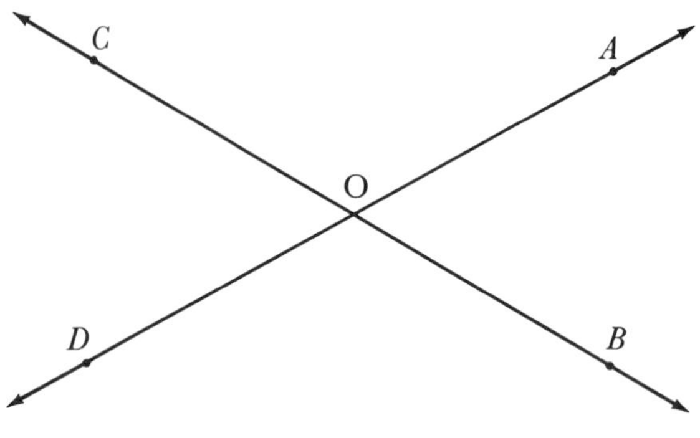

Figure 9-12

are not adjacent angles. The nonadjacent angles formed by the intersection of two lines are called *vertical angles.* Two angles are said to be *vertical* if their union is the union of two lines. Angles COD and AOB are said to be vertical angles.

Exercise Set 9-3

1. For each of the following sets of angles, tell which of the marked angles are adjacent.

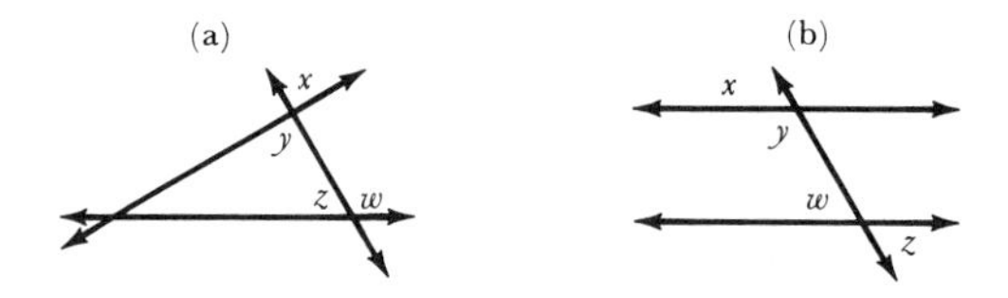

2. For Exercise 1, tell which angles are vertical.
3. Are angles x and y adjacent in the following figures? If not, explain.

(c)

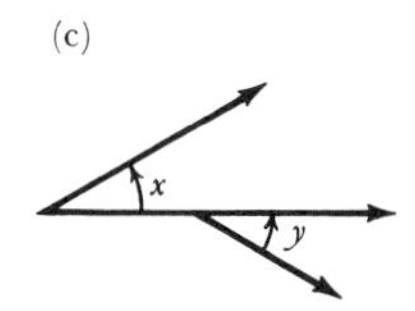

(d)

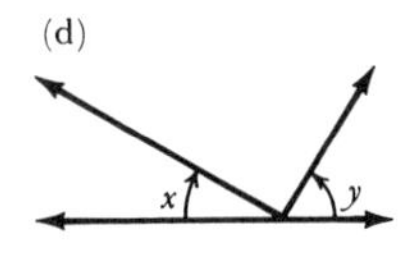

4. Make sketches that show the intersection of two angles as each of the following.
   (a) A segment (b) Empty
   (c) A ray (d) Exactly two points
   (e) One point (f) Ray and vertex of each (intersection of interiors empty)
5. Draw sketches of four possible intersections of a line and an angle.
6. For the figure given, name the following sets of points.
   (a) $\angle ABC \cap \overline{CA}$
   (b) $\overline{AB} \cap \angle ACB$
   (c) $\angle BAC \cap \angle ACB$
   (d) $\overline{AB} \cap \overline{CB}$
   (e) $\overrightarrow{BC} \cap \overline{CB}$

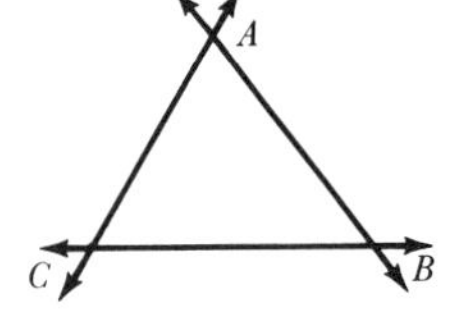

7. How many pairs of vertical angles are formed by
   (a) three concurrent lines? (b) four concurrent lines?
   (Concurrent lines contain a common point.)
8. Can two adjacent angles be a pair of vertical angles?
9. Can the angles of a pair of vertical angles be adjacent to one another?
10. Draw angle *CDE*. Label points *F* and *G* in the interior and *H* and *J* in the exterior.
    (a) Is every point on $\overline{FG}$ in the interior?
    (b) Is every point on $\overline{HJ}$ in the exterior?
    (c) Can you find points *H* and *J* in the exterior such that $\overline{HJ} \cap \angle CDE$ is not empty?
    (d) Can $\overline{FH} \cap \angle CDE$ be empty?

*11. Consider the intersection of three lines in space.

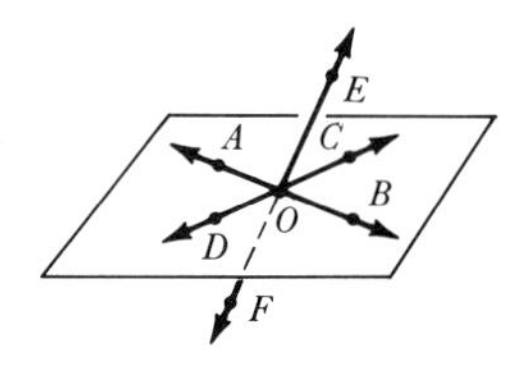

Two of these lines, $\overleftrightarrow{AB}$ and $\overleftrightarrow{CD}$, lie in the same plane; the third line,

$\overleftrightarrow{EF}$, does not lie in their plane.
(a) How many angles are formed? Name them.
(b) Name six pairs of adjacent angles.
(c) Which angles are vertical? Name them.

*12. Suppose you are given four points in space.
(a) What is the greatest number of lines determined by them? The greatest number of angles?
(b) Let the four points be in one plane. How many lines can be determined by them? Illustrate.

5

Simple Closed Curves

In this book, we intuitively consider a *curve* as the set of all points that can be represented by a pencil drawing such that the drawing is made without lifting the pencil and such that no portion of the curve is retraced other than single points. For example, take a line, a line segment, or a ray and deform it, without breaking it, into any other shape or position. The result is a *plane curve.* We agree that *curve* in this text will be used to mean plane curve. A *simple curve* passes through no point more than once. A *closed curve* is drawn so that the drawing starts and stops at the same point.

In Figure 9-13(a), the curve is not closed; (b) represents a simple closed curve; (c) and (d) are not simple closed curves. Why not?

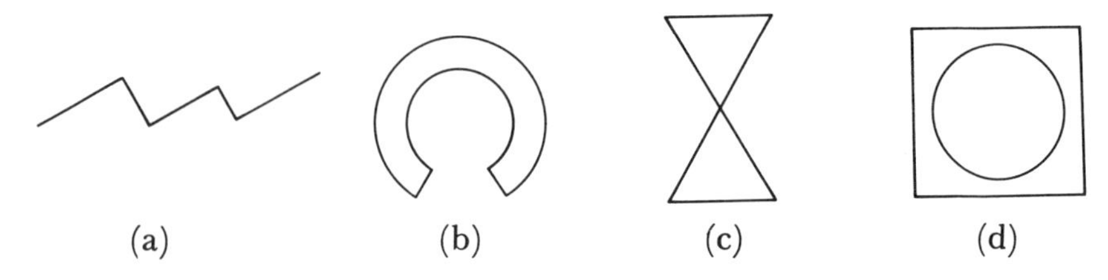

Figure 9-13

An important property of simple closed curves in a plane is that a simple closed curve separates the points of the plane into three sets: the set of points on the curve and two sets called the *interior* and the *exterior* of the curve. Although this result seems to be intuitively obvious, it is a well-known theorem that is extremely difficult to prove.

Theorem 9-4: *Jordan Curve Theorem.* Any simple closed curve c divides a plane into two regions, called the interior and exterior, for each of which c is the boundary.

Thus, for any simple closed curve c in the plane, the plane is the union of three sets, no two of which intersect. The three sets are: the set c, the interior of c, and the exterior of c. Any curve containing a point in the interior and a point in the exterior of a simple closed curve must intersect the simple closed curve. Such a curve is illustrated by the curve containing points A and B in Figure 9-14(a). Any two points in the interior (or any two points in the exterior) may be joined by a curve that does not intersect the simple closed curve. Look again at Figure 9-14(a) and note that curves between A and D can be in the interior of the simple closed curve and curves between E and B can be in the exterior of the simple closed curve.

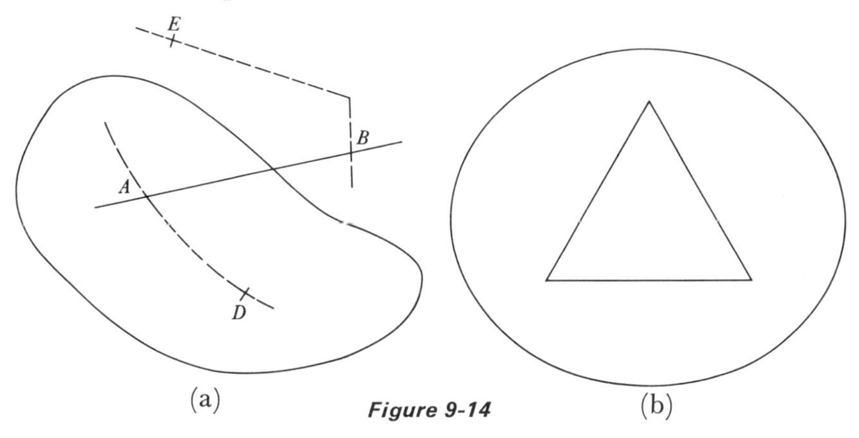

Figure 9-14

The interior of a simple closed curve is called a *region*. There are other types of sets in a plane which are also regions. In Figure 9-14(b), the part of the plane between the two simple closed curves is called a region. Generally, as defined here, a region (as a set of points) includes its boundary.

A set of points is said to be *convex* if a line segment $\overline{PQ}$ drawn between any two points P and Q of the set consists entirely of points of the given set. For example, the sets of points bounded by the first two closed curves in Figure 9-15 [(a) and (b)] are convex. However, neither of the sets of points bounded by the last two closed curves [(c) and (d)] is convex. Why?

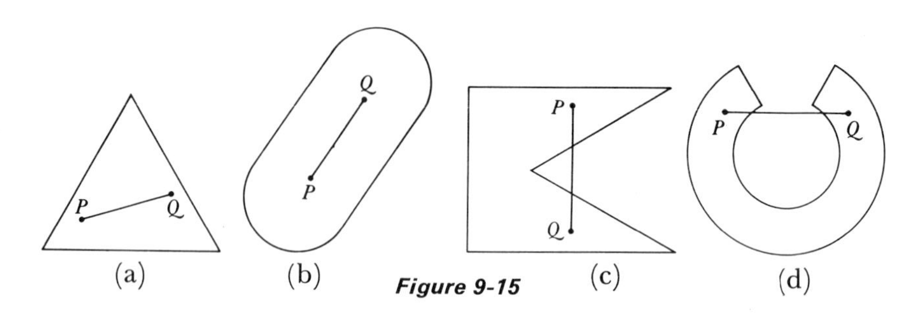

Figure 9-15

Certain simple closed curves play an important role in elementary geometry.

Definition 9-9: A *polygon* is a simple closed curve that is the union of three or more line segments, $\overline{AB}$, $\overline{BC}$, $\overline{CD}$, ... such that the points $A, B, C, D, \ldots$ are coplanar, and no three consecutively named points are collinear. (See Figure 9-16.)

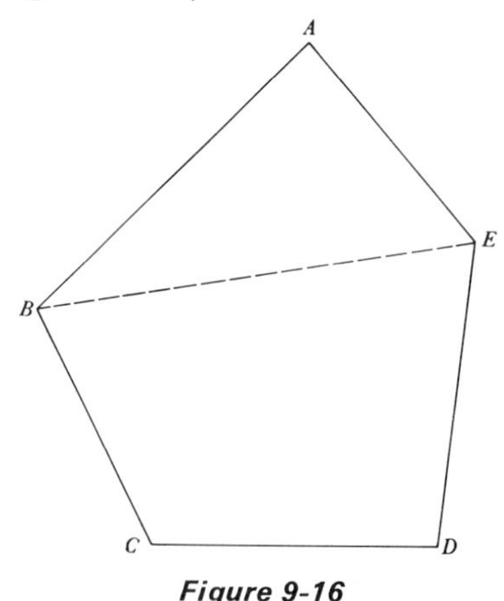

Figure 9-16

A polygon may be either convex or not convex. Can you sketch examples? However, in the remainder of this book, we will use the word polygon to represent a convex polygon.

For the polygon given in Figure 9-16, $\overline{AB}$, $\overline{BC}$, $\overline{CD}$, $\overline{DE}$, and $\overline{EA}$ are called the *sides* of the polygon $ABCDE$. The points A, B, C, D, and E are called the *vertices* of the polygon. *Adjacent vertices* are endpoints of the same side; and the *diagonals* of a polygon, such as $\overline{BE}$, are the line segments joining nonadjacent vertices.

Every point in the plane that is not on the polygon or in the interior of the polygon is said to be an external point. The union of the interior of a polygon and the polygon is called a *polygonal region.*

Definition 9-10: A *quadrilateral* is a polygon made up of four distinct line segments called sides.

Definition 9-11: A *triangle* is a polygon containing three distinct line segments called sides.

In everyday usage, the distinction between a polygon and a polygonal region is frequently not made. For example, many people use "triangle" to refer either to the boundary or to the triangular region. In nonmetric geometry, the distinction must be carefully maintained. Remember, a triangle is a polygon and not a region.

Let A, B, and C be three points not all on the same straight line; then triangle ABC, denoted by $\triangle ABC$, is $\overline{AB} \cup \overline{AC} \cup \overline{BC}$ (Figure 9-17). Just as $\overline{AB}$, $\overline{AC}$, and $\overline{BC}$ are the *sides* of the triangle, we speak of $\angle ABC$, $\angle ACB$, and $\angle BAC$ as the *angles* of the triangle. Since each side of a triangle is only part of the ray contained in the angle, it is intuitively obvious in Figure 9-17 that $\angle BAC$ includes much more than points of $\triangle BAC$. However, it is common practice to call $\angle BAC$ an angle of $\triangle BAC$.

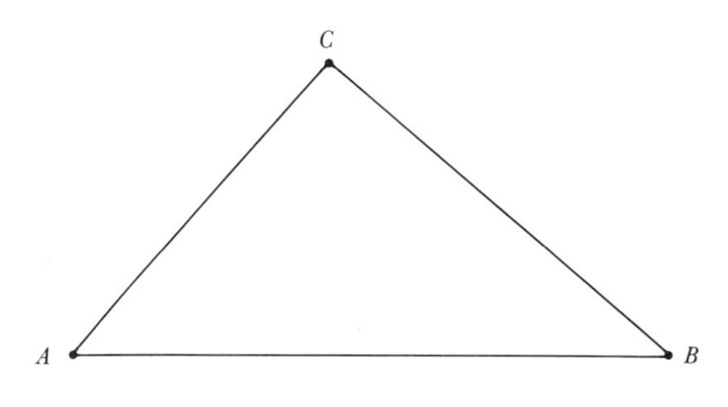

Figure 9-17

A *pentagon* is a polygon with five sides. It is formed by the union of five line segments making a simple closed curve. (See Figure 9-16.) Common names for other polygons are as follows:

Number of Sides	*Name of Polygon*
6	Hexagon
7	Heptagon
8	Octagon
9	Nonagon
10	Decagon
12	Dodecagon

Definition 9-12:

(a) A *parallelogram* is a quadrilateral whose opposite sides lie on parallel lines.
(b) A *trapezoid* is a quadrilateral in which exactly one pair of opposite sides lie on parallel lines.

Exercise Set 9-4

1. Identify the figures that are simple curves.

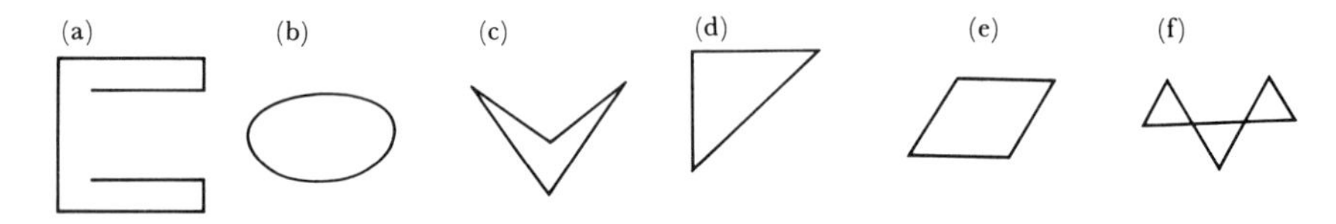

2. In Exercise 1, identify the figures that are closed curves.
3. Classify the figures in Exercise 1 that are closed as convex or not convex.
4. Draw a triangle XYZ. Name the following.
   (a) $\overline{XY} \cap \overline{XZ}$ (b) $\angle XYZ \cap \overrightarrow{XZ}$
   (c) $\overline{XZ} \cap \overleftrightarrow{YZ}$ (d) $\angle YZX \cap \overline{YZ}$
   (e) $\overline{XZ} \cap \overleftrightarrow{XZ}$ (f) $\overline{YZ} \cap \overline{ZX}$
   (g) $\overleftrightarrow{XZ} \cap \triangle XYZ$ (h) $\triangle XYZ \cap \angle XYZ$
5. Explain why an angle is not a simple closed curve.
6. Illustrate a line containing exactly one point of a simple closed curve. Two points. Three points. Four points.
7. Draw two triangles whose intersection is
   (a) one point. (b) two points. (c) three points.
   (d) four points. (e) five points. (f) six points.

8. Label three points not all on the same line as X, Y, and Z. Draw $\overleftrightarrow{XY}$, $\overleftrightarrow{XZ}$, and $\overleftrightarrow{YZ}$. Shade the Y side of $\overleftrightarrow{XZ}$. Shade the Z side of $\overleftrightarrow{XY}$. Shade the X side of $\overleftrightarrow{YZ}$. Has any part been shaded three times? Describe.
9. Give several examples of a convex polygon. What seems to be the smallest number of sides it can have? What about a polygon that is not convex?
10. Answer the following questions using the given figure.

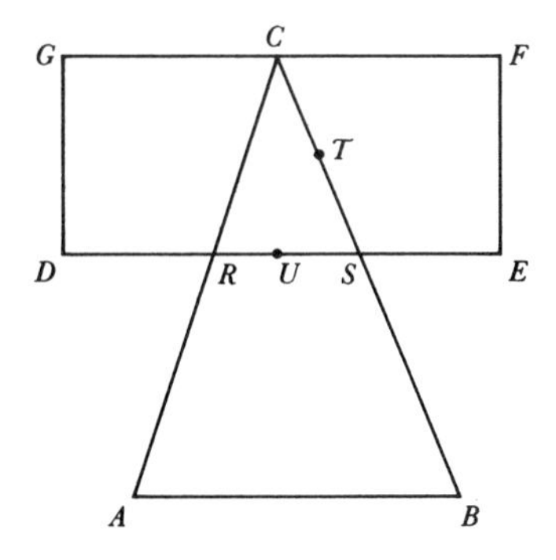

(a) Count the simple closed curves in the figure.
(b) Name a point on $\triangle ABC$ that is in the interior of $DEFG$.
(c) Name three points on curve ABC and also on curve $DEFG$.
(d) How are points T and B located with respect to curve $DEFG$?
(e) $ABSR \cap DEFG =$?

11. Let T be the union of $ABCD$ with its interior, the shaded part of

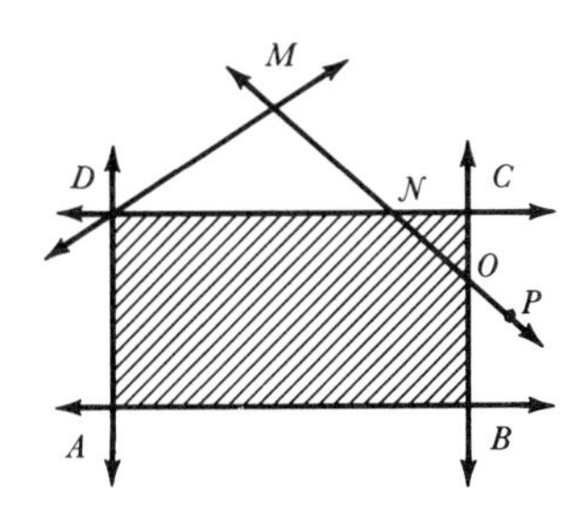

the figure. Decide which of the following are true.

(a) $\triangle NOC \cap ABCD = \{N, O\}$
(b) $\overrightarrow{MP} \cap T = \overline{NO}$
(c) $\overline{MO} \subset T$
(d) $\triangle NOC \subset T$
(e) $\angle DAB \subset T$
(f) D is in the interior of $\angle ABC$
(g) $M \in \overrightarrow{NO} \cup \overrightarrow{AD}$
(h) $\triangle MDN \cap T = \varnothing$

12. (a) Define the interior of $\triangle ABC$ in terms of the intersection of half-planes.
    (b) Define the exterior of $\triangle ABC$ in terms of the union of half-planes.
13. Each of the following polygons has how many diagonals?
    (a) Quadrilateral (b) Triangle (c) Pentagon
14. For the given figure, describe each of the following sets of points.

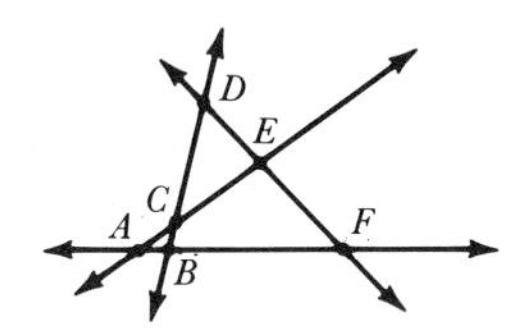

(a) $\triangle DBF \cap \overline{AC}$
(b) Exterior of $\triangle DBF \cap \overline{CE}$
(c) Interior of $\triangle DBF \cap \overleftrightarrow{CE}$
(d) Interior of $\triangle DBF \cap$ interior of $\triangle CDE$
(e) Interior of $\triangle DBF \cup$ interior of $\triangle CDE$
(f) Interior of $\angle EAF \cap$ interior of $\triangle CDE$

*15. Consider three houses (denoted by $\times$) in a row and three utilities (denoted by $\circ$) in a second row on a plane. Join each house to every utility by curves in the plane in such a way that no two curves cross or pass through houses or utilities except at their endpoints.

× × ×

• • •

*16. Consider the following figure.
    (a) Name three points in the exterior of $\triangle ABC$.
    (b) Name two points in the interior of figure *DEFGH*.
    (c) Determine if figures *DEFGH*, *ABC*, *DKLH*, *KBHD*, *FGH*, and *CAKEH* are simple closed convex curves.

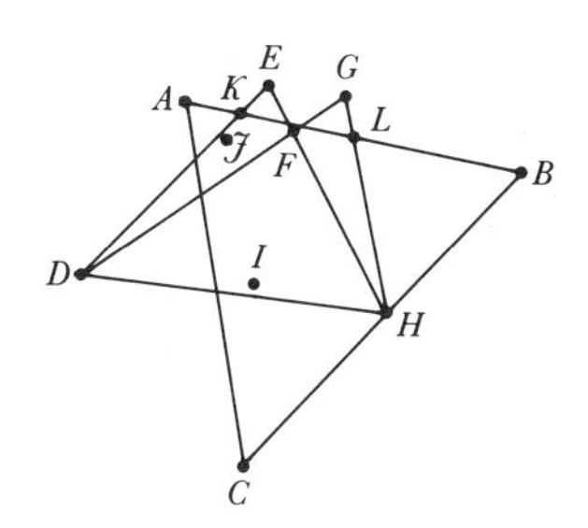

6

Simple Closed Surfaces

Simple closed surfaces are sets of points in space analogous to simple closed curves in a plane. Such surfaces separate the points of space into three disjoint sets of points: the set of points on the surface, the set of points interior to the surface, and the set of points exterior to the surface. We will consider the union of the points on the surface and the points interior to the surface as the *solid* bounded by the closed surface.

We introduce types of space surfaces by a discussion of space surfaces formed by polygonal regions.

Definition 9-13: A *polyhedron* is a simple closed surface formed by polygons and polygonal regions.

The union of points of space bounded by a polyhedron and the polyhedron will be called a *space region.* In Chapter 10, we will discuss the volume of a space region bounded by a polyhedron.

Intuitively, we consider two polygons to be *congruent* if they have the same shape. That is, if one of the polygons were superimposed upon the other, the two polygons would coincide point by point. (Congruent figures will be studied in Chapter 10.) A regular polyhedron is a polyhedron formed by congruent polygons. Polyhedrons with four, six, eight, twelve, and twenty faces or sides are called *tetrahedrons, hexahedrons, octahedrons, dodecahedrons,* and *icosahedrons,* respectively.

We will now consider various kinds of polyhedrons. The first is the prism.

Definition 9-14: A *prism* is a simple closed surface formed by two congruent polygonal regions in parallel planes along with three or more quadrilateral regions joining the two polygonal regions so as to form a closed space figure.

First of all we wish to study what are called *right* prisms. The sides (sometimes called lateral faces) of the right prisms are rectangles and the planes containing the sides are perpendicular to the planes containing the bases of the prisms. Any dihedral angle containing a side and a base is a right dihedral angle. (Rectangle, right angle, right dihedral angle, and perpendicular are metric terms that will be discussed in Chapter 10.)

The *right triangular prism*, Figure 9-18(a), resembles a wedge or a trough. The *rectangular prism*, Figure 9-18(b), could be a closed shoe box or simply a room. Certain types of chalk have surfaces that are shaped in the form of a *hexagonal prism*, Figure 9-18(d).

In the triangular prism in Figure 9-18(a), $\triangle ABC$ and $\triangle DFE$ are congruent and are in parallel planes. The two triangles are joined by three quadrilaterals, $ACED$, $ABFD$, and $BCEF$. The two triangles, $\triangle ABC$ and $\triangle DFE$, are called bases. The quadrilateral regions are called *lateral faces* of the prism. The *faces*, or *sides*, include both bases and lateral faces. The segments forming the boundaries of the faces of the

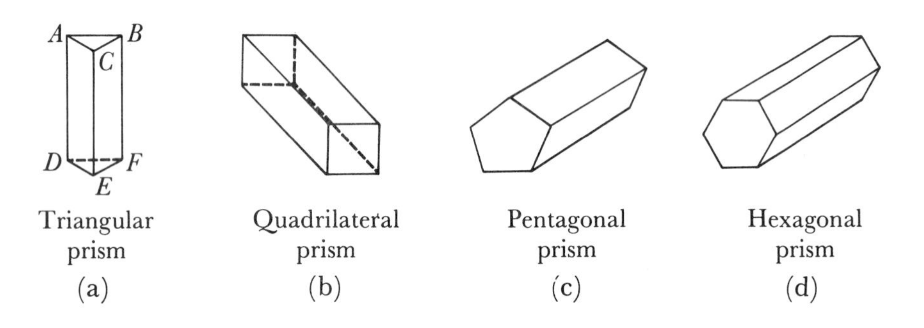

Figure 9-18

right prism, such as $\overline{DE}$, $\overline{CE}$, $\overline{AB}$, and so on in Figure 9-18(a), are called *edges*, and the corners, points A, B, C, D, E, and F, are called the *vertices* of the prism. Prisms are classified by the number of sides, edges, and vertices.

The Swiss mathematician Euler (1707–1783) discovered a relationship between the number of vertices V, edges E, and sides S of polyhedrons. This formula can be written as

$$V + S - E = 2.$$

Note in Figure 9-18 that the triangular prism has 6 vertices, 5 sides, and 9 edges. Since $6+5-9=2$, Euler's formula is satisfied. Count the number of sides, edges, and vertices in the quadrilateral prism and verify that Euler's formula is satisfied.

Another polyhedron of interest is the pyramid. A study of Egyptian history introduces the famous constructions of Egypt, which were built in the form of pyramids.

Definition 9-15: A *pyramid* is a simple closed surface formed by a simple closed polygonal region, a point not in the plane of the region, and the triangular regions joining the point and the sides of the polygonal region.

Pyramids are classified according to the polygonal regions forming the base, as in Figure 9-19.

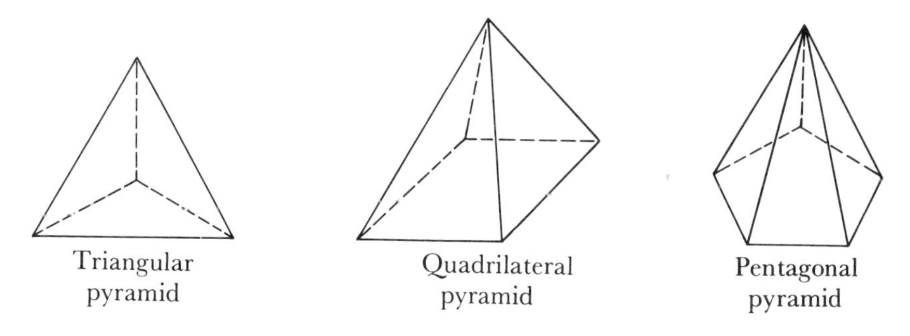

Figure 9-19

Regular solids are characterized by the fact that each face has the same number of edges; that the edges would coincide if one face were placed over another; that the angles would coincide if one were placed over another; and that each vertex is the intersection of the same number of edges. The following are examples of regular solids: a cube is formed by six identical quadrilaterals; a tetrahedron, by four identical triangles; an octahedron, by eight identical triangles; a dodecahedron, by twelve pentagonal regions; and an icosahedron, by twenty identical triangles. Are there other regular solids? Plato stated (correctly) that these are the only possible regular solids, so they are sometimes called the *Platonic solids*. If you will enlarge the patterns shown in Figure 9-20 and construct the regular solids, you will better understand this section.

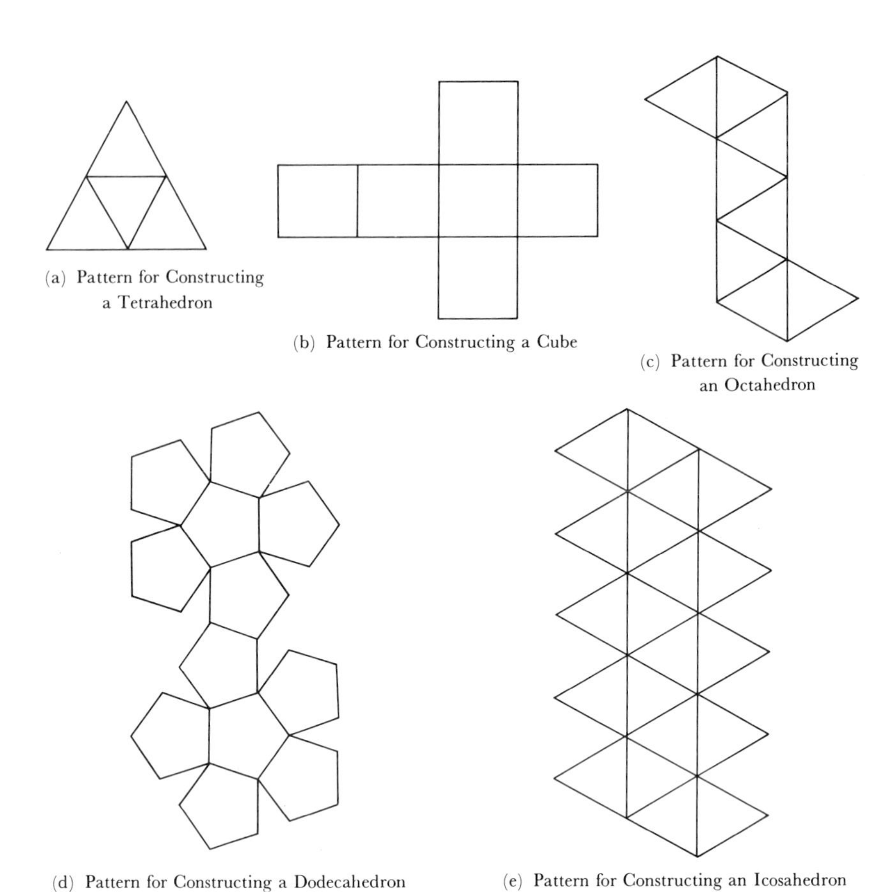

(a) Pattern for Constructing a Tetrahedron

(b) Pattern for Constructing a Cube

(c) Pattern for Constructing an Octahedron

(d) Pattern for Constructing a Dodecahedron

(e) Pattern for Constructing an Icosahedron

Figure 9-20

Exercise Set 9-5

1. Answer true or false.
   (a) A pyramid is a polyhedron.
   (b) A pyramid is a prism.
   (c) Every triangular prism is a right prism.
   (d) Some polyhedrons are prisms.
   (e) The bases of a prism lie in perpendicular planes.
   (f) The sides of a pyramid are parallelograms.
   (g) A pyramid with seven faces has a hexagon as the boundary of its base.
   (h) Euler's formula does not hold for pyramids.

2. (a) What is the smallest number of vertices of a polyhedron?
   (b) What is the smallest number of sides of a polyhedron?
   (c) What is the smallest number of edges of a polyhedron?
   (d) What is the smallest number of vertices of a prism?
   (e) What is the smallest number of sides of a prism?
   (f) What is the smallest number of edges of a prism?
3. Give the name for each space figure; denote the edges and vertices.

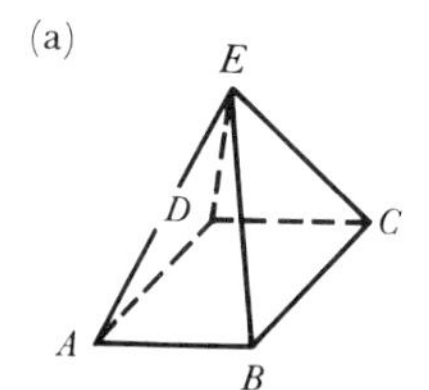

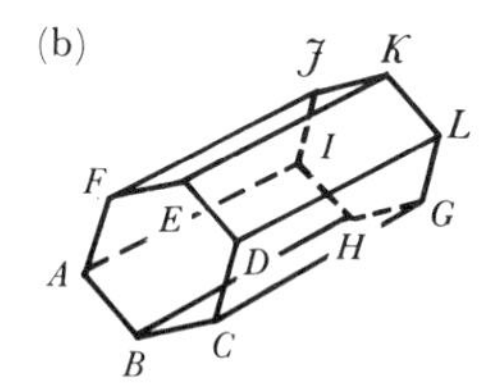

4. By counting the number of sides, vertices, and edges, verify Euler's formula for each space figure named below.
   (a) A pentagonal prism (b) A hexagonal prism
   (c) A quadrilateral pyramid (d) A pentagonal pyramid
5. Complete the table, indicating the number of vertices, edges, faces, and Euler's formula for each of the following regular solids.

Surface	V	E	S	$V+S-E$
Tetrahedron				
Cube				
Octahedron				
Dodecahedron				
Icosahedron				

*6. Count the number of vertices, edges, and sides on what is left of a quadrilateral prism after one corner is cut off. Does Euler's formula hold?

*7. A quadrilateral prism has a tunnel cut in it in the form of a pentagonal prism. Find V, E, S, and $V+S-E$.

*8. Every convex polyhedron has at least two more vertices than half the number of faces, or $V \geq 2 + \frac{1}{2}S$, where $V =$ number of vertices and $S =$ number of faces or sides. Establish the truth or falsity of this statement for the following figures.
   (a) Pentagonal pyramid (b) Quadrilateral prism
   (c) Hexagonal prism (d) Quadrilateral pyramid

Review Exercise Set 9-6

1. Classify each of the following as true or false. If a statement is false, explain why.
   (a) A line separates space.
   (b) A point separates a line.
   (c) For any two points A and B, $\overline{AB} = \overline{BA}$.
   (d) For any two points A and B, $\overrightarrow{AB} = \overrightarrow{BA}$.
   (e) For any two points A and B, $\overleftrightarrow{AB} = \overleftrightarrow{BA}$.
   (f) If $\overrightarrow{AB} = \overrightarrow{AD}$, then $B = D$.
   (g) If $\overline{AB} = \overline{AD}$, then $B = D$.
   (h) A segment separates a plane.
   (i) Closed curves are simple curves.
   (j) All straight lines are curves.
   (k) A polygon has more than three sides.
   (l) A plane region is a simple closed curve.
   (m) Three points, not all on the same line, determine a plane.
   (n) The union of two half-lines is a line.
   (o) The sides of an angle are line segments.
2. Describe each of the following sets of points with reference to the given figure.

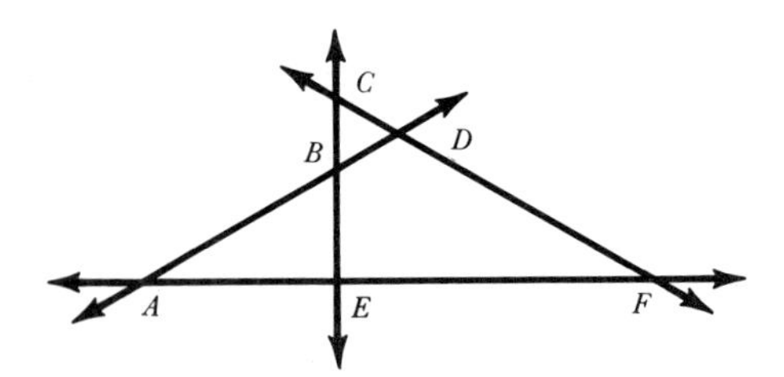

   (a) $\overrightarrow{AF} \cap \overline{AE}$
   (b) $\overline{CE} \cap \triangle ADF$
   (c) (interior $\triangle ADF$) $\cap \overrightarrow{CB}$
   (d) $\overline{AE} \cup \overrightarrow{EF}$
   (e) $\triangle BAE \cap \triangle ADF$
   (f) (interior $\triangle ABE$) $\cup$ (interior $\triangle ADF$)
   (g) $\overline{CD} \cap \overleftrightarrow{AD}$
   (h) $\overline{EF} \cup \overline{AE}$
3. (a) Draw a curve that is not a simple closed curve.
   (b) Draw a simple closed curve that is not convex.
   (c) Draw a simple closed curve that is convex.

4. Make sketches to indicate the intersection of two parallelograms such that the intersection is
   (a) an empty set. (b) exactly one point.
   (c) exactly two points. (d) exactly four points.
   (e) exactly five points. (f) exactly six points.
5. Sketch a curve that is the union of line segments when the curve is
   (a) simple but not closed. (b) simple and closed.
   (c) a convex polygon of four sides. (d) a convex polygon of five sides.
6. Sketch the following.
   (a) A plane parallel to line r and intersecting line s
   (b) Skew lines r and s
   (c) A plane containing line r, where r and s are skew lines
7. Can $\overline{AC} \cap \overline{BD} = \varnothing$ and $\overleftrightarrow{AC} \cap \overleftrightarrow{BD}$ be a point? A line? A segment? Describe in general the various possible relationships or comparisons that can exist between $\overline{AC} \cap \overline{BD}$ and $\overleftrightarrow{AC} \cap \overleftrightarrow{BD}$.
8. Is it possible for a ray to be a proper subset of another ray? Draw a figure to demonstrate your answer.

Suggested Reading

Points, Lines, Planes, Space: Allendoerfer, pp. 564, 567–569. Graham, pp. 293–299. Meserve, Sobel, pp. 171–175. Moore, Little, p. 285. Peterson, pp. 246–250. Smart, pp. 25–30, 218–222. Smith, pp. 289, 294, 297–304. Weaver, Wolf, pp. 94–97. Willerding, pp. 109–114.

Closed Curves—Polygons, Triangles, Circles: Allendoerfer, pp. 571–573. Graham, pp. 301–303. Meserve, Sobel, pp. 182–185, 193–196. Moore, Little, pp. 286–287, 325–326. Peterson, pp. 255–257, 259–262. Smart, pp. 63–68. Smith, pp. 290–295, 307–308. Weaver, Wolf, pp. 98–99. Willerding, pp. 122–123, 126–127, 132–133.

Angles: Allendoerfer, pp. 575-576. Graham, pp. 299–301. Peterson, pp. 252–253. Smart, pp. 30–34. Smith, pp. 308–310. Weaver, Wolf, pp. 96–98. Willerding, pp. 120–122.

Closed Surfaces: Allendoerfer, p. 579. Graham, pp. 303–304. Smart, pp, 229–233. Smith, pp. 296–297.

10

Informal Metric Geometry

1

Congruence and Measure

Having considered the more abstract qualities of geometry in Chapter 9, let us now turn to a geometry with which we are more familiar—the geometry of measurement. Before we begin our study, we need to stop a moment to consider a basic concept for measurement, the concept of congruence. Congruence is one of the important concepts of Euclidean geometry, like the concepts of "point," "line," and "between," and it is usually considered as an undefined term. Yet we intuitively consider two figures, or two sets of points, to be *congruent* if they have the same size and shape. That is, if one of the congruent figures were superimposed upon the other, the two would coincide point by point.

Example: Place a sheet of tracing paper over a figure and trace the figure. Remove the paper. Two congruent drawings (indicating sets of points) now exist. In Figure 10-1, (b) is the trace of (a). The two sets of points have the same size and shape. That is, these two sets of points are said to be congruent.

The first requirement for congruence is that one be able to set up a one-to-one correspondence between two sets of points. Glance back to the last section of Chapter 2 and review the material concerning one-to-one correspondence. Remember that a one-to-one correspondence exists between two sets when the elements of the two sets can be paired exactly. But congruence is more than a one-to-one correspondence; it also involves size and shape.

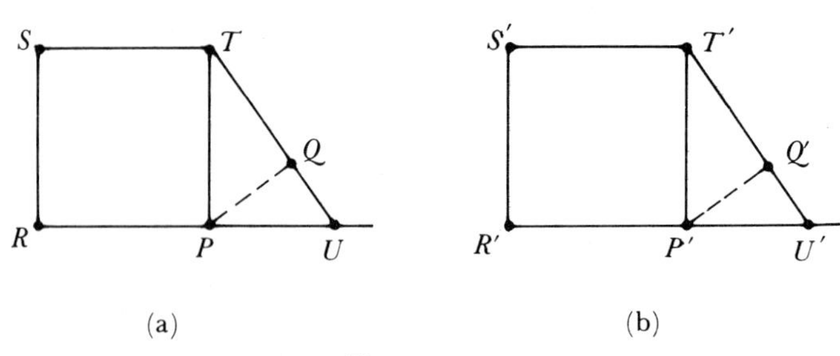

Figure 10-1

Example: Consider the two points P and Q in Figure 10-1(a). Let P' and Q' in (b) correspond to P and Q, respectively, in the one-to-one correspondence between these two sets. That is, in the tracing process, P' would coincide with P, and Q' would coincide with Q. Now consider segment $\overline{PQ}$. This segment is indicated by a dotted line because it is not a subset of the set of points in the figure. However, if Figure 10-1(b) is made to coincide with Figure 10-1(a), segment $\overline{P'Q'}$ will coincide with segment $\overline{PQ}$.

The two line segments $\overline{P'Q'}$ and $\overline{PQ}$ in the preceding example are said to be congruent, a relationship denoted by $\overline{P'Q'} \simeq \overline{PQ}$. Such line segments can be drawn using a compass, as illustrated by the following construction.

Construction 10-1: Construct a line segment of the same size as a given line segment.

Given: Line segment $\overline{CD}$ and ray $\overrightarrow{AF}$.

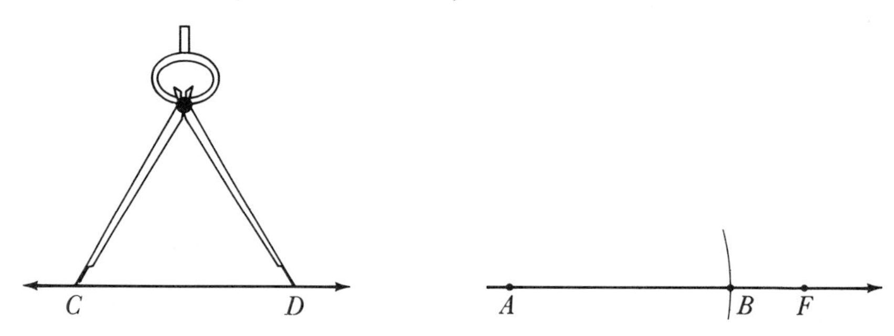

Figure 10-2

(a) Set your compass at C and at D.

(b) Without changing the setting of your compass, place one end at A and describe an arc, cutting the ray at B.

(c) $\overline{AB}$ is the same size as $\overline{CD}$.

Using congruence as a basic concept, we now move to the measurement of line segments. Man, seeking to answer the questions "How long?" and "How much?" as well as "How many?" probably began measuring about the same time he began counting. A measurement is a description of size. The word "measurement" refers both to the process of determining size and to the end result that describes that size. Before a measurement is made, a basic unit (such as an inch or a foot) is chosen. The measurement compares the set of points to the unit, which must be in the same category as the set of points being measured: to measure length we choose a unit of length; to measure area we choose a unit of area.

The concept of congruence, although it does not imply any special unit, is the mathematical basis for the theory of measurement of a set of points. For example, the inch ruler, one of the most common tools for measurement, consists of a number of congruent segments called inches. Note that each of these inches can be divided into congruent line segments so that there might exist a subdivision into four, eight, or even sixteen congruent, nonoverlapping segments.

But let's move now from the particular divisions of an inch ruler to the more general measurement of any line segment. First of all, it is necessary to select a line segment that will serve as a unit segment. For example, in Figure 10-3, suppose $\overline{RS}$ is selected to serve as a *unit segment* for the purpose of measurement. Now use $\overline{RS}$ to measure a line segment $\overline{AB}$.

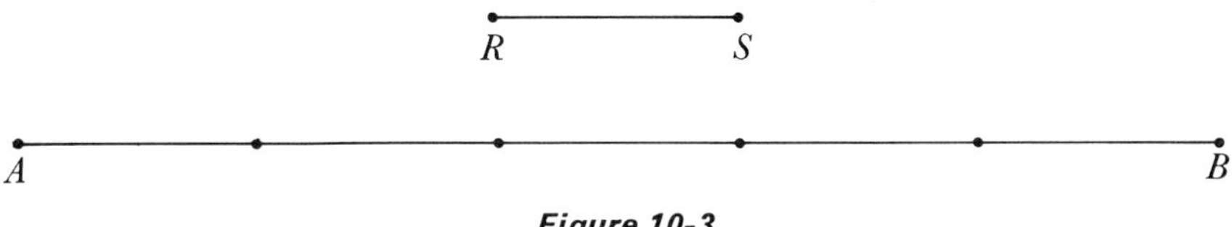

Figure 10-3

Starting at A, mark off line segments congruent to $\overline{RS}$. Note that these line segments have in common only an endpoint. If the union of n of these congruent segments is $\overline{AB}$, then we say that the *measure* of $\overline{AB}$ is n and denote this by $m(\overline{AB}) = n$.

Notice that the congruent segments do not overlap and that measure, as illustrated, is a positive integer. Of course, it is not always possible to obtain an integer for an answer if you use $\overline{RS}$ as the selected unit for measuring a given line segment $\overline{AB}$. That is, starting at A, an integral number of segments congruent to $\overline{RS}$ will not necessarily end at B. Thus, the measure of a segment is not always an integer.

In Chapter 8, we reasoned that there exists a one-to-one correspondence between the set of real numbers and the points on a number line. The real numbers that we associate with the points on the number line are called the *coordinates* of the points. This relationship was denoted by representations such as Figure 10-4. The line segments from -3 to $-2, \ldots,$ and 2 to 3 are congruent segments. Each of these segments may be used as a unit segment or a unit of measure. Each of the numbers $-3, -2, -1, 0, 1, 2,$ and 3 is a coordinate of a point. The point named by 0 in Figure 10-4 is called the *origin* of the coordinate system.

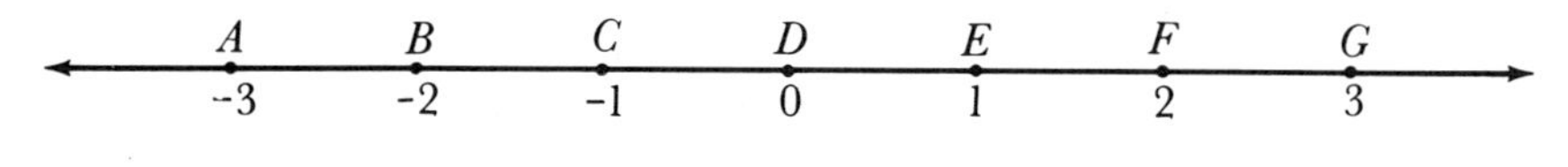

Figure 10-4

Now, let A and B be points on a number line with coordinates a and b, respectively, as shown in Figure 10-5. A and B are selected such that

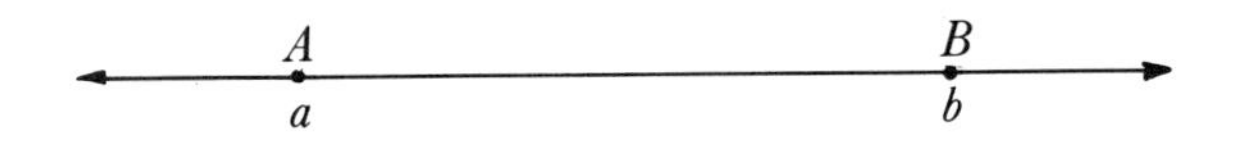

Figure 10-5

$a \leq b$. We now associate with every line segment a real number that we call the length or measure of the segment. Since $a \leq b$, then the measure of $\overline{AB}$, denoted by $m(\overline{AB})$, is $b - a$. Since $\overline{AB}$ and $\overline{BA}$ are congruent, we have a definition.

Definition 10-1: If A and B are any points on a number line with coordinates a and b, respectively, where $a \leq b$, then the measure of $\overline{AB}$ is given by $m(\overline{AB}) = m(\overline{BA}) = b - a$.

Notice that the measure of a line segment, as defined, is always a non-negative number, since a was defined to be less than or equal to b. If $A = B$, then $m(\overline{AB}) = a - a = b - b = 0$. Also, if $m(\overline{AB}) = 0$, then $A = B$.

Examples: Using Definition 10-1, the following measures can be obtained from Figure 10-4.

$m(\overline{EG}) = 3 - 1 = 2 \qquad m(\overline{DG}) = 3 - 0 = 3$

$m(\overline{FC}) = 2 - (-1) = 3 \qquad m(\overline{BD}) = 0 - (-2) = 2$

Of course, in Definition 10-1, a and b were not restricted to integers. The measure of $\overline{AB}$ in Figure 10-6 is $m(\overline{AB}) = 3\frac{7}{8} - 1\frac{1}{2} = 2\frac{3}{8}$.

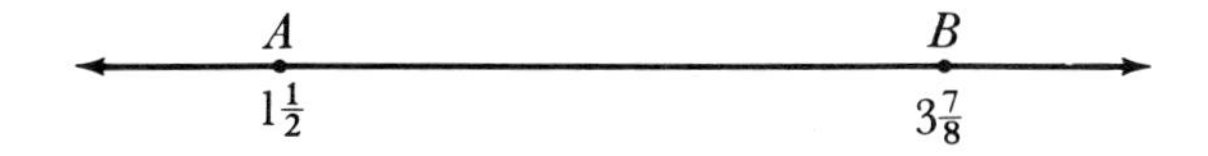

Figure 10-6

We can now relate congruence and measurement. It is evident that when two segments are congruent, their measures are equal; and if the measures are equal, then the two segments are congruent. When the segments $\overline{AB}$ and $\overline{CD}$ are congruent, we use the notation $m(\overline{AB}) = m(\overline{CD})$. Thus, $m(\overline{AB}) = m(\overline{CD})$ and "$\overline{AB}$ is congruent to $\overline{CD}$" mean the same thing. (*Note:* It is customary in many geometry books to shorten the notation for the measure of a segment. In the remainder of this book, we use both AB and $m(\overline{AB})$ to represent the measure of line segment $\overline{AB}$.)

The polygon represented in Figure 10-7 is the union of five line

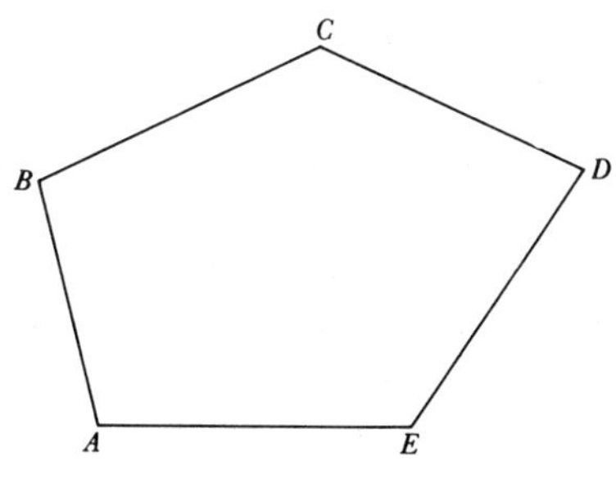

Figure 10-7

segments. The sum of the measures of the sides of the polygon is called the *perimeter* of the polygon. For example, suppose that

$$m(\overline{AB}) = m(\overline{BC}) = m(\overline{DE}) = 8, \quad m(\overline{EA}) = 6, \quad \text{and} \quad m(\overline{CD}) = 4.$$

Then the perimeter of the polygon is $8+8+8+6+4=34$.

Now let's consider our standard yardstick. Suppose you measure your desk with this yardstick and find that the width is 20 inches. Of course, all measurements are approximate, but if we treat this measurement as if it were exact, then

20 inches is the measurement,
20 is the measure, and
1 inch is the unit of measure.

A measurement, then, consists of two parts: the measure, which in this case is 20, and the unit of measure, which in this case is the inch.

Of course, if two people were to obtain the measurement of the width of a desk with different units, they would have difficulty comparing answers. Therefore, the need for a *standard unit* that is agreed upon and used by a majority of people is evident. Difficulties that have been encountered in reaching a decision on standard units revolve about a search for a better standard. Disagreements about units for linear measurement (measurement of line segments) became so common that at the beginning of the nineteenth century some French scientists called a meeting of scientists from many countries in an effort to establish an international set of measures. This group developed the *metric* system, of which the meter is the standard unit length. A meter is supposed to be one ten-millionth of the distance from the equator to the North Pole. Actually, the international standard of length is based on the orange radiation of the krypton isotope, which has an atomic weight of 86 (Kr_{86}). All other lengths are defined in terms of the standard meter. For example, a yard is approximately 0.9144 meters.

Of course, any one system that was universally adopted would have advantages over other systems. The metric system has advantages other than being widely adopted. Its main advantage comes from the fact that its subdivisions are multiples of ten, a feature that facilitates computation in the decimal number system. Since the United States has now taken steps to adopt the metric system, there is a probability that this system will be adopted for use world-wide. Check your ability to cope with a system that may soon replace our inches, feet, yards, miles, pounds, and quarts. (Table 10-1 should help!)

Metric Equivalents

Length	*Volume*	*Mass*
1 meter	1 liter	1 gram
= 10 decimeters	= 10 deciliters	= 10 decigrams
= 100 centimeters	= 100 centiliters	= 100 centigrams
= 1000 millimeters	= 1000 milliliters	= 1000 milligrams
= 0.1 dekameter	= 0.1 dekaliter	= 0.1 dekagram
= 0.01 hectometer	= 0.01 hectoliter	= 0.01 hectogram
= 0.001 kilometer	= 0.001 kiloliter	= 0.001 kilogram

English Units and Approximate Equivalents

12 inches = 1 foot	1 inch ≈ 2.54 centimeters	1 centimeter ≈ 0.4 inch
3 feet = 1 yard	1 yard ≈ 0.914 meter	1 meter ≈ 1.1 yards
5280 feet = 1 mile	1 mile ≈ 1.6 kilometers	1 kilometer ≈ 0.62 mile
320 rods = 1 mile	1 pound ≈ 0.45 kilogram	1 kilogram ≈ 2.2 pounds
	1 quart ≈ 0.95 liter	1 liter ≈ 1.05 quarts

Table 10-1

Examples:

2 yards ≈ 1.828 meters = 182.8 centimeters
4.2 quarts ≈ 4 liters = 4000 milliliters
16 miles ≈ 25.6 kilometers = 25,600 meters
6 pounds ≈ 2.7 kilograms = 2700 grams

Exercise Set 10-1

1. The following drawings represent congruences between figures. Enumerate one matching between points A, B, C, D and A', B', C', D' to obtain the congruence. For example, these two triangles can be matched as $A \leftrightarrow A'$, $B \leftrightarrow B'$, $C \leftrightarrow C'$.

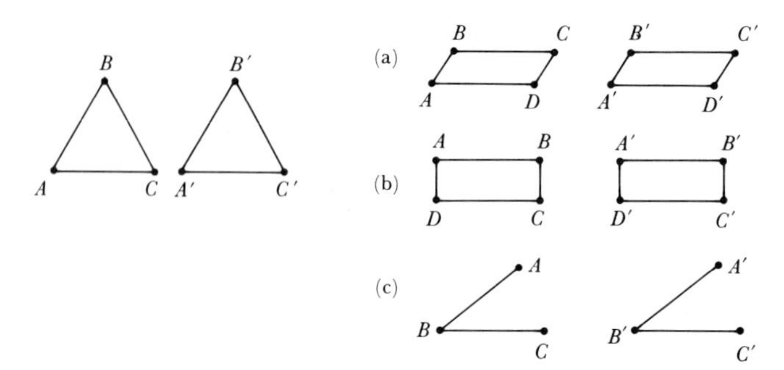

2. Classify the following statements as either true or false.
   (a) The perimeter of a triangle is greater than twice the measure of any side.
   (b) If $\overline{AB}$ is congruent to $\overline{AD}$, then B and D name the same point.
   (c) If $AD = CE$, then D and E name the same point.
   (d) The perimeter of a rectangle is greater than or equal to four times the measure of the shortest side.
   (e) If $m(\overline{AC}) = m(\overline{AD})$, then C and D name the same point.
   (f) The union of two different segments may have the same measure as one of the segments.
   (g) The union of two different segments may be congruent to one of the segments.
   (h) A yard is longer than a meter.
   (i) A centimeter is shorter than two inches.
   (j) Two congruent figures have the same shape.
   (k) If segments $\overline{AB}$ and $\overline{BC}$ are congruent, then they are equal.
   (l) If $\overline{AB} \simeq \overline{BC}$ and $\overline{BC} \simeq \overline{CD}$, then $\overline{AB} \simeq \overline{CD}$.
3. Classify each statement as either true or false.
   (a) 1 foot is longer than $\frac{1}{2}$ meter.
   (b) 1 kilometer is shorter than 1 mile.
   (c) 1 centimeter is longer than 2 inches.
   (d) 1 millimeter is shorter than 1 inch.
   (e) 2 centimeters is shorter than 1 inch.
   (f) 1 meter is longer than 39 inches.
4. Consider the line segments as given by the endpoints of the segments A, B, C, D, E, F, and G. Find each of the following measures.

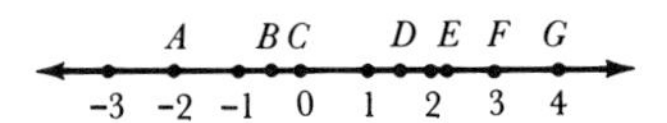

(a) $m(\overline{AB})$ (b) $m(\overline{CE})$ (c) $m(\overline{DB})$
(d) $m(\overline{GE})$ (e) $m(\overline{GA})$ (f) $m(\overline{CD})$

5. List the congruent line segments in Exercise 4.
6. In the following triplets of numbers, find a coordinate x so that $\overline{AB}$ is congruent to $\overline{BC}$.
   (a) A at $\frac{1}{2}$, B at 2, C at x (b) A at $\frac{3}{4}$, B at x, C at 2
   (c) A at -1, B at 3, C at x (d) A at x, B at 1, C at 4
7. Convert
   (a) 11 meters to yards. (b) 150 feet to centimeters.
   (c) 11 miles to kilometers. (d) 52 centimeters to inches.
   (e) 14 feet to meters. (f) 1 foot to yards.
   (g) 6 miles to yards. (h) 360 kilometers to miles.
8. Express the following measurements in three other ways, using different units.
   (a) 5 feet (b) 3 centimeters
   (c) 8 meters (d) 14 inches
9. Formulate rules for changing the following measurements.
   (a) Inches to feet
   (b) Inches to yards
   (c) Feet to miles
10. Joan's room in the dormitory is 5.6 yards long and 4.1 yards wide.
   (a) Convert the measurement of Joan's room to feet.
   (b) Convert to inches.
11. Find two possible coordinates for B so that $\overline{AB}$ is a unit segment, where A has the following coordinates.
   (a) $\frac{1}{2}$ (b) -3 (c) -4 (d) $\frac{7}{8}$ (e) -1 (f) 3

*12. The length of a meter has been defined as 1,650,763.73 wavelengths of the orange-red lines of the krypton isotope. Express the following in wavelengths.
   (a) Inch (b) Kilometer (c) Mile (d) Centimeter

*13. Consider two pairs of corresponding points $A \leftrightarrow A'$ and $B \leftrightarrow B'$ on congruent figures. Discuss the relationship between $\overline{AB}$ and $\overline{A'B'}$.

2

Measure of Angles

In the preceding chapter, we defined an angle to be a set of points formed by the union of two rays having the same endpoint. From this definition, we introduce the concept of congruent angles and then move on to the concept of angle measurement. Intuitively, two angles of the same size would be congruent because angle congruence is a special case of congruence in general and must satisfy the requirements for congruence. The two examples below illustrate congruent angles.

Example: Two congruent segments are segments with the same length. In Figure 10-8, $\overline{OA} \simeq \overline{O'A'}$ and $\overline{OB} \simeq \overline{O'B'}$. If $\angle AOB$ is congruent to $\angle A'O'B'$, then $\overline{AB} \simeq \overline{A'B'}$. This figure illustrates that corresponding line segments of congruent angles have the same size.

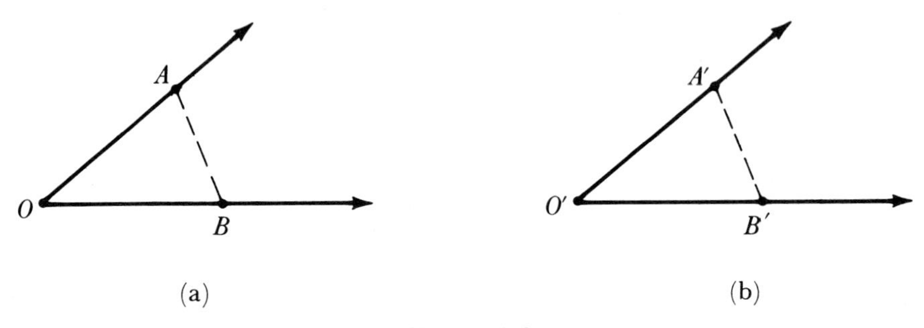

Figure 10-8

Example: Place a sheet of tracing paper over $\angle ABC$ in Figure 10-9. Make a trace of this angle, as indicated by $\angle DEF$ of Figure 10-9, so that D corresponds to A, E to B, and F to C. This work presents an intuitive meaning of the concept $\angle ABC \simeq \angle DEF$.

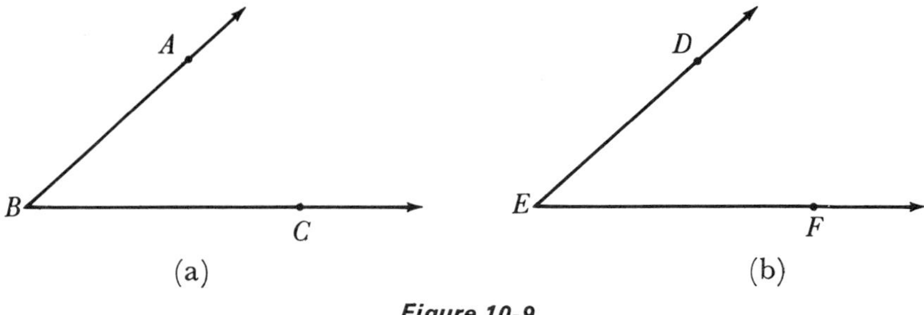

Figure 10-9

To find the measure of an angle, one determines a *unit angle* just as it was necessary to determine a unit segment for measuring a segment.

In Figure 10-10(a), suppose $\angle ABC$ is a unit angle. Then $\angle EFG$, in Figure 10-10(b), has a measure of 2, because $\angle EFG$ and its interior can be partitioned by $\overrightarrow{FH}$ into two adjacent angles, each of which is congruent to the unit angle, $\angle ABC$. That is, we assumed $\angle HFG \simeq \angle ABC$ and $\angle EFH \simeq \angle ABC$. Thus, the measure of $\angle EFG$ is 2.

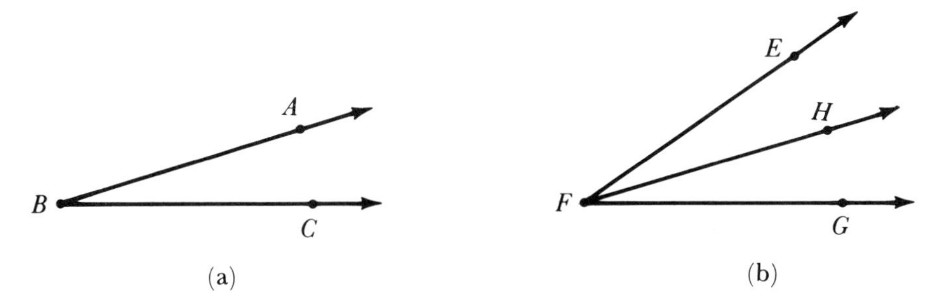

Figure 10-10

In Figure 10-11 the measure of $\angle EFG$ is 4 because $\angle EFG$ and its interior have been partitioned into four adjacent angles, each of which is congruent to the unit angle, $\angle ABC$.

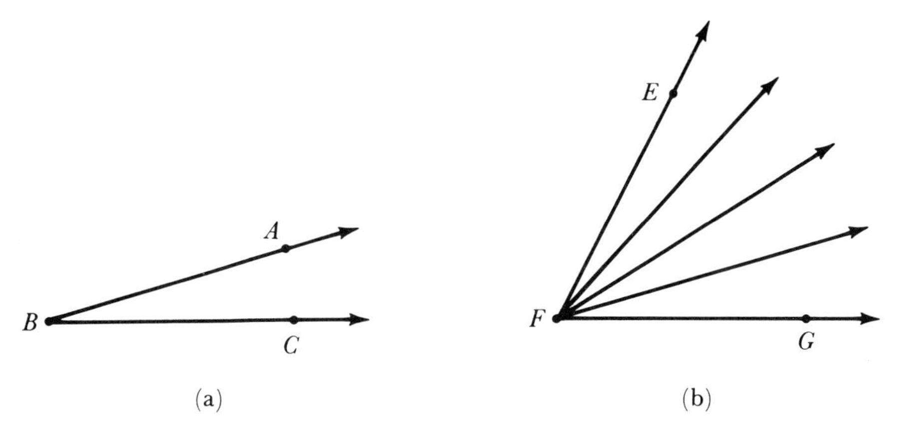

Figure 10-11

Just as a ruler, depending on the length of the unit segment, can be divided into two, four, eight, or more congruent segments, so an angle, depending on the size of its unit angle, might be divided into two, four, eight, or more adjacent, congruent angles. The rays of the congruent angles form what is called a ray coordinate system. Let us postulate that n congruent angles, where n is a counting number, can be placed in the interior of $\angle EFG$ to form a ray coordinate system. If we let one of these congruent angles be a unit, then $m(\angle EFG) = n$. We further postulate that there exists a correspondence that associates with each angle a real number n so that the measure of the angle is n.

In elementary mathematics, the unit of measurement most commonly used for angles is the *degree*, which is usually indicated by a small circular superscript. For example, 20 degrees is written 20°. For an understanding of the size of a unit angle, with a measure of one degree, form a ray coordinate system consisting of 180 congruent adjacent angles such that the outside rays on the first and last angles form a straight line. Call the measurement of each of these small angles a degree. Then the measurement of the large angle is 180°. The angle that contains within its interior the first 90 congruent angles has a measurement of 90° and is called a *right angle*. An angle that contains within its interior ten of these adjacent congruent angles is said to have a measurement of 10°.

Although angles may be measured in other units, in this text let's agree to use a degree as our standard unit for the measurement of an angle. A *protractor* is a device that has been developed to approximate the number of degrees in a given angle. A protractor, like an inch ruler, is a tool that has marked on it units of the ray coordinate system, a system involving 90 units, or degrees, within a right angle.

If the point marked O on the protractor is placed at the vertex of an angle, and if one side, $\overrightarrow{OA}$, of the angle coincides with the 0° line on the protractor, then the other side of the angle, $\overrightarrow{OB}$, will coincide with or be very close to a marked line on the protractor; and the measurement of the angle can be read from the protractor. In this text, we shall restrict our angles such that for any angle ABC, $0 \leq m(\angle ABC) \leq 180$.

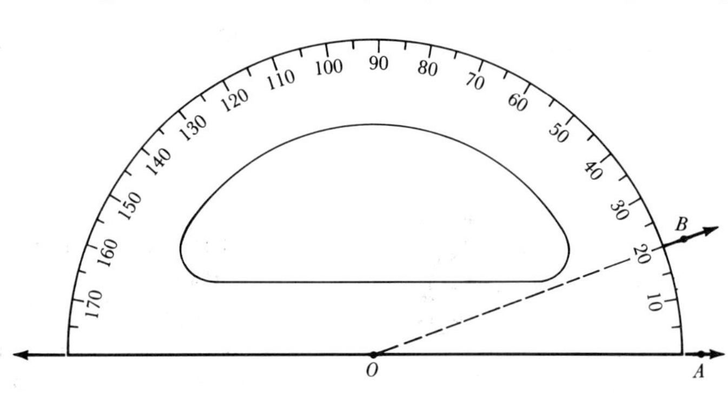

Figure 10-12

Examples: Determine the measurements of the following angles.

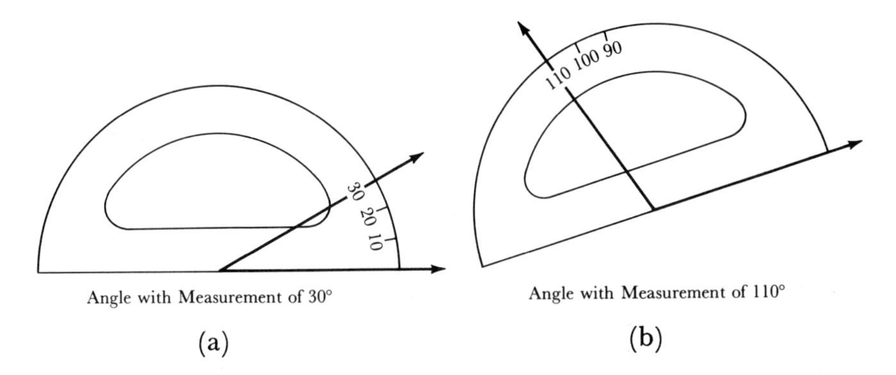

Angle with Measurement of 30°

(a)

Angle with Measurement of 110°

(b)

Figure 10-13

Certain special angles are classified according to the measurements of the angles. Two angles are *complementary* if the sum of the measurements of the angles is 90°. Two angles are *supplementary* if the sum of the measurements of the angles is 180°. Supplementary angles that are also adjacent form what is called a *straight angle.*

In Figure 10-14, $\angle ABC$ and $\angle ABD$ are both supplementary and adjacent. Rays $\overrightarrow{BD}$ and $\overrightarrow{BC}$ lie on the same line. We call $\angle DBC$ a *straight angle,* which has a measurement of 180°.

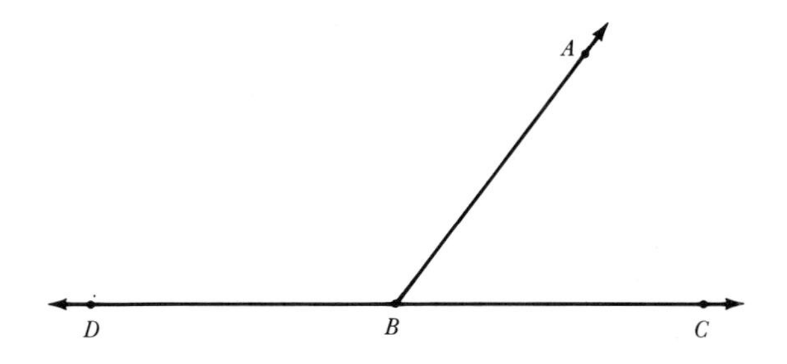

Figure 10-14

Sometimes, a degree is considered as being divided into 60 equal parts called *minutes* ($'$); and each minute, into 60 equal parts called *seconds* ($''$). Thus, $20° \; 41' \; 16''$ is read as 20 degrees, 41 minutes, 16 seconds.

Example: Express $15.4°$ in degrees and minutes. $0.4(60) = 24$. Thus $0.4° = 24'$. Then $15.4° = 15°\ 24'$.

In terms of measure, we can give meaning to terms that occur frequently in geometry.

(a) If the measure of an angle is less than 90, we say the angle is an *acute* angle; if the measure of an angle is more than 90, but less than 180, we say the angle is *obtuse*.

(b) The measure of a right angle is 90.

(c) The measure of a straight angle is 180.

(d) If the sum of the measures of two angles is 90, the angles are called *complementary*; if 180, they are *supplementary* angles.

(e) Two rays (or the lines containing the rays) forming a right angle are said to be *perpendicular*.

Exercise Set 10-2

1. Classify the following as true or false. If the statement is false, correct it.
   (a) If $\angle ABC \simeq \angle EFG$, then $B \leftrightarrow F$.
   (b) If $\angle ABC \simeq \angle EFG$, then $A \leftrightarrow E$.
   (c) If $\angle ABC \simeq \angle EFG$, then $\overline{BA} \simeq \overline{FE}$.
   (d) If $\angle ABC \simeq \angle EFG$, then $\overline{AB} \simeq \overline{EF}$.
   (e) If $\angle ABC \simeq \angle EFG$, then $\overline{BC} \simeq \overline{GF}$.
   (f) If $\angle ABC \simeq \angle EFG$, then $\angle ABC \simeq \angle GFE$.
   (g) Any two right angles are congruent.
   (h) Every angle is congruent to itself.
   (i) If $\angle ABC \simeq \angle EFG$, then $\angle ABC = \angle EFG$.
   (j) If $\angle ABC = \angle EFG$, then $\angle ABC \simeq \angle EFG$.
2. Make a tracing of a part of $\angle BAC$. See if you can place this tracing on $\angle BAC$ in another way. The part of the tracing obtained from $\overrightarrow{AB}$ now covers what?

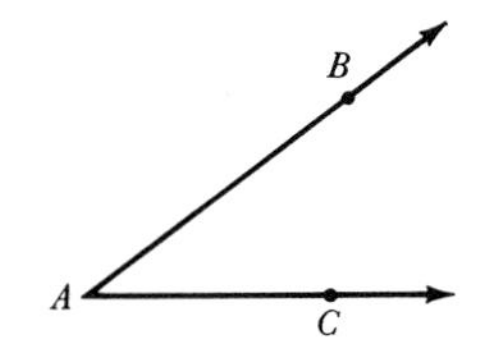

3. Make a conjecture on supplementary angles of congruent angles. Make tracings to test your conjecture.
4. Consider vertical angles $\angle AOB$ and $\angle COD$. Make a tracing of a part of $\angle AOB$ and see if it fits $\angle COD$. Does it seem to fit? Make a conjecture on vertical angles.

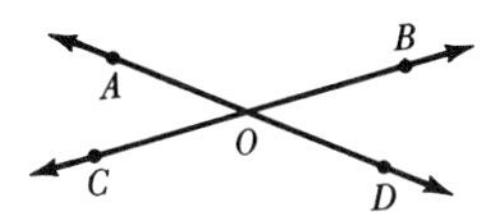

5. State which of the following have meaning in our development of the geometry of angles.
   (a) $\angle ABC \simeq \angle CBA$ (b) $\angle ABC = \angle CBA$
   (c) $\angle ABC + \angle DEF = 75$ (d) $m(\angle ABC) = 1$
   (e) $m(\angle ABC) = m(\angle CBA)$ (f) $\angle ABC + \angle CBD = \angle ABD$
   (g) $m(\angle ABC) + m(\angle CBD) = m(\angle ABD)$
6. If $m(\angle ABC) = 60$ and $m(\angle CBD) = 80$, what is $m(\angle ABD)$ if
   (a) C is in the interior of $\angle ABD$?
   (b) C is in the exterior of $\angle ABD$?
7. In the figure if $m(\angle COB) = 40$, find
   (a) $m(\angle AOB)$.
   (b) $m(\angle AOD)$.
   (c) $m(\angle COD)$.

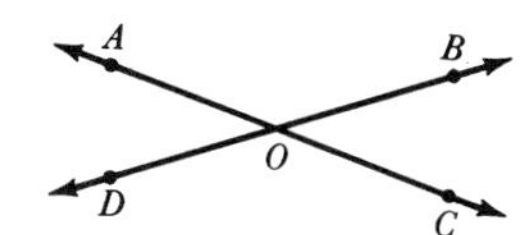

8. Find the measure of the supplement and the complement, if they exist, of the following angles.
   (a) $m(\angle A) = 70$ (b) $m(\angle B) = 2$
   (c) $m(\angle ABC) = 42$ (d) $m(\angle E) = 0$
   (e) $m(\angle F) = 114$ (f) $m(\angle DEF) = 14$
9. (a) Add $14° \, 26' \, 10''$ and $46° \, 22' \, 51''$.
   (b) From $25° \, 39' \, 17''$ subtract $14° \, 41' \, 27''$.
   (c) Find the complement of $18° \, 17' \, 42''$.
   (d) Find the supplement of $61° \, 19' \, 38''$.
10. (a) Are complementary angles necessarily adjacent, using our measure definition?
   (b) Do supplementary angles have to be adjacent?

11. Use a protractor to find the approximate measurement of the following angles.

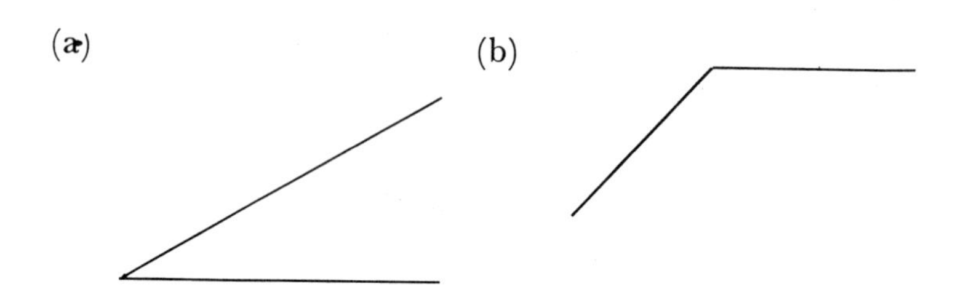

12. Add or subtract.

(a) $\begin{array}{r} 16^\circ\ 14'\ 17'' \\ +\ 8^\circ\ 50'\ 51'' \\ \hline \end{array}$ (b) $\begin{array}{r} 84^\circ\ 36'\ 41'' \\ +\ 17^\circ\ 41'\ 54'' \\ \hline \end{array}$

(c) $\begin{array}{r} 18^\circ\ 14'\ 17'' \\ -\ 8^\circ\ 50'\ 51'' \\ \hline \end{array}$ (d) $\begin{array}{r} 84^\circ\ 36'\ 41'' \\ -\ 17^\circ\ 41'\ 54'' \\ \hline \end{array}$

13. If the measure of the supplement of an angle is three times the measure of the complement of the angle, find the measure of the angle.
14. Draw a right triangle. Use the protractor to measure the acute angles of the triangle. Does the sum of the measures of the acute angles equal 90?
15. Construct a right angle and then bisect it. What is the measure of each of the two angles?

*16. (a) Draw three noncollinear points on your paper and label them *A*, *B*, and *C*. Connect the three with line segments. How many segments do you have?
(b) Measure $\angle ABC$, $\angle BCA$, and $\angle CAB$. What is the sum of the measurements of these three angles?
(c) Now place a fourth point *D* in such a position that no three of the four points are collinear and such that *D* is on the opposite side of $\overleftrightarrow{AC}$ from *B*. Draw line segments to *D* from *A*, *B*, and *C*. What type of figure is *ABCD*?
(d) Measure $\angle ABC$, $\angle BCD$, $\angle CDA$, and $\angle DAB$. What is the sum of the measurements of these angles?
(e) Make a conjecture as to the sum of the angles in the interior of a triangle. In the interior of a quadrilateral.

*17. Prove informally that vertical angles have the same measure and thus are congruent.

*18. Discuss the statement "Angles with different measures may be congruent."

*19. If the hands of a clock are such that the time is four o'clock, what is the measure of the angle formed by the hands of the clock? What if the time is three o'clock? Five o'clock?

3

Triangles

Another special case of general congruence is that of congruent triangles. When two triangles are congruent, a correspondence can be made between their vertices so that the sides and angles satisfy six segment and angular congruence relations, given in the following definition.

Definition 10-2: Two triangles, $\triangle ABC$ and $\triangle DEF$, are said to be congruent if the vertices A, B, C, and D, E, F, can be so paired that corresponding angles and corresponding sides are congruent. If A is paired with D, B with E, and C with F, then $\overline{AB} \simeq \overline{DE}$, $\overline{BC} \simeq \overline{EF}$, $\overline{CA} \simeq \overline{FD}$, $\angle ABC \sim \angle DEF$, $\angle BCA \sim \angle EFD$, and $\angle CAB \sim \angle FDE$.

It seems reasonable that one would not need to check all six congruence relations in order to say that two triangles are congruent. For example, suppose two sides and the included angle of one triangle were congruent to two sides and the included angle of a second triangle. It is evident that the other side and the other two angles would also be congruent. This fact is stated in the fundamental congruence axiom for triangles.

Axiom 10-1: If two sides and the included angle of one triangle are congruent to two sides and the included angle of a second triangle, the two triangles are congruent.

This axiom may be used to develop other conditions for determining whether or not two triangles are congruent. Two such examples are given in the following theorems.

Theorem 10-1: If two angles and the included side of one triangle are congruent to two angles and the included side of another triangle, the triangles are congruent.

Proof: Draw $\triangle ABC$ and $\triangle DEF$ with $\angle CAB \simeq \angle FDE$, $\angle ABC \simeq \angle DEF$, and $\overline{AB} \simeq \overline{DE}$ (Figure 10-15). There is a point X on $\overrightarrow{DF}$ such that $\overline{AC} \simeq \overline{DX}$. Then, by Axiom 10-1, $\triangle ABC \simeq \triangle DEX$. Why? Thus, $\angle ABC \simeq \angle DEX$. Hence $\angle DEX \simeq \angle DEF$. Therefore $\overrightarrow{EX}$ and $\overrightarrow{EF}$ are the same. This means that points F and X are the same, since two rays can intersect in only one point. Therefore, $\triangle ABC \simeq \triangle DEF$.

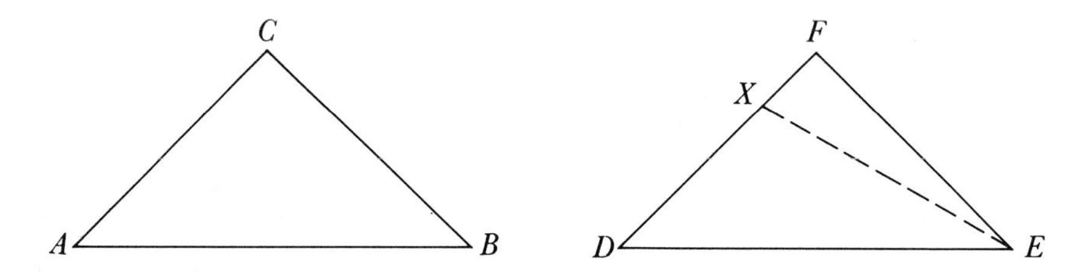

Figure 10-15

Theorem 10-2: If three sides of one triangle are congruent to three sides of another triangle, the triangles are congruent.

The proof is left as Exercise 9 in Exercise Set 10-3.

Congruence relations also define certain types of triangles.

Definition 10-3: (a) An *isosceles triangle* is a triangle having exactly two of its sides congruent.

(b) An *equilateral triangle* is a triangle having all three sides congruent.

(c) A *scalene triangle* is a triangle having no sides congruent.

Properties of isosceles and equilateral triangles are summarized as a theorem in order to indicate that the truth of these statements can be proved from the preceding theory.

Theorem 10-3: (a) If a triangle is isosceles, the angles opposite congruent sides are congruent.

(b) If two angles of a triangle are congruent, then the triangle is an isosceles triangle.

(c) If a triangle is equilateral, all three angles are congruent.

(d) If three angles of a triangle are congruent, then the triangle is an equilateral triangle.

A very special type of triangle that is used in many branches of engineering and science is called a right triangle.

Definition 10-4: If one angle of a triangle is a right angle, the triangle is called a *right triangle.*

In a right triangle, the side opposite the right angle is called the *hypotenuse* and the other two sides are called *legs.* The best known theorem involving right triangles is named for a Greek mathematician, Pythagoras.

Theorem 10-4: *The Pythagorean Theorem.* In a right triangle, the square of the measure of the hypotenuse is equal to the sum of the squares of the measures of the legs: $c^2 = a^2 + b^2$ (Figure 10-16).

Many proofs of the Pythagorean Theorem have been presented. In fact, the theorem was known to the Babylonians about a thousand years

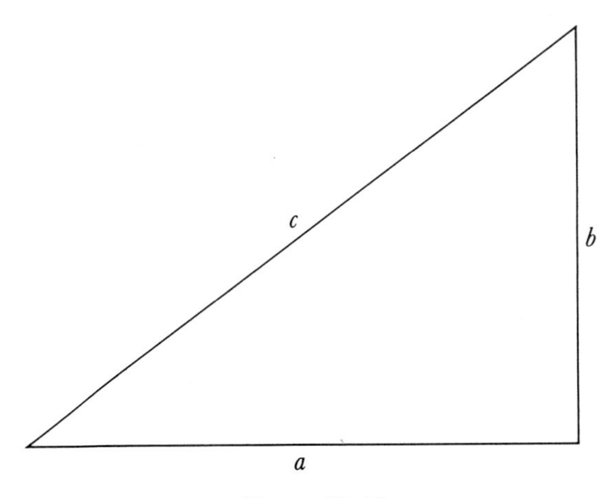

Figure 10-16

before Pythagoras, but the first general proof of the theorem is believed to have been given by Pythagoras about 525 B.C. There have been many conjectures as to the type of proof Pythagoras might have given; it is generally believed today that his proof was probably a dissection-type proof such as shown in Figure 10-17.

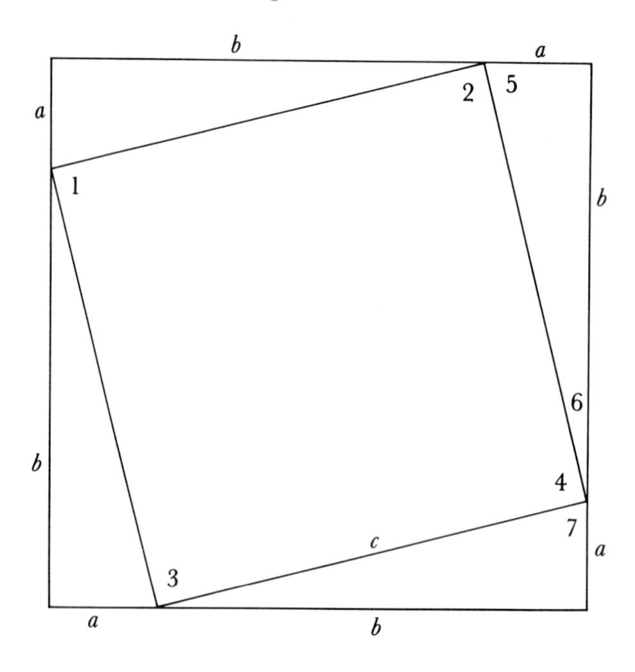

Figure 10-17

This figure is a square with sides that have measures of length $a+b$. In this square, we draw four right triangles with legs of measure a and b. The four right triangles are congruent by Axiom 10-1. Thus, the four hypotenuses are congruent. $m(\angle 6)+m(\angle 4)+m(\angle 7)=180$. Why? $m(\angle 5)+m(\angle 6)=90$. (See Exercise 14, Set 10-2.) Similarly, $m(\angle 7)=m(\angle 5)$. Why? Hence $m(\angle 7)+m(\angle 6)=90$, so $m(\angle 4)+90=180$, which means that $m(\angle 4)=90$. In a like manner, it can be shown that $m(\angle 1)=m(\angle 2)=m(\angle 3)=90$. Thus, the figure consisting of the four hypotenuses is a square.

The area of the large square is equal to the area of the smaller square plus the area of the four congruent right triangles. Although the concept of the area of plane regions will not be formally introduced until Section 6 of this chapter, it is assumed that you remember the formulas for the areas of the regions bounded by squares and triangles. The area of the large square is $(a+b)^2 = a^2 + 2ab + b^2$; also, the area of the small square is c^2. The total area of the four right triangles is $4(ab)/2 = 2ab$. Thus $a^2 + 2ab + b^2 = c^2 + 2ab$. Subtract $2ab$ from each side of this equation to obtain $a^2 + b^2 = c^2$, thus completing the proof.

Theorem 10-5: *Converse of the Pythagorean Theorem.* If the square of the measure of one side of a triangle is equal to the sum of the squares of the measures of the other two sides, the triangle is a right triangle.

Proof: Given triangle ABC with $c^2 = a^2 + b^2$. Now draw a right triangle DEF with legs of length a and b. Let d be the hypotenuse of this right triangle. Then $a^2 + b^2 = d^2$. Hence $c^2 = d^2$. Why? Thus, $c = d$. Why? Therefore, $\triangle ABC \simeq \triangle DEF$. Why? But $\triangle DEF$ is a right triangle. Hence $\triangle ABC$ is a right triangle, thus completing the proof.

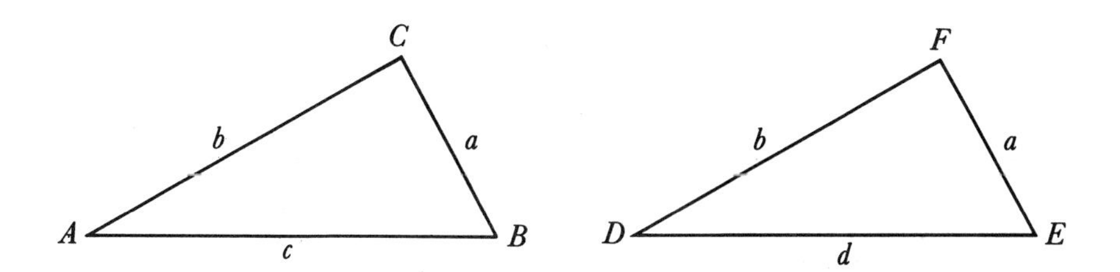

Figure 10-18

Example: A triangle has sides whose lengths are 5, 12, and 13. Prove that this triangle is a right triangle. Since $(13)^2 = 169$, and since $(12)^2 + (5)^2 = 144 + 25 = 169$, then $(13)^2 = (12)^2 + (5)^2$; hence, by Theorem 10-5, the given triangle is a right triangle.

Other interesting properties of right triangles are listed as theorems.

Theorem 10-6: Two right triangles are congruent if

(a) a leg (or hypotenuse) and an acute angle of the first are congruent to the corresponding leg (or hypotenuse) and acute angle of the second.

(b) the hypotenuse and a leg of the first are congruent to the hypotenuse and corresponding leg of the second.

(c) two legs of the first are congruent to the corresponding legs of the second.

Exercise Set 10-3

1. Are there congruent triangles in the following figures? If so, tell why. Like marks on the line segments indicate congruent line segments. Right angles are marked as ⌟ and equal angles are indicated by ∢ and ∢.

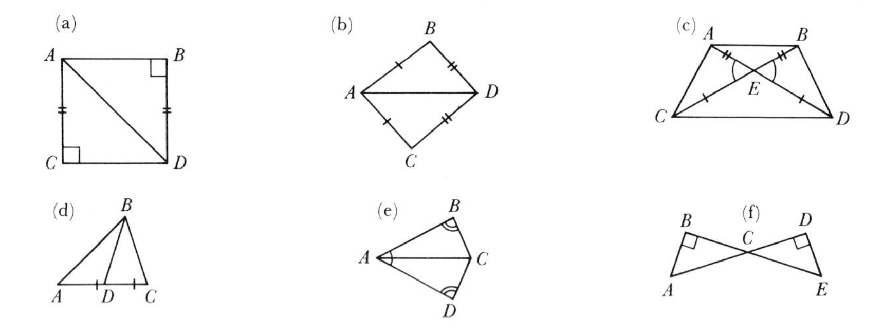

2. Draw a sketch of a triangle satisfying each set of conditions.
   (a) Obtuse and scalene (b) Acute and scalene
   (c) Acute and isosceles (d) Right and isosceles
   (e) Obtuse and isosceles (f) Acute and equilateral
3. Triangle ABC is an isosceles triangle with a right angle at A.
   (a) Name the two congruent sides.
   (b) Name the two congruent angles.
   (c) What is the measure of each of these congruent angles?
4. Suppose that $\triangle ABC$ and $\triangle DEF$ are given so that $\angle A \simeq \angle D$, $\angle C \simeq \angle F$, $\overline{CB} \simeq \overline{FE}$, and $\overline{AB} \simeq \overline{DE}$. Is $\triangle ABC \simeq \triangle DEF$? Explain your answer.
5. Show whether the triangles whose sides have the following measures are right triangles.
   (a) 6, 8, 10 (b) 5, 12, 13 (c) 7, 24, 25
   (d) 9, 40, 41 (e) 11, 60, 61 (f) 10, 24, 26
   (g) 28, 21, 35 (h) 40, 96, 104 (i) $a + b, a, b$

6. Find the perimeter of the following closed figures.

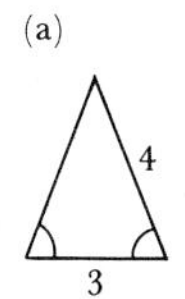

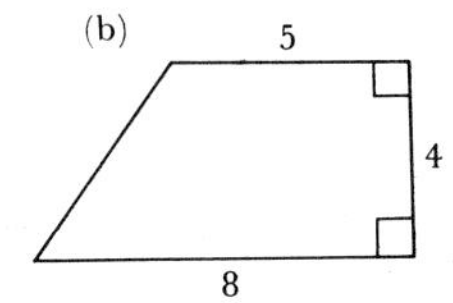

7. Experiment to determine the truth or falsity of the following statements about the relative lengths of the sides of a triangle.
   (a) The sum of the lengths of any two sides must be greater than the measure of the third side.
   (b) The difference in the measures of any two sides must be less than the measure of the third side.

*8. If the measure of the hypotenuse of a right triangle is $(m^2/4) + 2$ and one leg has measure $(m^2/4) - 2$, find the measure of the other leg.

Prove the theorems designated in Exercises 9–16.

*9. Theorem 10-2. *10. Theorem 10-3(a).
*11. Theorem 10-3(b). *12. Theorem 10-3(c).
*13. Theorem 10-3(d). *14. Theorem 10-6(a).
*15. Theorem 10-6(b). *16. Theorem 10-6(c).
*17. Show that if x and y are any numbers with $x > y$, then $x^2 + y^2$, $x^2 - y^2$, and $2xy$ are the sides of a right triangle.

4

Angles Formed by Transversals

In the previous chapter we pointed out that two lines may intersect at a point to form two pairs of vertical angles. Now consider what happens when a line intersects two or more lines. A line that intersects two or more lines in distinct points is called a *transversal* of these lines. In Figure 10-19, line r is a transversal relative to lines m and n. For this transversal, angles 1, 2, 3, and 4 are called *interior* angles, whereas angles 5, 6, 7, and 8 are called *exterior* angles. $\angle 1$ and $\angle 4$, and $\angle 2$ and $\angle 3$ are pairs of *alternate interior angles.* $\angle 5$ and $\angle 8$, and $\angle 6$ and $\angle 7$ are pairs of *alternate exterior angles.* $\angle 6$ and $\angle 2$, $\angle 4$ and $\angle 8$, $\angle 5$ and $\angle 1$, and $\angle 3$ and $\angle 7$ are called *corresponding angles.* Angles 3 and 6, 4 and 5, 1 and 8, and 7 and 2 are called *vertical angles.*

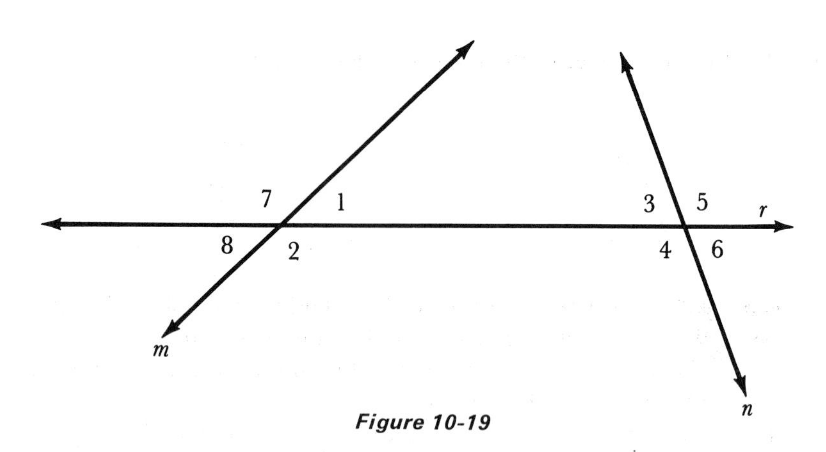

Figure 10-19

The following theorems use the idea of a transversal to state some important properties of parallel lines.

Theorem 10-7: (a) If lines m and n in a plane are intersected by a transversal in such a way that a pair of alternate interior angles are congruent, then m is parallel to n.

(b) Alternate interior angles formed by two parallel lines and a transversal are congruent.

Theorem 10-8: The opposite sides of a parallelogram are congruent.

Proof: Draw a diagonal connecting two opposite vertices of the parallelogram $ABCD$ (Figure 10-20). Now $\angle 1 \simeq \angle 4$, $\angle 3 \simeq \angle 2$, and $\overline{DB} \simeq \overline{BD}$. Why? Thus $\triangle ABD \simeq \triangle CDB$. Why? Thus $\overline{AB} \simeq \overline{CD}$ and $\overline{AD} \simeq \overline{CB}$. Why?

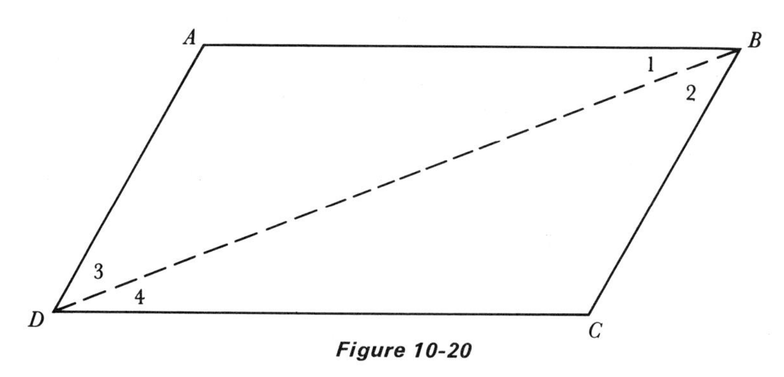

Figure 10-20

Three special parallelograms of interest are the *rhombus*, the *rectangle*, and the *square*. A *rhombus* is a parallelogram with all sides congruent. A *rectangle* is a parallelogram with right angles. A *square* is a rectangle with all sides congruent. Likewise, a square is a rhombus with four right angles.

5

Similar Geometric Figures

The term "similar" ordinarily means alike but not necessarily identical. Likewise, in mathematics we think of "similar" geometric figures as figures with the same shape but not necessarily the same size. Figure 10-21 illustrates two sets of four geometric figures; those in the first row are similar, respectively, to those in the second row.

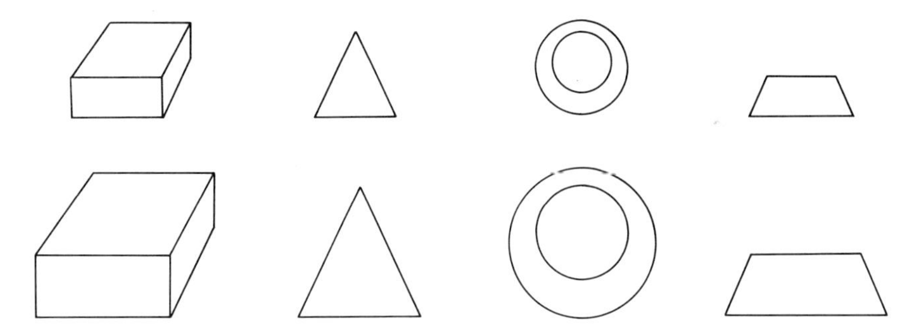

Figure 10-21

A familiarity with the word *ratio* is helpful in giving a mathematical definition of similar geometric figures. The quotient $(a \div b)$ is sometimes referred to as a ratio. Suppose we have 10 boys and 15 girls in our class. The ratio of boys to girls is "10 to 15," which may be expressed as 10/15. The ratio of a to b may be written as a/b.

Definition 10-5: Two geometric figures are said to be *similar* if there exists a one-to-one correspondence between their points, and if the ratio of the measures of any two line segments joining two pairs of points on one figure is equal to the ratio of the measures of the line segments joining corresponding points on the other figure.

In Figure 10-22, $AB/BC = DE/EF$; $BC/AC = EF/DF$; $AC/AB = DF/DE$ because the two triangles are similar. Since G corresponds to G' and H corresponds to H', $BG/GH = EG'/G'H'$ and $GH/AC = G'H'/DF$.

Theorem 10-9: Two triangles, $\triangle ABC$ and $\triangle DEF$, are similar (denoted by $\triangle ABC \sim \triangle DEF$) if and only if corresponding angles are congruent ($\angle CAB \simeq \angle FDE$, $\angle ABC \simeq \angle DEF$, and $\angle BCA \simeq EFD$).

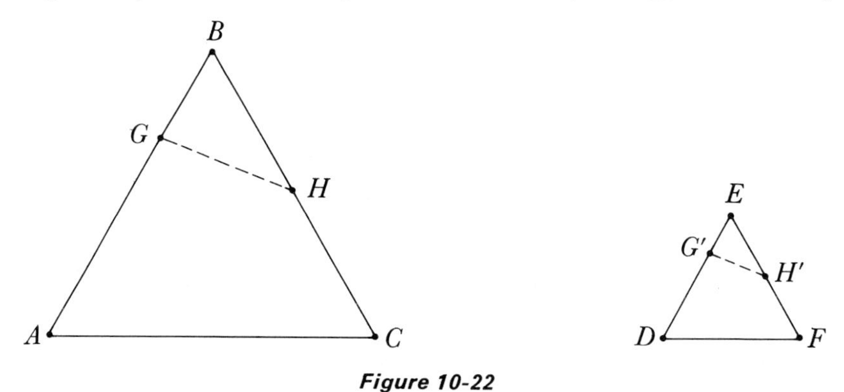

Figure 10-22

An important property of similar triangles is that the quotients of the measures of corresponding sides of similar triangles are equal. For example, in Figure 10-22

$$\frac{AB}{DE} = \frac{BC}{EF} = \frac{CA}{FD}.$$

(Remember that AB is another notation for $m(\overline{AB})$.)

Let us compare the similar triangles given in Figure 10-23. $\overline{AC}$ corresponds to $\overline{DF}$, $\overline{AB}$ corresponds to $\overline{DE}$, and $\overline{CB}$ corresponds to $\overline{FE}$.

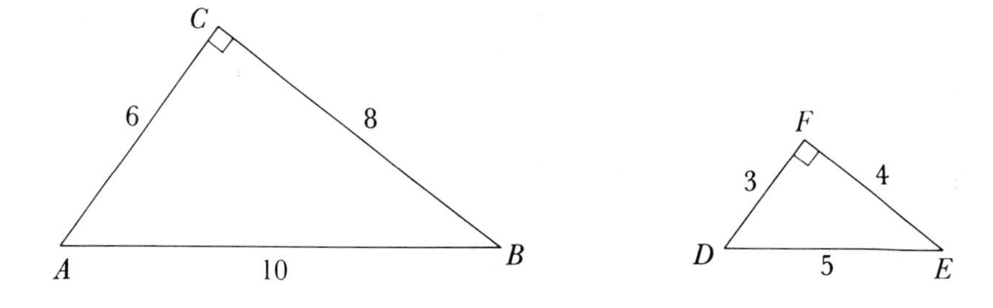

Figure 10-23

That is,

$$\frac{AC}{DF}=\frac{AB}{DE}=\frac{CB}{FE},$$

which is expressed numerically as

$$\frac{6}{3}=\frac{10}{5}=\frac{8}{4}=2.$$

Example: How high is a flagpole that casts a shadow of 80 feet, if a nearby post 8 feet high casts a shadow of 10 feet?

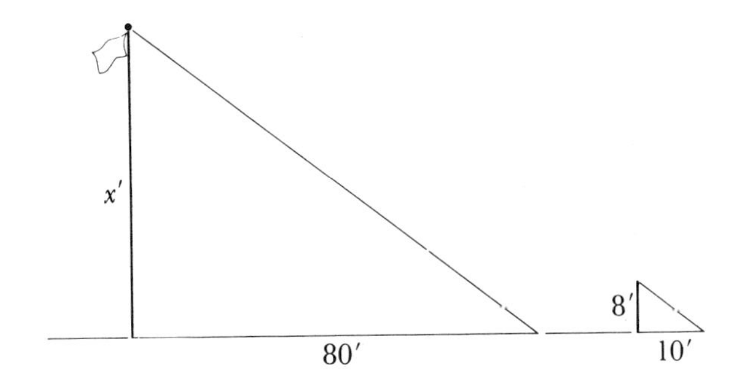

Figure 10-24

Since the sun rays make the same angles with the horizontal for the triangles involving the flagpole and the post, and since both the flagpole and the post are assumed to make right angles with the horizontal, the two triangles are similar, as indicated in Figure 10-24. Thus,

$$\frac{x}{8}=\frac{80}{10} \quad \text{or} \quad 10x=640 \quad \text{or} \quad x=64 \quad \text{or} \quad x=64 \text{ feet.}$$

Therefore the flagpole is 64 feet high.

Exercise Set 10-4

1. (a) Given $\triangle ABC \sim \triangle DEF$. If $m(\angle A)=70$ and $m(\angle E)=40$, find $m(\angle C)$ and $m(\angle F)$.

(b) Given $\triangle GHI \sim \triangle JKL$. If $GH = 5$, $GI = 7$, $JK = 8$, and $KL = 9\frac{3}{5}$, find HI and JL.
(c) Given $\triangle STU \sim \triangle VWX$, $TU = 8$, $SU = 8$, $VW = 4$, and $m(\angle TSU) = 60$, find ST, WX, and $m(\angle WVX)$.

2. Assume that the three triangles given are similar. Find the measure of the unknown sides.

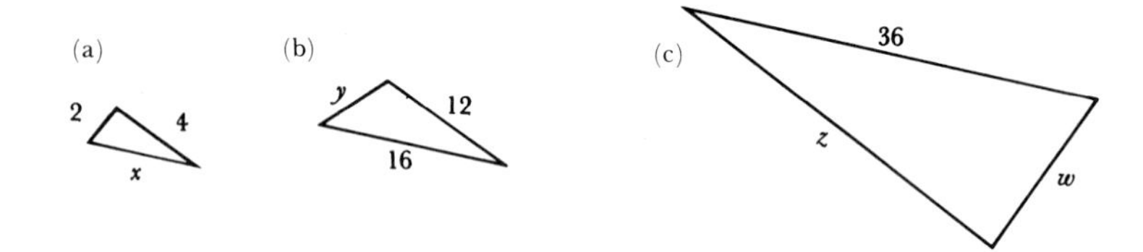

3. Find the perimeter of each triangle of the preceding exercise.
4. A snapshot is $2\frac{1}{4}$ inches wide and $3\frac{1}{4}$ inches long. It is enlarged so that it is 9 inches wide. How long is the enlarged picture? What is its perimeter?
5. At the same time that a yardstick held vertically casts a 4-foot shadow, a vertical flagpole casts a 24-foot shadow. How high is the flagpole?
6. On a certain map, $\frac{2}{3}$ inch is given as representing 60 miles. If the distance between two cities on the map is 4 inches, what is the distance in miles between the two cities?
7. Find x in the following, assuming in each exercise that the triangles are similar.

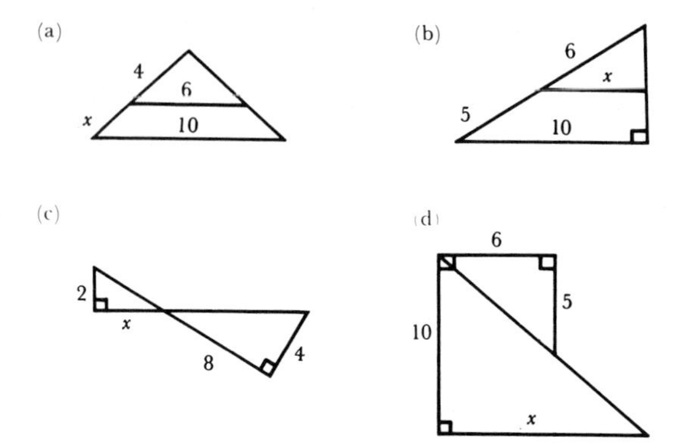

8. Given the figure below consisting of three parallel line segments cut by a transversal, list the angles congruent to $\angle 9$ and explain why each is congruent.

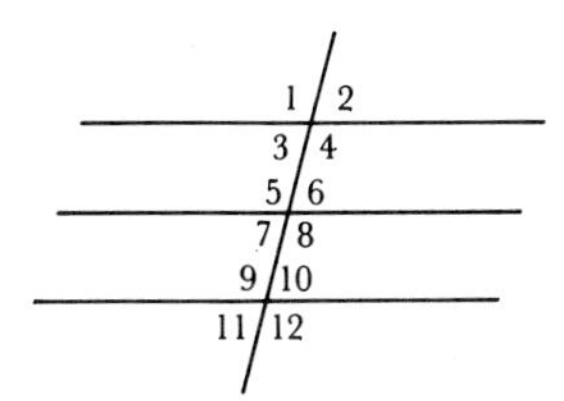

9. Find the height of the tree.

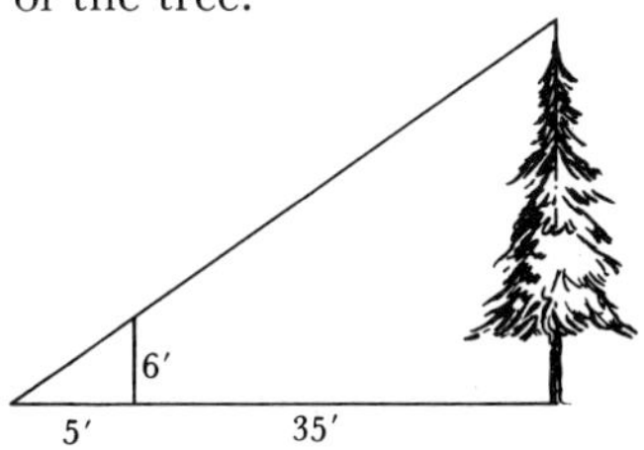

10. Suppose that $\triangle ABC \sim \triangle DEF$, where $\angle A \simeq \angle D$, $\angle B \simeq \angle E$, and $\angle C \simeq \angle F$. Let $m(\overline{AC}) = 7$, $m(\overline{DF}) = 14$, and $m(\overline{DE}) = 22$. If the perimeter of $\triangle ABC$ is 27, what is the perimeter of $\triangle DEF$?
11. In a triangle ABC, M is the midpoint of $\overline{AC}$ and N is the midpoint of $\overline{BC}$. What is the relationship between $\overline{MN}$ and $\overline{AB}$?
12. (a) Is it true that every equilateral triangle is similar to every other equilateral triangle? Why?
    (b) Is it true that every isosceles triangle is similar to every other isosceles triangle? Why?
    (c) Are congruent triangles always similar?
    (d) Are similar triangles always congruent?

*13. State the reflexive, symmetric, and transitive properties for similarity of triangles. (See Chapter 2.)

*14. Prove that the perpendicular from the hypotenuse to the opposite angle of a right triangle forms two triangles similar to each other and to the given triangle.

*15. In the accompanying figure, l is parallel to m, n is parallel to r, and $\overline{AB} \simeq \overline{AD}$.
    (a) Prove that quadrilateral $ABCD$ is a rhombus.
    (b) If n is perpendicular to l, prove that quadrilateral $ABCD$ is a square.

*16. In the figure, lines l and m are parallel while p is perpendicular to l. Prove that $\triangle ABC \sim \triangle ADE$.

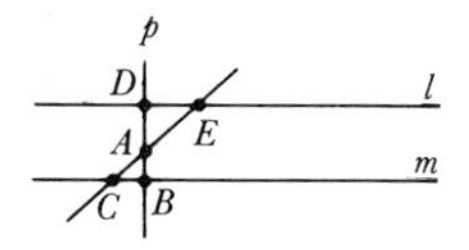

6

Area of Plane Figures

Now we turn from one-dimensional linear measure to two-dimensional plane measure. Let the word *area* refer to the measure of a region, where a region is the interior of a closed curve. In other words, "area" is the two-dimensional measure of the region bounded by a closed curve.

Consider, for example, the area bounded by a rectangle, one side of which has a measure of six units and the other side, three units (Figure 10-25). Thus, six congruent unit squares (the sides of which measure one unit) placed in a nonoverlapping position, will extend the entire length of the rectangle, while three unit squares will extend the width of the rectangle. The measuring process for two-dimensional regions is like that for one-dimensional figures: to find a measure, one determines how many nonoverlapping "units" fill the figure. In Figure 10-25, we

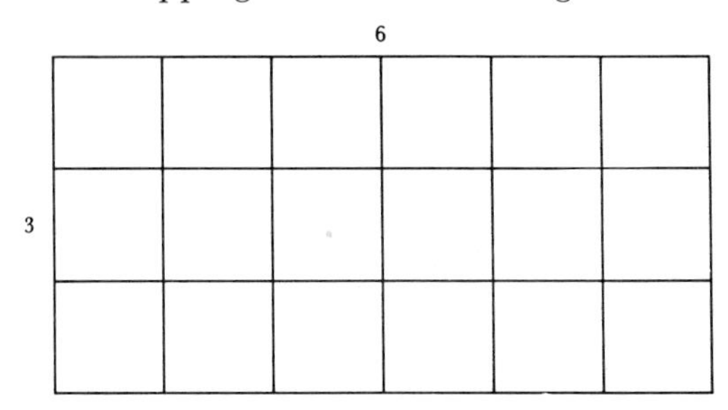

Figure 10-25

see that 18 unit squares completely cover the region of the rectangle. This work suggests that the unit square is quite reasonably designated as the unit for two-dimensional measure.

We postulate the existence of a two-dimensional area by considering the measure of a polygonal region. That is, we postulate the existence of a correspondence between a subset of real numbers and a polygonal region. There exists a one-to-one correspondence that associates with each convex polygonal region a unique positive real number, which we call *area*, with the following properties.

(a) *Area is additive.* That is, the area of the whole is equal to the sum of the areas of its nonoverlapping parts. If a region is decomposed into a finite number of regions such that any two regions have in common only line segments, curves, or points, then the area of the whole region is the sum of the areas of the regions into which it is decomposed.

(b) *If figure $A \simeq$ figure B, then the area bounded by A equals the area bounded by B.* If two regions are congruent, we can theoretically pick one up and put it down on the other so that they fit perfectly. That is, not only do congruent figures have the same size and shape, they also bound equal areas.

(c) *If a region is cut into parts and reassembled to form another region, then the two regions have the same area.* In Figure 10-26 the area of the rectangular region consisting of A and B in (a) is the same as the area of the triangular region composed of A and B in (b).

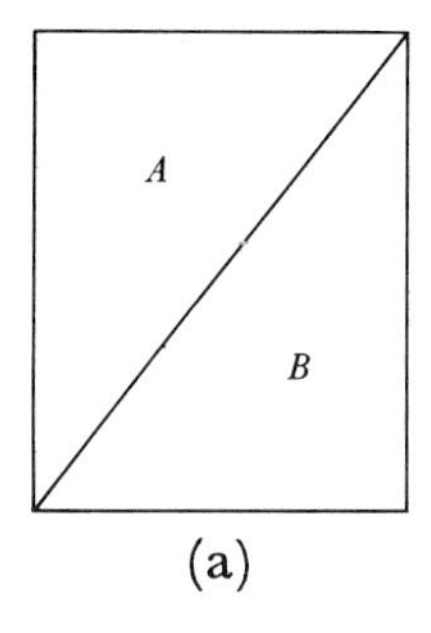

(a)

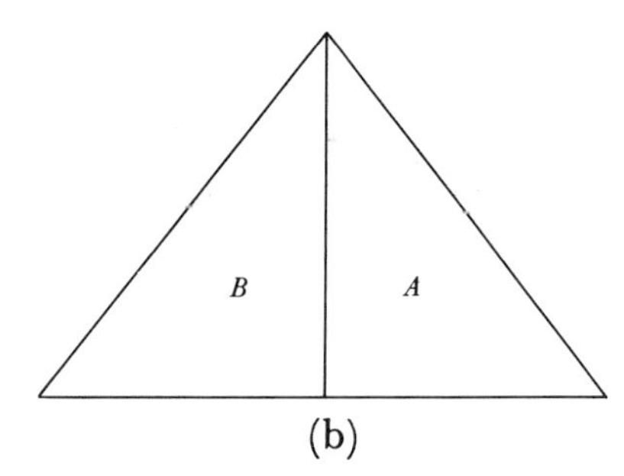

(b)

Figure 10-26

(d) *If l and w (called length and width) are the linear measures of two consecutive sides of a rectangle forming the boundary of a rectangular region, then the area of the rectangular region is the product of the length and width, denoted by $A = lw$.*

Example: Find the area of the region bounded by a rectangle that has a length of $6\frac{1}{2}$ and a width of 4.

$$\text{Area} = lw = (6\tfrac{1}{2})(4) = 26$$

Now let us apply the concept of area to parallelograms, triangles, and trapezoids.

Theorem 10-10: The area of the region bounded by a parallelogram is equal to the product of a (the altitude) and b (the corresponding base), denoted by Area $= ab$.

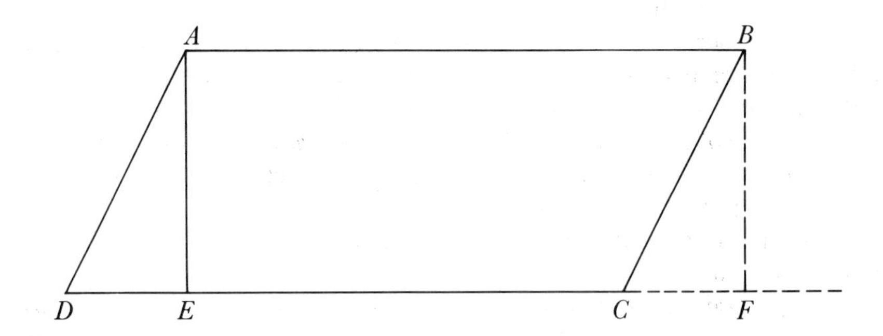

Figure 10-27

Proof: Consider parallelogram $ABCD$ in Figure 10-27. We agree that the angles are not right angles, so the figure is not a rectangle. We wish to develop a formula for the measure of the region bounded by the parallelogram. Draw line segment $\overline{AE}$ perpendicular to $\overleftrightarrow{DC}$. Now draw $\overline{BF}$ perpendicular to $\overleftrightarrow{DC}$. Quadrilateral $ABFE$ is a rectangle.

Now $\overline{AD} \simeq \overline{BC}$. Why? $\overline{AE} \simeq \overline{BF}$. Why? $\angle DEA \simeq \angle CFB$ by construction. Thus, $\triangle AED \simeq \triangle BFC$. Why? $\triangle AED$ has the same area as $\triangle BFC$. Thus the area of the region bounded by the parallelogram is the same as the area of the region bounded by the rectangle, and it can be written as $AE \cdot AB$ or $AE \cdot DC$. Hence, the area of the region bounded by the parallelogram equals $AE \cdot DC$. Let AE be called the *altitude* and DC the *base* of the parallelogram. Then Area $= ab$ where "a" denotes the altitude and "b" denotes the base. Notice that by definition an altitude is perpendicular to the base.

Example: Suppose the parallelogram in Figure 10-27 has $\overline{AE}$ of length 6 and $\overline{DC}$ of length 14. Find the area of $ABCD$.

Solution: Area $= ab = 6 \cdot 14 = 84$.

Theorem 10-11: The area of the region bounded by a triangle is one-half the product of the base and the altitude drawn to the base.

Proof: Consider $\triangle ABC$ in Figure 10-28, with base $\overline{CB}$ and altitude $\overline{AD}$. Through A, draw a line parallel to $\overleftrightarrow{CB}$; and through B, draw a line parallel to $\overleftrightarrow{CA}$. Let E be the intersection of these two lines. Then $AEBC$ is a parallelogram.

Consider $\triangle CAB$ and $\triangle EBA$. Now, $\overline{AE} \simeq \overline{BC}$ and $\overline{CA} \simeq \overline{EB}$. Why? Also, $\overline{AB} \simeq \overline{BA}$. Why? Thus $\triangle CAB \simeq \triangle EBA$. Why?

Hence the area of the region bounded by the parallelogram $AEBC$ is twice the area of the region bounded by the $\triangle CAB$. Therefore the area of the region bounded by $\triangle CAB = \frac{1}{2}$ the area of the region bounded by parallelogram $AEBC = \frac{1}{2}\, AD \cdot CB = \frac{1}{2}\, ab$.

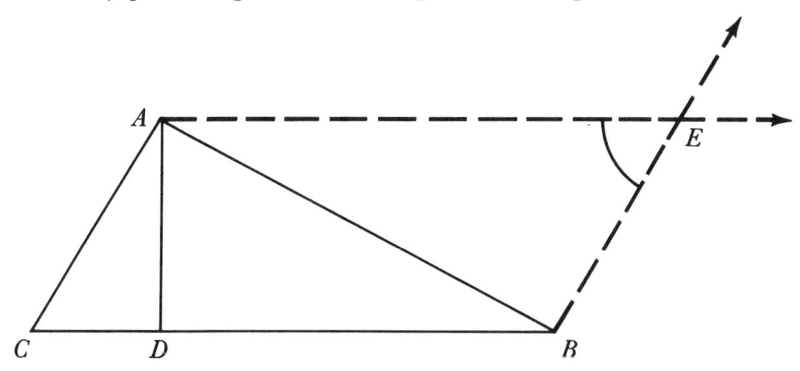

Figure 10-28

Example: Suppose $\triangle ABC$ in Figure 10-28 has $AD = 4$ and $CB = 12$. The area of the triangle is $(1/2) \cdot 4 \cdot 12 = 24$.

The area of the region bounded by a trapezoid may be considered as the sum of the areas of regions bounded by a rectangle and one or two triangles. For example, consider trapezoid $ABCD$ in Figure 10-29. The area of the region bounded by $ABCD$ = (area of the region bounded by rectangle $BCFE$) + (area of the region bounded by $\triangle ABE$) + (area of the region bounded by $\triangle FCD$) $= ca + \frac{da}{2} + \frac{ea}{2} = \left(c + \frac{1}{2}d + \frac{1}{2}e\right)a$ $= \left(\frac{2c + d + e}{2}\right)a = \left[\frac{c + (c + d + e)}{2}\right]a = \left(\frac{c + b}{2}\right)a$ where $b = c + d + e$. Thus, we have the following theorem.

Theorem 10-12: The area of the region bounded by a trapezoid is one-half the sum of the measures of the opposite parallel sides times the measure of the altitude; that is, $\frac{1}{2}(c+b)a$.

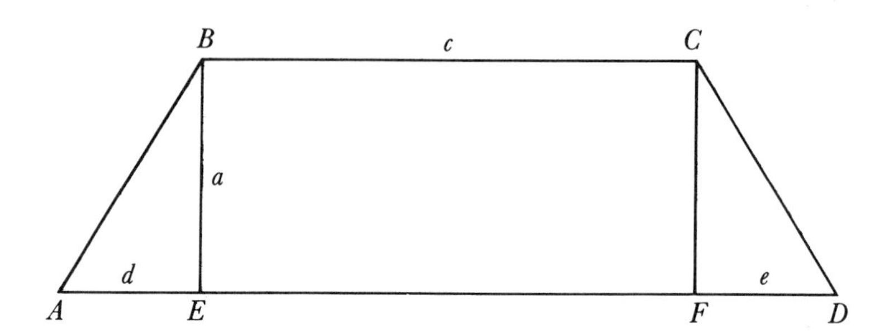

Figure 10-29

Just as in linear measure, we need to distinguish between the two-dimensional measure and the measurement of a region. Area measure is a positive real number assigned at this time to a polygonal region. The measurement of this region involves a standard unit. The measurement of a table may be 800 square inches. The area measure is 800 and the unit of measurement is square inches.

Example: Find the measurement of a rectangular region where the measurement of the length is 4 centimeters and of the width is 5 centimeters.

Measurement of area $= 4 \cdot 5 = 20$ square centimeters.

Exercise Set 10-5

1. Classify either as true or as false.
   (a) Every parallelogram is a trapezoid.
   (b) The "opposite sides" of a parallelogram are congruent.
   (c) No trapezoid has a pair of congruent sides.
   (d) If two triangles are congruent, then they have the same area.
   (e) Every parallelogram with one right angle is a square.
   (f) The union of a finite number of triangular regions which have only boundaries in common is always a polygon.
   (g) A rectangular region can be divided into two rectangular regions.

(h) There exists a square region with area 13.
(i) There exists a rectangular region with area $\sqrt{15}$.
(j) Parallelogram *ABCD* is a region.
(k) Parallelogram *ABCD* is a set of points.
(l) Parallelogram *ABCD* is the set $\{A, B, C, D\}$.
(m) Parallelogram *ABCD* is the set $\overline{AB} \cap \overline{BC} \cap \overline{CD} \cap \overline{DA}$.
(n) The area of parallelogram *ABCD* is base times altitude.
(o) All rectangles are squares.
(p) Some quadrilaterals are not trapezoids.
(q) Every altitude of a given trapezoid has the same measure.
(r) Every altitude of a given triangle has the same measure.

2. (a) How many square centimeters are in a square meter?
(b) How many square inches are in a square foot?
(c) How many square millimeters are in a square meter?
(d) How many square inches are in a square yard?
3. *ABCD* is a rectangle. Compute the area bounded by *ABCD* if *AB* and *BC* are given by the following pairs of values.
(a) 1.6, 0.21 (b) 7/2, 3/4 (c) 7, 3/2
(d) $1.0(10)^3$, $2.3(10)^{-5}$ (e) x^2y, yz^2 (f) 15 1/3, 17 4/5
4. Find the area of the region bounded by a trapezoid with altitude a and bases b and d if
(a) $a = 6$, $b = 10$, $d = 8$.
(b) $a = 4\frac{1}{2}$, $b = 6\frac{1}{2}$, $d = 7\frac{1}{2}$.
5. Find the measurement of the regions bounded by the figures shown, using the dimensions given.

(a)
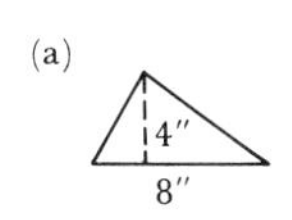

(b)
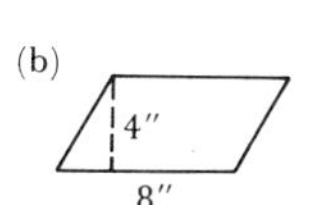

(c)
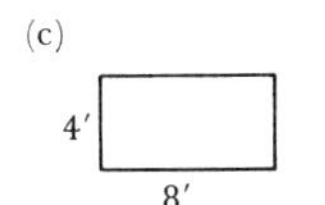

(d)
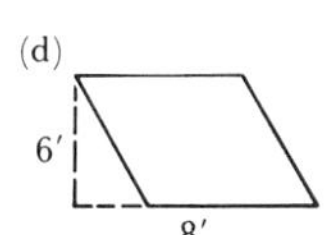

(e)
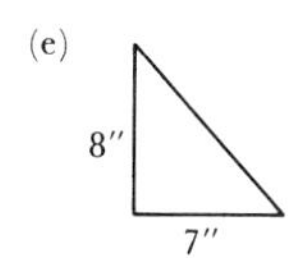

(f)
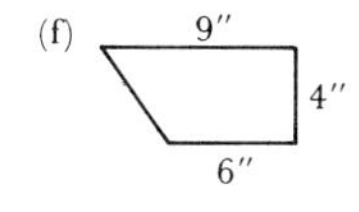

(g)
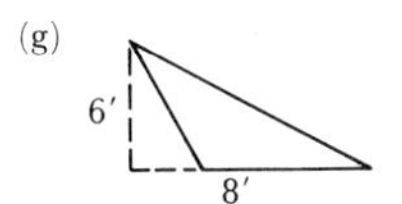

(h)
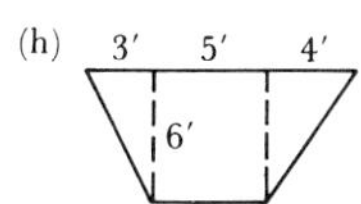

(i)
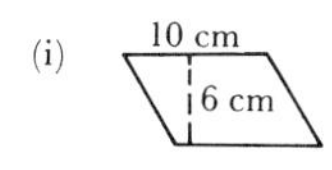

(j)
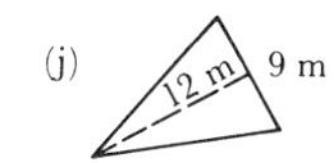

(k)
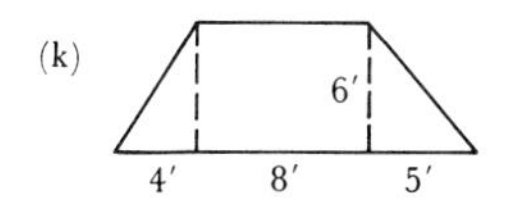

(l)
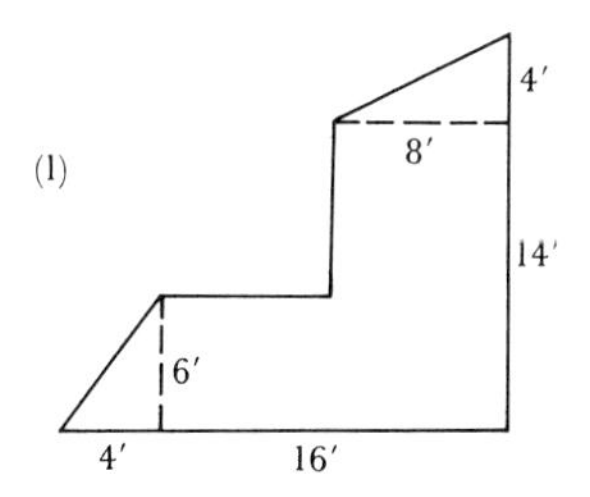

6. If the area of each of the following polygons is 56, find the missing quantities.
   (a) Triangle: base $= 14$, altitude $= ?$
   (b) Rectangle: length $= 28$, width $= ?$
   (c) Trapezoid: sum of parallel sides $= 56$, altitude $= ?$
   (d) Parallelogram: base $= 8$, altitude $= ?$
7. A rectangle is 4 units long and 3 units wide. If another rectangle twice as long has the same area, how wide is it?
8. For a triangular region ABC with altitude to base $\overline{AB}$ denoted by $\overline{CD}$ what happens to the area of the triangular region if
   (a) AB is doubled?
   (b) AB and CD are both doubled?
   (c) CD is tripled?
   (d) AB and CD are both tripled?
9. In trapezoid $ABCD$, $\overleftrightarrow{AB}$ is parallel to $\overleftrightarrow{CD}$ and $\overleftrightarrow{AD}$ is perpendicular to $\overleftrightarrow{DC}$. If $\overline{AB} = 6$, $\overline{AD} = 4$, and $\overline{DC} = 8$, find the area of the trapezoid region.

*10. The Jones plan to remodel parts of their home: the hall, living room dining room, kitchen, and den.

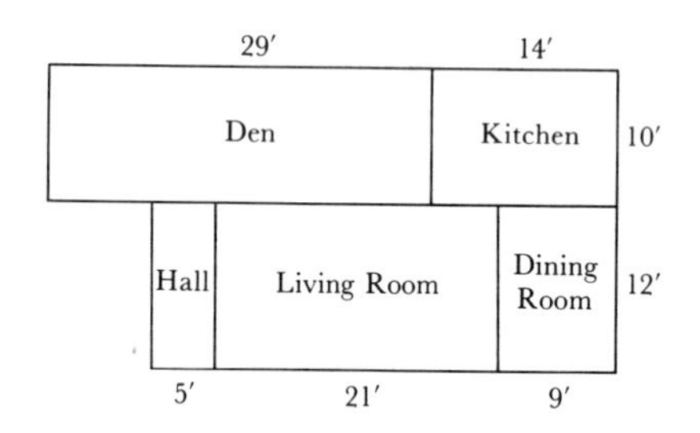

   (a) If the hall is to be carpeted in one pattern and the living and dining rooms in another, how many square yards of each carpet would the Jones need?
   (b) How much carpet would be needed if all three rooms in (a) were carpeted in the same pattern?
   (c) Mr. Jones wants to tile the den and kitchen himself. Discounting space that cabinets might cover, how much area must Mr. Jones tile in the kitchen? In the den?
   (d) Suppose that Mr. Jones plans to buy 9″ square tile. If the kitchen is one pattern and the den another, how many tiles of each pattern should he buy? How much tile should he buy if he uses the same pattern for both rooms?

*11. Prove or disprove the following statement: If two polygons have equal areas, then the polygons are congruent.

7

Circumference and Area of a Circle

A familiar class of simple closed curves is the set of circles. Examples of circular regions are coins, jar lids, dishes, wheels, and the like.

Definition 10-6: The set of all points in a plane at an equal distance from a fixed point in the plane is defined to be a *circle.*

In this definition of a circle, the fixed point is called the *center* of the circle, and any line segment from the fixed point to a point on the circle is called a *radius* of the circle. In Figure 10-30(a), a line is drawn through the center of the circle intersecting the circle in points A and B. The segment $\overline{AB}$ joining these two points is called a *diameter* of the circle, and $\overline{PA}$ is called a *radius* (so also is $\overline{PB}$ a radius). A line segment such as $\overline{BC}$ is called a *chord*, and $\overleftrightarrow{BC}$ is called a secant of the circle.

The *interior* of the circle is the set of all points of the plane whose distance from the center of the circle is less than the distance from the center of the circle to the circle. The *exterior* of a circle is the set of all points of the plane whose distance from the center of the circle is greater than the distance from the center of the circle to the circle.

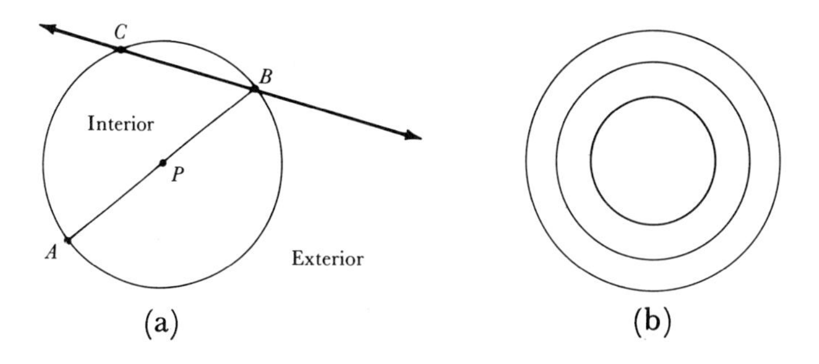

Figure 10-30

Two or more circles with the same center are called *concentric* circles (Figure 10-30(b)).

Figure 10-31 illustrates the intersection of a line and a circle. Note that in (a), the line intersects the circle in two distinct points, in (b), the intersection is one point, while in (c), the line and the circle do not intersect. If a line intersects a circle in only one point, the line is called a *tangent* and the common point is called a point of *tangency*.

Other interesting examples of intersections involve circles and polygons. In Figure 10-32(a), the intersection of the circle and the triangle

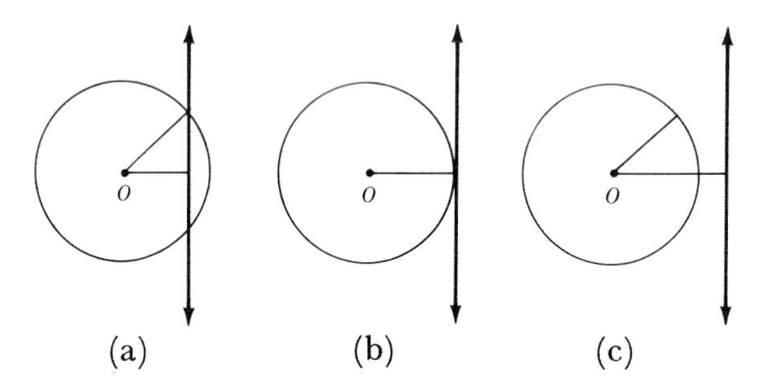

Figure 10-31

consists of three points. In this case, the triangle is said to be *inscribed* in the circle, or the circle is *circumscribed* about the triangle. In Figure 10-32(b), a quadrilateral is circumscribed about a circle, and the circle is inscribed in the quadrilateral. These two geometric figures have four common points of tangency.

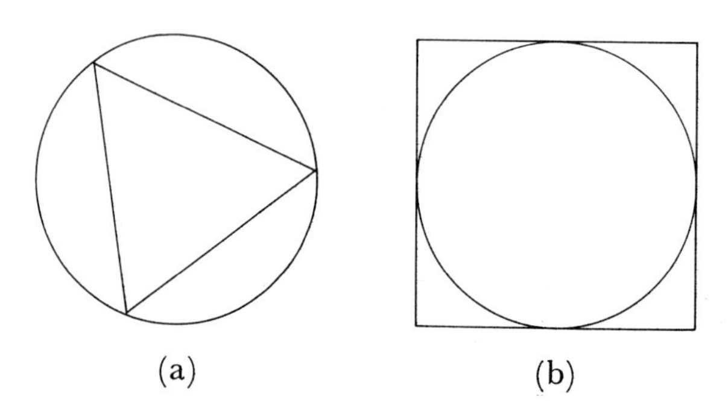

Figure 10-32

An arc of a circle is defined as the set of points consisting of two points on the circle and all the points on the circle between the two points. Actually, two points on a circle describe two arcs since the word " between " is not definite enough to describe the correct part of a circle. For example, in Figure 10-33, what is arc AC? To avoid this ambiguity, an arc is usually designated by three letters such as arc ABC, symbolized by $\widehat{ABC}$. In this notation, the first and third letters indicate the endpoints of the arc, and the middle letter indicates which of the two possible arcs is being described.

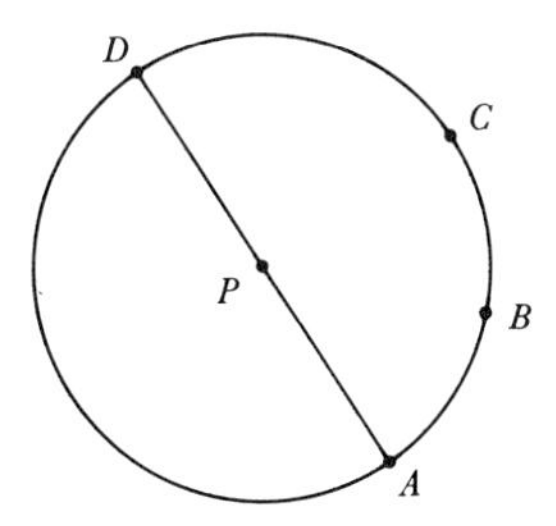

Figure 10-33

If the endpoints of an arc, such as $\widehat{ACD}$ in Figure 10-33, are the ends of a diameter of a circle, then the arc is called a *semicircle*. Thus, the endpoints of a semicircle and the center of a circle are collinear points.

In the preceding sections, we discussed the perimeter of a polygon. We now use this idea to give intuitive meaning to the circumference of a circle—that is, the distance around a circle.

Suppose one inscribes regular polygons inside a circle. In Figure 10-34 an equilateral triangle, a square, a pentagon, a hexagon, and a heptagon are inscribed inside circles. It is always possible to inscribe any such regular polygon inside a circle.

Now consider in Figure 10-35 square $ABCD$ inscribed within the circle. The perimeter of the square is given by $p_1 = AB + BC + CD + DA$. If c is the circumference of the circle, then $p_1 < c$ by inspection.

Now use the midpoint of each arc associated with a side of the square as a vertex of a new polygon. Call these points E, F, G, and H. Regular polygon $AEBFCGDH$ is an octagon with perimeter

$$p_2 = AE + EB + BF + FC + CG + GD + DH + HA.$$

Again it is obvious that $p_1 < p_2 < c$.

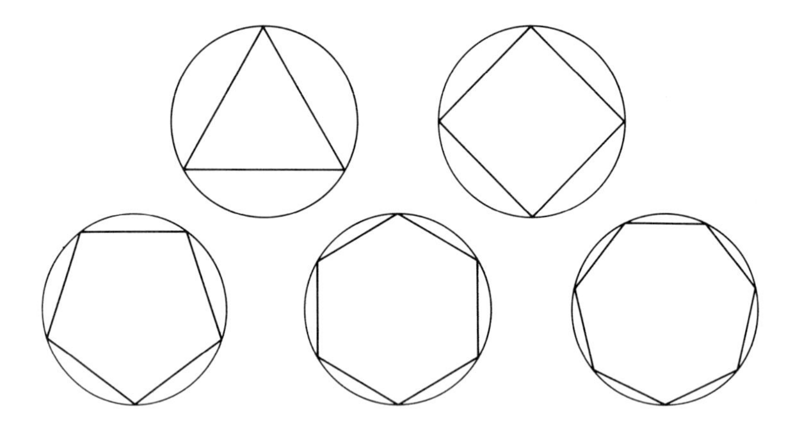

Figure 10-34

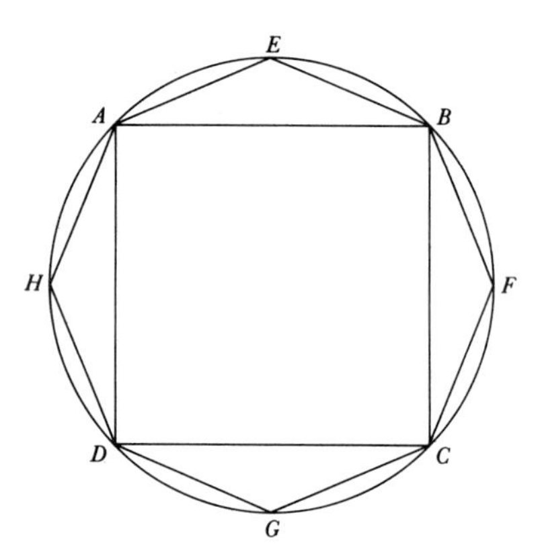

Figure 10-35

Now form a new polygon in the same manner with 16 sides. The perimeter of this new polygon would satisfy $p_1 < p_2 < p_3 < c$. Imagine this process as continuing with a 32-side polygon, a 64-side, or a 128-side polygon. Obviously the perimeters of the polygons are getting closer and closer to the circumference of the circle as the number of sides of the polygon increases. The *circumference* of the circle is the number that the perimeters get closer and closer to as more and more polygons are inscribed in the circle as indicated.

Definition 10-7: There is a real number π in terms of which the circumference of a circle is given by $c = 2\pi r$, where r represents the measure of the radius of the circle. π (read "pi") is an irrational number that can be approximated by a rational number to as many decimal places as desired; accurate to ten places, it is $\pi = 3.1415926534 \ldots$. Perhaps the most widely used approximation for π is 22/7, which agrees with the decimal representation of π to only two decimal places.

Example: Find the circumference of the circle with radius of measure 10.

$$\begin{aligned} c &= 2\pi r \\ &= 2\pi(10) \\ &= 20\pi \\ &= 20(22/7) = 440/7 \qquad \text{(approximately)} \end{aligned}$$

Often (without thinking) we use the phrase "area of a circle." Of course, a circle does not have area. What we mean is the area of the "closed circular region." We can approximate the area of the region bounded by a circle in many different ways. One way would be to inscribe regular polygons inside the circle (see Figure 10-35). Since regular polygons can be subdivided into triangles, and since the areas of triangular regions can be computed, one can easily obtain the area bounded by any regular polygon and thereby obtain an approximation to the area of a circular region. It is intuitively clear that as the number of sides of the inscribed regular polygon increases, the area of the polygonal region gets closer and closer to the area of the circular region. By continuing this process, it can be shown that the area of a circular region is πr^2, where r is the measure of the radius.

Examples: (a) Find the area of the circular region with a radius that measures 4. $A = \pi r^2 = (22/7) \cdot (4)^2 = 50.3$ (approximately).

(b) Find the area of the interior of a circle whose radius is 7. $A = (22/7) \cdot (7)^2 = 154$ (approximately).

Exercise Set 10-6

1. Discuss each of the following intersections.
   (a) Two circles (b) A line and a circle

2. Given that O is the center of the circle, let c and A represent the circumference and the area respectively.
   (a) If $DB = 6$, find c and A.
   (b) If $c = 8\pi$, find OB and A.
   (c) If $A = 16\pi$, find OB and c.
   (d) If $OB = \pi$, find c and A.

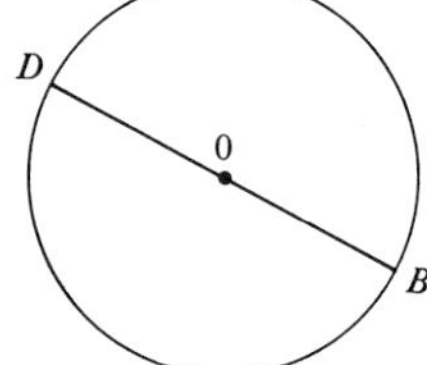

3. In the figure, name the following relative to the circle. (O is the center of the circle.)
   (a) $\overleftrightarrow{AB}$ (b) $\overline{FG}$
   (c) $\overline{OD}$ (d) $\overleftrightarrow{BC}$
   (e) Relationship between $\overleftrightarrow{AB}$ and $\overleftrightarrow{OD}$

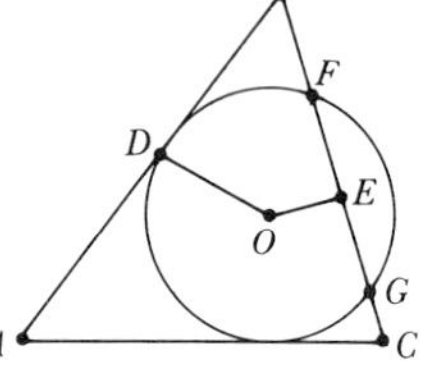

4. (a) Draw two circles whose intersection consists of only one point.
   (b) Draw three concentric circles.
5. Find the measurement of the area of each circular region for which the measurement of the radius is given. This time, approximate π by 3.1416.
   (a) 4 feet (b) 10 meters
   (c) 12 centimeters (d) 9 yards
6. Given that the circumference of a circle is 14π, find the measure of
   (a) the radius. (b) the diameter.
7. The figure below is a semicircle; the area of the interior of this closed curve is 24π. Find the length around the closed curve.

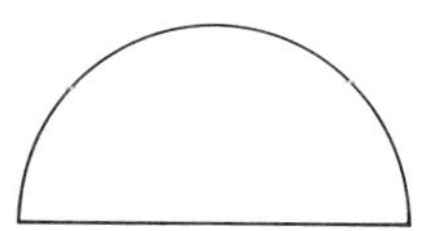

8. What is the area of the shaded region of the figure below if the measure of the radius of the circle is 2?

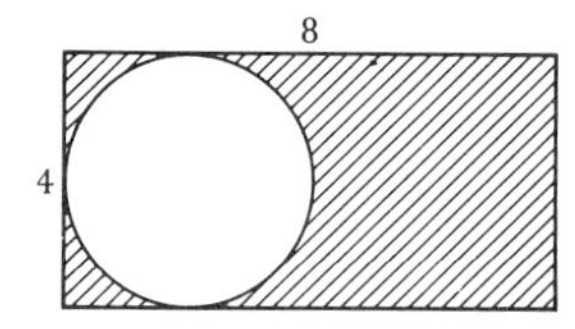

9. Compute the area of the region bounded as shown.

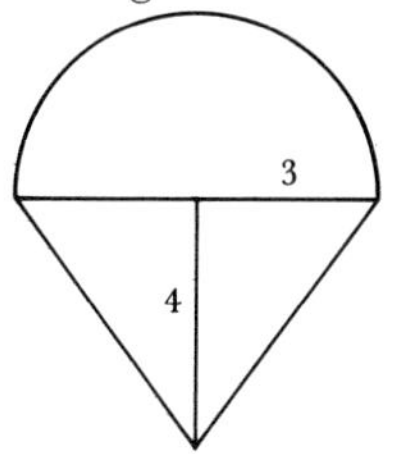

10. Consider the area bounded by two concentric circles. If $OP = 8$ and $OQ = 10$, what is the area of the region between the two circles?

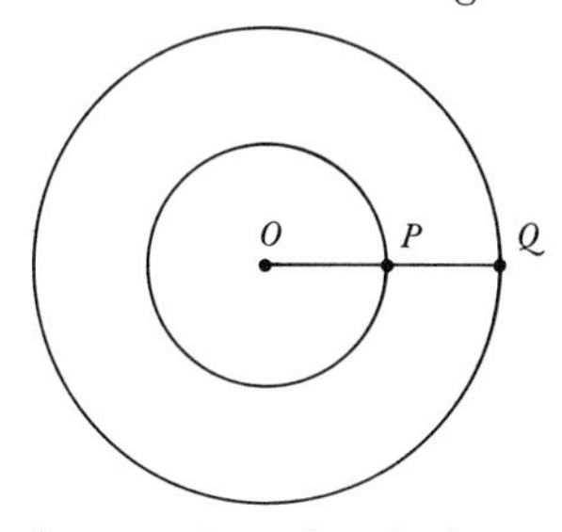

11. Let $E(c)$ represent the exterior of a circle and $I(c)$ the interior. Using the figure in Exercise 10, let c_1 represent the inside circle and c_2 the outside circle. What is the area of
    (a) $E(c_1) \cap I(c_2)$? (b) $I(c_1) \cap I(c_2)$?
    (c) $I(c_1) \cup I(c_2)$? (d) $I(c_1) \cap E(c_2)$?
12. If the circumference of a circle is doubled, how is the area changed?
13. Find the measure of the radius of the circle where the area of the circular region is double the circumference.
14. Write out the first four decimal places in 22/7. Compare this result with the approximation given for π in this book. Did you realize that 22/7 was such a poor approximation for π? How good an approximation to π is given by $\frac{10471}{3333}$?

*15. The formula for the area of a circle, $A = \pi r^2$, is given in terms of the radius. Express the formula for the area of a circle in terms of the circumference.

*16. If a point on a circular wheel traverses exactly the circumference of the wheel as the wheel is rotated, the wheel is said to have been rotated through one revolution. How many revolutions would be made in one mile by a car wheel with a diameter of 32″?

*17. If a car goes 50 mph, how many revolutions would a 30″ wheel turn in a half hour?

8

Volume

To obtain the measure of a space region, we use the same type of reasoning as was used for plane regions in the previous section. Associated with each closed space region is a number that we call the *volume,* the measure of a space region bounded by a closed space surface. Thus, volume has the same meaning for space regions as area has for plane regions.

The properties of volume are closely related to those of length and area. The concept "volume" has the following properties for space regions.

(a) *Volume is additive.* That is, the volume of the whole is equal to the sum of the volumes of its nonoverlapping parts. If space R and space S are nonoverlapping space regions (possibly with surfaces in common), then the volume of (space R) $\cup$ (space S) is equal to the volume of space R plus the volume of space S.

(b) *If space* $R \simeq$ *space* S, *then the volume of space* R *equals the volume of space* S.

(c) *If a space region is cut into parts and reassembled to form another space, then the two space regions have the same volume.*

The unit for the measurement of volume is a *unit cube.* A *cube* is a quadrilateral prism in which all six faces are regions bounded by congruent squares. In a unit cube, the side of each square measures one unit.

To demonstrate the use of a unit cube, consider the volume bounded by a *right rectangular prism* (sometimes called a rectangular parallelepiped). A right rectangular prism is a figure similar to a cube, except that its faces are rectangles. To find the volume of the region bounded by a rectangular prism, we must determine how many unit cubes can be fitted into the prism. Consider the rectangular prism in Figure 10-36. For this prism, *six* cubes can be placed on one edge (*the length*), *four* on a second edge (*the width*), and *three* on the third edge (*the height*). To generalize this result, we say that the volume of a right rectangular prism is the product of the length, the width, and the height; symbolically, $V = lwh$.

It is easy to see that the formula $V = lwh$ gives the number of unit cubes in a space region bounded by the right rectangular prism if l, w, and h are integers, since there would be l cubes along one direction, w

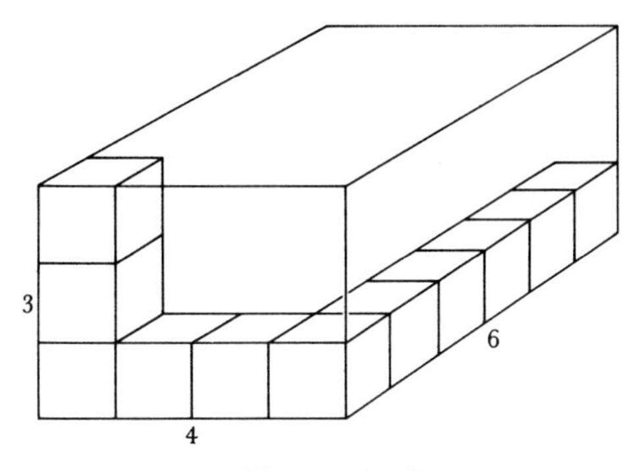

Figure 10-36

in the second direction, and h in the third direction. We postulate that the same formula holds by definition when l, w, and h are any real numbers.

Example: Find the volume bounded by the right rectangular prism which measures $3\frac{1}{2}$ by 6 by 8.

$$V = (3\tfrac{1}{2})(6)(8) = 168$$

Another measure associated with space surfaces is *surface area.* The *lateral* surface area of a space surface is the area of all the faces or sides (other than bases) of the space surface. The *total* surface area is the lateral area added to the area of the bases of the space surface.

Example: Find the total surface area of a right rectangular prism, where $l = 2$, $w = 4$, and $h = 5$.

$$\text{Total surface area} = 2(2 \cdot 4) + 2(2 \cdot 5) + 2(4 \cdot 5) = 76.$$

Figures 10-37 and 10-38 show two prisms with certain parts labeled. In Figure 10-37, a right section is drawn. A *right section* of a prism is the polygon formed by the intersection of the prism with a plane that is perpendicular to each of the lateral faces or sides of the prism. Figure 10-38 shows two parallel planes cutting a prism to obtain what we call the *bases* of the prism. The perpendicular distance between the two parallel planes is called the *altitude.*

In general for a prism we have:

Lateral surface area = (perimeter of right section) · (lateral edge of prism)
Total surface area = lateral surface area + area of bases
Volume = (area of a right section) · (lateral edge of prism)
Volume = (area of the base) · (altitude) = $b \cdot a$

Example: Consider a right triangular prism with an altitude of 10. The base and altitude of the triangle are 6 and 8, respectively. Find the volume of the space region bounded by the prism.

$$\text{Area of base} = \tfrac{1}{2}(6)(8) = 24$$
$$\text{Volume} = 24 \cdot 10 = 240$$

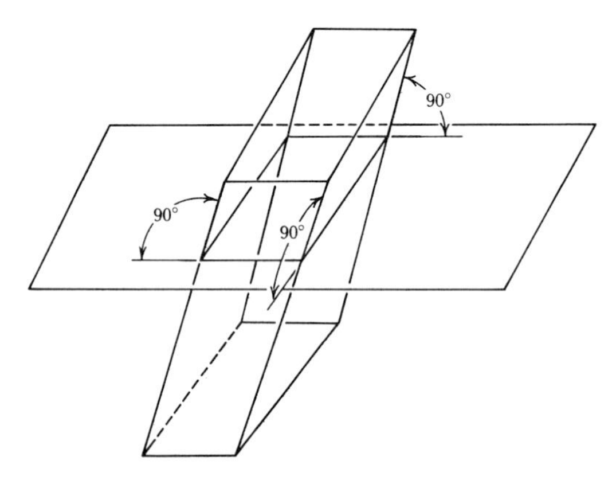

Figure 10-37

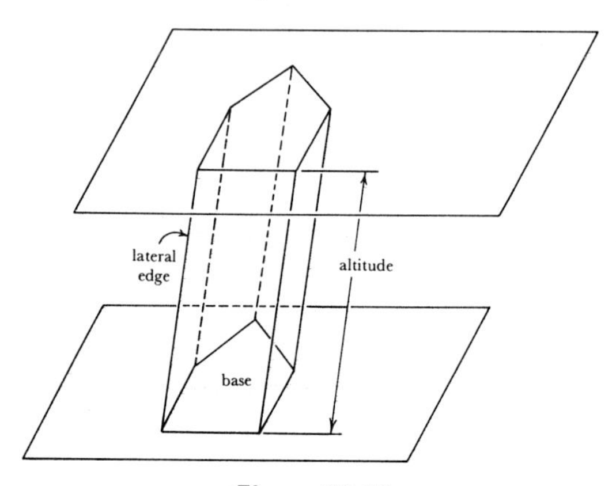

Figure 10-38

9

Volumes Bounded by Spheres, Cylinders, and Cones

Volumes other than those of right rectangular prisms are also of interest and of importance. For example, you might want to determine the volume of air in a basketball or the amount of food in a can. Others are interested in the surface area of the earth. Because of the need to determine volumes, we will consider in this section three additional simple closed surfaces: spheres, cylinders, and cones.

Can you visualize a set of points in space equidistant from a fixed point? This set of points would be more than a circle. In fact, it could be considered as the union of infinitely many circles. Such a space surface is called a *sphere.*

Definition 10-8: A set of points in space of equal distance from a fixed point is called a *sphere.*

The fixed point of the sphere is called the *center* of the sphere, and any line segment from the center to a point on the sphere is called the *radius* of the sphere.

Two or more spheres with the same center are called *concentric* spheres. A *chord* of a sphere is a line segment whose endpoints are on the sphere. If the chord contains the center of the sphere, it is called a *diameter* of the sphere.

The *interior of a sphere* is the set of all points whose distance from the center of the sphere is less than the distance from the center of the sphere to the sphere. The set of points whose distance from the center of the sphere is greater than the distance from the center of the sphere to the sphere is called the *exterior of the sphere.*

Consider the intersection of a sphere and a plane. If the intersection is not empty, it will be either one point or a circle. If the plane is tangent to the sphere, the intersection is but a single point. Such a situation is intuitively understood by visualizing a ball resting on a flat surface.

If the intersection of the plane and the sphere contains more than one point, then the intersection is a circle. Figure 10-39 shows two circles formed by the intersection of a plane and a sphere. A *great circle* on a sphere, illustrated in Figure 10-39(a), is any intersection of a sphere and a plane through the center of the sphere. All circles on a sphere

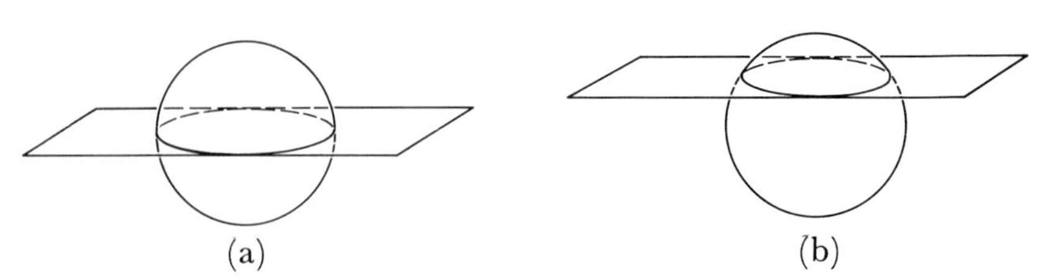

Figure 10-39

that are not great circles are called *small circles* [Figure 10-39(b)]. All great circles on a sphere have the same circumference, since their radii are equal. If we think of the earth as a sphere, and globes are used for such representation, the great circles through the North and South Poles are called *meridians.* The small circles formed by planes parallel to the plane of the equator are called *parallels of latitude.*

The next set of space points we wish to consider is called a *circular cylinder.* In Figure 10-40, we have three drawings that represent circular

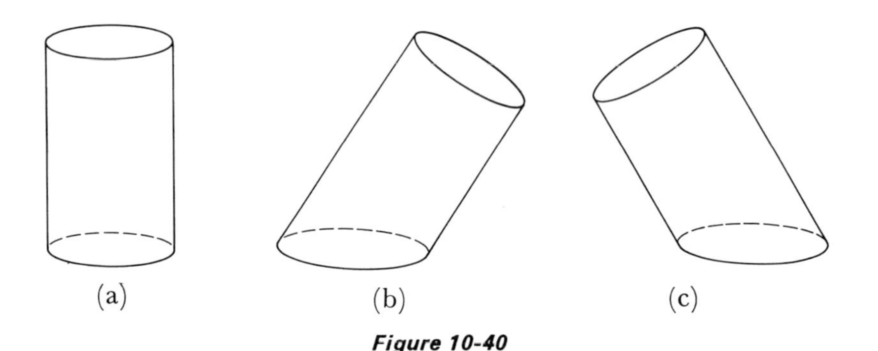

Figure 10-40

cylinders. Figure (a) is called a *right cylinder* and (b) and (c) are called *oblique cylinders.* The ends of a right circular cylinder (called bases) lie on parallel planes. Some common objects that suggest right circular cylinders are tin cans and iron pipes. Vertical segments of right circular cylinders, such as $\overline{AB}$ in Figure 10-41, are perpendicular to the plane of the base. The radius of the circular base is called the *radius* of the cylinder.

The third example of a simple closed surface is a *cone.* A *right circular cone,* shown in Figure 10-42, consists of a circular base and segments joining a fixed point to points on the base. Of course a general cone would be an oblique cone with any closed curve for a base.

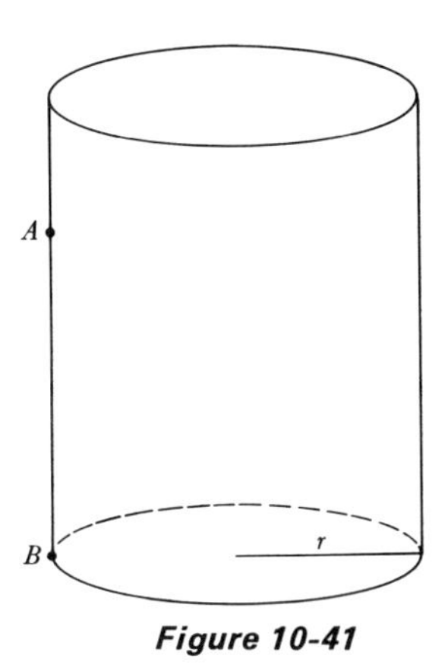

Figure 10-41

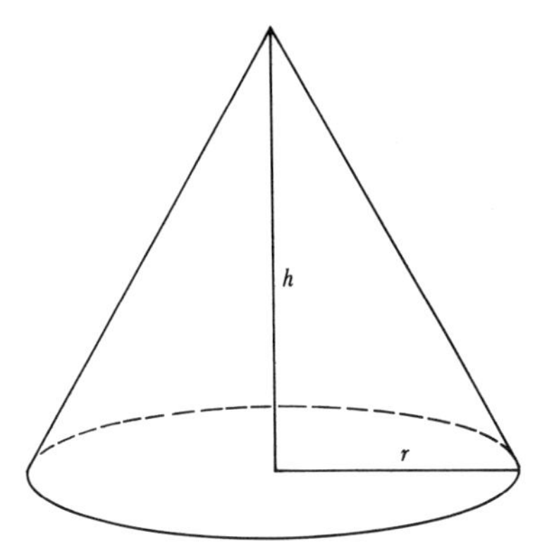

Figure 10-42

We now wish to consider two measures associated with a sphere: *surface area* and *volume*. The area of the surface and the volume of the region bounded by a sphere are given by

$$\text{area of the surface of a sphere} = 4\pi r^2,$$

$$\text{volume bounded by a sphere} = \frac{4}{3}\pi r^3.$$

Example: For a sphere whose radius is of measure 3, find the volume it bounds and its surface area.

$$V = \frac{4}{3}\pi(3)^3 \qquad\qquad A = 4\pi(3)^2$$

$$= 4\pi(3)^2 = 36\pi \qquad\qquad = 36\pi$$

As with a sphere, there are two fundamental measures associated with a cylinder. The *lateral surface area* of a right circular cylinder is composed of points of segments, each joining a point of the lower circle with a point directly above it on the upper circle. It should be intuitively clear that if we cut a cylinder without ends along one of these segments and flatten out the surface, then the right circular cylinder becomes a rectangular region. The length of this rectangle is simply the circumference of the circle, $2\pi r$, and the width is the height of the cylinder, h. Thus the lateral surface area of the cylinder will be the same as the area of the rectangular region, so

$$\text{lateral area of a cylinder} = 2\pi rh.$$

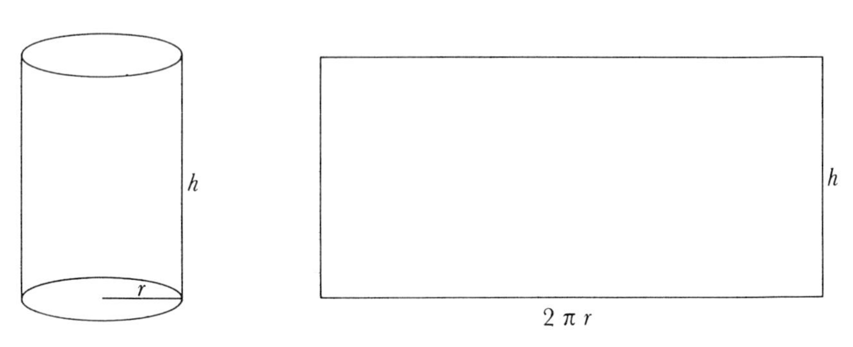

Figure 10-43

Since both the top and bottom of a cylinder are circles, the total surface area of a right circular cylinder is given by

$$2\pi rh + 2\pi r^2.$$

Since the volume of a region bounded by a cylinder could be considered as approximated by an inscribed regular prism, it seems reasonable to expect the formula for volume to be the same. The formula for the volume of the interior of a regular prism was given as the product of the area of the base and the altitude, or height. Since the area of the base of a cylinder is πr^2, the volume is given by

$$\text{volume of the interior of a cylinder} = \pi r^2 h.$$

(Think of the amount of soup in a tin can.)

Consider now the volume bounded by the *circular cone* illustrated in Figure 10-42. We consider only right circular cones where the height, or altitude, of the cone, represented by line h, is perpendicular to the base.

$$\text{volume of the interior of a cone} = \tfrac{1}{3}\pi r^2 h.$$

$$\text{lateral surface area of a cone} = \pi r\sqrt{r^2 + h^2}.$$

Exercise Set 10-7

1. Find the volume of the interior of a cone if the radius of the base is 5 and the altitude is 6.
2. Find the volume of the interior and total surface area of a cylinder whose diameter is 8 and whose altitude is 10.

3. What would happen to the volume in Exercise 2 if the radius were doubled? If the altitude were doubled? If both were doubled?
4. Find the volume bounded by and the total surface area of the right rectangular prisms with the following measurements.
   (a) $l = 4$, $w = 6$, and $h = 5$
   (b) $l = 1.2$, $w = 0.4$, and $h = 3.7$
   (c) $l = \frac{3}{4}$, $w = \frac{7}{8}$, and $h = 3\frac{1}{2}$
5. Find the volume of the following space regions (all prisms are right prisms).

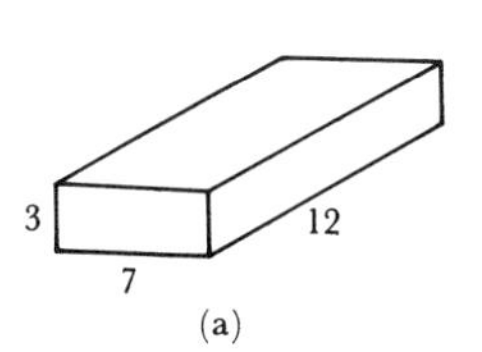

(a)

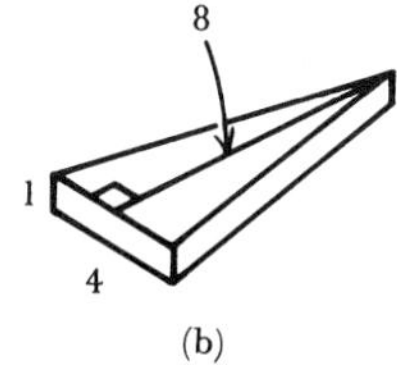

(b)

6. Find the total surface area of the figure in Exercise 5(a).
7. Find the measurement of the volume of a pipe if the measurement of the inner radius is 3 inches, the outer radius, $3\frac{1}{4}$ inches, and the length, 4 feet.
8. (a) What is the effect on the volume bounded by a cone if the altitude is doubled? (b) If the measure of the radius of the base is doubled?
9. A sphere where the measure of the radius is 12 is cut by a plane such that the distance of the plane from the center of the sphere is 4. What is the measure of the radius of the circle of intersection?
10. What is the total surface area of the right circular cone whose circumference is 8 and slant height is 6?
11. Find the volume of the interior of a cone if the circumference of the circular base is 6π and the altitude is 4.
12. How many gallons of paint (to be purchased in one-gallon cans) would one need to paint a cylindrical tank if the radius of the tank is 10 feet and the altitude of the tank is 12 feet? Assume that one gallon of paint will cover 300 square feet.
13. How many gallons of paint (to be purchased in one-gallon cans) would be needed to paint a spherical tank 10 feet in diameter? (See Exercise 12.)
14. Find the total surface area and the volume of the figures below.

(a)

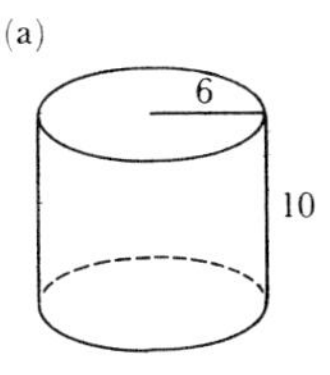

(b)

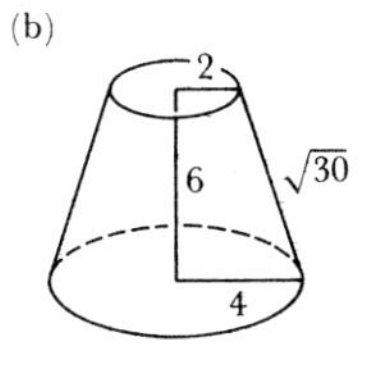

15. A right rectangular prism has edges that measure 16, 24, and 36 inches. If a gallon contains 231 cubic inches of water, how many gallons will the container hold?
16. A container in the shape of a right prism is 44 inches high and contains one gallon of water. How many square inches are there in the base? (See Exercise 15.)
*17. A cylinder and a cone have congruent radii and congruent altitudes. If the cone is placed inside the cylinder, what is the volume of the cylinder not occupied by the cone?
*18. The bases of a prism are equilateral triangles and the faces of the prism are squares. If the lateral surface area is 27, what is the measure of a side of the triangular bases?
*19. Find a side of the cube in which the volume equals the total surface area.
*20. Assume that the diameter of the earth is 12,756 kilometers. What is its surface area? Volume?

10

Geometric Transformations

The geometric figures that we have studied may be *transformed* into other geometric figures in a number of ways. To introduce transformations we consider a more general concept, mapping. A *mapping* of one set of points onto another is a pairing of the elements of the two sets such that each element of the first set is paired with exactly one element of the second set and each element of the second set is paired with at least one element of the first. In Figure 10-44 there is a mapping from set A, $\{a, b, c\}$, to set B, $\{d, e\}$. Mapping in geometry is equivalent to function in algebra. Recall that a function is a set of ordered pairs such that no two pairs have the same first element. In Figure 10-44, d is said to be an *image* of both a and b.

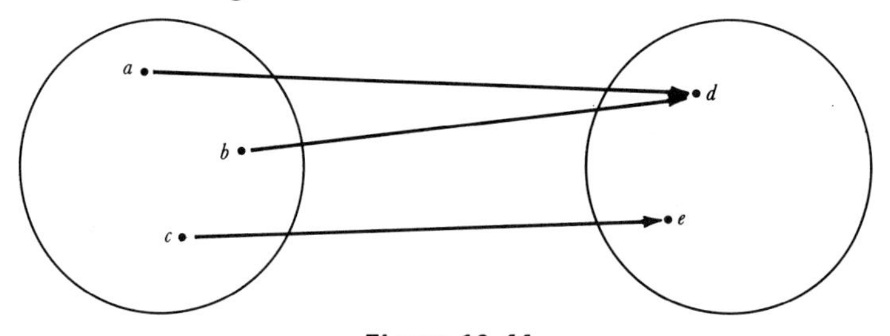

Figure 10-44

A *transformation* is a mapping of one set onto another such that each element of the second set is an image of only one element in the first set. An example of a transformation is given in Figure 10-45.

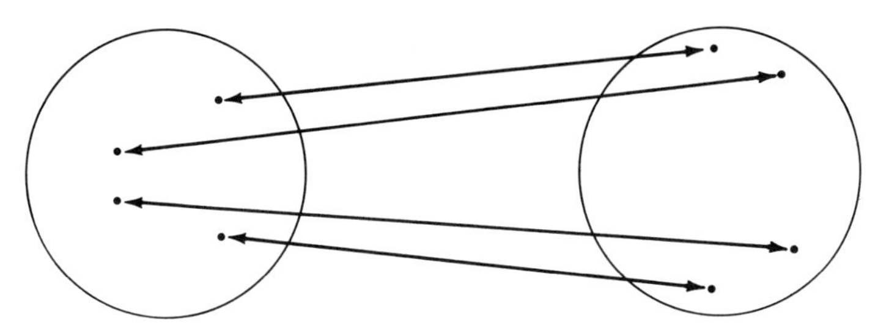

Figure 10-45

Example: Consider line l in Figure 10-46 with points A and B. $\overline{AB}$ is a defined line segment. Suppose we wish to move or transform $\overline{AB}$ to

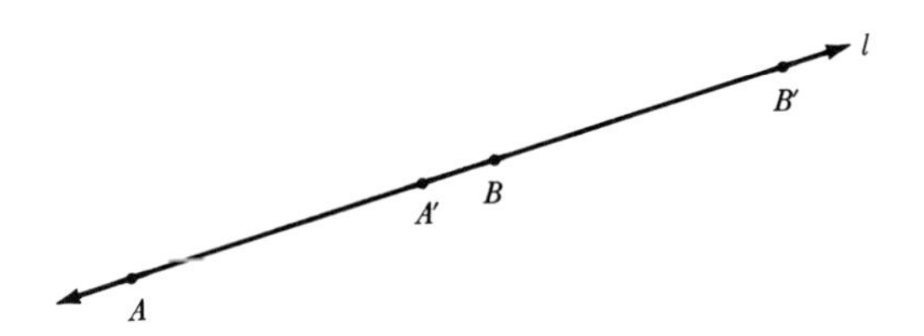

Figure 10-46

another position on line l. Pick a point A' one inch from point A to correspond to point A. For each point on $\overline{AB}$, select a corresponding point one inch away until point B' is named to correspond to B. Thus, $\overline{A'B'}$ is the image of $\overline{AB}$ under the transformation.

Also, $\overline{AB} \simeq \overline{A'B'}$ because of the way we have obtained the correspondence between points on $\overline{AB}$ and on $\overline{A'B'}$. A transformation that yields a congruent set of points is called a *rigid motion* or an *isometry*.

Definition 10-9: A transformation is a *rigid motion* if it preserves the distance between each pair of corresponding points. Two sets of points that correspond in a rigid motion are congruent.

Let's now consider specific types of rigid motion transformations. One such example is a *translation*, a transformation in which each of the points of a geometric figure may be thought of as being moved a certain number of units in a given direction.

Example: Move a given line segment $\overline{MN}$, as in Figure 10-47, two inches to the right in the direction of ray l. First, determine a point M'

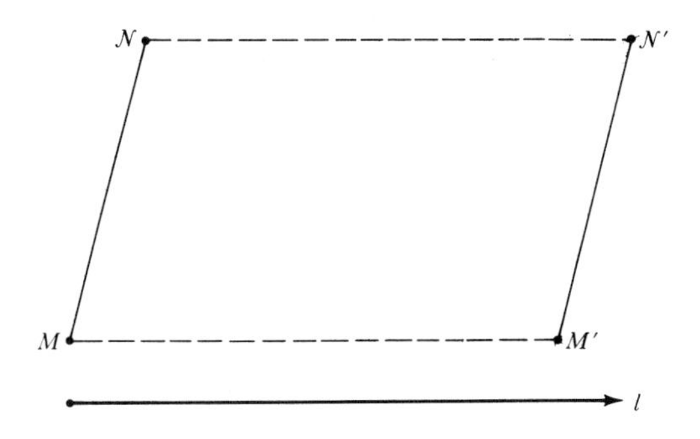

Figure 10-47

two inches from point M on a line parallel to ray l. In fact, translate each point on $\overline{MN}$ parallel to l two inches from its original position. The point N' is two inches from N, and $\overline{NN'}$ is parallel to l. Notice that the segments such as $\overline{MM'}$ and $\overline{NN'}$ are of equal length. Connecting the points M' and N' with a line segment determines the segment $\overline{M'N'}$, a segment congruent to $\overline{MN}$.

Definition 10-10: A *translation* in the direction of ray l is a rigid motion transforming points P into P' and Q into Q' such that the segments $\overline{PP'}$ and $\overline{QQ'}$ are parallel to a ray l and such that $\overline{PP'} \simeq \overline{QQ'}$.

Now shift your attention from translating points to reflecting points. Study the example of reflection and relate it to the definition that will be given.

Example: Taking line segment $\overline{PQ}$, as in Figure 10-48, reflect it about a selected line l in the following manner: From point P construct a line

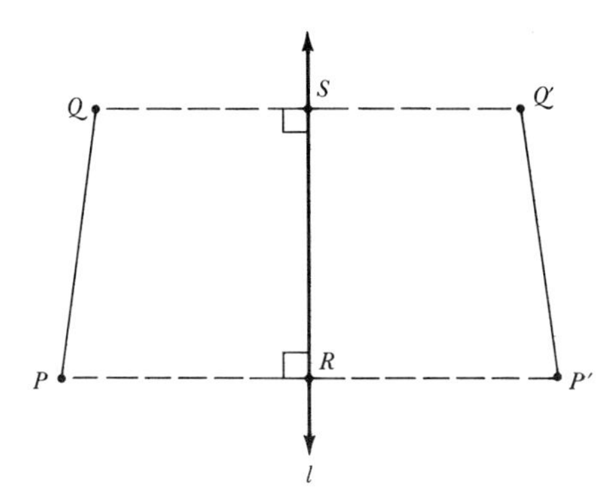

Figure 10-48

perpendicular to l, calling the point of intersection R. To determine a point P' to correspond to P, construct a segment $\overline{RP'}$ from R on $\overleftrightarrow{PR}$ such that $\overline{PR} \simeq \overline{RP'}$. For each point on $\overline{PQ}$, determine the perpendicular to l and a congruent segment on that perpendicular line. Thus, $\overline{QS} \simeq \overline{SQ'}$. The images of points on $\overline{PQ}$ are on the segment $\overline{P'Q'}$, a segment congruent to $\overline{PQ}$ and a reflection of it about line l.

Definition 10-11: A *reflection* is a rigid motion transforming each point P into a point P' on the opposite side of a line l so that the segment $\overline{PP'}$ is perpendicular to l at some point O and so that $\overline{PO} \simeq \overline{OP'}$.

Intuitively, reflections are obtained by flipping a geometric figure about a line.

Example: In Figure 10-49, $\triangle A'B'C'$ is the reflection of $\triangle ABC$ about line l.

A reflection may transform a geometric figure into itself, as illustrated by the circle in Figure 10-49. In this case we say the geometric figure is *symmetric about the line l* and that l is *the line of symmetry.*

Another rigid motion transformation, rotation, is introduced intuitively by the following example.

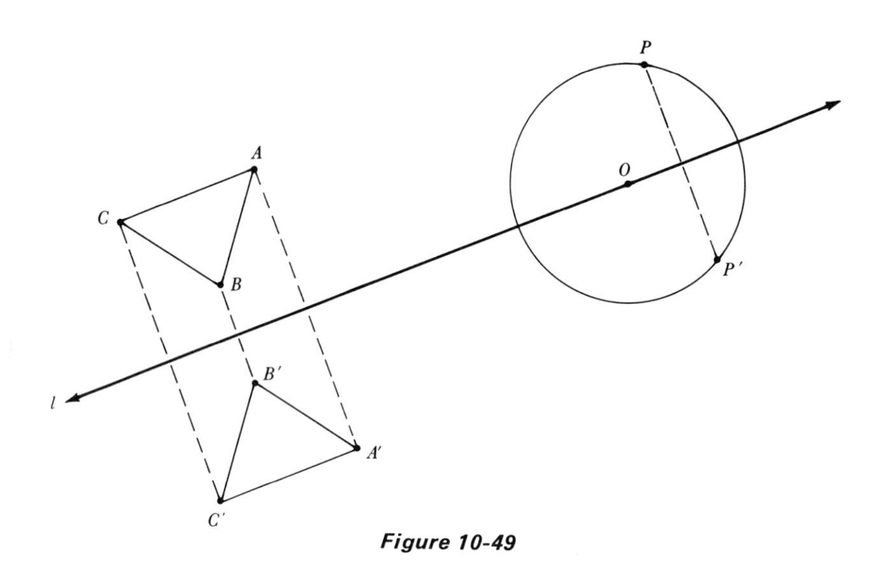

Figure 10-49

Example: Rotate the given segment $\overline{XY}$, shown in Figure 10-50(a), 45° in a counterclockwise direction about X. To perform the rotation, simply establish X as the one fixed point, the vertex, and draw an angle of 45°. Define point Y' such that $\overline{XY} \simeq \overline{XY'}$.

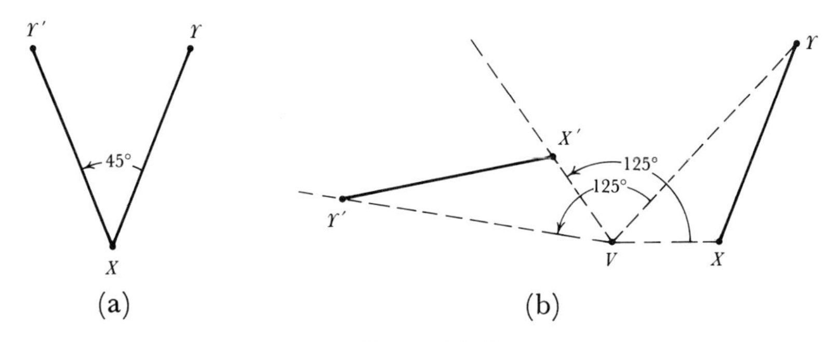

Figure 10-50

The rotation about point V in Figure 10-50(b) is more complex since $\overline{XV}$ and $\overline{YV}$ are on different lines. Rotate segment $\overline{XV}$ counterclockwise 125° about V. That is, with V as vertex and $\overrightarrow{VX}$ as one ray, construct an angle of 125°. Define X' on the new ray so that $\overline{VX} \simeq \overline{VX'}$. Similarly, rotate $\overline{VY}$ 125° about V, defining point Y' such that $\overline{VY} \simeq \overline{VY'}$. Segment $\overline{X'Y'}$ is congruent to segment $\overline{XY}$ and is a 125° rotation of $\overline{XY}$ about V.

Definition 10-12: A *rotation* is a rigid motion transforming point P about a point R onto point P' in either a counterclockwise or clockwise direction such that $\angle PRP'$ is of a given degree and $\overline{RP} \simeq \overline{RP'}$.

Other types of reflections are evident in everyday experience. For example, in Figure 10-51, we have a reflection from a surface such as a mirror. If A is a source of light, then $\angle AOB \simeq \angle DOC$.

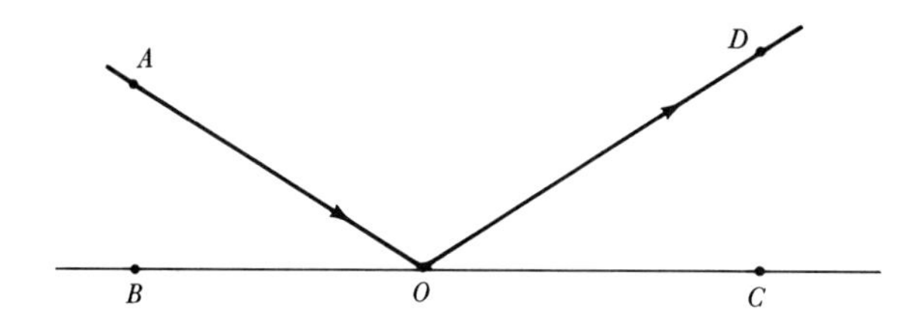

Figure 10-51

Exercise Set 10-8

1. Given the geometric figure to the right and the following images, match the transformations with the images.
   (1) Translation
   (2) Reflection about the middle bar
   (3) Reflection about the left leg
   (4) 90° rotation in a counterclockwise direction about P

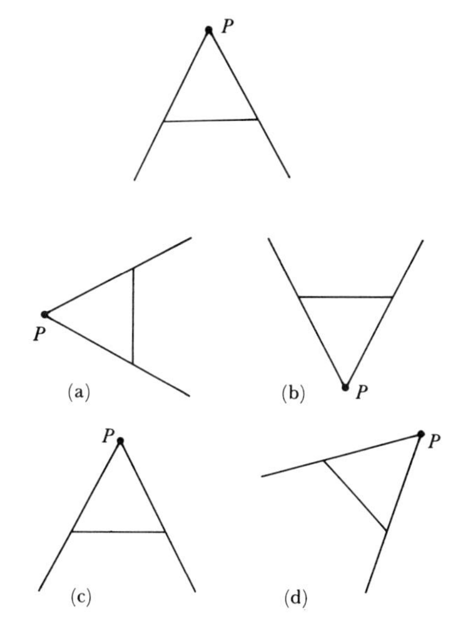

2. Trace each of the following figures on your own paper and then translate each traced figure two inches to the right.

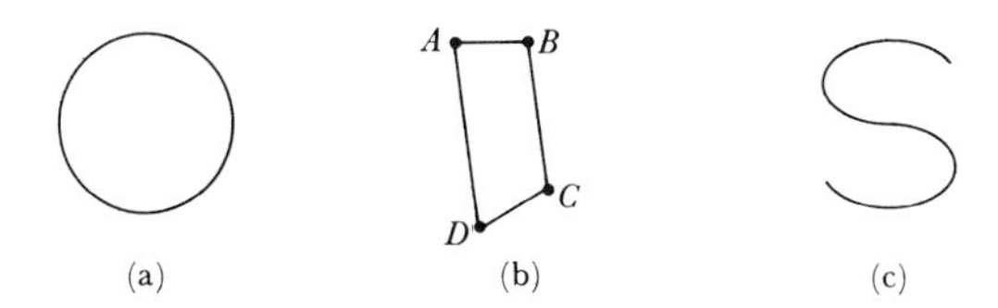

(a) (b) (c)

3. Using each of the following figures, construct a reflection about line l.

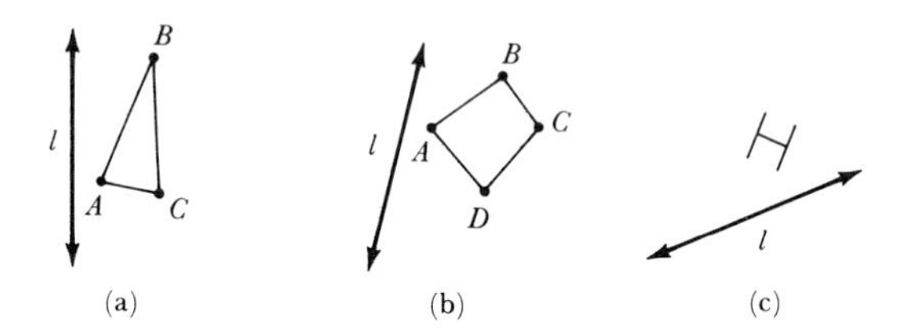

(a) (b) (c)

4. Rotate each of the following figures 120° counterclockwise about point A.

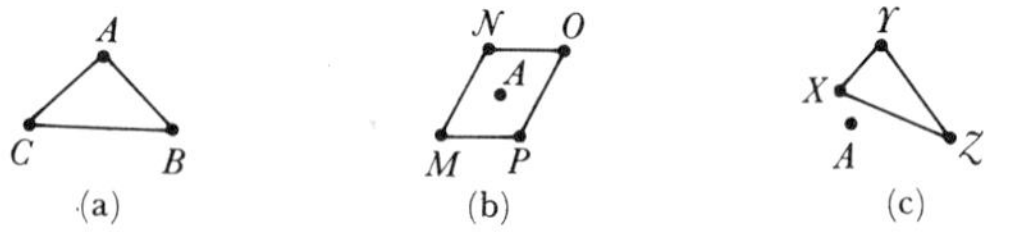

(a) (b) (c)

5. Construct an image for the letter H with the following transformations.

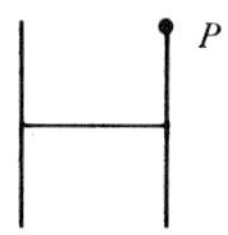

(a) Translate P down the length of the leg
(b) Reflect P about the opposite leg
(c) Rotate H 180° about P
(d) Reflect H about the middle bar

6. Determine whether translation, reflection, or rotation has been performed in the figures below.

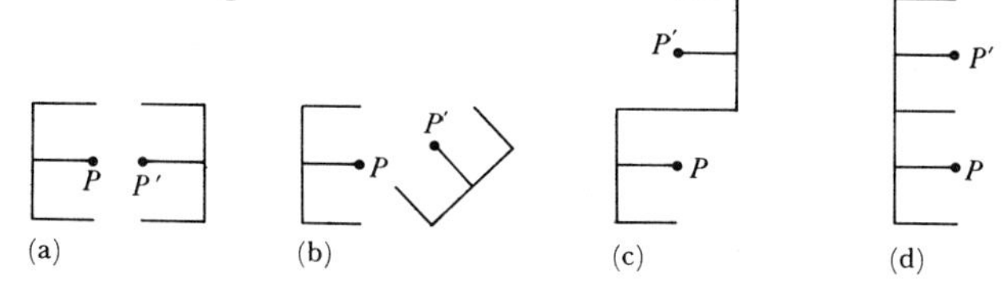

(a) (b) (c) (d)

7. Find all possible lines of symmetry, if any, in the following figures.

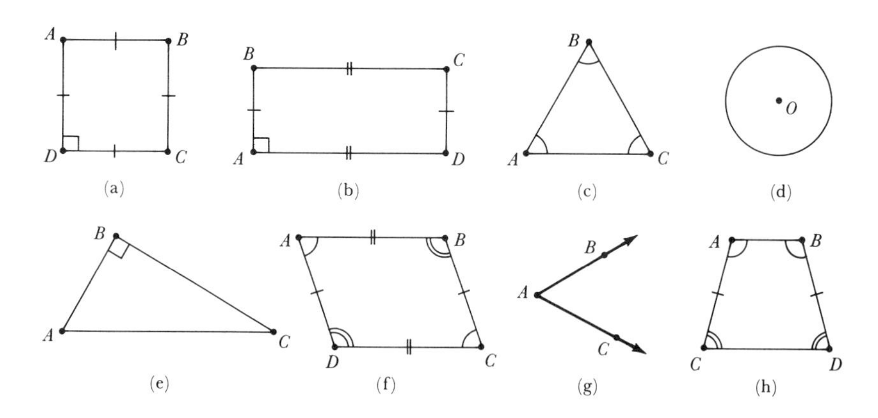

*8. Consider a transformation of a square into itself. Describe the following as translations, rotations, or reflections. $\begin{pmatrix} ABCD \\ BCDA \end{pmatrix}$ means A is transformed into B, B into C, C into D, and D into A.

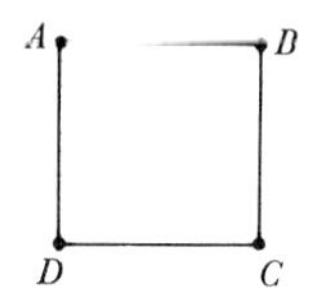

(a) $\begin{pmatrix} ABCD \\ BADC \end{pmatrix}$ (b) $\begin{pmatrix} ABCD \\ DCBA \end{pmatrix}$ (c) $\begin{pmatrix} ABCD \\ CDAB \end{pmatrix}$

(d) $\begin{pmatrix} ABCD \\ BCDA \end{pmatrix}$ (e) $\begin{pmatrix} ABCD \\ CBAD \end{pmatrix}$ (f) $\begin{pmatrix} ABCD \\ DABC \end{pmatrix}$

*9. (a) Draw a geometric figure to show that a translation may be equivalent to two reflections.
(b) Draw a geometric figure to show that a rotation may be equivalent to two reflections.

*10. (a) Draw a geometric figure. Perform on this figure a rotation followed by a translation. Show that the same result may be obtained by three reflections.
(b) Reverse the rotation and translation and work (a).

Review Exercise Set 10–9

1. Relative to the drawing given, which of the following are true?
   (a) $\overline{AB} = \overline{AE}$
   (b) $\overline{CB} \simeq \overline{BD}$
   (c) $m(\overline{CB}) = m(\overline{CE})$
   (d) $\overline{CD} > \overline{BD}$
   (e) $m(\overline{CB}) < m(\overline{AC})$
   (f) $\overline{BD} = \overline{DB}$
   (g) $m(\overline{AB}) - 2 = \overline{BD}$

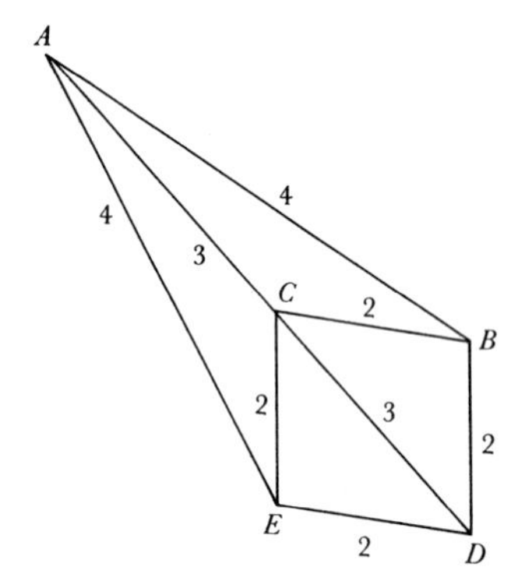

2. Mark each of the following statements as true or false.
   (a) All the points of one cylinder can lie in the interior of another cylinder.
   (b) All the points of a conc can lic in the exterior of a sphere.
   (c) The intersection of a sphere and a line, if the intersection is not null, is a line segment.
   (d) Concentric circles do not intersect.
   (e) Every plane intersects a right circular cone in a circle.
   (f) If the intersection of a plane and a sphere is not null or is not one point, then it is a circle.
   (g) Every plane parallel to a base that intersects a right circular cylinder intersects the cylinder in a circle.
   (h) The intersection of a cylinder and a plane may be a line.
   (i) The intersection of a cylinder and a plane may be a point.
   (j) A translation transforms a triangle into a congruent triangle.
   (k) Some equilateral triangles are right triangles.
   (l) A circle is a polygon.
   (m) If the areas of two rectangles are equal, the rectangles are congruent.
   (n) All rectangles are parallelograms.
   (o) All radii of a circle are congruent line segments.
   (p) If the perimeters of two squares are equal, the areas are equal.
   (q) Some right triangles are isosceles.
   (r) The center is a point in the set of points that form a circle.

(s) If a line segment has the center of a circle for one endpoint and a point on the circle for the other endpoint, it is called a diameter.

3. Find the area of the figure given.

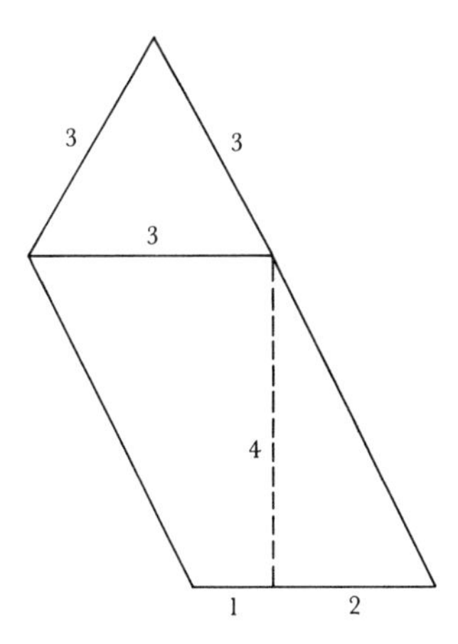

4. Find the volume of the interior of a cone if the circumference of the circle is 8π and the altitude is 3.
5. What happens to the volume of the interior of a right circular cylinder if both the circumference of the base and the altitude are doubled?
6. If the perimeter of a square is equal to the circumference of a circle, compare the areas of the bounded regions.
7. The area of a circular region is equal to the circumference of the circle. Find the measure of the radius of the circle.
8. If a 32″ diameter wheel on a car traveling 25 mph turns 225 revolutions, how long has it been traveling?
9. The altitude and radius of a right circular cone determine a plane that intersects the cone. If the altitude is 7 and the radius is 5, find the area of the polygon that represents the intersection of the plane and the cone.

Suggested Reading

Congruent Figures: Allendoerfer, pp. 600–605. Graham, pp. 310–312. Moore, Little, pp. 289–295. Smith, pp. 313–314, 322–325.

Transversals, Parallelograms: Allendoerfer, pp. 590–591. Graham, pp. 322–323. Moore, Little, pp. 280–284, 306–311. Smart, pp. 60–61, 151–152. Weaver, Wolf, pp. 109, 144.

Triangles: Allendoerfer, pp. 590–591, 595, 600–606. Graham, pp. 323–324. Moore, Little, pp. 289–295. Smart, pp. 69–76, 89–92, 152–154. Smith, p. 327. Weaver, Wolf, pp. 143–145.

Measurement: Allendoerfer, pp. 583–609. Graham, pp. 307–310, 314–315. Moore, Little, pp. 340–345. Smart, pp. 41–62, 101–103, 109–112, 245–264. Smith, pp. 316–318.

Area, Volume of Prisms: Allendoerfer, pp. 588–593. Graham, pp. 320–327, 328–332. Moore, Little, pp. 295–300, 347–350. Peterson, Hashisaki, pp. 267–271, 273–275. Smart, pp. 146–163, 245–253. Smith, pp. 329–331. Weaver, Wolf, pp. 140–142.

11

Concepts of Probability

1

Introduction to Probability

In the middle of the seventeenth century, a French gambler, the Chevalier de Méré, had a siege of bad luck which led him to inquire as to why the dice had seemingly turned against him. He contacted a noted mathematician, Blaise Pascal, who later discussed the dice problems of Chevalier de Méré with another noted mathematician, Pierre de Fermat. From studies by these two mathematicians, a new branch of mathematics, probability, was born.

You may be surprised to find that mathematicians today are still studying the science of gambling—dice throwing, coin tossing, and other games of chance—for possible applications in various areas of life. For example, the sale of insurance, the purchase of stocks, the reliability of equipment, the prediction of weather, the significance of medical research, and the study of warfare each involve an aspect of gambling.

It is often said that only two things in life are certain: death and taxes. This saying, of course, emphasizes the fact that life is full of uncertainties, especially life today. A scientific, mathematical study of these uncertainties involves what is known as *probability theory.* Every day, one hears such words as *probability*, *chance*, *odds*, and *likelihood.* However, it is not always clear what is meant by the different uses of these terms. For example, three common questions are: What is the probability that it will rain today? What are the chances that our football team will win? What are the odds that I will make money on this particular business deal? Such questions are examples of how probability might be construed as a measure of personal belief or personal confidence. In general, the probability concept relates to the degree of confidence one has in the occurrence of a particular event.

In another context, probability may be considered as that which may be expected to happen in repeated trials. The probability that a head will show when an ordinary coin is tossed once is 1/2. This is true because if the coin were tossed a large number of times, the percentage of times that a head would be observed should approach (that is, be close to) 50%. Thus, one might say that probability is applicable to events that can be repeated over and over under much the same conditions each time.

In this chapter, we give precise definitions for probability in terms of numbers, and we develop methods for computing and using these numbers. However, before doing this, we need to examine the fundamental mathematical concept of what is called *sample space*, on which our theory of probability will be based.

Let us consider an *experiment* as a process by which an observation or measurement is made. The result need not be numerical, but most of the time in this book we consider only those cases in which it *is* numerical. When performing an experiment, one is interested in the *outcome* of the observation. The examples that follow indicate possible outcomes of experiments.

Example: Suppose we have a needle fixed to a circular piece of cardboard so that it will spin and so that it is as equally likely to stop at one place as another. (We will call this construction a "spinner" and will use it in many examples in this chapter.) First of all, we consider a spinner divided into four equal sections, colored red, black, white, and green (Figure 11-1).

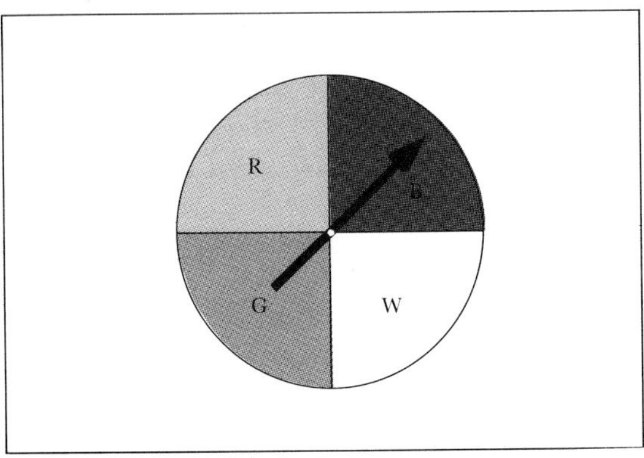

Figure 11-1

Let an experiment consist of spinning the needle and observing the color of the region where the needle stops. (If the needle stops on a line separating two regions, we agree to designate this situation as the color of the region the needle would move into if rotated clockwise.) The four possible outcomes of this experiment are not numbers but colors:

{red, black, white, green}.

Example: Consider an experiment of drawing one card from six cards marked by numbers 1 through 6 and observing the number appearing on the card. The possible outcomes are

{1, 2, 3, 4, 5, 6}.

Example: In another possible experiment with the six cards, we might observe whether the number appearing on a card is even or odd. Thus, this experiment would have only two possible outcomes, an even number or an odd number.

Definition 11-1: The set of possible outcomes of an experiment is called the *sample space* of the experiment if the set is finite and has the following properties.

(a) Every element of the sample space is a possible outcome of the experiment.
(b) For any performance of the experiment, one and only one outcome must occur.

The set of outcomes meeting the conditions of this definition are said to be *exhaustive* (nothing can occur that is not listed) and *mutually exclusive* (no two outcomes will describe the same result).

Definition 11-2: Every element in the set of the sample space is called a *sample point.*

Example: Suppose two coins are tossed. The possible outcomes of the experiment are two heads, HH, two tails, TT, a head on the first coin and a tail on the second coin, HT, or a tail on the first coin and a head on the second coin, TH. Thus, the sample space is the set {HH, HT, TH, TT} with four sample points.

Example: Draw one card each from two sets of six cards numbered 1 through 6 and observe the numbers on the cards. Find the sample space.

The answer is the set {(1, 1), (1, 2), (1, 3), (1, 4), (1, 5), (1, 6), (2, 1), (2, 2), (2, 3), (2, 4), (2, 5), (2, 6), (3, 1), (3, 2), (3, 3), (3, 4), (3, 5), (3, 6), (4, 1), (4, 2), (4, 3), (4, 4), (4, 5), (4, 6), (5, 1), (5, 2), (5, 3), (5, 4), (5, 5), (5, 6), (6, 1), (6, 2), (6, 3), (6, 4), (6, 5), (6, 6)}. There are thirty-six sample points.

Example: It is possible for an experiment to be such that its outcomes do not comprise a sample space. Can you make up an example of an experiment with this characteristic?

Let the experiment consist of drawing a card from a standard deck of cards. Suppose the space is enumerated as {spades, hearts, clubs, diamonds, aces}. This set does not constitute a sample space, because the trial yielding the ace of hearts would give two elements of the set. This example emphasizes that the requirements of the definition must be satisfied for the outcomes of an experiment to be a sample space.

Many times, our interest is not restricted to only one sample point in a sample space but includes the occurrence of any one of several possible outcomes. For example, in drawing a card from a set of six, we may be interested in the sample points that are even numbers, 2, 4, and 6. We note that these points are a subset of the sample space {1, 2, 3, 4, 5, 6}.

Definition 11-3: An *event* is a subset of a sample space.

Example: Consider an experiment which consists of drawing a card from our set of six and observing the number that appears on the card.

Some events will be:	The subsets are:
(a) observing a 1,	{1},
(b) observing a 3,	{3},
(c) observing a 6,	{6},
(d) observing an even number,	{2, 4, 6},
(e) observing an odd number,	{1, 3, 5},
(f) observing a number divisible by 3,	{3, 6},
(g) observing a number less than 4,	{1, 2, 3},
(h) observing a number greater than 4,	{5, 6},
(i) observing a number greater than 17.	$\varnothing$.

Of course, these are not all the events that could be listed relative to the experiment. Certain distinct characteristics of the events listed are apparent. Events (a), (b), and (c) differ from the remaining events listed in that each contains only one point of the whole sample space, whereas events (d) through (h) all involve more than one point, and (i) is empty.

When an event comprises only one point of the sample space, it is called a *simple event*; a *compound event* involves more than one sample point.

There are only six simple events, such as observing a 3, associated with the preceding experiment of drawing a card. Can you name other simple events? In a like manner, there may be many compound events, such as the numbers less than 5 or the even integers, associated with this experiment. Can you name other compound events?

An event may be represented in a Venn diagram by circling, or enclosing, the sample points in the diagram. For example, consider the sample space of the outcomes of tossing three coins. The sample points comprising the event characterized by getting at least two heads are enclosed in the diagram of Figure 11-2.

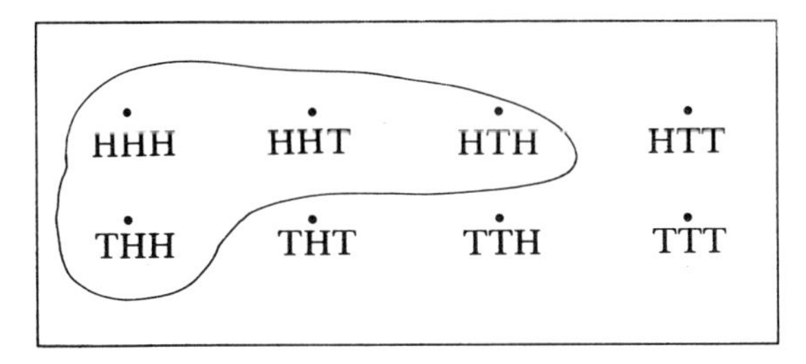

Figure 11-2

Sometimes we are interested in the sample points that are *not* a part of the event. These points are in the subset defined by the complement of the set of points representing the event. By the notation of Chapter 2, the complement of event E is denoted by $\overline{E}$. The complement of getting at least two heads is the subset

$$\{\text{TTT, THT, HTT, TTH}\}.$$

Compound events involving several sample points are necessary and very useful in probability theory. These compound events often involve unions and intersections of other compound events.

Let A and B be two events in a sample space S. $A \cap B$ represents the event A and B. ($A \cap B$ is the event that occurs if and only if A occurs and B occurs.) Likewise, $A \cup B$ represents event A or event B or both. ($A \cup B$ is the event that occurs if and only if A occurs or B occurs or both occur.) Thus, we note that in probability theory "and" means intersection and "or" denotes union.

Example: Consider the experiment described in a previous example where two cards were drawn one each from two sets of six cards.

Let A be all the sample points where the sum is 7 and B all the sample points where at least one card shows a 5. $A \cap B$ is the set

$$\{(2, 5), (5, 2)\},$$

and $A \cup B$ is

$$\{(1, 6), (2, 5), (3, 4), (4, 3), (5, 2), (6, 1), (5, 5), (1, 5), (3, 5), (4, 5), (6, 5), (5, 1), (5, 3), (5, 4), (5, 6)\}.$$

Exercise Set 11-1

1. Give the sample space for each of the four spinners.

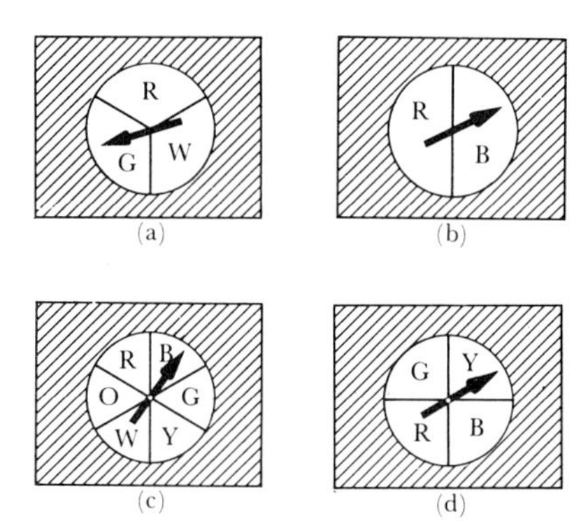

(a) (b) (c) (d)

2. An experiment consists of spinning two of the spinners in Exercise 1. List the sample space if the two spinners are
   (a) (a) and (b)
   (b) (c) and (a)
   (c) (a) and (d)
   (d) (b) and (c)
   (e) (d) and (b)
   (f) (c) and (d)

3. List the points in the sample space for the simultaneous toss of a coin and drawing a card from a set of six numbered 1 through 6. (*Hint:* There are twelve points in the space.)
4. An experiment consists of an ordinary coin being tossed three times. What is the sample space?
5. A box contains three red balls and four black balls. Let R represent a red ball and B a black ball. Tabulate the sample space if
   (a) one ball is drawn at a time.
   (b) two balls are drawn at a time.
   (c) three balls are drawn at a time.
6. Ed cannot decide which automobile to purchase. Thus, he writes the names of seven cars {Ford, Chevrolet, Pontiac, Oldsmobile, Buick, Plymouth, and Dodge} on slips of paper and distributes the pieces of paper in a box. Ed is planning to buy the car named on the slip of paper drawn from the box.
   (a) What are the possible outcomes?
   (b) If Ed decided to purchase two automobiles, what are the possible outcomes?
7. For the following pair of spinners, list the number pairs of possible outcomes of spinning both spinners.

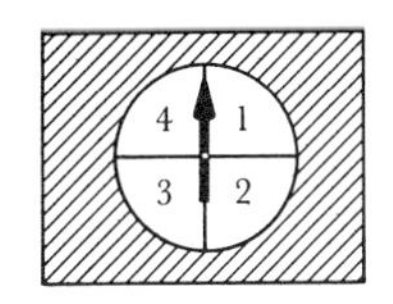

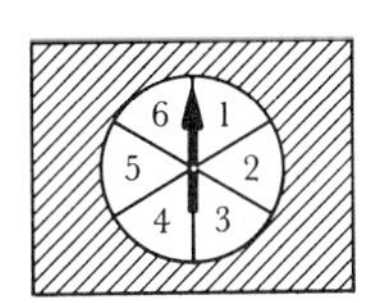

8. Three coins are tossed, and the number of heads is recorded. Which of the following sets is a sample space for this experiment? Why do the other sets fail to qualify as sample spaces?
   (a) {1, 2, 3} (b) {0, 1, 2}
   (c) {0, 1, 2, 3, 4} (d) {0, 2, an odd number}
   (e) $\{x \mid x$ is a whole number less than 2 and greater than $1\}$
   (f) {0, 1, 2, 3}
9. A sociologist is interested in a comparative study of the problems associated with slums in five major cities. He has selected for his study Chicago, Detroit, Philadelphia, Los Angeles, and Miami. However, he decides to use only four cities in his study. List the sample space of possible alternatives.

2

The Concept of Probability

Consider an experiment of drawing one card each from two sets of six cards (numbered 1 through 6). Let S represent the sample space and let event A be a subset of S. Suppose A is the set of number pairs such that the sum of the numbers on the two cards is 7.

Now imagine that we perform the experiment a number of times, say N, and count the number of times f (the frequency) that the event A actually occurs. For example, in this experiment, suppose that the number 7 occurs 150 times in 900 repetitions of the experiment. The quotient

$$\frac{f}{N} = \frac{150}{900} = \frac{1}{6}$$

is called the *relative frequency* of event A.

If we were to perform the experiment over and over again, we would find that the relative frequency would vary a great deal for various values of N, especially for the cases where N is small. But as N gets larger and larger, actual experience would reveal that the relative frequency, f/N, tends to stabilize. For example, if we were to repeat a large number of times the experiment of tossing an ordinary coin, the relative frequency of obtaining a head would approach 1/2. This experimentation suggests that we may associate with an event a number P which the relative frequency seems to approach for larger and larger N. This number P, associated with or assigned to an event, is called the *probability of the event* A and is denoted by $P(A)$.

Example: If a student were selected from Lamor University (see Table 11-1), 176/600 would be

	Number of Students at Lamor University
School A	176
School B	244
School C	180
Total	600

Table 11-1

assigned as the probability that a student is enrolled in School A. The probability that a student is in School C is 180/600.

Consider again the experiment of tossing a coin. We assume that one is justified in thinking that it is *equally likely* that the coin will show a head or a tail. The idea of "equally likely" is essentially an intuitive idea indicating "equally probable" or "having equal chances." It seems reasonable to assume that the coin is just as likely to fall heads as tails. Thus, there are two ways the coin can fall; however, there is but one way for the coin to show a head. Thus, the probability that a head will show when we flip a coin is 1/2. Note that the important quantity is not the number of ways that the coin can show a head, but the number of ways it can show a head divided by the total number of equally likely outcomes when flipping a coin. A definition for the assignment of probabilities in an experiment involving a finite number of equally likely events may be stated as follows.

Definition 11-4: If a sample space S consists of $n(S) = N$ simple, equally likely events, and if A is a subset consisting of $n(A) = r$ simple, equally likely events, then the *probability of A* is

$$P(A) = \frac{n(A)}{n(S)} = \frac{r}{N}.$$

Sometimes this rule is given as the classical definition of probability. Suppose there are N possible, equally likely outcomes of an experiment. If r of these have a particular characteristic, so that these outcomes can be classified as a success, then the probability of a success is defined to be r/N.

Example: Consider the experiment of drawing a card from the set of six (numbered 1 through 6). There are six equally likely possible outcomes of the experiment of drawing a card; let two of these, a 3 and a 6, represent a success, E. Thus,

$$P(E) = \frac{2}{6} = \frac{1}{3}.$$

Example: Consider an experiment of tossing three coins. Equally likely outcomes may be listed as {HHH, HHT, HTH, HTT, THT,

THH, TTH, TTT} Consider a success as getting exactly two tails. Then $E = \{HTT, THT, TTH\}$. Therefore, $P(E)$, the probability of E, is 3/8, and we write $P(E) = 3/8$.

Example: Consider the two spinners in Exercise 7 of Exercise Set 11-1. If both the spinners are activated, find the probability of a total of 5 points.

In Exercise 7 of Exercise Set 11-1, it was found that there were twenty-four possible outcomes of spinning both spinners. Four of these,

$$\{(1, 4), (2, 3), (3, 2), (4, 1)\},$$

give a total of 5 points. Thus,

$$P(5 \text{ points}) = \frac{4}{24} = \frac{1}{6}.$$

Example: A poll is taken at Lamor University to determine whether or not the students wish to abolish grades. Table 11-2 indicates the results of this poll when the students are divided into three groups, (A) lower-division students, (B) upper-division students, and (C) graduate students.

	Favor Abolishing Grades	*Do Not Favor Abolishing Grades*	*No Opinion*
Group A	150	50	10
Group B	100	80	8
Group C	30	70	2

Table 11-2

(a) What is the probability that a student at Lamor University selected at random from Group A would be in favor of abolishing grades? (To say that a student is selected *at random* means that each student has the same probability of being selected, or has an equally likely opportunity of being selected.)

(b) What is the probability that a student selected at random from Group B has no opinion?

(c) What is the probability that a student selected at random will be in Group C?

(d) What is the probability that a student selected at random will be in Group B and will favor abolishing grades?

(e) What is the probability that a student selected at random has no opinion?

(f) What is the probability that a student selected at random favors abolishing grades?

The answers are:

(a) P (student selected from Group A favors abolishing grades) $= 150/210$.
(b) P (student selected at random from Group B has no opinion) $= 8/188$.
(c) P (student in Group C) $= 102/500$.
(d) P (student is in Group B and favors abolishing grades) $=$ $100/500$.
(e) P (student has no opinion) $= 20/500$.
(f) P(student favors abolishing grades) $= 280/500$.

Exercise Set 11-2

1. Consider the following record of the number of incorrect answers on a background test in history administered to the freshman class at Lamor University.

Incorrect Answers	Number of Students
0	20
1	80
2	120
3	250
4	260
5	190
6	80

(a) Assign a probability to the event of obtaining three incorrect answers.
(b) Assign a probability of obtaining four or five incorrect answers.
(c) Assign a probability of getting all answers correct.
(d) Assign a probability of getting less than three incorrect answers.

2. One card is drawn from each of two sets of four cards (each set numbered 1 through 4). Note that the two cards can be drawn in sixteen different ways. What is the probability that the sum of the numbers is
   (a) greater than four? (b) equal to nine?
   (c) equal to seven? (d) greater than seven?
3. Suppose there is an equally likely probability that the spinner will stop at any one of the six numbered sections for the given spinner.

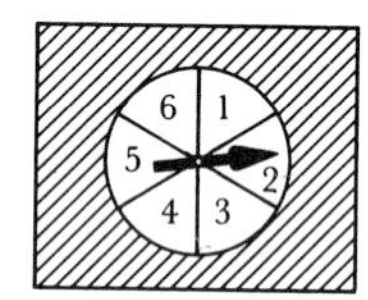

   (a) What is the probability of stopping in an even-numbered section?
   (b) What is the probability of stopping in a section that is a multiple of 3?
   (c) What is the probability of stopping in an even-numbered section or in a section that is a multiple of 3?
   (d) What is the probability of stopping at an even-numbered or odd-numbered section?
   (e) What is the probability of stopping at a number less than 4?
4. A box contains four black, seven white, and three red balls. If one ball is drawn, what is the probability that it is
   (a) black? (b) red?
   (c) white? (d) red or white?
5. A medical survey of the cause of death among a group of 110 adult males was categorized as to the cause of death, and the age of the subject at the time of death.

Cause of Death	*21-40*	*41-60*	*61-80*
Heart Disease	4	9	14
Cancer	2	4	8
Stroke	1	2	5
Flu and Pneumonia	0	1	2
Diabetes	1	0	1
Tuberculosis	0	1	0
Other	17	14	24

If one of these subjects is chosen at random, what is the probability that he
(a) died of cancer?
(b) was 35 years old when he died and did not die from any of the diseases listed by name?
(c) was over 61 years old when he died?
(d) died of tuberculosis at the age of 61?
(e) died of a stroke or of heart disease?

6. Assume that the spinner with three equally likely sections is spun along with the spinner in Exercise 3. What is the probability that the sum of the numbers is
(a) greater than seven?
(b) equal to eight?
(c) equal to three?
(d) less than five?

7. Two coins are tossed.
(a) What is the probability of getting two heads?
(b) What is the probability of getting exactly one head? At least one head?

8. A card is drawn from an ordinary deck of cards.
(a) What is the probability of getting a heart?
(b) What is the probability of drawing an ace?
(c) What is the probability of drawing the jack of spades?
(d) What is the probability of drawing a red card?

3

Counting Schemes

Sometimes it is difficult to compute the number of sample points in an experiment. This section will present counting schemes that should assist in this computation. First of all, we consider experiments that can be partitioned into subexperiments.

Example: A coin is tossed and at the same time the spinner in Exercise 3, Exercise Set 11-2, is activated. How many outcomes can you list for this experiment?

The coin can fall in two ways, H or T. The spinner may stop at one of six places, 1, 2, 3, 4, 5, or 6. The sample space has twelve possible outcomes: {(H, 1), (H, 2), (H, 3), (H, 4), (H, 5), (H, 6), (T, 1), (T, 2), (T, 3), (T, 4), (T, 5), (T, 6)}.

The preceding example may be considered as an experiment partitioned into two subexperiments, tossing a coin and activating a spinner. There are two possible outcomes for tossing the coin and six possible outcomes for the spinner. Note that $2 \cdot 6 = 12$, the total number of outcomes for the experiment.

In general, if event A has m distinct outcomes and event B has n distinct outcomes, and if the occurrence or nonoccurrence of event A does not affect the possible outcome for event B, then an experiment that consists of both event A and event B has mn distinct outcomes.

Example: A student plans a trip from New York to Miami to Tallahassee. From New York to Miami he can travel by bus, train, airplane, or boat. However, from Miami to Tallahassee he can travel only by bus or airplane. In how many different ways can he make the trip?

He can travel from New York to Miami in four different ways and from Miami to Tallahassee in two different ways. Therefore, he can make the entire trip in $4 \cdot 2 = 8$ different ways.

Sometimes the preceding statement on outcomes needs to be stated in the following manner. If event A has m distinct outcomes and if, after event A has occurred, event B has n distinct outcomes, then an experiment that is a composite of A and B has mn distinct outcomes. The phrase "after event A has occurred" in the preceding statement is very important, as the following example indicates.

Example: Assume a two-toned car can be selected with black, red, or white as either the body color or the top color. Then there are three choices for the body (or top) color; but once a choice has been made, there are only two selections left for a top (or body) color. Thus, there are $3 \cdot 2 = 6$ possibilities for two-toned cars with three possible colors.

Example: A psychologist places three rats in three cages named A, B, and C, as in Figure 11-3. Cage A can be filled with any one of the three rats. After Cage A has been filled, Cage B can be filled by either of the two remaining rats. Therefore, Cages A and B can be filled $3 \cdot 2$ ways. After Cages A and B have been filled, the remaining rat must

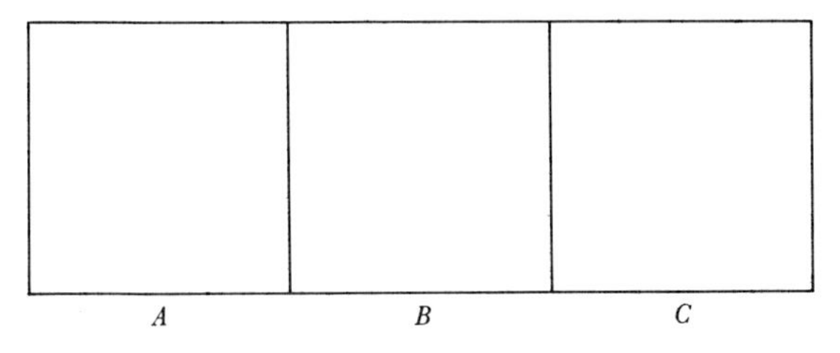

Figure 11-3

occupy Cage C; or, Cage C can be filled in only one way. Therefore, Cages A, B, and C can be filled in $3 \cdot 2 \cdot 1$ ways. We denote this multiplication by $3 \cdot 2 \cdot 1 = 3!$

Likewise,

$$6! = 6 \cdot 5 \cdot 4 \cdot 3 \cdot 2 \cdot 1,$$

and

$$8! = 8 \cdot 7 \cdot 6 \cdot 5 \cdot 4 \cdot 3 \cdot 2 \cdot 1.$$

In general, the product $m \cdot (m-1) \cdot (m-2) \cdot \cdots \cdot 3 \cdot 2 \cdot 1$, denoted by $m!$, is called *m-factorial.*

Example:

$$6! = 6 \cdot 5 \cdot 4 \cdot 3 \cdot 2 \cdot 1 = 720$$
$$5! = 5 \cdot 4 \cdot 3 \cdot 2 \cdot 1 = 120$$
$$4! = 4 \cdot 3 \cdot 2 \cdot 1 = 24$$
$$3! = 3 \cdot 2 \cdot 1 = 6$$
$$2! = 2 \cdot 1 = 2$$

To complete the definition above, we define both $1!$ and $0!$ to be 1. The statement that $0! = 1$ may seem unusual, but you will learn later in your work with factorials that this definition is reasonable and consistent with the factorial idea for positive whole numbers.

Suppose we have five cities—A, B, C, D, and E—under consideration for an experiment, but we learn that we have money available for only two cities. In how many different ways could the two cities be selected from the five? To obtain the answer we list the possible outcomes:

$$\{(A, B), (A, C), (A, D), (A, E), (B, C), (B, D), (B, E), (C, D), (C, E), (D, E)\}$$

Thus, two cities can be selected from five in ten different ways. Note that if we take the number of different ways of selecting the first city, 5, times the number of ways of selecting the second city after the first city has been selected, 4, the number of ways of selecting the two cities in a given order would be $5 \cdot 4$. Since we are not interested in order, (A, C) and (C, A) are the same selections. Thus, we divide $5 \cdot 4$ by 2 to obtain the answer 10.

$$\frac{5 \cdot 4}{2} = 10$$

However, in order to make the numerator into a factorial we write

$$\frac{5 \cdot 4 \cdot (3 \cdot 2 \cdot 1)}{(2 \cdot 1) \cdot (3 \cdot 2 \cdot 1)} = \frac{5!}{2!\,3!}.$$

Definition 11-5: The number of ways r objects can be chosen from among n objects is called the number of *combinations* of n things taken r at a time. The number is given by the formula

$${}_nC_r = \frac{n!}{r!(n-r)!}.$$

Example: Consider the preceding problem of selecting two cities from five. Now $n = 5$ and $r = 2$. Therefore

$${}_5C_2 = \frac{5!}{2!(5-2)!} = \frac{5 \cdot 4 \cdot 3 \cdot 2 \cdot 1}{(2 \cdot 1) \cdot (3 \cdot 2 \cdot 1)} = 10.$$

This result is in agreement with our enumeration in the preceding discussion.

Example: In how many ways can a committee of three be selected from five people whom we name A, B, C, D, and E? The answer is

$${}_5C_3 = \frac{5!}{3!2!} = \frac{5 \cdot 4 \cdot 3 \cdot 2 \cdot 1}{3 \cdot 2 \cdot 1 \cdot 2 \cdot 1} = 10.$$

This answer may be verified by enumeration to be

$$\{ABC, ABD, ABE, ACD, ACE, ADE, BCD, BCE, BDE, CDE\}.$$

Example: In the preceding example, in how many ways can a committee of five people be selected?

Obviously, five people can be selected from five people in one and only one way. However, using the formula given in Definition 11-5,

$$_5C_5 = \frac{5!}{5!0!}.$$

Setting $5!/5!0!$ equal to 1 and simplifying, one obtains

$$\frac{5!}{5!0!} = 1,$$

or

$$\frac{5 \cdot 4 \cdot 3 \cdot 2 \cdot 1}{5 \cdot 4 \cdot 3 \cdot 2 \cdot 1 \cdot 0!} = 1,$$

or

$$\frac{1}{0!} = 1,$$

or

$$0! = 1.$$

This result is in agreement with our earlier statement that $0! = 1$.

Combination notation is especially useful in certain types of probability problems, as indicated by the following examples.

Example: A box contains seven white balls and three red balls. Three balls are drawn at random. What is the probability of drawing two white balls and one red ball?

The number of sample points in the experiment would be $_{10}C_3$, since three balls must be drawn from ten. Two white balls must be drawn from seven; this can happen in $_7C_2$ ways. One red ball must be drawn from three; this can happen in $_3C_1$ ways. Therefore,

$$P(\text{2 white balls and 1 red ball}) = \frac{_7C_2 \cdot {_3C_1}}{_{10}C_3}.$$

Since $_7C_2 = 21$, $_3C_1 = 3$, and $_{10}C_3 = 120$,

$$P(\text{2 white balls and 1 red ball}) = \frac{21 \cdot 3}{120} = \frac{63}{120} = \frac{21}{40}.$$

Example: A history professor requires each of his students to write research papers on two of fifteen topics. Of the fifteen topics, five are easy to research, seven are difficult to research, and three are practically impossible to research. One student picks his two topics at random. What is the probability of his getting

(a) two easy topics?
(b) one difficult topic and one practically impossible topic?
(c) two practically impossible topics?

The solutions are:

(a) $P(2 \text{ easy topics}) = \dfrac{{}_5C_2 \cdot {}_7C_0 \cdot {}_3C_0}{{}_{15}C_2} = \dfrac{10 \cdot 1 \cdot 1}{105} = \dfrac{2}{21}.$

(b) $P(1 \text{ difficult and } 1 \text{ impossible topic})$

$$= \frac{{}_7C_1 \cdot {}_3C_1 \cdot {}_5C_0}{{}_{15}C_2} = \frac{7 \cdot 3}{105} = \frac{1}{5}.$$

(c) $P(2 \text{ impossible topics}) = \dfrac{{}_7C_0 \cdot {}_5C_0 \cdot {}_3C_2}{{}_{15}C_2} = \dfrac{1 \cdot 1 \cdot 3}{105} = \dfrac{1}{35}.$

Exercise Set 11-3

1. Evaluate.
   (a) ${}_{10}C_6$ (b) ${}_{10}C_0$ (c) ${}_{15}C_1$
   (d) ${}_4C_2$ (e) ${}_rC_2$ (f) ${}_rC_{r-1}$
2. In how many ways can three personality tests be administered to a class?
3. In how many ways can five speakers be arranged on a program?
4. A student body president is asked to appoint a committee consisting of five boys and three girls. He is given a list of twelve boys and ten girls from which to make the appointments. How many different committees can be selected?
5. How many different hands consisting of five cards can be drawn from an ordinary pack of cards?
6. A special committee of three men must be selected from a board of directors involving twelve men. In how many ways can the committee be selected?
7. A teacher has prepared three multiple-choice examinations and five examinations involving discussion questions. Four different examinations are randomly selected from the eight for final examinations for four classes. What is the probability that two finals are multiple choice and two involve discussion questions?

8. A monthly draft call for a given district consists of 10 men. There are 12 men classified as eligible above 21 years of age and 15 eligible young men 21 years of age or less.
   (a) What is the probability that all men drafted will be above 21 years of age?
   (b) What is the probability that all men drafted will be less than 22 years of age?
   (c) What is the probability that five of the men drafted will be from the group above 21 years of age and five will be in the other group?
9. A file contains 20 good sales contracts and five canceled contracts. If four contracts are selected at random, what is the probability that none has been canceled?
10. A class consists of 20 boys and 10 girls. The class has four complimentary tickets to a local dance. Students receive the tickets by drawing names from a hat. What is the probability that the winners could go to the dance as partners (boy and girl)?

*11. A hat contains twenty slips of paper numbered 1 to 20. If three are drawn without replacement, what is the probability that all three are numbered less than 10?

*12. Why is ${}_6C_2$ the same as ${}_6C_4$? Generalize for ${}_nC_r = ?$.

*4

Properties of Probability

In the preceding section, probability was considered as a relative frequency and as a ratio for a finite number of equally likely events. These concepts will be used to develop important properties of probability.

One way to observe that probability is a number between, and possibly including, 0 and 1 is to consider probability as r/N. Since the number of equally likely outcomes of an event can never be greater than the total number of possible equally likely outcomes, r/N can never be greater than 1. That the probability can equal 1 is illustrated by the following example. The probability of getting a number less than or equal to six in the toss of a die is $6/6 = 1$. Events for which the probability is 1 are said to be sure, or *certain*. In like manner, the probability of getting a seven in the toss of a die is 0/6, because it is impossible to get a seven. When there is no way that an event can occur, set A representing the event is the null set. Thus $P(A) = 0$ when $A = \varnothing$.

Theorem 11-1: For any two events A and B, the probability of A or B is given by

$$P(A \cup B) = P(A) + P(B) - P(A \cap B).$$

Example: Consider an experiment of tossing two dice. One die is colored red and the other, black. Let event A represent the event in the sample space defined by the upper face of the red die being greater than 3 with no requirement on the black die; let event B represent the event defined by the upper face of the black die being less than 5 with no requirement on the red die. There are, of course, thirty-six possible outcomes of tossing two dice. Eighteen of these satisfy the requirements for event A, and twenty-four satisfy the requirements for event B. Since $18 + 24 > 36$, some of the simple outcomes in A and B must be the same. These may be enumerated as

$$\{(4, 1), (4, 2), (4, 3), (4, 4), (5, 1), (5, 2), (5, 3), (5, 4), (6, 1), (6, 2), (6, 3), (6, 4)\}.$$

Thus, $P(A \cap B) = 12/36$. Therefore,

$$\begin{aligned} P(A \text{ or } B) &= P(A \cup B) \\ &= P(A) + P(B) - P(A \cap B) = \frac{18}{36} + \frac{24}{36} - \frac{12}{36} = \frac{30}{36} = \frac{5}{6}. \end{aligned}$$

In Figure 11-4, it is evident that $A \cup B$ consists of 30 outcomes. Thus,

$$P(A \cup B) = \frac{30}{36} = \frac{5}{6}.$$

Thus, Theorem 11-1 is verified for this example.

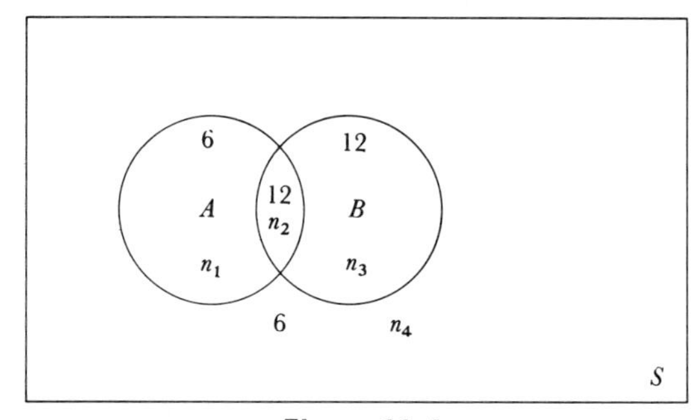

Figure 11-4

We may generalize this reasoning by considering a sample space composed of N simple events. For example, suppose there are n_1 sample points in A and not in B, n_2 sample points in $A \cap B$, n_3 sample points in B but not in A, and n_4 sample points in neither A nor B, where

$$N = n_1 + n_2 + n_3 + n_4 .$$

Now

$$P(A) = \frac{n_1 + n_2}{N},$$

$$P(B) = \frac{n_2 + n_3}{N},$$

$$P(A \cap B) = \frac{n_2}{N},$$

$$P(A \cup B) = \frac{n_1 + n_2 + n_3}{N}.$$

Note that Theorem 11-1 is satisfied by these probabilities in that

$$P(A \cup B) = P(A) + P(B) - P(A \cap B)$$

since

$$\frac{n_1 + n_2 + n_3}{N} = \frac{n_1 + n_2}{N} + \frac{n_2 + n_3}{N} - \frac{n_2}{N}$$

$$= \frac{n_1 + n_2 + n_3}{N}.$$

Example: In the drawing of a card from a set of eight cards (numbered 1 through 8), what is the probability of getting a number less than five or an even number?

Let A represent the event of getting a number less than 5 and B the event of getting an even number.

$$P(A) = 4/8 = 1/2$$
$$P(B) = 4/8 = 1/2$$

Now there are two numbers that are both even and less than five.

$$P(A \cap B) = 2/8 = 1/4$$

Thus,

$$\begin{aligned} P(A \cup B) &= P(A) + P(B) - P(A \cap B) \\ &= 1/2 + 1/2 - 1/4 = 3/4. \end{aligned}$$

If two events are mutually exclusive, the occurrence of one of them excludes the occurrence of the other, since they have no sample points in common. In the tossing of a die, let event A consist of getting an odd number. Let event B consist of getting a 4 or a 6. Then these two events are mutually exclusive, since in the tossing of a die, if event A occurs, event B could not occur because neither 4 nor 6 is odd. If event C is getting a number divisible by 3, then event B and event C are not mutually exclusive, because both contain the sample point 6.

When $A \cap B = \varnothing$ (A and B are mutually exclusive), $P(A \cap B) = 0$. Thus,

$$\begin{aligned} P(A \cup B) &= P(A) + P(B) - P(A \cap B) \\ &= P(A) + P(B) - 0 \\ &= P(A) + P(B). \end{aligned}$$

Theorem 11-2: If A and B are mutually exclusive, then

$$P(A \cup B) = P(A) + P(B).$$

Example: If you toss two dice, what is the probability of getting a sum of 7 or 8? Since getting a 7 or getting an 8 involves mutually exclusive events,

$$P(7 \text{ or } 8) = P(7) + P(8) = \frac{6}{36} + \frac{5}{36} = \frac{11}{36}.$$

Example: Find the $P(7 \leq x \leq 10)$, where x is the sum of the number of dots on the upper faces of two randomly tossed dice. Since the events of getting a 7, 8, 9, or 10 are mutually exclusive,

$$\begin{aligned} P(7 \leq x \leq 10) &= P(7) + P(8) + P(9) + P(10) \\ &= \frac{6}{36} + \frac{5}{36} + \frac{4}{36} + \frac{3}{36} = \frac{1}{2}. \end{aligned}$$

Let us now make an observation concerning the probability that an event does *not* occur. The probability of getting a 6 on the toss of a die is

1/6. What is the probability of *not* getting a 6? There are five equally likely ways of not getting a 6, namely, getting either a 1, 2, 3, 4, or 5. Thus, the probability of *not* getting a 6 is 5/6. Note that $5/6 = 1 - 1/6$. In general, if $P(A)$ is the probability that event A occurs, then $P(\overline{A})$ is the probability that A does not occur, and

$$P(\overline{A}) = 1 - P(A).$$

Theorem 11-3: The probability of the complement of event A, $P(\overline{A})$, is given by $P(\overline{A}) = 1 - P(A)$.

Proof: $A \cup \overline{A} = S$, which is the sample space. By Theorem 11-2,

$$P(S) = P(A) + P(\overline{A}),$$

since A and $\overline{A}$ are mutually exclusive. Since $P(S) = 1$, $P(A) + P(\overline{A}) = 1$. Therefore, $P(\overline{A}) = 1 - P(A)$.

Example: Consider again the probability of not getting a 6 in the toss of a die. Let A represent the event of getting a 6; that is, $P(A) = 1/6$. Now let $\overline{A}$ represent the event of *not* getting a 6. Then

$$P(\overline{A}) = 1 - P(A) = 1 - 1/6 = 5/6.$$

When two events are related in such a way that the occurrence or nonoccurrence of one of them does not affect the probability of the occurrence or nonoccurrence of the other, we say the events are *independent.* For example, the tossing of two coins gives independent events. The outcome of the second coin is not affected by the first coin, and vice-versa.

Definition 11-6: If two events A and B are independent, then the probability of A and B is equal to the probability of A multiplied by the probability of B, or $P(A \cap B) = P(A) \cdot P(B)$.

The events defined by getting a head in the toss of a coin and getting a 6 in the roll of a die are independent events. Thus, the probability of both occurring is the product of the individual probabilities. That is,

$$P(\text{a head and a 6}) = P(\text{head}) \cdot P(6) = \frac{1}{2} \cdot \frac{1}{6} = \frac{1}{12}.$$

Example: From a box containing four red balls and three black balls, one ball is draw, inspected, and returned to the box. Then a second ball is randomly drawn. What is the probability of drawing two red balls?

Let A represent getting a red ball on the first draw and B represent getting a red ball on the second draw. Since the ball was returned after the first draw, A and B are independent events. Thus,

$$P(A \cap B) = P(A) \cdot P(B) = \frac{4}{7} \cdot \frac{4}{7} = \frac{16}{49}.$$

We consider the concept of *conditional probability* from the standpoint of assigning probabilities in a sample space. Consider an example relative to a poll to determine whether or not the five-hundred hourly-wage earners of a company favor a strike. The results of the poll are tabulated into three groups, X, Y, and Z according to salary.

	Favor a Strike	*Do Not Favor a Strike*	*No Opinion*
Group X	150	50	10
Group Y	100	80	8
Group Z	30	70	2

Table 11–3

Suppose an hourly-wage earner is selected at random from Group X. Obviously, the probability that he favors a strike is 150/210. This differs from the probability, 100/188, that an hourly-wage earner in Group Y favors a strike. Thus, the probability depends on the group from which the employee is selected. This is called *conditional probability* and is denoted by P(favors a strike | group X) or P(favors a strike | group Y). The vertical bar is read "given" or "knowing that." Thus, $P(A | B)$ is the probability of A occurring given that B has occurred. P(an employee does not favor a strike | group Z) $= 70/102$. Likewise, P(an employee is in group Y | he is in the set that favors a strike) $= 100/280$.

From our assignment of probabilities associated with this example, we note an interesting relationship between $P(A | B)$, $P(A \cap B)$, $P(A)$, and $P(B)$. Let A represent the fact that an employee favors a strike, and let B represent the fact that an employee is in group Y. Then $P(A|B) = 100/188$. The probability that an employee selected at random is in group Y and simultaneously favors a strike is

$$P(A \cap B) = 100/500.$$

Finally, note that $P(B) = 188/500$. We now observe that for this example,

$$P(A|B) = \frac{P(A \cap B)}{P(B)},$$

since

$$\frac{100}{188} = \frac{100/500}{188/500}.$$

It is interesting to consider the case in which A and B are interchanged. We investigate to determine whether or not it is true that

$$P(B|A) = \frac{P(A \cap B)}{P(A)}.$$

Now $P(B|A)$, the probability that an employee is in group Y, given that he favors a strike, is equal to 100/280, and $P(A) = 280/500$. Therefore,

$$P(B|A) = \frac{P(A \cap B)}{P(A)},$$

since

$$\frac{100}{280} = \frac{100/500}{280/500}.$$

Consider now conditional probability relative to a sample space composed of N simple events. For example, consider two events A and B in the sample space depicted in Figure 11-4. Suppose there are n_2 sample points in $A \cap B$, n_1 sample points in A and not in B, n_3 in B but not in A, and n_4 sample points in neither A nor B, where

$$N = n_1 + n_2 + n_3 + n_4.$$

Obviously,

$$P(A) = (n_1 + n_2)/N, \quad P(B) = (n_3 + n_2)/N, \quad \text{and} \quad P(A \cap B) = n_2/N.$$

Now, if it is given that the sample points are in B, then

$$P(A|B) = \frac{n_2}{n_2 + n_3}.$$

Likewise,

$$P(B|A) = \frac{n_2}{n_1 + n_2}.$$

Therefore,

$$P(A|B) = \frac{P(A \cap B)}{P(B)},$$

because

$$\frac{P(A \cap B)}{P(B)} = \frac{n_2/N}{(n_3 + n_2)/N} = \frac{n_2}{n_3 + n_2} = P(A|B).$$

Likewise,

$$P(B|A) = \frac{P(A \cap B)}{P(A)},$$

because

$$\frac{P(A \cap B)}{P(A)} = \frac{n_2/N}{(n_1 + n_2)/N} = \frac{n_2}{n_1 + n_2} = P(B|A).$$

Theorem 11-4: If A and B are any events, then

$$P(A|B) = \frac{P(A \cap B)}{P(B)}.$$

Again we emphasize that $P(A|B)$ is the conditional probability that A will occur once B has occurred.

Example: Suppose we take six cards and mark them with 1, 2, 3, X, Y, Z.

(a) What is the probability of drawing a card with a letter?

(b) What is the probability of drawing a card with a letter on a second draw if we know that the first card drawn had a letter on it?

(c) What is the probability of drawing a letter on the second card if it is known that the first card drawn had an X on it?

(d) What is the probability of drawing a letter on a second card if it is known that one card has been drawn with a number on it?

Answers:

(a) $P(\text{card with letter}) = 3/6 = 1/2$.

(b) $P(\text{letter on second card} | \text{letter on first card}) = 2/5$ since there are left in the deck of cards after a letter card has been drawn two cards with letters and three cards with numbers.

(c) $P(\text{letter on second card} | X \text{ on first card}) = 2/5$ since the deck of cards contains two letter cards and three cards with numbers after an X has been drawn.

(d) $P(\text{letter on second card} | \text{number on first card}) = 3/5$ since the deck of cards contains three letter cards and two cards with numbers after a card with a number is drawn.

The formula for conditional probability, as given in Theorem 11-4, can be rewritten as a product relation by multiplying by the denominator.

$$P(A \cap B) = P(A) \cdot P(B|A)$$

or

$$= P(B) \cdot P(A|B).$$

The product formula may be extended to the occurrence of three events in the following manner.

$$P(A \cap B \cap C) = P(A) \cdot P(B|A) \cdot P(C|A \cap B)$$

Example: An urn contains four red marbles and six white marbles. Two marbles are drawn, one at a time, without replacement. What is the probability of drawing two red marbles?

$P(2R) = P(R) \cdot P(R|R)$, so the probability of two red marbles is the probability of a red marble on the first draw times the probability of a red marble on the second draw, after a red marble has been drawn on the first draw.

$$P(2R) = \frac{4}{10} \cdot \frac{3}{9} = \frac{2}{15}$$

The second probability is 3/9, because only three red marbles in a total of nine marbles remain in the urn after the first red marble is drawn.

Example: In the preceding example, what is the probability of drawing a red marble and then a white marble in the order given?

$$P(R \cap W) = P(R) \cdot P(W|R) = \frac{4}{10} \cdot \frac{6}{9} = \frac{4}{15}$$

Exercise Set 11-4

1. In a certain college, 30% of the freshmen failed mathematics, 20% failed English, and 15% of these groups failed both mathematics and English.
   (a) If a freshman failed English, what is the probability that he failed mathematics?
   (b) If a freshman failed mathematics, what is the probability that he failed English?
   (c) What is the probability that he failed mathematics or English?
2. An experiment consists of tossing a coin seven times. Describe the complement of each set below.
   (a) Getting at least two heads
   (b) Getting three, four, or five tails
   (c) Getting one tail
   (d) Getting no heads
3. A box contains six different-colored balls: red, white, blue, black, green, and yellow. If two balls are drawn at random, one at a time, and replaced, what is the probability of getting
   (a) a yellow ball followed by a red ball?
   (b) a red ball followed by a blue ball?
   (c) a yellow and red ball? (*Hint:* This can happen in more than one way.)
4. If A and B are events in a sample space such that

$$P(A) = 0.6, \quad P(B) = 0.2, \quad \text{and} \quad P(A \cap B) = 0.1,$$

   compute the following.
   (a) $P(\overline{A})$ (b) $P(\overline{B})$ (c) $P(A \cup B)$
5. If A and B are events with $P(A \cup B) = 5/8$, $P(A \cap B) = 1/3$, $P(\overline{A}) = 1/2$, compute the following probabilities.
   (a) $P(A)$ (b) $P(B)$ (c) $P(\overline{B})$ (d) $P(\overline{A} \cup B)$
6. The probability that John will live twenty additional years is 1/5, and the probability that his wife will live twenty additional years is 1/4. Find the probability that
   (a) both will live twenty additional years.
   (b) at least one will live twenty additional years.
   (c) only John will live twenty additional years.
   (d) neither will live twenty additional years.
7. A card is drawn from a standard deck of cards. What is the probability that it is a jack, given that it is a face card (K, Q, J)?

8. From an urn containing five red balls and three white balls, two balls are drawn successively at random. What is the probability that the first is white and the second is red?
9. In a recent survey, it was found that 60% of the people in a given community drink Lola Cola and 40% drink other soft drinks; 15% of the people interviewed indicated that they drink both Lola Cola and other soft drinks. What percent of the people drink either Lola Cola or other soft drinks?
10. Assume that two cards are drawn from a standard deck of playing cards. What is the probability that a king is drawn followed by an ace
    (a) if the first card is replaced before the second is drawn?
    (b) if the first card is not replaced before the second is drawn?
11. An urn contains the following balls: five colored red and white, three black and white, four green and white, six red and black, four red and green, and five black and green.
    (a) Given that you have drawn a ball that is partly green, what is the probability that it is partly white?
    (b) Given that the ball you have drawn is partly white, what is the probability that it is partly red?

Review Exercise Set 11-5

1. In a certain town, 40% of the families have incomes exceeding $8000 a year, 30% of the families have two or more automobiles, and 15% have both two or more automobiles and incomes exceeding $8000 a year. A family is selected at random from this town.
    (a) If the family has two or more automobiles, what is the probability that the family income exceeds $8000 a year?
    (b) If the family income exceeds $8000 a year, what is the probability that the family has two or more automobiles?
    (c) What is the probability that a family has an income exceeding $8000 a year or has two or more automobiles?
2. (a) What is the probability of drawing four cards that are all hearts from an ordinary deck of cards?
    (b) What is the probability of drawing the ace, king, queen, and jack of hearts?

3. A number x is selected at random from a set of numbers,

$$\{1, 2, 3, \ldots, 8\}.$$

What is the probability that

(a) x is less than 5?
(b) x is prime?
(c) x is even?
(d) $x \geq 2$?
(e) x is less than 5 and even?
(f) $x < 7$?
(g) x is less than 5 or even?
(h) $x^2 > 3$?

4. From a bag containing six red, four black, and three green balls, one ball is drawn. What is the probability that it is

(a) red or black?
(b) blue?
(c) red or black or green?
(d) not red and not green?
(e) not black?
(f) green?
(g) not red or not black?
(h) red and black?
(i) not red or not green?
(j) not green?

5. A die is tossed three times. If the throws are assumed to be independent, what is the probability that on the first throw the die will show an even number; on the second throw, a number divisible by 3; and on the third throw, a one?

6. A shipment contains ninety-six good items and four defective items. Threc items are drawn, one at a time, from the shipment.

(a) What is the probability of three defective items?
(b) What is the probability of no defective items?
(c) What is the probability of two good items and one defective item?
(d) What is the probability in (a) if the items are replaced after each drawing?
(e) What is the probability in (b) if the items are replaced after each drawing?

7. Three young men of a common age form a partnership. An insurance company estimates that the probability that men of such an age will live to be 65 is 70%.

(a) What is the probability that all of them will live to be 65 years of age?
(b) What is the probability that at least one will live to be 65 years of age?
(c) What is the probability that exactly one will live to be 65 years of age?
(d) What is the probability that none of them will live to be 65 years of age?

Suggested Reading

Sample Space and Events: Freund, pp. 129–134. Lewis, pp. 85–88. Mack, pp. 47–50. Mendenhall, pp. 44–51, 55–62. Mode, pp. 5–16, 27–37. Nichols, Swain, pp. 318, 321–326. Spector, pp. 376–379. Weaver, Wolf, pp. 213–215. Wheeler, Peeples, pp. 165–171.

Probability: Adams, p. 22. Freund, pp. 134–138, 149–152. Graham, pp. 363–364. Huntsberger, pp. 75–84. Lewis, pp. 90–95. Mack, pp. 51–62. Mendenhall, pp. 62–67. Mode, pp. 40–50. Nichols, Swain, pp. 318–333. Spector, pp. 362–383. Weaver, Wolf, pp. 209–212, 216–228. Wheeler, Peeples, pp. 171–220.

12

Introduction to Statistics

1

Tables and Graphs

If you asked the man on the street for his interpretation of statistics, he could well answer "Statistics is a hocus-pocus of numbers." Or he might comment that some people use statistical statements in an attempt to "prove" a statement that may or may not be true. This is a normal reaction. We are bombarded continuously with the misuse of statistics until some of us have calloused our thinking against this field. Yet, when we ourselves lack adequate information, we often confuse our opponents with a mumbo-jumbo of statistics. These tactics would not be possible if an adequate knowledge of the use of statistics were a part of everyone's education.

H. G. Wells, more than eighty years ago, remarked that statistical thinking would one day be as necessary for efficient citizenship as the ability to read and write. Yes, statistics can be used by those who wish "to tell a lie," but it is also an efficient tool for those who are searching for truth. We need to be able to distinguish between these extremes.

Admittedly, one cannot become proficient in the use of statistics in seven short lessons—no matter how compact the material. However, this introduction to the field of statistics will hopefully make you aware of the importance of the subject and will encourage additional individual study so that you will be able to recognize obvious misuses of statistics.

Statistics is involved with the collection of numerical facts—as boring as this may be to some. Yet, statistics is much more than this. First of all, we concern ourselves with the display or representation of numerical facts. No matter how accurate the statistical facts, we can still be confused by tables and charts unless we have some knowledge of this field.

Statistical data may be presented in a variety of ways. If data are recorded in a *table*, they are usually easy to read and understand. Consider the grades in Table 12-1. It is easy to see that 36 out of 60 students earned a C grade.

Final Exam Grades

A	4
B	15
C	36
D	3
F	2
Total	60

Table 12-1

Table 12-2 illustrates that tables may also be arranged horizontally.

Number of Graduates: Lovin College

Year	1961	1962	1963	1964	1965	1966
Number of Graduates	152	163	197	185	201	196

Year	1967	1968	1969	1970	1971	1972
Number of Graduates	210	189	195	205	200	180

Table 12-2

Tables are useful in determining a single item, such as the category with the largest or smallest entry, as illustrated by the C category or the F category in Table 12-1. However, tables are sometimes not as useful as graphs for making comparisons.

Statistical data in this form do not facilitate immediate interpretation. It is often necessary to group the data into *classes* before general characteristics can be detected and measured. Group the data in Table 12-5 into five classes, 15–19, 20–24, 25–29, 30–34, and 35–39, by making a mark or tally in Table 12-6 for each value that occurs in a class. The total of the tallies for each class results in a figure that is the frequency for the class. (Recall that the term frequency was used in the preceding section.) The result of the tabulation involving the classes and the

Class	*Tallies*	*Frequency*
15–19	||||	4
20–24	~~||||~~ ||	7
25–29	~~||||~~	5
30–34	||	2
35–39	||	2

Table 12-6

frequencies is called a *frequency distribution* or *group frequency distribution.* Note that certain interesting facts can be obtained easily from the frequency distribution. For example, the largest number of entries (seven) falls in the class 20–24. Note also that the number of observations per class increases up to the seven and then starts decreasing.

Since the observations in Table 12-5 seem to be recorded to the nearest whole number, the class interval 15–19 theoretically includes all measurements from 14.5 to 19.5. Numbers indicating intervals such as 14.5–19.5 are called *class boundaries.* The number 14.5 is a lower class boundary and 19.5 is an upper class boundary. The class boundaries may be obtained by adding the upper class limit and lower class limit of consecutive classes and dividing by 2. For example, the class boundaries for Table 12-6 could be written as 14.5–19.5, 19.5–24.5, 24.5–29.5, 29.5–34.5, and 34.5–39.5.

Table 12-7 gives the lengths of engagements (by the number of months) of 30 newly married college seniors. It is difficult to see a trend or pattern in the data. However, by grouping, certain statistical facts become evident.

The process of grouping or tabulating data into classes consists of three steps. First, one determines the number and size of the classes and chooses the class limits. Second, the data are divided or tabulated into

10	2	9	6	11
17	4	10	7	3
1	4	11	6	3
8	15	12	9	12
8	18	12	6	10
8	18	12	6	9

Table 12-7

the selected classes; finally, a count is made of the number of entries in each class.

It is difficult to provide general instructions for separating data into classes, since the grouping of data is essentially arbitrary and depends in many cases on the ultimate use of the grouped data. However, certain questions always arise, such as "How many classes?" "How long should the classes be made?" "What numbers should be used as class limits?" Consequently, some discussion of a general nature is given for grouping data into classes.

First, note in Table 12-7 that the shortest length of engagement is 1 month and the longest is 18 months. The difference in the longest and the shortest periods (called the *range* of the data) is 17. To decide on the number of classes, a practical rule may be followed which suggests that there are few occasions requiring fewer than five classes or more than twenty classes. We arbitrarily select six classes for our grouping. Since 17 ÷ 6 is 2.833, the length of the classes (if the classes are of equal length) must be more than 2.833 in order to include all the data in six classes. Whenever feasible, it is desirable to have classes of equal length. Thus, we arbitrarily select the following class limits: 1–3, 4–6, 7–9, 10–12, and so on.

The tallies shown in the middle column of Table 12-8 indicate the number of entries in each class, which when tabulated give the last column, the frequency of the class.

Class Intervals	*Tallies*	*Frequency*
1–3	||||	4
4–6	卌 |	6
7–9	卌 ||	7
10–12	卌 ||||	9
13–15	|	1
16–18	|||	3

Table 12-8

In the preceding section, we studied graphical procedures for displaying or representing frequency distributions. Note in the histogram in Figure 12-7 that each side of a rectangle terminates at a class boundary.

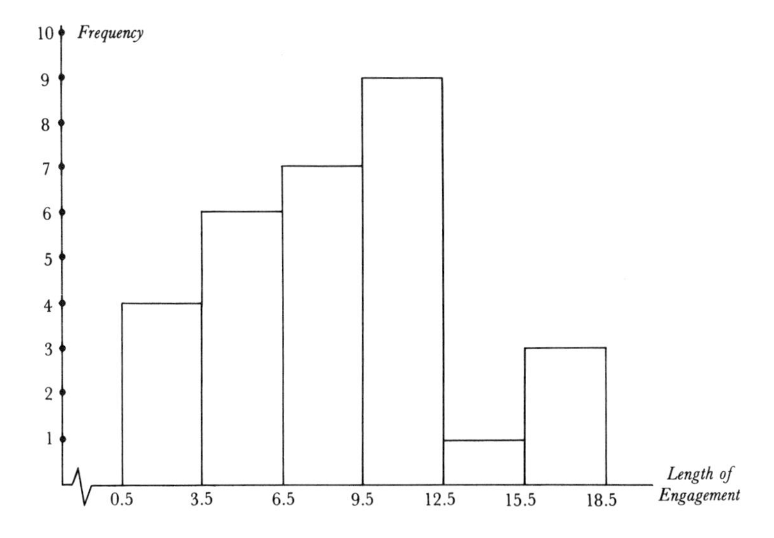

Figure 12-7

The *length of a class interval* can be found by taking the difference in the boundaries of a class or the difference in consecutive lower (or upper) class limits. The mid-value of a class interval is called the *class mark.* It is obtained by computing the sum of the boundaries of a class and dividing that sum by 2. It can also be computed as the sum of a lower class boundary and one-half of the class interval. The class mark is important in drawing frequency polygons and in computing certain statistics such as the mean and standard deviation, which will be discussed later.

The class marks for the ages of the mothers in Table 12-6 are 17, 22, 27, 32, and 37. The class marks for the lengths of engagements (Table 12-8) are 2, 5, 8, 11, 14, and 17.

Class marks are used in constructing a line frequency polygon, discussed in the preceding section. Figure 12-8 presents a frequency polygon for the grouped frequency distribution in Table 12-8.

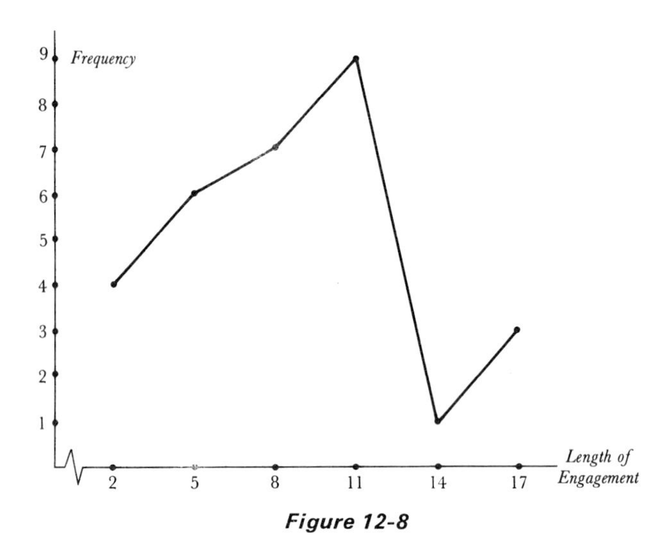

Figure 12-8

Exercise Set 12-2

For Exercises 1 through 6, group the data in the manner outlined in each exercise. Then use the tabulated data to complete parts (a) through (d).

(a) Make a frequency distribution.
(b) Find the class mark for each class.
(c) Construct a histogram.
(d) Construct a frequency polygon.

Save the frequency distributions obtained in this exercise set for use in problems in the remaining exercise sets of this chapter.

1. The following are the amounts rounded to the nearest \$1 that a sample of 50 freshmen spent on textbooks during a fall semester.

33	41	35	53	42	47	41	31	38	37
30	38	37	33	41	35	42	50	41	38
39	42	41	40	40	38	37	41	45	48
35	36	35	38	33	39	40	40	47	38
37	38	37	34	35	44	44	46	40	39

Using six intervals (of minimum integral length), make a frequency distribution, with the first class beginning at 30.

2. The grades of 60 students in a mathematics course were recorded as follows.

96	71	43	77	74	73	87	81	91	79
78	72	82	81	87	93	95	53	64	66
83	71	58	97	53	74	61	68	67	63
55	81	62	87	76	74	71	65	93	71
74	71	77	83	85	84	94	56	63	65
75	48	89	84	75	76	75	61	91	90

Make a frequency distribution, starting the first class at 43 and using seven intervals of minimum integral length.

3. The following are the IQ scores of 30 first-grade students in one classroom.

128	133	100	115	82	99
107	142	98	112	152	100
105	78	114	84	86	110
96	93	101	94	86	124
120	100	102	107	94	128

Group these scores into six intervals (of minimum integral length), starting the first class at 75.

4. A small foreign country each year exports 20 main products, ranging from iron ore, to toy medical kits, to surgical instruments. The value of each export in millions of dollars is given in the table below. Group the values into six intervals of length 70, starting the first class limit at 60.

86	62	239	290	207
285	232	214	131	195
424	343	476	140	398
363	348	156	222	370

5. The 25 scores below were achieved by a group of college freshmen on a mathematics placement test. Collect the following information into five groups, the first beginning with the class limit of 450.

477	485	527	483	582
567	513	609	596	525
566	540	451	519	530
576	656	525	621	603
648	555	535	528	546

6. The daily air particulate count per cubic meter of air in a smog-laden city for the months of July and August is given below.

156	195	420	465	191	225	159	171
205	145	461	407	163	275	101	223
225	159	508	395	159	255	114	191
185	210	509	388	175	235	151	223
160	265	565	388	151	220	163	227
149	307	593	305	184	254	171	229
145	333	535	263	198	240	172	
187	393	515	207	202	234	171	

Make the frequency distribution starting the first class at 100, using ten intervals of minimum integral length.

*7. Given the following histogram, construct a frequency distribution and find the class marks.

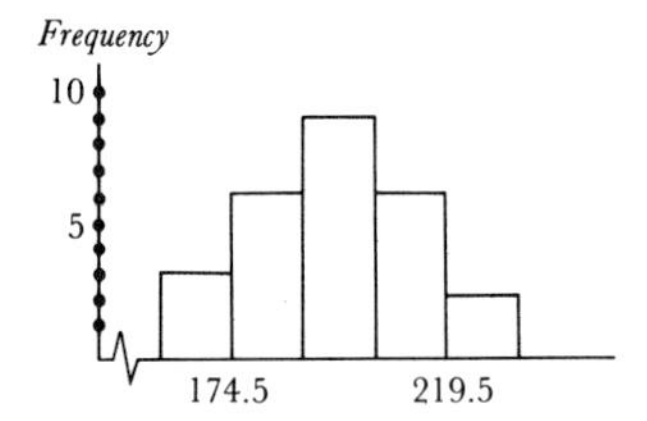

*8. Given the following frequency polygon, tabulate a frequency distribution.

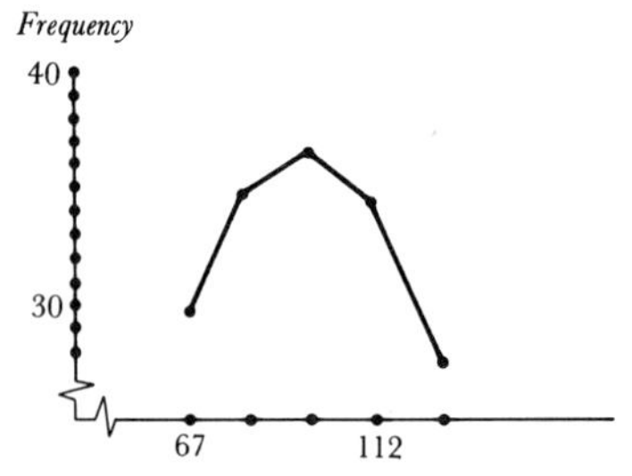

3

A Statistical Average: The Arithmetic Mean

The graphical methods discussed in Section 1 of this chapter are useful in presenting a general description of collected data. However, there are limitations to the use of graphical techniques for describing and analyzing data. Many of these limitations can be overcome by the use of certain statistics, called "averages." There are five different averages in general use—the arithmetic mean, the median, the mode, the geometric mean, and the harmonic mean. In this book, we will study only the arithmetic mean, the median, and the mode.

One of the most common averages is the *arithmetic mean* (sometimes called arithmetic average). The arithmetic mean of a set of n measurements is the sum of the measurements divided by the number of measurements.

Definition 12-1: Consider n observations $x_1, x_2, x_3, \ldots, x_n$. The formula for the *arithmetic mean*, denoted by $\bar{x}$, is given by

$$\bar{x} = \frac{x_1 + x_2 + x_3 + \cdots + x_n}{n}.$$

Example: Find the arithmetic mean of 8, 16, 4, 12, and 10.

$$\bar{x} = \frac{8 + 16 + 4 + 12 + 10}{5} = 10$$

Example: Find the arithmetic mean of 25, 25, 25, 25, 30, 30, 30, 40, 40, 40, 40, 50.

$$\bar{x} = \frac{25 + 25 + 25 + 25 + 30 + 30 + 30 + 40 + 40 + 40 + 40 + 50}{12}$$

$$= \frac{25(4) + 30(3) + 40(4) + 50(1)}{4 + 3 + 4 + 1}$$

$$= \frac{400}{12} = 33\tfrac{1}{3}$$

In the preceding example, consider the 4, 3, 4, and 1 as the frequencies of the 25, 30, 40, and 50, respectively. If we generalize the formula for finding the arithmetic mean to include frequencies as illustrated in the preceding example, then

$$\bar{x} = \frac{x_1 f_1 + x_2 f_2 + x_3 f_3 + \cdots + x_m f_m}{f_1 + f_2 + f_3 + \cdots + f_m},$$

where f_i is the frequency of x_i.

If data are presented in a frequency table, we may have no way of knowing the distribution of the data within a class. We therefore assume that the data are uniformly distributed within a class interval about the class mark or that all of the data within a class interval are located at the class mark. Thus, we use the formula for $\bar{x}$ given in the preceding paragraph, where x represents the class mark, f the frequency of each class and m the number of class intervals.

Example: Find the arithmetic mean for the data in Table 12-9.

Class Mark x	*Frequency* f	xf
17	4	68
22	7	154
27	5	135
32	2	64
37	2	74
Total	20	495

Table 12-9

Answer: $\bar{x} = \frac{495}{20} = 24.75.$

We should note that the arithmetic mean of a frequency distribution is not exactly equal to the mean obtained directly from the observations. This discrepancy is due to the fact that, in finding the arithmetic mean of a frequency distribution, we assume that the values of all the observations in each class are equal to the class mark, or are equally spaced about the class marks.

Example: Using the first three columns of Table 12-10, we find the arithmetic mean of the lengths of engagements as given previously in Table 12-8.

Class Mark	*Frequency*	*xf*	$y = \frac{x-2}{3}$	*yf*
2	4	8	0	0
5	6	30	1	6
8	7	56	2	14
11	9	99	3	27
14	1	14	4	4
17	3	51	5	15
Total	30	258		66

Table 12-10

Answer: $\bar{x} = \frac{258}{30} = 8.6.$

The computation involved in finding the arithmetic mean for examples similar to the preceding example may be greatly simplified by the following theorem.

Theorem 12-1: If $y = \frac{x-a}{b}$ so $x = by + a$, where a and b are any real numbers, then the mean of the x's is the sum of b times the mean of the y's and a, or $\bar{x} = b\bar{y} + a$.

This theorem is illustrated using the preceding example. Let x represent the set of original class marks. Subtract the first class mark, 2, from each class mark and divide the result by the width of the class, 3. Call the result y. This operation is symbolized by the equation

$$y = \frac{x-2}{3}.$$

When $x = 2, y = 0$; when $x = 5, y = 1$; when $x = 8, y = 2$; and so forth, as seen in the fourth column of Table 12-10. Using the sum of the fifth and second columns,

$$\bar{y} = \frac{66}{30} = 2.2.$$

By Theorem 12-1,

$$\begin{aligned}\bar{x} &= 3\bar{y} + 2,\\ &= 3(2.2) + 2,\\ &= 8.6.\end{aligned}$$

Proof of Theorem 12-1:

$$\begin{aligned}\bar{x} &= \frac{x_1f_1 + x_2f_2 + x_3f_3 + \cdots + x_mf_m}{f_1 + f_2 + f_3 + \cdots + f_m}\\ &= \frac{(by_1 + a)f_1 + (by_2 + a)f_2 + (by_3 + a)f_3 + \cdots + (by_m + a)f_m}{f_1 + f_2 + f_3 + \cdots + f_m}\\ &= \frac{by_1f_1 + af_1 + by_2f_2 + af_2 + by_3f_3 + af_3 + \cdots + by_mf_m + af_m}{f_1 + f_2 + f_3 + \cdots + f_m}\\ &= \frac{(by_1f_1 + by_2f_2 + by_3f_3 + \cdots + by_mf_m) + (af_1 + af_2 + af_3 + \cdots + af_m)}{f_1 + f_2 + f_3 + \cdots + f_m}\\ &= \frac{b(y_1f_1 + y_2f_2 + y_3f_3 + \cdots + y_mf_m)}{f_1 + f_2 + f_3 + \cdots + f_m} + \frac{a(f_1 + f_2 + f_3 + \cdots + f_m)}{f_1 + f_2 + f_3 + \cdots + f_m}\\ &= b\bar{y} + a.\end{aligned}$$

Exercise Set 12-3

1. Compute the arithmetic mean of the following sets of data.
   (a) 2, 1, 7, 3, 4, 6, 10, 5, 9
   (b) 3, 1, 1, 4, 2, 2, 7, 9, 6
   (c) 3, 8, 7, 3, 8, 6, 5, 5
   (d) 8, 16, 5, 4, 6, 6, 10, 7, 9, 10, 13, 14
2. The following is the distribution of scores on a test administered to freshmen at Laneville College.

Score	*Frequency*
140–149	3
130–139	4
120–129	8
110–119	13
100–109	4
90–99	2
80–89	0
70–79	1

(a) Find the arithmetic mean.
(b) Find the arithmetic mean by subtracting the last class mark from each class mark.
(c) Find the arithmetic mean by subtracting 114.5 from each class mark.
(d) Find the arithmetic mean by subtracting 114.5 from each class mark and by dividing the result by 10.

3. Consider the following data.

Class	*Frequency*
100–119	2
120–139	3
140–159	4
160–179	6
180–199	21
200–219	18
220–239	20
240–259	18
260–279	20
280–299	8

(a) Compute the arithmetic mean by subtracting the first class mark from each class mark.
(b) Compute the mean by subtracting 189.5 from each class mark.
(c) Compute the mean by subtracting 189.5 from each class mark and dividing the result by 20.

4. The following table indicates the length in days of 25 labor strikes. Divide the table into six classes of minimum integral length, beginning the first class at 10. Then calculate the arithmetic mean.

28	37	55	22	55
32	14	61	44	45
23	26	20	16	47
15	27	33	36	26
33	63	64	15	37

Compute the arithmetic mean for the data presented in Exercises 5 and 6.

5. A state-wide survey revealed that 48 industrial plants were discharging large amounts of waste into the state water systems. The following table gives the number of plants discharging certain millions of gallons of waste.

Millions of Gallons	*Number of Plants*
.5– 3.5	6
3.5– 6.5	9
6.5– 9.5	13
9.5–12.5	10
12.5–15.5	7
15.5–18.5	3

6. An ethnologist investigated the life span of a tribe of natives on a remote South Pacific island. During the year that he lived with the tribe, a total of 35 deaths occurred. He developed the following table giving the age in years of each person at the time of death. (*Hint:* Before finding the arithmetic mean, divide the table into five classes of minimum integral length, starting the first class at 0.)

39	39	1	10	43	2	21
45	4	45	52	1	18	35
47	33	45	18	26	58	22
50	18	26	63	15	$\frac{1}{2}$	39
6	3	1	$\frac{1}{2}$	39	14	41

7. Compute the arithmetic means for data presented in the following exercises in Exercise Set 12-2.
   (a) Exercise 1 (b) Exercise 2
   (c) Exercise 3 (d) Exercise 4
   (e) Exercise 5 (f) Exercise 6

*8. A Biblical scholar studied the lengths of the words in the Old Testament. Why don't you do the same for the first 100 words of the King James version by tabulating a frequency distribution?

*9. Repeat Exercise 8 for some modern version of the Bible.

4

Averages: The Median and the Mode

The *median* of a set of observations is a number such that half of the observations are larger than or equal to it and half of the observations are smaller than or equal to it. If a set of observations is arranged in order of increasing or decreasing magnitude, it is easy to determine the median.

Definition 12-2: If an odd number of observations are arranged in order of decreasing or increasing magnitude, the *median* of the set of numbers is the value of the middle number in the array. For an even number of observations, the median is half-way between the two middle numbers.

Example: Consider the set of five measurements 7, 1, 2, 1, 3. Arranged in increasing order of magnitude, they may be written as 1, 1, 2, 3, 7. Hence, the median of the set of numbers is 2.

Example: The array

$$25, 2, 5, 6, 5, 23, 7, 10, 22, 15, 21, 23,$$

which can be arranged in decreasing order as

$$25, 23, 23, 22, 21, 15, 10, 7, 6, 5, 5, 2,$$

includes twelve observations, so there is no middle number of the array. The median is

$$\frac{15 + 10}{2} = 12.5.$$

The median for data expressed as a frequency distribution is a bit more involved. When the distribution is represented by a histogram, the median is a point such that the areas enclosed by the histograms on each side of the point are equal. Analytically, the process involves assuming that the values in the class that contains the median are uniformly distributed across the interval. Then the class interval is divided into two parts so that each part contains the frequency necessary for the frequency above and the frequency below the dividing point to be one-half of the total frequency.

Example: Find the median of the data in Table 12-11, which illustrates the results of a survey of weekly take-home pay (denoted by x) for 100 employees.

x	f
60–69	4
70–79	10
80–89	18
90–99	24
100–109	14
110–119	10
120–129	9
130–139	7
140–149	4
	$N = 100$

Table 12-11

The first step in finding the median is to find 1/2, or 50 %, of the total frequency. In this case, 50 % of the total frequency is $(1/2)(100) = 50$. The total frequency in the first three classes is $4 + 10 + 18 = 32$, and the total frequency in the first four classes is $4 + 10 + 18 + 24 = 56$. Thus, the class 90–99 contains the median.

There are 32 weekly wages below this class. Therefore, $50 - 32 = 18$ wages are needed from this class below the median in order to have 50 % of the wages below the median. The class boundaries of the interval containing the median are 89.5–99.5. Since 18 wages are needed from the 24 in this class (assumed to be spread uniformly throughout the interval), these 18 wages should be in the first 18/24 of the interval. Now the interval is of width 10; hence, the length of the interval below the median should be $(18/24)10 = 7.5$. Thus, the median, m, is

$$m = 89.5 + 7.5 = 97.0.$$

This relationship can be expressed in a formula as

$$m = L + \left[\frac{N/2 - F_b}{F_i}\right] i,$$

where

L is the lower boundary of the class containing the median,
N is the total frequency,
F_i is the frequency of the class containing the median,
F_b is the total frequency below the class containing the median,
i is the length of the class interval.

Thus, in the preceding example,

$$m = 89.5 + \left[\frac{100/2 - 32}{24}\right]10,$$

$$= 89.5 + (18/24)10,$$

$$= 89.5 + 7.5 = 97.0.$$

The median of a set of data is a special case of a more general statistic called a *percentile*. The tenth percentile, P_{10}, is a value such that one-tenth of the observations are less than it. Nine-tenths of the observations are less than the ninetieth percentile, P_{90}. Thus, the median is the fiftieth percentile, P_{50}.

Example: Find the tenth percentile and the ninetieth percentile for the following set of observations: 8, 14, 12, 64, 7, 9, 42, 84, 76, 92, 41, 15, 17, 26, 47, 16, 21, 22, 23, 24. There are twenty observations; $(1/10)(20) = 2$, or two of the observations are less than the tenth percentile. Arranged in increasing order, the second observation is 8 and the third observation is 9. The value of the tenth percentile is a number between 8 and 9. Under our assumption that statistical data are evenly spread, we arbitrarily select a point half-way between 8 and 9. Thus, $P_{10} = 8 + .5(9 - 8) = 8.5$. To find P_{90}, note that since $(9/10)20 = 18$, P_{90} is located between the eighteenth and nineteenth observations (76 and 84, respectively). Thus,

$$P_{90} = 76 + .5(84 - 76) = 80.$$

In this example we found the ninetieth percentile to be 80. We can also state that the *percentile rank* of score 80 is 90.

We use a formula very similar to the formula for the median to find percentiles for frequency distributions.

$$P = L + \frac{(pN - F_b)i}{F_i}$$

where

L is the lower class boundary of the class containing the percentile.
N is the total frequency.
p is the percentage of the observations less than and equal to the percentile.
F_b is the total frequency below the class containing the percentile.
F_i is the frequency of the class containing the percentile.
i is the class interval.

Example: Consider the following frequency distribution.

Class Limits	*Frequency*
80–99	4
100–119	16
120–139	24
140–159	36
160–179	14
180–199	4
200–219	2

(a) Find the tenth percentile.

$$P_{10} = 99.5 + \left[\frac{(1/10)100 - 4}{16}\right]20 = 107.00$$

(b) Find the ninetieth percentile.

$$P_{90} = 159.5 + \left[\frac{(9/10)100 - 80}{14}\right]20 = 173.79$$

Definition 12-3: The *mode* for a set of observations is the observation that occurs most frequently, if such exists.

Example: For the set of observations, 2, 3, 8, 5, 3, 6, 5, 3, 6, 3, three is the mode because it occurs four times and no other observation occurs this many times.

When the data are tabulated as a frequency distribution, the class with the greatest frequency is known as the modal class. In Table 12-6, the modal class is the class designated as 20–24. The modal class for the data in Table 12-11 is the class 90–99.

Now that we have introduced three averages (mean, median, and mode), the question naturally arises "Which average should one use?" The only reasonable answer to this question is to use the average that best fulfills the purpose at hand. For example, a business executive who is attempting to estimate the average cost of a wage increase would compare mean salaries before and after the increase. On the other hand, a census taker who desires to describe a standard of living might quote a median salary. A purchasing agent for a clothing store would definitely have use for the mode.

Of course, given only the original observations (before grouping), the median is somewhat easier to compute than the mean. If the data have been grouped as a frequency distribution, the median and the modal class are still somewhat easier to calculate than the mean. For cases where the first class or the last class are open-ended (such as "less than 100" or "more than 5000"), the mean cannot be calculated without assuming values for the class marks in the open-ended classes.

The mode and the median are affected less by extreme observations than the mean. For example, consider the average salary of a set of five salaries: $6000, $7000, $8000, $9000, and $250,000. The average salary given by the arithmetic mean is $56,000, whereas the average salary given by the median is $8000. In this case, the $8000 median average is more representative of a typical salary than the mean average of $56,000.

When a change in averages is of interest, one should use the arithmetic mean instead of the median because a change in an arithmetic mean is equal to the arithmetic mean of the changes in the variable. The median lacks this useful property.

Example: Consider two sets of observations, x and y.

$$x = 4,\ 8,\ 12,\ 30,\ 20,\ 10$$

$$y = 10,\ 14,\ 20,\ 44,\ 30,\ 14$$

$$y - x = 6,\ 6,\ 8,\ 14,\ 10,\ 4$$

$$\bar{x} = \frac{4 + 8 + 12 + 30 + 20 + 10}{6} = 14$$

$$\bar{y} = \frac{10 + 14 + 20 + 44 + 30 + 14}{6} = 22$$

$$\overline{y - x} = \frac{6 + 6 + 8 + 14 + 10 + 4}{6} = 8$$

Note that $\overline{y - x} = \bar{y} - \bar{x}$ since $8 = 22 - 14$. Now for the median

$$m(\text{of } x) = 11, \quad m(\text{of } y) = 17, \quad m(\text{of } y - x) = 7,$$
$$m(\text{of } y - x) \neq m(\text{of } y) - m(\text{of } x) \text{ since } 7 \neq 17 - 11.$$

Generally speaking, the mean is used more than the other two averages in that it plays a very important role in the advanced study of statistics. In conclusion, the mean can be described as the center of a set of observations in the sense that the sum of the differences in the mean and each observation greater than the mean is equal to the sum of the differences in the mean and each observation smaller than the mean.

Example:

$$\bar{x} = \frac{4 + 8 + 12 + 30 + 20 + 10}{6} = 14.$$

$$\begin{array}{rcl} 14 - 4 = 10 & \qquad & 30 - 14 = 16 \\ 14 - 8 = 6 & & 20 - 14 = 6 \\ 14 - 10 = 4 & & \overline{22} \\ 14 - 12 = 2 & & \\ \overline{22} & & \end{array}$$

The median can be described as the middle value of a set of observations in the sense that there are as many observations greater than the median as there are observations smaller than the median.

The mode, provided it exists, can be used as a typical value when one is interested in the most common value.

Exercise Set 12-4

1. Compute the median and the mode for the following sets of data.
   (a) 3, 6, 2, 6, 5, 6, 4, 1, 1
   (b) 7, 1, 3, 1, 4, 6, 5, 2
   (c) 21, 13, 12, 6, 23, 23, 20, 19
   (d) 18, 13, 12, 14, 12, 11, 16, 15, 21
2. For each of the given sets of observations, find the tenth and ninetieth percentiles.
   (a) 16, 14, 12, 13, 15, 18, 24, 8, 10, 4
   (b) 18, 47, 64, 32, 41, 92, 84, 27, 14, 12

Find the modal class and the median for the data given in Exercises 3 through 5.

3. A study on the educational expenditures of local governments is summarized by the following table.

Amount Spent	*Number of Cities*
$100,000–$150,000	24
$150,000–$200,000	33
$200,000–$250,000	41
$250,000–$300,000	33
$300,000–$350,000	62
$350,000–$400,000	41
$400,000–$450,000	33
$450,000–$500,000	13

4. A case worker, in reviewing his files for a period of one year, found that the ages of the people with whom he had worked varied as follows.

Age	*Frequency*
1–10	31
11–20	16
21–30	15
31–40	23
41–50	3
51–60	3
61–70	4
71–80	2
81–90	1
91–100	2

5. From a study of overpopulation in certain underdeveloped countries, the number of births in one year in each of 50 countries is presented in the following table.

Number of Births	*Number of Countries*
5,000–15,000	11
15,000–25,000	8
25,000–35,000	7
35,000–45,000	2
45,000–55,000	4
55,000–65,000	5
65,000–75,000	5
75,000–85,000	2
85,000–95,000	4
95,000–105,000	2

6. The arithmetic mean salary for 99 people is given as \$8,000. How much will the addition of one salary of \$60,000 increase the average of the salaries (arithmetic mean)?
7. The following data have been collected on the expenses (excluding travel) of six trips made by teachers in the mathematics department at Snelling College.

Number of Days on Trip	*Total Expense*	*Expense per day*
.5	\$13.50	\$27.00
2.5	12.00	4.80
3	21.00	7.00
1	9.50	9.50
8	32.00	4.00
5	60.00	12.00
20	\$148.00	\$64.30

Let

$$\bar{x} = \frac{\$64.30}{20} = \$3.22 \text{ average expense per day}$$

or

$$\bar{x} = \frac{\$148.00}{20} = \$7.40 \text{ average expense per day}$$

or

$$m = \frac{\$13.50 + \$21.00}{2} = \$17.25 \text{ average expense per day.}$$

Which result is realistic? Why are the others not applicable?

8. The weight in pounds of the members of the Lo-Ho College football squad are as follows: 165, 140, 162, 204, 158, 180, 172, 174, 176, 189, 195, 198, 207, 180, 218, 201, 196, 190, 184, 173, 180, 192, 194, 196, 199, 184, 160, 148, 154, 180, 192, 196, 188, 174, 184, 194, 172, 176, 164, 180.
   (a) What is the mean weight of the football squad?
   (b) What is the median weight?
   (c) Find the mode of this set of observations.
   (d) If you were a sportswriter assigned to do a story on this squad, how would you describe the (average) weight?

9. Group the data of Exercise 8 into classes 140–149, 150–159, 160–169, and so forth.
   (a) What is the mean weight of the football squad?
   (b) What is the median weight?
   (c) Find the modal class.
   (d) Compare answers in Exercises 8 and 9.

5

Standard Deviation

No average by itself gives information about what might be called the dispersion or the scattering of observations. For certain purposes, it is good to know the average salary of the heads of households in a given community; however, the social scientist is also interested in the distribution of family incomes or the variability of salaries.

There are several ways to measure dispersion or scattering of observations. The easiest measurement to calculate is the *range* (defined earlier to be the difference in the largest and smallest values in a set of observations). For example, for the set of data, 7, 3, 1, 15, 41, 74, 35, the range is $74 - 1 = 73$. The range is not usually a good measure of dispersion because it often varies with the number of observations in a set of observations. Note also that the range can change greatly if just one observation at either end is changed. For example, suppose the 74 in the set of observations listed above was miscopied and listed as 24 instead. Note that the range changes from 73 to 40.

Since the range is affected by extreme values, other measures of dispersion are preferable. In this section, we consider two measures of dispersion, *mean deviation* and *standard deviation*. Mean deviation (denoted by *m.d.*) is the arithmetic mean of the absolute value of the deviations of each variate x from the mean $\bar{x}$. That is,

$$m.d. = \frac{|x_1 - \bar{x}| + |x_2 - \bar{x}| + \cdots + |x_n - \bar{x}|}{n}.$$

Now, are you wondering why we complicate this definition by inserting absolute values rather than simply averaging the deviations from the mean? The following example should answer this question for you.

Example: Find the mean deviation of the following data: 6, 8, 10, 11, 13, and 15.

$$\bar{x} = \frac{6 + 8 + 10 + 11 + 13 + 15}{6} = 10.5$$

Now before computing the mean deviation, we will compute the average deviation of each variate from the mean, 10.5.

Average deviation

$$= \frac{(6 - 10.5) + (8 - 10.5) + (10 - 10.5) + (11 - 10.5) + (13 - 10.5) + (15 - 10.5)}{6}$$

$$= \frac{-4.5 + (-2.5) + (-.5) + .5 + 2.5 + 4.5}{6} = 0$$

Thus, the average deviation is 0. In fact, this is true for all distributions. Consequently, since the average deviation tells nothing about scattering, we can see the necessity of taking the absolute value.

m.d.

$$= \frac{|6 - 10.5| + |8 - 10.5| + |10 - 10.5| + |11 - 10.5| + |13 - 10.5| + |15 - 10.5|}{6}$$

$$= \frac{4.5 + 2.5 + .5 + .5 + 2.5 + 4.5}{6} = 2.5$$

Because mean deviation does involve absolute value, it is not as important in statistics as are other measures of scattering called variance and standard deviation. *Variance* and *standard deviation* also measure scattering from the mean, but in the computation of variance and standard deviation we use the squares of the deviations instead of their absolute values.

Variance can be obtained in three easy steps:

(a) Take the difference or deviation of each observation from the arithmetic mean.
(b) Square each deviation.
(c) Obtain the mean of the squared deviations.

These computations are indicated by the following formula for variance.

Definition 12-4: *Variance*, denoted by s^2, is defined by

$$s^2 = \frac{(x_1 - \bar{x})^2 + (x_2 - \bar{x})^2 + (x_3 - \bar{x})^2 + \cdots + (x_n - \bar{x})^2}{n}.$$

Of course, when data are given as a frequency distribution, one considers the frequency of each data point (in this case, class marks). Thus, the formula for variance for a grouped frequency distribution may be written as

$$s^2 = \frac{(x_1 - \bar{x})^2 f_1 + (x_2 - \bar{x})^2 f_2 + (x_3 - \bar{x})^2 f_3 + \cdots + (x_m - \bar{x})^2 f_m}{f_1 + f_2 + f_3 + \cdots + f_m}.$$

Definition 12-5: *Standard deviation*, denoted by s, is defined to be the square root of variance.

Example: Find the standard deviation for the data: 5, 7, 1, 2, 3, and 6, using Table 12-12.

	x	$x - \bar{x}$	$(x - \bar{x})^2$
	5	1	1
	7	3	9
	1	-3	9
	2	-2	4
	3	-1	1
	6	2	4
Total	24		28

Table 12-12

$$\bar{x} = 24/6 = 4$$
$$s^2 = 28/6 = 4.67$$

The standard deviation is $\sqrt{4.67}$, which is approximately 2.16.

Example: Find the variance and standard deviation of the distribution as tabulated in Table 12-13.

x	f	xf	$x - 14.8$	$(x - 14.8)^2$	$(x-14.8)^2 f$
4	10	40	-10.8	116.64	1166.4
8	10	80	-6.8	46.24	462.4
12	20	240	-2.8	7.84	156.8
16	30	480	1.2	1.44	43.2
20	20	400	5.2	27.04	540.8
24	10	240	9.2	84.64	846.4
	100	1480			3216.0

Table 12-13

$$\bar{x} = 1480/100 = 14.8$$
$$s^2 = 3216/100 = 32.16$$

The standard deviation is $\sqrt{32.16}$, or approximately 5.67.

Theorem 12-2: If $y = (x - a)/b$, so $x = by + a$, where a and b are any real numbers, then the variance of the x's equals b^2 times the variance of the y's, or

$$s_x{}^2 = b^2 s_y{}^2.$$

The proof of this theorem is a straightforward calculation and is left as an exercise.

The computation of the variance in the preceding example can be simplified by using Theorem 12-2. Let

$$y = \frac{x - 4}{4}.$$

x	$y = \frac{x-4}{4}$	f	yf	$y - \bar{y}$	$(y - \bar{y})^2 f$
4	0	10	0	-2.7	72.9
8	1	10	10	-1.7	28.9
12	2	20	40	- .7	9.8
16	3	30	90	.3	2.7
20	4	20	80	1.3	33.8
24	5	10	50	2.3	52.9
		100	270		201.0

Table 12-14

$$\begin{aligned} \bar{y} &= 270/100 = 2.7 \\ s_y{}^2 &= 201/100 = 2.01 \\ s_x{}^2 &= 16s_y{}^2 = 16(2.01) = 32.16 \\ s_x &= 5.67 \end{aligned}$$

One advantage of the standard deviation over other measures of dispersion is that when sets of observations are combined in various ways, the standard deviation of the combined data can be calculated from the means and the standard deviations of the original data. We note also in the next section that standard deviation plays a very special role in the utilization of what is called the normal distribution.

Exercise Set 12-5

1. For each of the given sets of observations, find the standard deviation.
   (a) 16, 14, 12, 13, 15, 18, 24, 8, 10, 4
   (b) 18, 47, 64, 32, 41, 92, 84, 27, 14, 12
   (c) 9, 7, 16, 14, 12, 13, 14, 18, 24, 8, 10, 4
   (d) 10, 17, 18, 47, 64, 32, 41, 92, 84, 27, 14, 12
2.

x	10	14	18	22
f	4	6	8	2

   (a) Compute $s_x{}^2$.
   (b) Calculate $s_x{}^2$ by first finding $s_y{}^2$ from

$$y = \frac{x - 10}{4}.$$

3. Find the standard deviations for the following problems.
   (a) The following table shows the distribution of scores on a test administered to freshmen at Laneville College.

Score	*Frequency*
140–149	3
130–139	4
120–129	8
110–119	13
100–109	4
90–99	2
80–89	0
70–79	1

(b) A poll taker tabulated the ages of 30 users of a vitamin pill designed to make one feel young. The results are shown in the table below.

Age	*Frequency*
20–29	1
30–39	2
40–49	4
50–59	5
60–69	9
70–79	6
80–89	3

(c) The distance in miles that students at Alabaster College live from home (discounting those who live under 100 miles) is given by the following table.

Distance in Miles	*Frequency*
100–119	2
120–139	3
140–159	4
160–179	6
180–199	21
200–219	18
220–239	20
240–259	18
260–279	20
280–299	8

(d) Alabaster College has 1426 students. The following table indicates their ages, recorded to the nearest year.

Age	*Number of Students*
15–19	562
20–24	450
25–29	350
30–34	58
35–39	6

(e) The following table gives the frequency of the number of absences per semester of students in a psychology course.

Number of Absences	*Frequency*
0	200
1	180
2	86
3	49
4	34
5	14
6	13
7	12
8	8
9	3
10	1

4. Work Exercise 3(a) using $y = (x - 114.5)/10$.
5. Work Exercise 3(b) using $y = (x - 54.5)/10$.
6. Work Exericse 3(c) using $y = (x - 189.5)/20$.
7. Work Exercise 3(d) using $y = (x - 17)/5$.
8. Using $y = x - 5$, work Exercise 3(e).
9. Prove Theorem 12-2.
10. Find the standard deviations for the following problems in Exercise Set 12-2.
    (a) Exercise 1 (b) Exercise 2
    (c) Exercise 3 (d) Exercise 4
    (e) Exercise 5 (f) Exercise 6

*6

The Normal Frequency Curve

If one constructs several frequency polygons from data accumulated in the physical, social, and biological sciences, it is evident that many of the figures obtained are "bell-shaped" and are somewhat symmetrical about the mean. These distributions have the form of the *normal distribution*, a well-known distribution in the field of statistics.

The standard normal distribution has a graph shaped like the curve in Figure 12-9. We use t to represent the standard normal variable and

y to represent the frequency. The maximum value of the curve is attained at $t = 0$. The normal curve has perfect symmetry. (See Figure 12-10.) Because of this characteristic, the mean, median, and

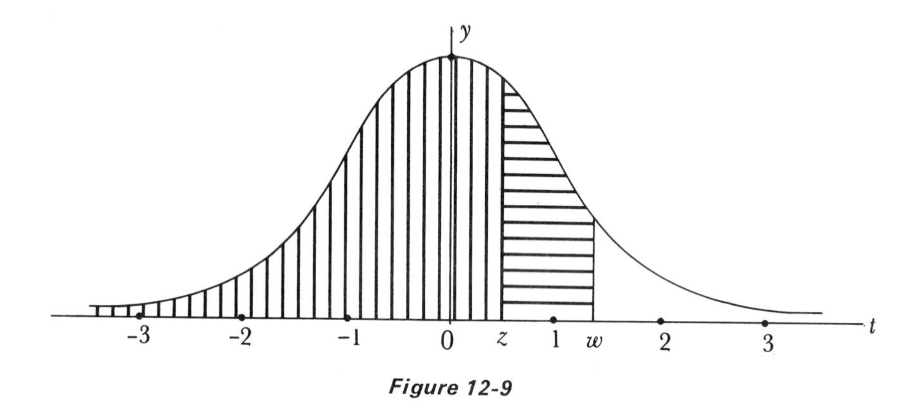

Figure 12-9

mode of the distribution will all have the same value, m. The range is not defined because there are values occurring as far out as you wish to go—that is, the curve never comes down to the axis.

Standard deviation is very important relative to the normal curve. Figure 12-10 indicates that 68% of the values of the variable are within one standard deviation of the mean, 95% are within two standard deviations, and 99% are within three standard deviations.

To gain a better understanding of continuous curves, such as the standard normal distribution, consider the set of data in Table 12-15. We divide each frequency by the total frequency, 20, to obtain what is called the *relative frequency*.

Class	*Frequency*	*Relative Frequency*
10–14	2	.10
15–19	4	.20
20–24	7	.35
25–29	5	.25
30–34	2	.10

Table 12-15

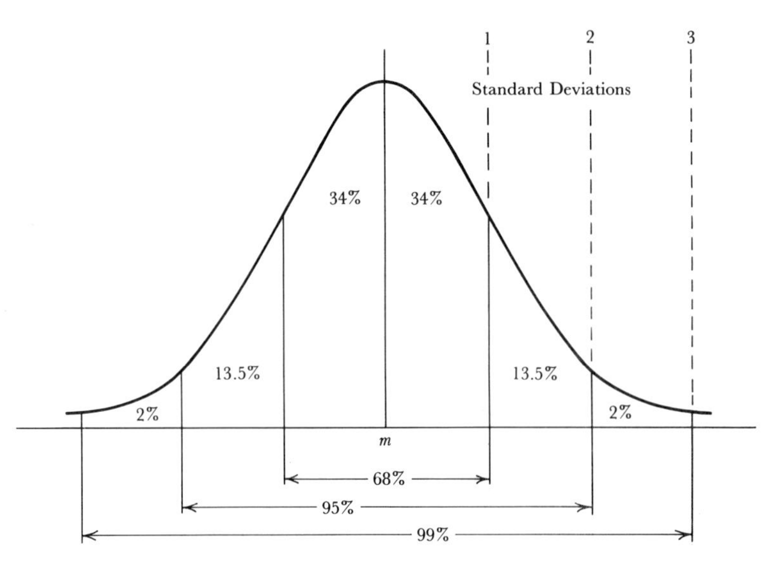

Figure 12-10

Using the relative frequency concept of probability as discussed in the preceding chapter, the probability that an observation is in the class 20–24 would be .35. The probability of an observation being in class 15–19 would be .20, and so on.

Now we will draw a histogram representing the relative frequency in Table 12-15. Suppose we select the horizontal scale such that the width of the rectangle associated with each class is one; the height of each rectangle is the relative frequency of a class (Figure 12-11).

Note that the area of each rectangle is the probability that an observation occurs in a given class, and note also that the sum of the areas of all the rectangles is 1.

Now draw a continuous curve to approximate the histogram in Figure 12-11. Assume that the area under this curve is likewise 1. Since the shaded area under the continuous curve is an approximation of the area of rectangle *ABCD*, and since the area of rectangle *ABCD* gives the probability that an observation falls in class 15–19, the shaded area under the curve gives an approximation to the probability that an observation falls in class 15–19. This intuitive approach should assist in understanding some of the important properties of the standard normal distribution.

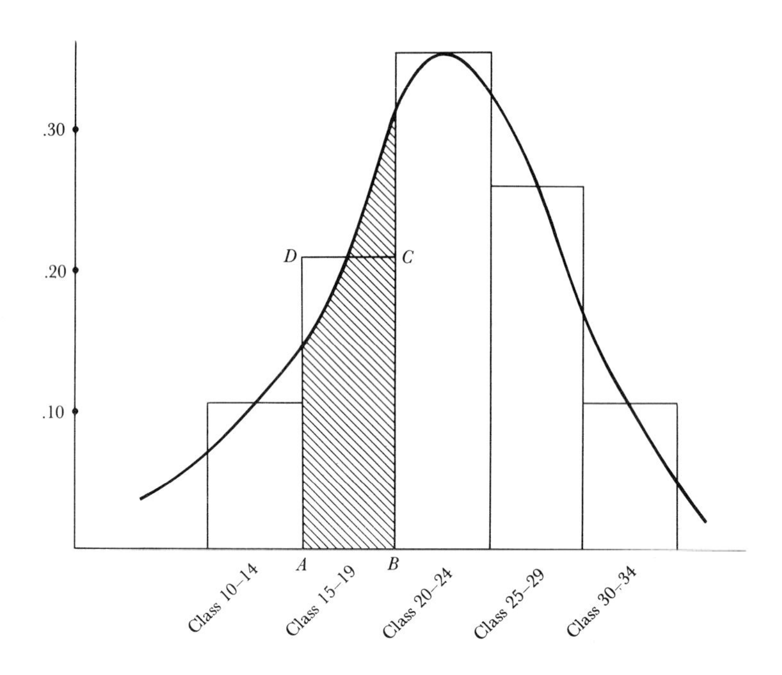

Figure 12-11

The area under the standard normal curve is 1; thus, to find the probability that t is between z and w, one obtains the area under the curve between z and w (horizontally shaded region in Figure 12-9). Table 12-16 gives the area under the normal curve less than or equal to z (vertically shaded in Figure 12-9). A property of this curve is that the area less than z is the same as the area less than or equal to z. The A, or area, in this table is actually the probability that t is less than or equal to z, which is the same as the probability that t is less than z.

To obtain the probability that $t \leq z$ for $z > 0$, one obtains from Table 12-16 the area under the curve for $t \leq z$. For example, to find $P(t \leq 2)$, one notes in the table that this probability is 0.9772. In a like manner, $P(t \leq 1.7) = 0.9554$. Remember that $P(t < 1.7)$ is also 0.9554.

Example: Find $P(t \leq 1.2)$.

From Table 12-16 we read that the value of A for $z = 1.2$ is 0.8849. Thus, $P(t \leq 1.2) = 0.8849$.

Area under Normal Curve Less Than or Equal To z

z	A	z	A
0.1	0.5398	1.6	0.9452
0.2	0.5793	1.7	0.9554
0.3	0.6179	1.8	0.9641
0.4	0.6554	1.9	0.9713
0.5	0.6915	2.0	0.9772
0.6	0.7257	2.1	0.9821
0.7	0.7580	2.2	0.9861
0.8	0.7881	2.3	0.9893
0.9	0.8159	2.4	0.9918
1.0	0.8413	2.5	0.9938
1.1	0.8643	2.6	0.9953
1.2	0.8849	2.7	0.9965
1.3	0.9032	2.8	0.9974
1.4	0.9192	2.9	0.9981
1.5	0.9332	3.0	0.9987

Table 12-16

Recall that the normal curve is symmetrical about $t = 0$. This fact is important as we discuss areas under the curve. For example, the area under the curve to the right of $t = 2$ is the same as the area under the curve to the left of $t = -2$.

Example: Find $P(t \geq -1)$.

Since the curve is symmetrical, the area above $t = -1$ is the same as the area below $t = 1$. Thus, $P(t \geq -1) = P(t \leq 1) = 0.8413$.

Since the total area under the curve is 1, the area above $t = a$ would equal 1 minus the area below $t = a$ for all real a.

Example: Find $P(t > 2.3)$.

$$\begin{aligned} P(t > 2.3) &= 1 - P(t \leq 2.3) \\ &= 1 - 0.9893 \\ &= 0.0107 \end{aligned}$$

Example: From Table 12-16, find $P(t \leq -0.4)$.

$$\begin{aligned} P(t \leq -0.4) &= P(t \geq 0.4) \\ &= 1 - P(t < 0.4) \\ &= 1 - 0.6554 \\ &= 0.3446 \end{aligned}$$

Sometimes one needs to compute the probability that t is in a certain range, say between .4 and 1.4. This is indicated by $P(0.4 < t < 1.4)$. This can be obtained by finding the probability that t is less than 1.4 and subtracting from this the probability that t is less than 0.4.

$$\begin{aligned} P(0.4 < t < 1.4) &= P(t < 1.4) - P(t < 0.4) \\ &= 0.9192 - 0.6554 \\ &= 0.2638 \end{aligned}$$

The standard normal curve has no units such as pounds or inches. For normal curves involving units, a transformation is necessary to change the variable to the variable of the standard normal curve.

If x is a measurement having a mean $\bar{x}$ and a standard deviation s, then $t = (x - \bar{x})/s$ gives the value in what is called *standard units.* Thus, if x is normally distributed with mean $\bar{x}$ and standard deviation s, then $t = (x - \bar{x})/s$ is the variable of the standard normal curve.

Example: Find the probability that the normal variable x, with mean 175 and standard deviation 20, is less than or equal to 215.

$$t = \frac{215 - 175}{20} = 2$$

$$P(x \leq 215) = P(t \leq 2) = 0.9772$$

Example: The grades on a certain test are known to be normally distributed with mean 74 and standard deviation 8. What is the probability that a student will make less than 58 on this test?

$$t = \frac{58 - 74}{8} = -2$$

$$\begin{aligned} P(x < 58) &= P(t < -2) \\ &= P(t > 2) \\ &= 1 - P(t \leq 2) \\ &= 1 - 0.9772 \\ &= 0.0228 \end{aligned}$$

Thus, 2.28% of the students will make less than 58 on the test; equivalently, the probability that a student chosen at random will make less than 58 is 0.0228.

Example: Let x be normally distributed with mean 10 and variance 4. Find the probability that x lies between 11 and 14.

First, we find the t that corresponds to 11 and 14.

$$t = \frac{x - \bar{x}}{s} = \frac{11 - 10}{2} = 0.5$$

$$t = \frac{x - \bar{x}}{s} = \frac{14 - 10}{2} = 2$$

Thus,

$$\begin{aligned} P(11 < x < 14) &= P(0.5 < t < 2) \\ &= P(t < 2) - P(t < 0.5) \\ &= 0.9772 - 0.6915 \\ &= 0.2857. \end{aligned}$$

If x lies between $\bar{x} - s$ and $\bar{x} + s$, then

$$t = \frac{\bar{x} - s - \bar{x}}{s} = -1 \quad \text{and} \quad t = \frac{\bar{x} + s - \bar{x}}{s} = 1.$$

$$\begin{aligned} P(-1 \le t \le 1) = P(t \le 1) - P(t \le -1) &= P(t \le 1) - [1 - P(t < 1)] \\ &= 0.8413 - 0.1587 = 0.6826. \end{aligned}$$

Since $P(-2 \le t \le 2) = 0.9544$ and $P(-3 \le t \le 3) = 0.9974$, the following summary may be made for any normal distribution.

If measurements are normally distributed with mean $\bar{x}$ and standard deviation s, then

68.26% of the measurements deviate less than 1 s from $\bar{x}$.
95.44% of the measurements deviate less than 2 s from $\bar{x}$.
99.74% of the measurements deviate less than 3 s from $\bar{x}$.

Example: It is found that the test grades on a certain test follow a normal curve. If grades of A or F are given to students whose scores deviate more than 1.6 standard deviations from the mean, what proportion of the students receive grades of B, C, or D?

The dividing line between A's and B's is 1.6 standard deviations above the mean, or $\bar{x} + 1.6s$, and the dividing line for F's is $\bar{x} - 1.6s$.

$$t = \frac{\bar{x} + 1.6s - \bar{x}}{s} = 1.6 \quad \text{and} \quad t = \frac{\bar{x} - 1.6s - \bar{x}}{s} = -1.6.$$

$$\begin{aligned} P(-1.6 < t < 1.6) &= P(t < 1.6) - P(t < -1.6) \\ &= 0.9452 - P(t > 1.6) \\ &= 0.9452 - [1 - P(t \leq 1.6)] \\ &= 0.9452 - [1 - 0.9452] \\ &= 0.8904. \end{aligned}$$

Hence, 89% of the grades will be B's, C's, or D's and 11 % of the grades will be A's or F's.

Example: The semester grades given by the mathematics department at Rowdy College follow a normal curve. If 0.838 of the grades are B's, C's, or D's, how many standard deviations from the mean give the dividing lines between A's and B's and between D's and F's?

Since the middle area (symmetrical) under the standard normal curve is 0.838, the sum of the two tails would be $1 - 0.838 = 0.162$. Since the two tails have equal areas (symmetrical curve), each tail is of area 0.081. The area of the left tail plus the center area would be $0.838 + 0.081 = 0.919$. $P(t \leq 1.4)$ is 0.9192. Thus, 1.4 standard deviations above the mean and 1.4 standard deviations below the mean are the dividing lines between A's and B's and between D's and F's, respectively.

Exercise Set 12-6

1. Find the following probabilities from Table 12-16.
   (a) $P(t \leq 1.5)$ (b) $P(t \leq 2.6)$
   (c) $P(t \leq -.4)$ (d) $P(t < -1.6)$
   (e) $P(t > 1.5)$ (f) $P(t > 2.4)$
   (g) $P(t > -2.1)$ (h) $P(t > -1.8)$
   (i) $P(1.3 \leq t \leq 2.4)$ (j) $P(2.1 < t < 2.8)$
   (k) $P(-1.2 \leq t \leq .3)$ (l) $P(-2.6 \leq t \leq 1.4)$
2. If x is a variable having a normal distribution with $\bar{x} = 12$ and $s = 4$, find the probability that
   (a) $x \leq 16$ (b) $x \geq 10$ (c) $10 \leq x \leq 14$
   (d) $8 \leq x \leq 16$ (e) $x \geq -4$

3. If x is normally distributed with $\bar{x} = 1$ and $s = 2$, find the following.
   (a) $P(x > 4)$ (b) $P(0 \leq x \leq 3)$
4. If x is normally distributed with $\bar{x} = 1$ and $s = 2$, find a number a for (a)–(f) below.
   (a) $P(x > a) = 0.1151$ (b) $P(x > -a) = 0.8159$
   (c) $P(x < a) = 0.9965$ (d) $P(x < -a) = 0.2743$
   (e) $P(1 - a < x < 1 + a) = 0.8664$
   (f) $P(1 - a < x < 1 + a) = 0.3830$
5. It is known from experience that the number of telephone orders made daily to a company approximates a normal curve with mean 350 and standard deviation 20. What percentage of the time will there be more than 400 telephone orders per day?
6. A large set of measurements is closely approximated by a normal curve with mean 30 and standard deviation 4.
   (a) Find what percentage of the measurements can be expected to lie in the interval from 20 to 32.
   (b) Find the probability that a measurement will differ from the mean by more than 5.
7. The grades in a certain class are normally distributed with mean 76 and standard deviation of six. 61 is the lowest D, 70 the lowest C, 82 the lowest B, and 91 the lowest A. What percentage of the class will make A's? What percent B's? What percent C's? D's? F's?
8. It is decided to change the grading scale in this class so as to have 4.5 percent A's. 9.1 percent B's, 72.9 percent C's, 11.7 percent D's, and 1.8 percent F's. What is the new grading scale?

*7

Correlation Coefficient

Consider the ordered pairs of scores in Table 12-17. These scores exhibit a relation between the sets of scores on a mathematics aptitude test (M) and the scores on a final examination in statistics (F).

M	40	30	30	30	20	20	20	10
F	100	80	85	84	72	60	65	30

Table 12-17

A casual glance at the table suggests that a high score on the mathematics aptitude test is associated with a rather high score on the final examination; a low score on the aptitude seems to correspond to a low score on the final examination.

We need a more precise method of determining whether or not there is a connection or relationship between two variables beyond that which seems apparent from a glance at or study of a table. The *correlation coefficient* provides a means for determining whether a linear relationship exists between two variables. That is, if the ordered pairs of points are plotted in the same manner that points were plotted for the frequency polygon, the correlation coefficient measures whether or not there is some straight line such that the plotted points are located reasonably close to the line in contrast to being scattered away from the line.

How do we determine this indicator of the linear relationship between two variables x and y, when the linear relationship is indicated by pairs of values (x, y) of observations? The measure of this linear relationship, denoted by r, can be obtained from the following formula.

Definition 12-6: The *correlation coefficient* of a set of observations (x, y) is given by

$$r = \frac{(x_1 - \bar{x})(y_1 - \bar{y})f_1 + (x_2 - \bar{x})(y_2 - \bar{y})f_2 + \cdots + (x_m - \bar{x})(y_m - \bar{y})f_m}{Ns_x s_y}$$

where $N = f_1 + f_2 + f_3 + \cdots + f_m$.

If the frequency for each pair of variables is 1, this formula can be written as

$$r = \frac{(x_1 - \bar{x})(y_1 - \bar{y}) + (x_2 - \bar{x})(y_2 - \bar{y}) + \cdots + (x_n - \bar{x})(y_n - \bar{y})}{ns_x s_y}$$

For the pairs of scores given in Table 12-17, the correlation coefficient may be computed as shown in Table 12-18.

$$\bar{x} = \frac{200}{8} = 25 \qquad \bar{y} = \frac{576}{8} = 72$$

$$s_x^2 = \frac{600}{8} = 75 \qquad s_y^2 = \frac{3118}{8} = 389.75$$

$$s_x = 8.66 \qquad s_y = 19.74$$

$$r = \frac{1310}{8(8.66)(19.74)} = 0.90$$

x	y	$x-25$	$y-72$	$(x-25)^2$	$(y-72)^2$	$(x-25)(y-72)$
40	100	15	28	225	784	420
30	80	5	8	25	64	40
30	85	5	13	25	169	65
30	84	5	12	25	144	60
20	72	-5	0	25	0	0
20	60	-5	-12	25	144	60
20	65	-5	-7	25	49	35
10	30	-15	-42	225	1764	630
200	576			600	3118	1310

Table 12-18

The correlation coefficient is a very useful characteristic that can be obtained from a set of observations (x, y). Without proof, we make some comments concerning this statistical concept. The correlation coefficient is always less than or equal to 1 and is also greater than or equal to -1. If r is 1 or -1, all the points, as plotted, lie on some straight line. If r is 0, the points are so scattered that the relationship between x and y is not linear. When r is positive, y increases with x; when r is negative, y decreases with increasing x.

The values of r falling between 0 and 1 or between 0 and -1 are more difficult to explain. Some people erroneously think that an r of 0.90 is "twice as good" as an r of 0.45 and "three times as good" as an r of 0.30. An r of 0.90 means that $(0.90)^2 = 0.81$ of the scattering in the y's is due to changes in the x's. An r of 0.45 means that $(0.45)^2$, or approximately 0.20, of the scattering in the y's is due to changes in the x's. Thus, an r of 0.90 indicates that four times as much of the scattering of the y's is accounted for by changes in the x's as is accounted for by an r of 0.45.

The correlation coefficient is widely used, and it is also widely abused. Many people use the correlation coefficient as a measure of a cause-effect relationship. x does not necessarily cause y to vary in a given manner. r simply describes the linear relation (points scattered about a straight line) between the two variables.

Example: Consider the following pairs of x and y that occur with given frequencies in a set of observations. Find the correlation coefficient.

(x, y)	(1, 20)	(2, 23)	(3, 30)	(4, 33)	(5, 34)
f	7	8	9	6	8

x	y	f	xf	yf	$x-\bar{x}$	$y-\bar{y}$	$(x-\bar{x})(y-\bar{y})f$	$(x-\bar{x})^2f$	$(y-\bar{y})^2f$
1	20	7	7	140	-2	-8	112	28	448
2	23	8	16	184	-1	-5	40	8	200
3	30	9	27	270	0	2	0	0	36
4	33	6	24	198	1	5	30	6	150
5	34	8	40	272	2	6	96	32	288
		38	114	1064			278	74	1122

Table 12-19

$$\bar{x} = 114/38 = 3 \quad \text{and} \quad \bar{y} = 1064/38 = 28$$

$$s_x^2 = \frac{74}{38} = 1.95 \qquad s_y^2 = \frac{1122}{38} = 29.53$$

$$s_x = 1.40 \qquad s_y = 5.43$$

$$r = \frac{278}{38(1.40)(5.43)} = 0.96$$

Since r is 0.96, there is a linear relationship between the variables.

Exercise Set 12-7

For the pairs of values in Exercises 1–3, compute
(a) $\bar{x}$, (b) $\bar{y}$, (c) s_x^2, (d) s_y^2, (e) r.

1.

x	6	8	10	12	14	16
y	1	8	10	15	20	24

2.

x	6	9	10	14
y	100	102	107	109

3.

x	18	40	72	100
y	140	200	250	300

4. For the following data, find
   (a) $\bar{x}$, (b) $\bar{y}$, (c) $s_x^{\,2}$, (d) $s_y^{\,2}$, (e) r.

x/y	20	30	40
1	1	6	8
5	8	10	12
9	2	4	5

(The numbers inside the table represent frequencies. For example, the $x = 1, y = 20$ has a frequency of 1; $x = 5, y = 30$ has a frequency of 10; and $x = 9, y = 40$ has a frequency of 5.)

5. Obtain the correlation coefficient for the relationship between the scores on a mathematics aptitude test and the scores on a final examination in statistics, as given in Table 12-17, by subtracting 10 from each aptitude score and 60 from each final examination score.

*6. Find r for the data given in the following tables.

(a)

	0–49	50–99	100–149
0–4	1	1	
5–9		2	3
10–14		3	4
15–19			5

(b)

	10–16	17–23	24–30	30–36
0–4	1			
5–9	1	3	4	
10–14		4	5	1
15–19				4

*7. In Exercise 2, subtract 10 from each x and 100 from each y. Find a new r. How does it compare with the old r?

*8. In Exercise 4, let $u = \dfrac{x - 5}{4}$ and $v = \dfrac{y - 20}{10}$. Find r_{uv} (relationship between u and v). How does it compare with r_{xy}, found in Exercise 4?

*9. If $u = \dfrac{x - c}{a}$ and $v = \dfrac{y - d}{e}$, find a relationship between r_{xy} and r_{uv}.

8

Misuses of Statistics

We conclude this short chapter on an introduction to statistics with some examples of the misuse of statistics. You probably have heard the statement "There are liars, damn liars, and there are statisticians." This statement emphasizes the point that some people who quote statistics give results that are misleading. Some of these abuses are indicated by Darrell Huff in his book, *How to Lie with Statistics.* Consider the following examples of the misuse of statistics.

One misuse of statistics involves the shifting of the definition to suit your purpose. For example, in 1949 a Russian assertion stated that there were 14 million unemployed in the United States. This number far exceeded the number of unemployed as given by the United States Bureau of Labor. What was the discrepancy? The Russian figure involved everyone who worked less than 40 hours a week, while the American figure represented those who were not employed. Thus, by changing a definition one can manipulate a statistic to prove an argument. In accurate statistical statements, all terms should be completely defined.

A second misuse of statistics is due to inaccurate measurement or tabulation of data. Many opinion-polling organizations have difficulty in securing personnel to take opinions and record data accurately.

A third misuse is due to the method of selecting the group from which the information is to be obtained. For example, each child in a school system was asked how many children are in his family. From these data, the mean number of children in a family was obtained. The answer was much too large. Why? The answer was too large because a family with seven in school occurred seven times in the survey, whereas a family of one child with only one in school occurred only once. The method of taking the survey was weighted in favor of large families.

Undoubtedly, the most prevalent misuse of statistics involves inappropriate comparisons. A toothpaste commercial states that a group using Droat toothpaste has 25% fewer cavities than a group using other toothpastes. Who knows the composition of the other toothpaste preparations? Maybe they include ingredients which will cause cavities. Or maybe the people using Droat toothpaste have fewer teeth, hence fewer cavities.

Another misuse is due to the failure to include representative groups within a sample. Ronald Clodhopper polled his neighbors and found the average salary was $40,000. We certainly would not conclude that the average salary in the United States was $40,000—especially since Ronald lived in an exclusive suburb of a prosperous metropolitan area.

Of course, we have already learned that an average without some indication of dispersion can be misleading. The mean temperature in a particular city is 86 degrees. Want to move there ? We hope not, because it is exceedingly hot in summer and exceedingly cold in winter. Don't be misled by averages!

If you get nothing else out of this chapter, we hope you take time to analyze statements involving statistics. Such analysis is the answer of an educated person to the misuse of statistics.

Review Exercise Set 12-8

1. A city survey was made to determine the average family earnings. This information was obtained from all the employers' records in the city. Would the result be misleading? Why?
2. The president of a university quoted that the enrollment was up 4%. However, the registrar said it was down 1%. Give one reason for this discrepancy.
3. One-third of the women students at Lonesome University became pregnant last year. Can you think of reasons why this statement might be misleading?
4. The mean salary of all employees at the Brown Corporation is $30,000. Make up an example to show how this statistic may be misleading.
5. Find the mean, median, and standard deviation for the following observations: 104, 106, 101, 102, 110, 108, 111, 116.
6. Find the mean, median, standard deviation, and modal class for the following frequency distributions.

x	10	20	30	40	50
f	4	6	8	4	3

7. Find r for the following distribution.

x	1	2	3	4	5	6	7	8	9	10
y	4	7	8	10	12	14	18	18	20	22

*8. Find r for the data given in the following table.

y/x	8	9	10
1	7	10	4
2	6	0	8
3	10	11	5

Suggested Reading

Averages: Cooley, Wahlert, pp. 358–361. Fujii, pp. 470–472. Graham, pp. 251, 358–363. Jacobs, pp. 406–410. Moore, Little, pp. 148–154. Wheeler, Peeples, pp. 438–443. Wren, pp. 298–300.

Correlation Coefficient: Cooley, Wahlert, p. 374. Moore, Little, pp. 164–166. Wheeler, Peeples, pp. 460–463.

Dispersion: Cooley, Wahlert, pp. 362–366. Fujii, pp. 474–479. Jacobs, pp. 415–417. Moore, Little, pp. 156–163. Wheeler, Peeples, pp. 444–448. Wren, pp. 300–303.

Frequency Distributions and Graphs: Cooley, Wahlert, pp. 355–357. Graham, pp. 347–357. Jacobs, pp. 392–398, 422–430. Moore, Little, pp. 153–154. Wheeler, Peeples, pp. 432–437. Wren, pp. 348–353.

Normal Distribution: Cooley, Wahlert, pp. 367–369. Jacobs, pp. 418, 439. Moore, Little, pp. 167–170. Wheeler, Peeples, pp. 472–476.

13

Mathematical Systems

1

Introduction

Throughout the preceding chapters, references have been made to *mathematical systems*. You recall that the set of whole numbers was defined along with the operations of addition and multiplication. From these definitions, the properties of the whole numbers were developed. Thus, we said that the whole numbers, together with the defined operations and developed properties, constitute a mathematical system.

The integers also constitute a mathematical system. To obtain the integers, we started with a set of elements and then defined some operations, later developing a number of interesting properties. In Chapter 7, we discussed the system of rational numbers that we called a field; and in Chapter 8, attention was given to the real number system. Now we wish to give additional attention to the meaning of a mathematical system and to discuss the structure of a system. In this chapter, we will emphasize *finite* mathematical systems, or systems having a finite number of elements.

2

Clock Arithmetic

To help us understand the abstract idea of a mathematical system, we now introduce an example of a finite mathematical system called *clock arithmetic*.

The numbers 1, 2, 3, ..., 12 on the face of a clock are used to tell us the time of day. We classify these numbers as *clock numbers*. When one considers future time, one adds a fixed number of hours to the present time. The time is now 9:00 A.M. and you have an appointment in four hours. Four hours after 9:00 A.M. is 1:00 P.M. Thus, 9:00 A.M. $\oplus$ 4 hours = 1 : 00 P.M., where $\oplus$ is used as the symbol for clock addition.

In the same way, past events are denoted by subtraction. We had breakfast three hours ago. The time is now 8:00 A.M. At what time did we have breakfast? Letting the symbol $\ominus$ denote clock subtraction, $8 \ominus 3 = 5$. Thus, we had breakfast at 5:00 A.M.

If we were unfamiliar with time as given by a 12-hour clock, this new type of arithmetic might be difficult. However, we can easily verify that each of the following is correct.

Example: Add 27 hours to 7:00 A.M. $7 \oplus 27 = 10$. The clock time is 10:00 A.M. Did you get this answer?

Example: It takes 7 hours to drive to your house. You wish to arrive by 3:00 P.M. What time should you start? $3 \ominus 7 = 8$, or you should start by 8:00 A.M.

Let us now consider the numbers 1 through 12 as the elements of a set and consider addition to be based on counting in a clockwise direction. Forgetting any A.M. or P.M. designations, we make a table of addition facts for a 12-hour clock, as shown in Table 13-1. Note that

$\oplus$	1	2	3	4	5	6	7	8	9	10	11	12
1	2	3	4	5	6	7	8	9	10	11	12	1
2	3	4	5	6	7	8	9	10	11	12	1	2
3	4	5	6	7	8	9	10	11	12	1	2	3
4	5	6	7	8	9	10	11	12	1	2	3	4
5	6	7	8	9	10	11	12	1	2	3	4	5
6	7	8	9	10	11	12	1	2	3	4	5	6
7	8	9	10	11	12	1	2	3	4	5	6	7
8	9	10	11	12	1	2	3	4	5	6	7	8
9	10	11	12	1	2	3	4	5	6	7	8	9
10	11	12	1	2	3	4	5	6	7	8	9	10
11	12	1	2	3	4	5	6	7	8	9	10	11
12	1	2	3	4	5	6	7	8	9	10	11	12

Table 13-1

no matter where we start on the clock, we are always at the starting position 12 hours later. Thus, for any element b of the set, $b + 12 = b$.

The following examples provide interesting illustrations of clock arithmetic.

Example: Solve $t + 8 = 2$. In the addition table, we note that $6 + 8 = 2$; therefore, $t = 6$.

Example: $6 - 2t = 8$. To find t, we first find what number added to 8 gives an answer of 6. The answer is 10. Thus, $2t = 10$, and $t = 5$.

The construction of a finite mathematical system can also be illustrated by a four-minute clock. The sketch in Figure 13-1 represents the face of a four-minute clock. Such a clock might be used to time the rounds and intermissions in a boxing match.

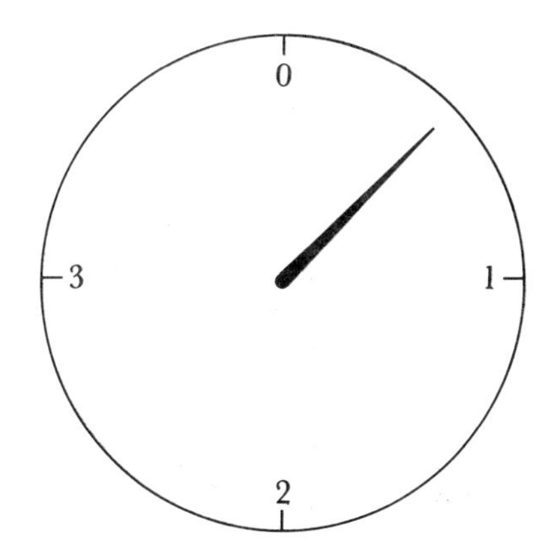

Figure 13-1

Notice that this mathematical system will contain only four numbers—0, 1, 2, and 3. Addition in this system (denoted by $\triangle$) may seem a little strange at first, for addition is defined and interpreted to mean that the hand of the clock moves the same number of positions as the number to be added. If the hand is at 1 and moves for 2 minutes, then it will be at 3, so $1 \triangle 2 = 3$. In a similar manner, if the hand is at 2 and it moves for 2 minutes, then it will be at 0, so $2 \triangle 2 = 0$. Finally, suppose the hand is at 3 and it moves for 2 minutes. Then it will be at 1, so $3 \triangle 2 = 1$.

A complete definition of the operation of addition for this system can be displayed easily by a table. You should check the entries in Table 13-2 by visualizing the movement of the hand of the four-minute clock.

$\triangle$	0	1	2	3
0	0	1	2	3
1	1	2	3	0
2	2	3	0	1
3	3	0	1	2

Table 13-2

Since the body of the table contains no entries that are not members of the set {0, 1, 2, 3}, and since there is an entry for every position in the table, then the set is closed under the operation $\triangle$. Stated differently, the sum of any two of the four elements contained in this set is one of the four elements of the set.

The commutative property of addition is satisfied by this mathematical system. A statement of the commutative property may be given as $a \triangle b = b \triangle a$ for all elements a and b of the set. As an example of this property, compare $2 \triangle 3$ and $3 \triangle 2$. The first sum is determined by starting at 2 and proceeding three units in a clockwise direction. The second sum starts at 3 and proceeds two units in a clockwise direction. In each case, the sum is 1.

$(a \triangle b) \triangle c = a \triangle (b \triangle c)$ for all elements a, b, and c of the set. As an example, compare $(2 \triangle 3) \triangle 1$ and $2 \triangle (3 \triangle 1)$.

$$(2 \triangle 3) \triangle 1 = 1 \triangle 1 = 2$$
$$2 \triangle (3 \triangle 1) = 2 \triangle 0 = 2$$

Thus,

$$(2 \triangle 3) \triangle 1 = 2 \triangle (3 \triangle 1).$$

As another example,

$$(2 \triangle 1) \triangle 2 = 3 \triangle 2 = 1$$

and

$$2 \triangle (1 \triangle 2) = 2 \triangle 3 = 1$$

so that

$$(2 \triangle 1) \triangle 2 = 2 \triangle (1 \triangle 2).$$

These examples show that the associative property of addition holds for the selected elements of the set. By continuing this process, one could test all 64 possible combinations of triples of elements of the set and thus verify that the associative property of addition holds.

Note that 0 plays the same role in this system as in the system of integers because $0 \triangle n = n \triangle 0 = n$, where $n = 0$, 1, 2, and 3. Because of this relationship, the number 0 will be called the additive identity of the system. A statement of the identity property of this set of elements may be given as follows:

There exists an element a such that for all elements b of the set,

$$a \triangle b = b \triangle a = b.$$

In general, it is easy to discover the identity element in a table that defines an operation. The (horizontal) row that has the identity element for a heading will be identical to the (vertical) column headings of the table. Similarly, the column that has the identity element for a heading will be identical to the row headings of the table. Note that 12 is the identity in Table 13-1.

In a similar manner, the inverses of the elements of a set relative to an operation are easy to discover from a table. Since the operation involving an element and its inverse produces the identity, one can find the identities in the body of a table and then determine the two elements that when paired by the operation yield the identity for an answer. For the example given, the additive identity is 0; and the pairs of numbers that may be added to give 0 are 3 and 1, 2 and 2, and 0 and 0. Thus 3 is the additive inverse of 1, and 1 is the additive inverse of 3; 2 is its own inverse; and 0 is its own inverse.

Hence, each element of this set has an additive inverse, defined as follows:

If a is an element of a mathematical system, then b is the inverse of a with respect to the operation $\triangle$ *if* $a \triangle b$ *equals* $b \triangle a$ *equals the identity of the system.*

The five properties satisfied by the operation of addition as defined for the four-minute clock characterize an important category of mathematical systems, and the study of such systems is an important branch of mathematics. Many mathematical systems are similar to clock arithmetic systems with one operation; we will now study the properties of this class of systems. First, we shall give a name to a mathematical system having these properties. This classification of mathematical systems is called a *group*.

Definition 13-1: A *group* is a mathematical system consisting of a set of elements and one binary operation satisfying the following properties.

(a) The set is closed under the operation.

(b) The elements of the set satisfy the associative property for the operation.

(c) There is a unique identity element for the operation on the set.

(d) Every element in the set has a unique inverse in the set relative to the operation.

It should be noted that the system may well have other properties in addition to these and still be a group. Many systems have operations that satisfy all the requirements to be a group and, in addition, satisfy the commutative property. Such groups are often called *Abelian groups* for the famous Norwegian mathematician, Niels Henrik Abel; however, they are also simply called *commutative groups*.

Let us now discuss the mathematical systems already developed. For example, the 12-hour clock and the four-minute clock are not only groups, they are also Abelian groups. Looking back to the mathematical systems discussed in the first eight chapters of this book, we find that the natural numbers cannot be classified as a group with respect to either addition or multiplication as an operation. Under addition, there is closure and associativity but no identity or inverses; under multiplication, there is closure, associativity, and an identity, but not all elements have inverses. Approximately the same statements could be made for the whole numbers. An exception would be the existence of the identity for addition.

The integers do constitute a group under the operation of addition. Under addition, there is closure, associativity, an additive identity, and additive inverses for all members of the set. In fact, integers form a commutative group under addition, since pairs of elements commute with respect to addition. The integers do not constitute a group under the operation of multiplication, since not all elements have inverses.

The rational number system is obviously a commutative group under the operation of addition. Do the rational numbers constitute a group under the operation of multiplication? The answer is *no*, since 0 has no inverse. However, we sometimes consider sets with one or more elements omitted. If we consider the rational numbers without the element 0, then this new set of numbers under the operation of multiplication does constitute a commutative group.

To conclude this discussion on clock arithmetic, it should be obvious now that we can construct a system with many different time periods—

a five-minute clock, a seven-minute clock, and so on. The tables representing the operations of these systems are similar to the one we constructed to define addition on a four-minute clock. Let us consider multiplication as defined on a four-minute clock. This operation is defined in Table 13-3.

$\triangle$	0	1	2	3
0	0	0	0	0
1	0	1	2	3
2	0	2	0	2
3	0	3	2	1

Table 13-3

For your own satisfaction, you will want to discuss the structure of this mathematical system relative to the defined operation of "clock-type" multiplication. Is the operation commutative and associative on the set? Do identities and inverses exist? Can this system be classified as a group?

This new kind of arithmetic we have been discussing is sometimes called *modular arithmetic* or *congruence arithmetic*. In the system associated with Tables 13-2 and 13-3, the number 4 is called the *modulus of the system*, and we say the operation on the system is defined modulo 4, written (mod 4). Of course, addition and multiplication tables can be written for (mod 3), (mod 7), and so forth. We can use any natural number as a modulus of a system, as we shall indicate in Section 3.

Exercise Set 13-1

1. Make an addition and multiplication table for a three-minute clock.
2. Consider the system defined by Table 13-3.
   (a) Is the operation commutative?
   (b) Is the operation associative?
   (c) Find the identity.
   (d) What is the inverse of 0? Of 1? Of 2? Of 3?
   (e) Can this system be classified as a group? A commutative group?

3. Construct addition and multiplication tables for a five-minute clock, and use the tables to answer the following questions.
   (a) Is the set of numbers closed under the operation of multiplication? Addition?
   (b) Is addition commutative? Associative?
   (c) Is multiplication commutative? Associative?
   (d) Find the additive identity; the multiplicative identity.
   (e) Find the additive inverses of 1, 2, and 3.
   (f) Find the multiplicative inverses of 2, 3, and 4.
   (g) Is the sum of two even numbers always even?
   (h) Is the sum of an even and odd number always odd?
   (i) Can this system be classified as a group under addition?
   (j) Can this system be classified as a group under multiplication?
4. Work Exercise 3 for a six-minute clock.
5. Work Exercise 3 for a seven-day clock. In other words, if today is Sunday, or the first day, then nine days from now will be day three, or Tuesday.
6. Find t for the following, where the answer is some numeral on a 12-hour clock.

   (a) $t = 8 + 7$ (b) $t = 5 - 8$
   (c) $t = 9 - 11$ (d) $t = (4)(7)$
   (e) $t = (3)(9)$ (f) $t = 2 - 7$
   (g) $t + 9 = 5$ (h) $2 - t = 11$
   (i) $\frac{t}{2} + 4 = 8$ (j) $(3t) - 4 = 5$
   (k) $t = 9 - 1$ (l) $t = 5 - 9$
   (m) $t = 3 \cdot 7$ (n) $t = 4 \cdot 5$
7· Work Exercise 6 for a 24-hour clock.
8. Determine if the set of real numbers is a group under addition; under multiplication. Also, determine if, in either case, it is an Abelian group.
*9. Determine if p, q, and r, with the operations $\odot$ and $*$ defined in the tables below, are groups or Abelian groups.

(a)

$\odot$	p	q
p	p	q
q	q	p

(b)

$*$	p	q	r
p	p	q	r
q	q	r	p
r	r	p	q

*10. Consider all rational numbers x such that $0 \leq x \leq 1$, with the operation $x * y = x + y - xy$. Is this a group? An Abelian group?

3

Modular Arithmetic

You should understand that the choice of the four-minute clock in Section 2 was purely arbitrary. Any clock could have served the same purpose. Actually, once an operation on a mathematical system is completely defined by a table, the physical model may be discarded; and the system is then studied from the table. Similar tables could have been constructed with or without reference to the clock. Mathematical systems of this sort are known as *modular arithmetics* or *congruences*.

Once again, this new type of number system probably seems very strange to you. Yet it is the kind of mathematics that we need to use for many machines with dials or controls. In addition, it is useful in any situation involving a finite set of numbers where the numbers may be repeated.

In this new type of mathematical system, emphasis is placed on the fact that two integers differ by a multiple of some natural number. Numbers that differ by multiples of a given natural number are said to be congruent modulo the natural number.

Definition 13-2: Two integers a and b are *congruent modulo m* (where m is a natural number) if and only if $m \mid (a - b)$.

This relationship is usually denoted by $a \equiv b \pmod{m}$ and is read "a is congruent to b (mod m)." Thus, two numbers are congruent modulo m if their differencc is divisible by m.

Examples:

$$10 \equiv 3 \pmod 7 \text{ because } 7 \mid (10 - 3)$$
$$14 \equiv 8 \pmod 6 \text{ because } 6 \mid (14 - 8)$$
$$1 \equiv 5 \pmod 4 \text{ because } 4 \mid (1 - 5)$$
$$10^2 \equiv 2 \pmod 7 \text{ because } 7 \mid (10^2 - 2)$$
$$6 \not\equiv 4 \pmod 3 \text{ because 3 does not divide } (6 - 4)$$

Theorem 13-1: Integers a and b are congruent modulo m if their remainders are equal when divided by m.

Proof: Let r and t be the respective remainders when a and b are divided by m. Then, by the division algorithm, $a = q_1 m + r$ and

$b = q_2 m + t$, where $0 \leq r < m$ and $0 \leq t < m$. Suppose $r \geq t$. Then

$$a - b = (q_1 - q_2)m + (r - t).$$

Since $0 \leq r < m$ and $0 \leq t < m$, then $0 \leq (r - t) < m$. Now, if $m \mid (a - b)$, then since $m \mid m$ we obtain $m \mid (r - t)$. But $0 \leq (r - t) < m$, so if $m \mid (r - t)$, then $r - t$ must be equal to 0, so $r = t$; thus, the remainders are the same if $m \mid (a - b)$. Conversely, if $r = t$, then $r - t = 0$, and so

$$a - b = (q_1 - q_2)m,$$

which implies that $m \mid (a - b)$. Thus, the definition of $a \equiv b \pmod{m}$ and the statement that the remainders are the same when a and b are divided by m are equivalent.

Examples: (a) $17 \equiv 35 \pmod{3}$ because 17 divided by 3 leaves a remainder of 2 and 35 divided by 3 also produces a remainder of 2.

(b) $701 \equiv 7001 \pmod{7}$ because, when the numbers are divided by 7, the remainders are each 1.

In particular, $a \equiv 0 \pmod{m}$ means that $m \mid a$; and, conversely, if $m \mid a$, then $a \equiv 0 \pmod{m}$.

Congruence is obviously a relation between two numbers. The question arises "Is it an equivalence relation?" Let us test the required properties to answer this question.

Theorem 13-2: Congruence modulo m is an equivalence relation.

Proof: (a) Reflexive: $a \equiv a \pmod{m}$ because $a - a = 0$ is obviously divisible by m.

(b) Symmetric: If $a \equiv b \pmod{m}$, then $b \equiv a \pmod{m}$ because if $a - b$ is divisible by m, then $b - a$ is divisible by m.

(c) Transitive: If $a \equiv b \pmod{m}$ and $b \equiv c \pmod{m}$, then $a \equiv c \pmod{m}$ because if a and b have the same remainders when divided by m and if b and c have the same remainders, then a and c must have the same remainders.

Addition in a system modulo m is clearly the same as addition of whole numbers except when the sum is greater than or equal to m. When the sum is greater than or equal to m, we divide the sum by m and use the remainder in place of the ordinary sum. Consider some examples involving addition (mod 5).

Example: $2 + 1 \equiv 3 \pmod 5$, $2 + 2 \equiv 4 \pmod 5$, $2 + 3 \equiv 0 \pmod 5$, and $2 + 4 \equiv 1 \pmod 5$. Did you check these additions? All the answers are the same as would be obtained using a five-minute clock.

Table 13-4 is a mod 5 addition table. You will recognize that the entries in the table are the same as the entries in the addition table in Exercise 3 of the preceding exercise set.

Addition Mod 5

+	0	1	2	3	4
0	0	1	2	3	4
1	1	2	3	4	0
2	2	3	4	0	1
3	3	4	0	1	2
4	4	0	1	2	3

Table 13-4

Multiplication in a system modulo m is the same as multiplication of whole numbers except when the product is greater than or equal to m; then it is reduced by the congruence relationship $a \equiv b \pmod m$ to a number less than m. Table 13-5 illustrates multiplication mod 5.

Multiplication Mod 5

×	0	1	2	3	4
0	0	0	0	0	0
1	0	1	2	3	4
2	0	2	4	1	3
3	0	3	1	4	2
4	0	4	3	2	1

Table 13-5

Once we have the table defining an operation (mod m), we can immediately discuss the properties of the system. By inspection, we see that both addition and multiplication (mod 5) are commutative and associative. In fact, multiplication is distributive over addition. In (mod 5), the additive identity is 0; and the additive inverse of each element is defined: 4 is the additive inverse of 1; 3, the additive inverse of 2; 2, of 3; and 1, of 4. The multiplicative identity is 1; the multiplicative inverses of 1, 2, 3, and 4 are 1, 3, 2, and 4, respectively. Thus, you can see that the operations of addition and multiplication (excluding 0) in modulo 5 form two Abelian groups.

Many properties of equality carry over to congruence. The following theorem is important in the next section.

Theorem 13-3: If $a \equiv b \pmod{m}$ and $c \equiv d \pmod{m}$, then

(a) $a + c \equiv b + d \pmod{m}$,
(b) $a - c \equiv b - d \pmod{m}$,
(c) $ka \equiv kb \pmod{m}$, k any integer,
(d) $ac \equiv bd \pmod{m}$.

Examples: We know that $16 \equiv 2 \pmod{7}$ and $38 \equiv 17 \pmod{7}$ because $7|(16 - 2)$ and $7|(38 - 17)$. Using these congruences, we have the following relations.

(a) $16 + 38 \equiv 2 + 17 \pmod{7}$. This can be verified by checking that $7|(54 - 19)$.

(b) $38 - 16 \equiv 17 - 2 \pmod{7}$.

(c) $(16)(38) \equiv (2)(17) \pmod{7}$ or $608 \equiv 34 \pmod{7}$.

Proof of Part (a): That $a + c \equiv b + d \pmod{m}$ is an immediate result of Definition 13-2. If $a \equiv b \pmod{m}$, then $m|(a - b)$. Similarly, if $c \equiv d \pmod{m}$, then $m|(c - d)$. Thus $m|[(a - b) + (c - d)]$, so it is clear that $m|[(a + c) - (b + d)]$. Then, by Definition 13-2,

$$a + c \equiv b + d \pmod{m}.$$

The proofs of the other parts of this theorem appear in the exercises.

Exercise Set 13-2

1. Label each of the following as true or false.
   (a) $6 \cdot 5 \equiv 4 \pmod{6}$ (b) $18 - 5 \equiv 1 \pmod{12}$
   (c) $6 \equiv 5 \pmod{3}$ (d) $79 \equiv -17 \pmod{4}$
   (e) $7 + 4 \equiv 5 \pmod{6}$ (f) $6 \equiv 3 \pmod{8}$
   (g) $8 \equiv 7 + 1 \pmod{5}$ (h) $34 \equiv -7 \pmod{13}$
2. Without tables, perform the following operations and, in each case, reduce the answer to a whole number less than the modulus.
   (a) $2 + 3 \pmod{4}$ (b) $2 \cdot 4 \pmod{5}$
   (c) $1 - 5 \pmod{3}$ (d) $15 - 2 \pmod{7}$

(e) $3 + (2 + 5) \pmod 6$ (f) $1 - 13 \pmod 4$
(g) $3(-2) \pmod 5$ (h) $(-2)(5) \pmod 7$
(i) $15 - 4 \pmod{11}$ (j) $-2(4 + -1) \pmod 7$
(k) $5^3 \pmod 9$ (l) $4^3 \pmod{12}$
(m) $(22)(46) \pmod 7$ (n) $(624)(589) \pmod 9$

3. By inspection, find a number x that will make each of the following statements true.
   (a) $x \equiv 5 \pmod 3$ (b) $x + 10 \equiv 9 \pmod 4$
   (c) $x + 5 \equiv 7 \pmod 9$ (d) $x + 20 \equiv 13 \pmod 3$
   (e) $x + 4 \equiv 6 \pmod 7$ (f) $x + 5 \equiv 7 \pmod 5$
   (g) $x + 3 \equiv 9 \pmod{11}$ (h) $x + 4 \equiv 3 \pmod 7$
4. (a) Is $4 - 7 \equiv 1 - 4 \pmod 8$?
   (b) In what modulus is $1 - 3 \equiv 5$?
   (c) For what modulus is $5 - 9 \equiv 10$?
   (d) Is $6 - 2 \equiv 3 - 9 \pmod{10}$?
5. Demonstrate that arithmetic (mod 7) satisfies the distributive property for multiplication over addition by using at least two specific examples.
6. Show that the set {0, 1, 2, 3, 4} forms a group under addition (mod 5).
7. Show that the set {1, 2, 3, 4, 5} does not form a group under addition (mod 6).
8. Determine if the set {1, 2, 3} with multiplication (mod 4) forms a group. The table, for example, would be as shown below.

·	1	2	3
1	1	2	3
2	2	0	2
3	3	2	1

9. Determine if the set {1, 2, 3, 4, 5, 6} forms a group under each operation below.
   (a) Addition (mod 7) (b) Multiplication (mod 7)
   (c) Addition (mod 8) (d) Multiplication (mod 8)

*10. Prove or disprove that a set of integers under multiplication (mod m) is a group if m is a composite number.

*11. Prove that modular numbers cannot be ordered so that if $x < y$ and $a < b$, then $ax < by$.

*12. Is the converse of Theorem 13-3(c) true? Why or why not?

*13. If $a \equiv b \pmod m$ and $c \equiv d \pmod m$, prove that
   (a) $a - c \equiv b - d \pmod m$.
   (b) $ka \equiv kb \pmod m$.
   (c) $ac \equiv bd \pmod m$.

4

Casting Out 9's and 11's

Some kind of a partial check for the correctness of computations is found in every elementary textbook. Two favorites are "casting out 9's" and "casting out 11's." Either of these checks may be applied to addition, subtraction, multiplication, or division. The check by casting out 9's consists of finding a representative for each entry in the problem, a check that is accomplished by adding the digits of each entry and finding a number to which the sum is congruent (mod 9). Generally this procedure is stated: "Add together all the digits in each entry, divide by 9, and represent the entry by the remainder." The same operations are performed on the representatives of the entries as are performed on the entries. If the computation is correct, the sum of the digits in the answer of the problem must be congruent (mod 9) to the answer obtained using the representatives.

Before doing the examples below, note that $5864 = 5(1000) + 8(100) + 6(10) + 4$. Now $1000 \equiv 1 \pmod 9$, since $9 \mid 1000 - 1$; $100 \equiv 1 \pmod 9$, since $9 \mid (100 - 1)$; and $10 \equiv 1 \pmod 9$, since $9 \mid (10 - 1)$. By Theorem 13-3, $5(1000) \equiv 5 \pmod 9$, $8(100) \equiv 8 \pmod 9$, $6(10) \equiv 6 \pmod 9$, and $4 \equiv 4 \pmod 9$; hence

$$5864 \equiv 5 + 8 + 6 + 4 \pmod 9.$$

Thus, every number is congruent (mod 9) to the sum of its digits.

Examples: (a) Addition

$$\begin{array}{rcrl}
5864 \equiv 5 + 8 + 6 + 4 \equiv 23 \equiv & 5 & \pmod 9 \\
3219 \equiv 3 + 2 + 1 + 9 \equiv 15 \equiv & 6 & \pmod 9 \\
\underline{4691} \equiv 4 + 6 + 9 + 1 \equiv 20 \equiv & \underline{2} & \pmod 9 \\
 & 13 & \pmod 9 \\
13{,}774 \equiv 1 + 3 + 7 + 7 + 4 \quad \equiv & 22 & \pmod 9
\end{array}$$

Since $22 \equiv 13 \pmod 9$, the answer 13,774 checks by this test.

(b) Subtraction

$$\begin{array}{rcrl}
82{,}726 \equiv 8 + 2 + 7 + 2 + 6 \equiv 25 \equiv & 7 & \pmod 9 \\
\underline{51{,}914} \equiv 5 + 1 + 9 + 1 + 4 \equiv 20 \equiv & \underline{2} & \pmod 9 \\
 & 5 & \pmod 9 \\
30{,}811 \equiv 3 + 0 + 8 + 1 + 1 \quad \equiv & 13 & \pmod 9
\end{array}$$

Since $13 \not\equiv 5 \pmod{9}$, the answer 30,811 is clearly incorrect. Why?

(c) Subtraction

$$\begin{array}{llr}
43{,}705 \equiv 4+3+7+0+5 \equiv 19 \equiv 1 \equiv & 10 & \pmod{9} \\
\underline{24{,}830} \equiv 2+4+8+3+0 \equiv 17 \equiv 8 \equiv & \underline{8} & \pmod{9} \\
 & 2 & \pmod{9} \\
18{,}875 \equiv 1+8+8+7+5 & \equiv 29 & \pmod{9}
\end{array}$$

Now $29 \equiv 2 \pmod{9}$, so the answer is probably correct.

(d) Multiplication

$$\begin{array}{llr}
345 \equiv 3+4+5 \equiv 12 \equiv & 3 & \pmod{9} \\
\underline{26} \equiv 2+6 \qquad \equiv 8 \equiv & \underline{8} & \pmod{9} \\
2070 & 24 & \pmod{9} \\
\underline{690} & & \\
8970 \equiv 8+9+7+0 \equiv 24 & & \pmod{9}
\end{array}$$

$24 \equiv 24 \pmod{9}$, so the multiplication is quite likely correct.

You probably noticed in the preceding examples that the statement was made that the answers were quite likely correct. If a mistake in a computation happened to be a multiple of 9, show with an example that casting out 9's would not indicate the mistake.

In a similar manner, we can verify the usual rule used for checking computations by casting out 11's. In casting out 11's, every other digit starting with the units digits is added and the other digits are subtracted. A check by casting out 11's consists of finding a representative for each entry in the problem by adding and subtracting the digits as indicated above, then finding a number to which the sum is congruent (mod 11). The same operations are performed on the representatives as on the original entries. The answer from the original entries must be congruent (mod 11) to the answer obtained from the remainders.

Consider 5864 as $5(1000) + 8(100) + 6(10) + 4$. Now $1000 \equiv -1 \pmod{11}$, since $11|(1000 + 1)$; $100 \equiv 1 \pmod{11}$, since $11|(100 - 1)$; and $10 \equiv -1 \pmod{11}$, since $11|(10 + 1)$. Thus, $5(1000) \equiv -5 \pmod{11}$, $8(100) \equiv 8 \pmod{11}$, $6(10) \equiv -6 \pmod{11}$, and $4 \equiv 4 \pmod{11}$. Hence $5864 \equiv -5 + 8 + -6 + 4 \pmod{11}$.

Examples: (a) Addition

$$\begin{array}{lrl}
5864 \equiv -5+8+-6+4 \equiv & 1 & \pmod{11} \\
\underline{3219} \equiv -3+2+-1+9 \equiv & \underline{7} & \pmod{11} \\
 & 8 & \pmod{11} \\
9083 \equiv -9+0+-8+3 \equiv & -14 & \pmod{11}
\end{array}$$

Now $-14 \equiv 8 \pmod{11}$; thus, the answer 9083 is quite likely correct.

(b) Subtraction

$$\begin{array}{rl} 856 \equiv 8 + -5 + 6 \equiv & 9 \pmod{11} \\ \underline{294} \equiv 2 + -9 + 4 \equiv & \underline{-3} \pmod{11} \\ & 12 \pmod{11} \\ 562 \equiv 5 + -6 + 2 \equiv & 1 \pmod{11} \end{array}$$

$1 \equiv 12 \pmod{11}$; thus, 562 is probably a correct answer.

(c) Multiplication

$$\begin{array}{rll} 345 \equiv 3 + -4 + 5 & \equiv & 4 \pmod{11} \\ \underline{13} \equiv -1 + 3 & \equiv & \underline{2} \pmod{11} \\ 1035 & & 8 \pmod{11} \\ \underline{345} & & \\ 4485 \equiv -4 + 4 + -8 + 5 \equiv -3 \pmod{11} & & \end{array}$$

Since $-3 \equiv 8 \pmod{11}$, quite likely the multiplication is correct.

Exercise Set 13-3

1. Perform the following operations and check by casting out 9's.
   (a) $583 + 427$ (b $583 \cdot 427$
   (c) $4371 \cdot 287$ (d) $4371 + 287$
   (e) $385 - 156$ (f) $7937 - 6065$
   (g) $43{,}590 + 11{,}825$ (h) $117{,}426 + 587{,}421$
   (i) $1746 - 1428$ (j) $(264)(876)$
   (k) $988 \cdot 234$ (l) $814 \cdot 721$
2. Check all the parts of the preceding exercise by casting out 11's.
3. The problem $69 \cdot 73 = 5048$ checks by casting out 11's but does not check by casting out 9's. Why?
4. Find a check for computation in base twelve similar to casting out 9's in base ten. Check the computation on the following problems by this procedure.
   (a) 4TE + T0E (b) (831)(173)
   (c) 80T − 1E4 (d) 68E · TE
5. Problems in different number bases may be checked by casting out a number one less than the base. Devise a rule similar to "casting out nines" and check each of the following problems.
   (a) $465_{\text{eight}} + 741_{\text{eight}}$ (b) $(41_{\text{five}})(34_{\text{five}})$
   (c) $(78_{\text{nine}})(64_{\text{nine}})$ (d) $(21_{\text{three}})(12_{\text{three}})$

*6. Devise a test for casting out one more than the base. Verify this test for base five.

*7. Use the test in Exercise 6 to check the computations in Exercise 5.

*8. Use the concept of modular arithmetic to develop the test for divisibility by the following.

(a) 2 (b) 5

(c) 9 (d) 11

*9. Prove that if $a \equiv b \pmod{m}$, then $a^{10} \equiv b^{10} \pmod{m}$.

*5

Linear Congruences

Two additional theorems are needed as we develop the theory of solving for x in linear congruences of the form $kx \equiv b \pmod{m}$. The first question that must be answered is "Under what conditions can one deduce that if $ka \equiv kb \pmod{m}$, then $a \equiv b \pmod{m}$?" This conjecture is obviously not true in all cases because $6 \cdot 7 \equiv 6 \cdot 1 \pmod{12}$ and $7 \not\equiv 1 \pmod{12}$. The answer to our question is given in the following theorem.

Theorem 13-4: If $ka \equiv kb \pmod{m}$ and if k and m are relatively prime, then $a \equiv b \pmod{m}$.

In other words, we can divide a common factor from each side of a congruence provided the common factor and the modulus are relatively prime. The preceding theorem is a special case of the following more general statement.

Theorem 13-5: If $ar \equiv br \pmod{m}$ and if d is the g.c.d. of m and r, then $a \equiv b \pmod{m/d}$.

Examples: (a) $5 \cdot 6 \equiv 8 \cdot 6 \pmod{9}$, so $5 \equiv 8 \pmod{3}$, since the g.c.d. of 6 and 9 is 3.

(b) $15 \cdot 8 \equiv 21 \cdot 8 \pmod{24}$, so $15 \equiv 21 \pmod{3}$, since the g.c.d. of 8 and 24 is 8. Furthermore, $5 \equiv 7 \pmod{1}$. Why?

(c) $7 \cdot 5 \equiv 21 \cdot 5 \pmod{7}$; therefore, $7 \equiv 21 \pmod{7}$, since the g.c.d. of 5 and 7 is 1. Furthermore, $1 \equiv 3 \pmod{1}$.

Proof: Since $m \mid (ar - br)$, $km = ar - br = (a - b)r$. Since $d \mid m$ and $d \mid r$, $k(m/d) = (a - b)(r/d)$. Thus, $(m/d) \mid (a - b)(r/d)$. Now, g.c.d. $(m/d, r/d) = 1$. Therefore, $(m/d) \mid (a - b)$ or $a \equiv b \pmod{m/d}$.

The numbers 2, 7, 12, 17, -3, -8, and so on, are mutually congruent modulo 5. For example, $2 \equiv 17 \pmod 5$ because $5 \mid (2 - 17)$; $12 \equiv -3 \pmod 5$ because $5 \mid [(12 - -3)]$. Now, consider the set of all integers congruent to 2 modulo 5; this set is called a *residue class modulo* 5.

In a like manner, consider the set of all integers congruent to 0 modulo 5, the set congruent to 1 (mod 5), congruent to 3 (mod 5), and, finally, the set congruent to 4 (mod 5).

Now, every integer n is in one of these classes because by the division algorithm involving n and 5, $n = 5 \cdot q + r$, where $0 \leq r < 5$. Thus, the remainder r is either 0, 1, 2, 3, or 4. In this way, the set of integers is partitioned into five sets of numbers called the *residue classes modulo* 5.

Of course, there is nothing special about the selection of residue classes modulo 5. When dealing with congruences modulo a fixed integer m, the set of all integers may be partitioned into m classes, called the *residue classes modulo m*, such that any two elements in the same class are congruent. Elements from different classes are not congruent. For the purposes of this course, we will generally use the smallest non-negative element from a class to represent the class. We now select an element from each residue class to form a complete residue system.

Definition 13-3: A complete *residue system* modulo m $(m > 1)$ is a set of integers with the following properties:

(a) No two elements of the set are congruent.
(b) If x is any integer, then x is congruent to some element of the set.

Examples of a complete residue system (mod m) are $\{0, 1, 2, 3, \ldots, m - 1)\}$, $\{1, 2, 3, \ldots, m\}$, and $\{m + 1, m + 2, \ldots, 2m\}$. No complete residue system (mod m) can contain fewer than m elements. Neither can it have more than m elements. Thus, a complete residue system modulo m is always a set consisting of exactly m different elements, no two of which are congruent.

The following complete residue system is of interest because it is useful in solving linear congruences.

Theorem 13-6: If $\{a_1, a_2, \ldots, a_m\}$ is a complete residue system (mod m) and g.c.d. $(k, m) = 1$, then $\{ka_1, ka_2, \ldots, ka_m\}$ is a complete residue system (mod m).

Proof: The set $\{ka_1, ka_2, \ldots, ka_m\}$ certainly contains m elements, since $\{a_1, a_2, \ldots, a_m\}$ contains m elements. To prove that the set is a complete residue system, we must prove that no two of the elements are congruent. Assume some $ka_i \equiv ka_j \pmod{m}$, where $i \neq j$. Since g.c.d. $(k, m) = 1$, then by Theorem 13-4, $a_i \equiv a_j \pmod{m}$. This is a contradiction, since a_i and a_j are elements of a complete residue system. Thus, $ka_i \not\equiv ka_j \pmod{m}$, and so $\{ka_1, ka_2, \ldots, ka_m\}$ is a complete residue system.

The preceding theorem enables us to discuss the solution of a congruence of the form $kx \equiv b \pmod{m}$, where g.c.d. $(k, m) = 1$. Since k is relatively prime to m, the set $\{ka_1, ka_2, \ldots, ka_m\}$ is a complete residue system, and thus one element of the set must be congruent to b modulo m. Consequently, some a_i is a solution of the linear congruence. Consider $2x \equiv 5 \pmod{7}$. Now, $\{0, 1, 2, 3, 4, 5, 6\}$ is a complete residue system (mod 7). Hence, 2 times one element of this set is congruent to 5 (mod 7). By trial and error, we find that $2(6) \equiv 5 \pmod{7}$, since $7 | (12 - 5)$. Therefore, $x \equiv 6 \pmod{7}$ is a solution to the congruence $2x \equiv 5 \pmod{7}$. The discussion in the last paragraph proves the following theorem.

Theorem 13-7: The linear congruence, $kx \equiv b \pmod{m}$, where g.c.d $(k, m) = 1$, always has an integral solution for x.

Example: Solve $3x \equiv 4 \pmod{5}$. Since g.c.d. $(3, 5) = 1$, then, by Theorem 13-7, this linear congruence has a solution. The solution will be one of the elements of any complete residue class modulo 5. One such complete residue system is $\{0, 1, 2, 3, 4\}$. By trial, we find that $x \equiv 3 \pmod{5}$ is a solution. Of course, any other element in the residue class containing 3 would also be a solution—for example, 8, 13, 18, 23, ...

Example: Solve $3x \equiv 5 \pmod{11}$. Since g.c.d. $(3, 11) = 1$, then, by Theorem 13-7, the linear congruence has a solution. A solution can be found in $\{0, 1, 2, 3, 4, 5, 6, 7, 8, 9, 10\}$. We find a solution to be $x \equiv 9 \pmod{11}$. Naturally, any number congruent to 9 (mod 11) will also be a solution.

Theorem 13-8: The congruence $kx \equiv b \pmod{m}$ has a solution if and only if g.c.d. (k, m) divides b. If the congruence has a solution, the number of distinct solutions is equal to g.c.d. (k, m).

Example: Solve $4x \equiv 5 \pmod 6$. No solution exists, for the g.c.d. of 4 and 6 does not divide 5.

Example: Solve $4x \equiv 10 \pmod 6$. Now g.c.d. $(4, 6) = 2$, and $2|10$; therefore, the congruence has a solution. In fact, since the g.c.d. $(4, 6) = 2$, the equation has two solutions. $2x \equiv 5 \pmod 3$ has a solution $x = 1$. The other solution may be obtained by adding the new modulus 3. Thus, $4x \equiv 10 \pmod 6$ has two solutions, $x \equiv 1 \pmod 6$ and $x \equiv 4 \pmod 6$.

Exercise Set 13-4

1. Determine the equivalence class (mod 9) to which each of the following integers belongs, where we adopt {0, 1, 2, 3, 4, 5, 6, 7, 8} as a system of residues.
   (a) -14 (b) 37 (c) 27
   (d) -86 (e) 50 (f) -126
2. Find all the nonnegative solutions less than the modulus for each of the following congruences.
   (a) $33x \equiv 42 \pmod{105}$ (b) $4x \equiv 13 \pmod 9$
   (c) $15x \equiv 60 \pmod{105}$ (d) $5x \equiv 3 \pmod{22}$
   (e) $4x + 1 \equiv 8 \pmod{11}$ (f) $25 + x \equiv 3 \pmod 8$
   (g) $73 + x \equiv 5 \pmod 9$ (h) $33 + x \equiv 6 \pmod 7$
   (i) $6x \equiv 11 \pmod 8$ (j) $6x \equiv 12 + 2x \pmod 8$
   (k) $5x \equiv 21 \pmod{15}$ (l) $2x \equiv 7 \pmod{10}$
3. Find a replacement for x, if it exists, that will satisfy the following.
   (a) $x \cdot x \equiv 1 \pmod 3$ (b) $x^2 \equiv 1 \pmod 5$
   (c) $4x \equiv 3 \pmod 5$ (d) $4x \equiv 5 \pmod 7$
   (e) $x^2 + 2x \equiv 0 \pmod 9$ (f) $x^2 + x \equiv 1 \pmod 9$

*4. Let $\boxplus$ be an operation on two elements a and b of a set such that $a \boxplus b = a + b + ab$. Complete a table for this operation on integers {1, 2, 3, 4, 5} in mod 6. Does it form a group?

*5. (a) Prove Theorem 13-4.
   (b) Prove Theorem 13-8.

*6

Other Systems

In most of the mathematical systems considered in this chapter, the elements of the sets were numbers. We now exhibit a mathematical

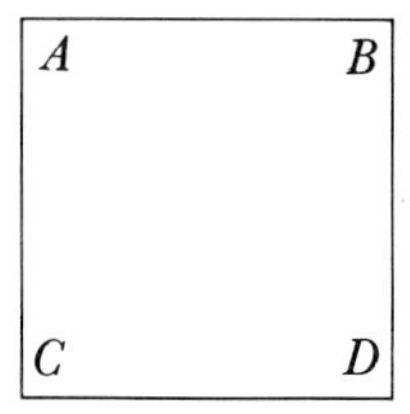

Figure 13-2

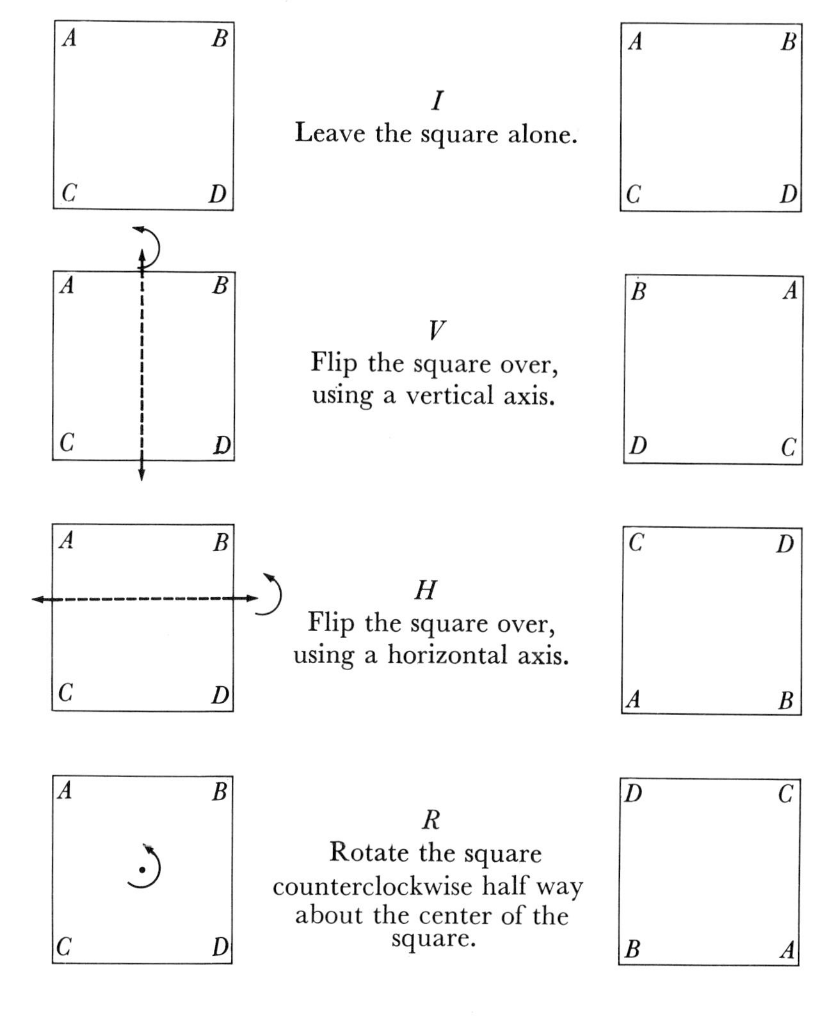

Figure 13-3

system without numbers for the purpose of showing how such a system may be generated.

What is needed in order to invent a mathematical system? For a start, we need a set of things. Let us consider an example in which the elements of a set will be *four movements of a square.* Take a square piece of cardboard and label its four corners A, B, C, and D, as indicated in Figure 13-2. Be sure to label the back of the card in the same manner as the front. Now we define four elements of a set to be movements relative to the square. These elements, which we shall call I, V, H, and R, are defined in Figure 13-3.

Thus, we have a set of elements I, V, H, R. To have a mathematical system, we now need an operation. We define an operation, denoted by $\odot$, to be successive applications of the motion to the square. For example,

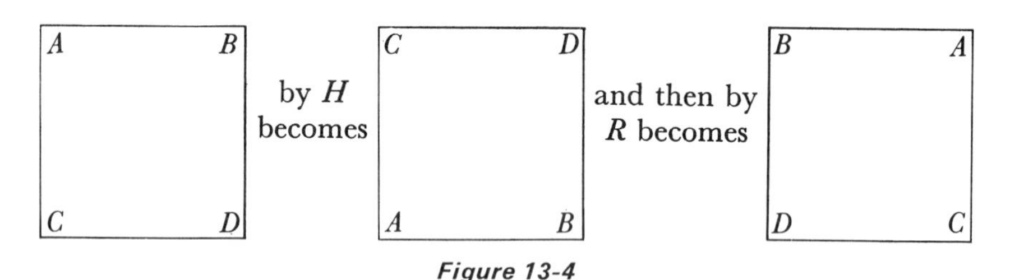

Figure 13-4

Thus, $H \odot R = V$. Can you define the operation relative to other elements of the set? Check the results in Table 13-6. Determine why this operation forms a commutative group.

$\odot$	I	V	H	R
I	I	V	H	R
V	V	I	R	H
H	H	R	I	V
R	R	H	V	I

Table 13-6

As we saw in the foregoing sections, mathematical systems arise from varying situations. Not only do they come from many different sources, but they also have different properties. In this chapter, we have become concerned with systems having certain common properties. Actually, this concept of characterizing systems is a part of what we might call "modern algebra." Since so many mathematical systems are similar, we can classify many mathematical systems by considering three well-known classifications—a *group*, a *ring*, and a *field*.

In Section 2 of this chapter, we defined a group and discussed whether or not the previously studied mathematical systems constituted groups.

It seems natural now to consider a mathematical system that can be classified as an Abelian group under two operations. Also, the elements of this system will satisfy the distributive law. In other words, this mathematical system will satisfy eleven properties. Can you name them? When this is accomplished, we obtain one of the most important and useful systems to be considered. It is called *field*.

Definition 13-4: A *field* is a mathematical system consisting of a set of elements (at least two) and two binary operations, $\oplus$ and $\odot$, satisfying the following properties:

(a) The mathematical system is a commutative group with respect to $\oplus$.

(b) The set of elements, with the identity for $\oplus$ excluded, is a commutative group with respect to $\odot$.

(c) The elements of the set satisfy the distributive law for $\odot$ over $\oplus$.

We have already discussed the fact that the rational numbers constitute a field. Rational number arithmetic as it is studied in elementary school is actually a study of field properties. Addition, multiplication, subtraction, and division are often called the four basic field operations. Of course, the real number system is also a field.

Another way to classify mathematical systems is through what are called *rings*.

Definition 13-5: A *ring* is a mathematical system consisting of a set of elements and two binary operations, $\oplus$ and $\odot$, satisfying the following properties:

(a) The mathematical system is a commutative group with respect to $\oplus$.

(b) The set is closed with respect to $\odot$.

(c) The elements of the set satisfy the associative property for $\odot$.

(d) The elements of the set satisfy the distributive property for $\odot$ over $\oplus$.

The most common example of a ring is the system of integers with the operations of addition and multiplication. With respect to addition, the integers have been previously classified as a commutative group. Also under multiplication, there are closure and associativity, and multiplication is distributive over addition; thus, the integers satisfy all the requirements necessary to be classified as a ring. Since the elements of the set are also commutative with respect to multiplication, the integers are called a commutative ring.

If a commutative ring happens to have an identity element for both operations and if $ab = 0$ implies either $a = 0$ or $b = 0$, it is then called an *integral domain*. Notice that the second operation on the ring does not necessarily form a group from the nonzero elements of the ring. What is missing? Also, notice that the ring of integers is an integral domain.

Review Exercise Set 13-5

1. You are given the geometric transformations I, X, Y, W, V, and Z as defined below.

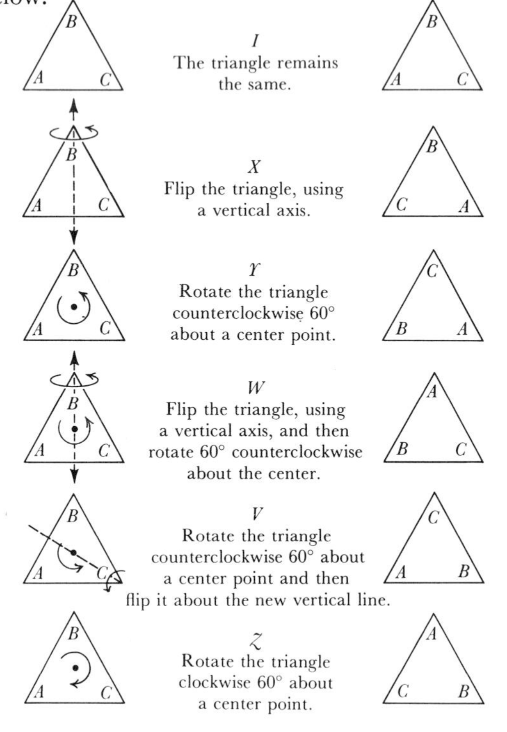

Define operation $\odot$ to be successive applications of these geometric transformations. Now perform the operation and answer the following questions.

(a) Make a table for the operation $\odot$ defined by the movement of a triangle.
(b) Is the set closed under the operation?
(c) Is the operation commutative?
(d) Is the operation associative?
(e) Is there an identity element for the operation?
(f) Does each element have an inverse for the operation?
(g) Is the system a group?

2. True or false?
(a) Every field is an integral domain.
(b) Every field is a ring.
(c) Every ring is an integral domain.
(d) Every integral domain is a field.
(e) Every integral domain is a ring.
(f) Every system that is commutative is associative.
(g) For each element of a group under the operation $\otimes$, each element $\otimes$ its inverse equals the identity of the set.
(h) For every element of a group, the inverse of the inverse of an element a is a.

3. The operation $\otimes$ is defined to be the smaller of two numbers if they are unequal and one of the numbers if they are equal.
(a) Make a table defining the operation for the set $\{1, 2, 3, 4, 5\}$.
(b) Is the operation commutative?
(c) Is there an identity element?
(d) Is the operation associative?
(e) Do all the elements of the set have an inverse with respect to the operation?

4. The operation $\triangle$ is defined to be the g.c.d. of two numbers. Work problems (a) through (d) of Exercise 3.

5. The operation $\otimes$ is defined to be the least common multiple of two numbers Work problems (a) through (d) of Exercise 3.

6. Construct addition and multiplication tables (mod 2).
(a) Does this system form a group under addition?
(b) Under multiplication with the inverse of 0 deleted?

7. Which of the following are groups?
(a) The nonnegative integers, operation addition
(b) The odd integers, operation addition
(c) The even integers, operation addition

(d) The positive even integers, operation addition
(e) Mod 7 multiplication
(f) Mod 4 addition
(g) The positive rationals, operation multiplication
(h) $\{n\sqrt{2} \mid n \text{ is a whole number}\}$, operation addition
(i) Even integers, operation multiplication

8. Which of the following modular mathematical systems, with 0 deleted, can be classified as groups under multiplication?
(a) mod 5 (b) mod 8 (c) mod 9
(d) mod 7 (e) mod 10 (f) mod 12

9. Why is it that a modular mathematical system under multiplication cannot form a group unless 0 is deleted?

10. Explain why each of the following tables does not define an operation so that the mathematical system is a group.

(a)

$\otimes$	x	y	z
x	x	y	z
y	y	x	z
z	z	z	z

(b)

$\odot$	a	b
a	a	b
b	b	b

11. Which of the following is a ring under the operations of addition and multiplication?
(a) Arithmetic (mod 5) (b) Arithmetic (mod 4)
(c) The even integers (d) The odd integers
(e) The positive rationals (f) $\{n\sqrt{2} \mid n \text{ is a whole number}\}$

12. Consider the set $T = \{A, B, C, D, E, F\}$ and the table below.
(a) Is λ a binary operation on T? Why?
(b) Is T closed with respect to λ? Why?
(c) Is λ commutative?
(d) Is there an identity for λ?
(e) What is the inverse of B? Of C?

λ	A	B	C	D	E	F
A	A	A	A	A	A	A
B	A	B	C	D	E	F
C	A	C	E	A	C	E
D	A	D	A	D	A	D
E	A	E	C	A	E	C
F	A	F	E	D	C	B

13. Consider the system of integers with addition and multiplication (mod n), and classify whether such a system is an integral domain.
    (a) $n = 12$ (b) $n = 5$ (c) $n = 6$
    (d) $n = 4$ (e) $n = 7$ (f) $n = 10$

Suggested Reading

Clock Arithmetic, Modular: Allendoerfer, pp. 383–389. Graham, pp. 191–192, 198–203. Meserve, Sobel, pp. 80–88. Peterson, pp. 324–336. Weaver, Wolf, pp. 249–250.

Casting Out 9's and 11's: Allendoerfer, pp. 389–390. Graham, pp. 203–205. Meserve, Sobel, p. 7.

Linear Congruences: Peterson, pp. 336–341.

Groups, Fields, Rings, Integral Domains: Graham, pp. 193–197. Peterson, pp. 349–354. Weaver, Wolf, pp. 252–268.

14

Algebra and Geometry

1

A Review of Equations and Inequalities

Many algebraic and geometric ideas have been introduced in earlier chapters of this book. However, in this chapter, we consider several additional topics concerning algebra and geometry, topics that may be useful in future study and work. In this course, we cannot present all of the significant topics of a good algebra or analytic geometry course, but we attempt to introduce here the language and the approach to such courses. Although we consider isolated topics, an effort will be made to present these topics as applications of logical mathematics rather than mere manipulation.

By the end of this chapter, we will see that the real number system is not sufficient for our demands; thus, an extension of the real numbers is made to obtain what are called *complex numbers*. The complex number system is defined, and the real number system is viewed as a subset of the complex numbers. We begin our study of algebra with a review of equations and inequalities. Consider the following examples.

Example: Solve $3x + 5 = 11$.

Add $^-5$ to both members (or subtract 5 from both sides of the equation).

$$(3x + 5) + (^-5) = 11 + (^-5) \qquad \text{Why?}$$
$$3x + (5 + {}^-5) = 11 + {}^-5 \qquad \text{Why?}$$
$$3x + 0 = 6 \qquad \text{Why?}$$
$$3x = 6 \qquad \text{Why?}$$

Multiply both members by 1/3 (or divide both members by 3).

$$\frac{1}{3}(3x) = \frac{1}{3}(6)$$

$$\left[\frac{1}{3}(3)\right]x = 2 \qquad \text{Why?}$$

$$1 \cdot x = 2 \qquad \text{Why?}$$

$$x = 2$$

Thus, 2 is a solution, or root, of the equation. A solution can be checked by substituting in the original equation:

$$3(2) + 5 = 6 + 5 = 11.$$

Example: Find the solution set of $(x/3) - 2 < 4$.

$$\frac{x}{3} + {}^{-}2 < 4. \qquad \text{Why?}$$

Add 2 to both members.

$$\left(\frac{x}{3} + {}^{-}2\right) + 2 < 4 + 2$$

$$\frac{x}{3} + ({}^{-}2 + 2) < 6 \qquad \text{Why?}$$

$$\frac{x}{3} < 6 \qquad \text{Why?}$$

Multiply both members by 3.

$$3\left(\frac{x}{3}\right) < 3 \cdot 6$$

$$3\left(\frac{1}{3} \cdot x\right) < 3 \cdot 6 \qquad \text{Why?}$$

$$\left[3\left(\frac{1}{3}\right)\right]x < 18 \qquad \text{Why?}$$

$$x < 18$$

The solution set consists of all real numbers less than 18.

The procedures used in the preceding examples may also be used for solving equations and inequalities involving only letters that represent numbers.

Example: Solve $y = mx + b$ for x.

Add $-b$ to both members.

$$y + (-b) = (mx + b) + (-b)$$
$$y + (-b) = mx + (b + -b)$$
$$y - b = mx$$

Multiply both members by $1/m$.

$$(1/m)(y - b) = (1/m)mx$$
$$(1/m)(y - b) = [(1/m)(m)]x$$
$$(1/m)(y - b) = 1 \cdot x$$
$$(1/m)(y - b) = x$$

$$\frac{y - b}{m} = x \quad \text{or} \quad x = \frac{y - b}{m}$$

We encounter mathematical problems in various phases of our everyday activity. Most of these problems are stated in verbal form rather than in mathematical terms. Solutions do not always exist for these problems. When solutions do exist, you can generally proceed as follows to solve the problems.

(a) Translate the verbal statements into mathematical equations.
(b) Solve the equations.
(c) Interpret the solutions in terms of the real-life problems at hand.

Before translating a verbal statement into a mathematical form, you should enumerate all of the variables involved in the problem. In general, these variables will be the desired answers for the problem. Let x (or some other letter) represent one of the unknowns. Then, if possible, express all other unknowns in terms of x. When all of the variables in the problem have been expressed in terms of x, you can write the verbal statement relating the variables in equation form.

Example: Twice John's weight added to 50 pounds is equal to 300 pounds. Find John's weight.

Solution: (a) Let x represent John's weight. Twice John's weight would be represented by $2x$, and twice John's weight added to 50 pounds would be $2x + 50$. From the verbal statements, $2x + 50$ is equal to 300 pounds.

(b)

$$\begin{aligned} 2x + 50 &= 300 \\ 2x &= 250 \\ x &= 125 \end{aligned}$$

(c) John's weight is 125 pounds.

Example: Paul's grade is double that of Sue. Three times Sue's grade less one-fourth of Paul's grade is equal to 100. Find the grades of Paul and Sue.

Solution: (a) Let x represent Sue's grade. Then $2x$ would represent Paul's grade. Three times Sue's grade would be $3x$, and one-fourth of Paul's grade would be $2x/4 = x/2$. Three times Sue's grade less one-fourth of Paul's grade would be $3x - x/2$. This quantity is equal to 100.

(b)

$$\begin{aligned} 3x - \frac{x}{2} &= 100 \\ 6x - x &= 200 \\ 5x &= 200 \\ x &= 40 \end{aligned}$$

(c) Sue's grade is 40. Paul's grade is $2(40) = 80$.

Example: A candy store has two grades of the same type of candy, one that costs \$0.80-a-pound and one that costs \$0.50-a-pound. How much of each should the store manager use in order to have a 100-pound mixture that costs \$0.74 per pound?

Solution: (a) Let x represent the number of pounds of \$0.80 candy to be put in the mixture. Then $100 - x$ will represent the number of pounds of \$0.50 candy. The cost and weight of each kind of candy leads to the following equation.

(b)

$$\begin{aligned} 0.80x + 0.50(100 - x) &= 0.74(100) \\ 0.30x + 50 &= 74 \\ 0.30x &= 24 \\ x &= 80 \end{aligned}$$

(c) Thus, the manager should mix 80 pounds of \$0.80 candy with $100 - 80 = 20$ pounds of \$0.50 candy to obtain a mixture of 100 pounds of \$0.74 candy.

Exercise Set 14-1

1. Solve for the variable and check.
   (a) $x + 4 = -6$ (b) $x + (-3) = -7$
   (c) $2x - (-5) = 6 - x$ (d) $4x + (-7) = 3x - 1$
   (e) $\frac{x}{5} - \frac{1}{3} = \frac{x}{3} + \frac{1}{5}$ (f) $\frac{x}{2} - \frac{1}{4} = \frac{3x}{4} + 1$
2. Solve for the variable x.
   (a) $3 + (-x) > 0$ (b) $x + (-5) < -2$
   (c) $-x + (-4) < -7$ (d) $-2x - 3 < 5$
   (e) $\frac{x}{3} + 2 < -5$ (f) $-3x - 4 < -5$
3. Solve for the variable indicated.
   (a) $A = P + Prt$; r (b) $A = P + Prt$; t
   (c) $y = mx + b$; b (d) $y = mx + b$; m
   (e) $S = \frac{a}{1 - r}$; r (f) $A = P + Prt$; P
4. Ed, David, and Bill painted the fence around Mrs. Jones's garden. David painted half as much as Ed, and Bill painted a fifth as much as Ed. How much of the fence did each boy paint?
5. Mrs. Cone has nickels, dimes, and quarters in her purse. The number of nickels is three times the number of dimes, and the number of quarters is one-fifth the number of dimes. If the total amount of money is \$4.50, how many coins of each kind does she have?
6. The sum of two integers is 62. The second integer is 11 more than twice the first. What are the integers?
7. Tom has twice as many books as Joe. Together they have 75 books. How many does each boy have?
8. Twenty pounds added to four times Sue's weight is 500 pounds. How much does Sue weigh?
9. Mrs. Brown and Mrs. Bishop compared grocery bills and found that Mrs. Brown's was half as much as three times Mrs. Bishop's. If the sum of the two bills was \$65.50, what was each woman's bill?

10. A new portable refrigerator cost the Jones family \$297.75. At the time of this purchase, the family also bought a new sailboat at twice the cost of the refrigerator. Since their cabin on the river was not well equipped, the family decided to purchase three new sleeping bags. These three cost one-fifth the price of the refrigerator. What was the total amount spent (discounting any tax)? What was the price of the sailboat and the price of each sleeping bag?
11. A grocer mixes 40 pounds of \$.60-a-pound nuts with 60 pounds of \$.45-a-pound nuts. If he wants to receive at least the same amount of money as he did when he sold the nuts separately, how much should he charge for the mixture?

*12. A matrix is a rectangular array of numbers. Two 2×2 matrices,

$$\begin{pmatrix} a & b \\ c & d \end{pmatrix} = \begin{pmatrix} w & x \\ y & z \end{pmatrix},$$

are defined to be equal only if $a = w$, $b = x$, $c = y$, and $d = z$. Solve for the unknowns in the following matrices.

(a) $\begin{pmatrix} w+7 & 2(4x-3) \\ 3y-2 & 5z \end{pmatrix} = \begin{pmatrix} -2 & 10 \\ -1 & 25 \end{pmatrix}$

(b) $\begin{pmatrix} 8 & 1/2-(2/3)x \\ 1/3(33) & 6z+1/2 \end{pmatrix} = \begin{pmatrix} w-5 & 4 \\ (1/2)y & (1/6)z+3 \end{pmatrix}$

*13. Dime and Dollar, two coin collectors, decided to swap some of their coins. Dime agreed to swap a certain number of coin A, three times the number of A of coin B, and one-fourth the number of A of coin C. Dollar gave Dime twice the original number of coin A, one-third of Dime's number of A in coin B, and four times Dime's number of A in coin C. If we judged the value of the swap simply on the number of coins (not their value), who made the better deal? If the total number of coins which Dime exchanged was 204, how many coins each of A, B, and C did he relinquish? If the total number of coins which Dollar exchanged was 304, how many coins each of A, B, and C did he relinquish?

2

Graphs of Linear Equations

In this section we introduce the Cartesian coordinate system and learn to draw graphs of linear equations and inequalities. Most of this material is a review of what you learned in high school. However, we shall refresh your memory by starting at an elementary level and then moving rapidly through the material.

In a given plane draw two perpendicular lines intersecting at point 0. (See Figure 14-1.) In the typical case, one line $0x$ is horizontal while the other line $0y$ is vertical. The two lines are called *coordinate axes*. Traditionally, the horizontal line is called the *x axis* and the vertical line is called the *y axis*. The point of intersection of the two lines is called the *origin*. The plane in which the two axes lie is called the *coordinate plane*. The four sections into which the axes divide the plane are called *quadrants*. The four quadrants are labeled I, II, III, and IV, as in Figure 14-1.

Arbitrarily select a unit for measuring distance as the basis for number scales on the x and y axes. In Figure 14-1, the number scales have been selected such that the intersection of the two lines is represented by zero. Although the number scales were selected to be the same units on the x and y axes, they need not be equal. Note that positive numbers are to the right on the x axis and up on the y axis, and negative numbers are to the left and down on the x axis and y axis, respectively.

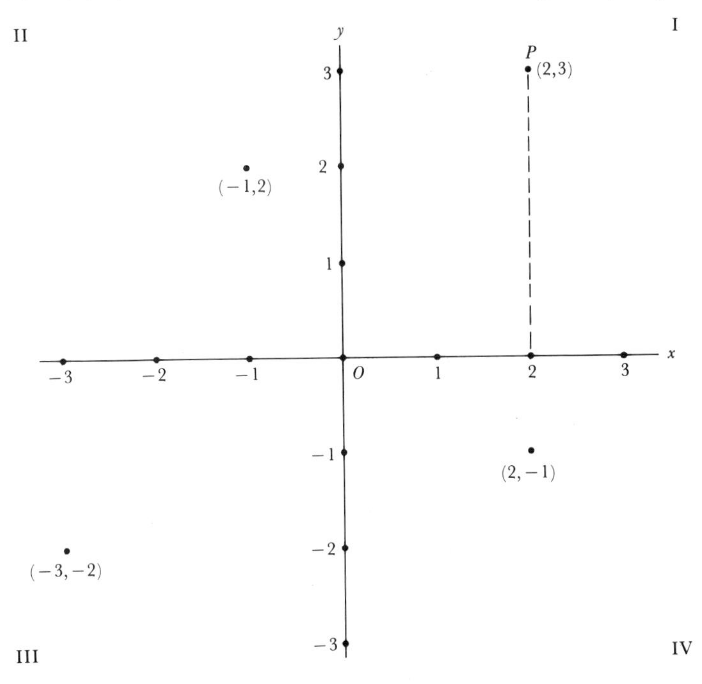

Figure 14-1

Once the integers have been positioned, we can locate points representing rational and irrational numbers. We conceive of each axis as representing a line where each point on the line corresponds to a real number.

Let P be any point in the coordinate plane. We define an ordered pair of coordinates (x, y) for P as follows:

The *abscissa*, or horizontal coordinate of P, is the distance x measured from the vertical axis, or y axis, to P. If x is positive, the distance is measured to the right; and if x is negative, the distance is measured to the left.

The *ordinate*, or vertical coordinate of P, is the distance y measured from the horizontal axis, or x axis, to P. If y is positive, the distance is measured up; and if y is negative, the distance is measured down.

Example: Plot the point $P = (2, 3)$. At 2 on the x axis, in Figure 14-1, erect a perpendicular to the x axis. Go up three units on this perpendicular to reach P. P can also be located by constructing a perpendicular to the y axis at 3 and then moving two horizontal units to the right.

If an ordered pair of numbers (a, b) gives the coordinates of a point, we may refer to it as "the point (a, b)."

Example: Locate, in Figure 14-1, the points $(-1, 2)$, $(2, -1)$, and $(-3, -2)$.

Consider the equation

$$x + y = 4.$$

Let a and b be two real numbers. We say that the ordered pair (a, b) *satisfies* $x + y = 4$, or is a *solution* of $x + y = 4$, if the equation becomes a true statement when a is substituted for x and b is substituted for y. For example, $(2, 2)$ is a solution of $x + y = 4$, and so are $(3, 1)$, $(1, 3)$, $(-1, 5)$, $(4, 0)$, and $(0, 4)$. In fact, there are infinitely many solutions for this equation in the real number system.

If we plot all the points of the solution set, the resulting figure is called the *graph* of the solution set or the *graph* of the equation. Of course, the solution set is infinite so we cannot possibly plot every ordered pair of the solution set. We can, however, get an idea of what a graph looks like by plotting several representative points. For example, we use five solutions of $x + y = 4$ to obtain the graph in Figure 14-2. All five points seem to lie on the same straight line.

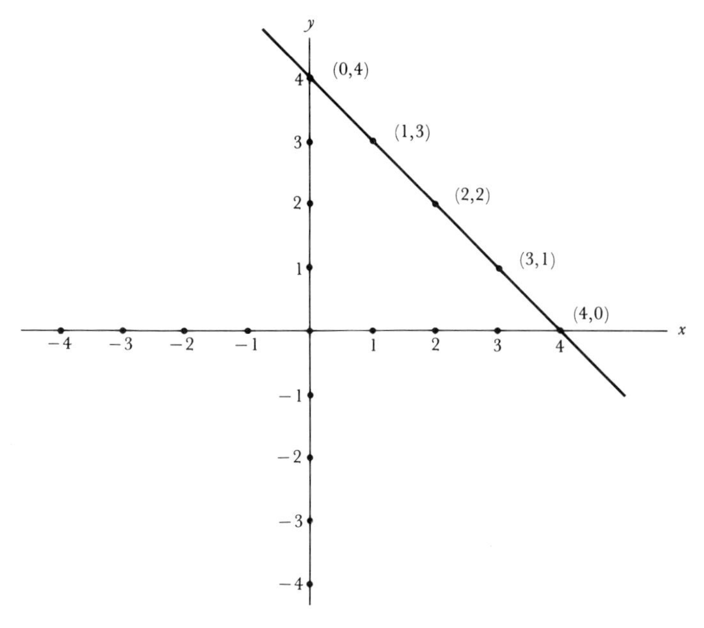

Figure 14-2

Example: Suppose x and y are related in such a way that $y = 2x + 3$. Some pairs of numbers that satisfy this relationship are (0, 3), (1, 5), (-1, 1), (-2, -1), (-3, -3), as plotted in Figure 14-3. Note also that fractional points ($\frac{1}{2}$, 4), ($\frac{1}{4}$, $3\frac{1}{2}$) satisfy the equation. If one were able to

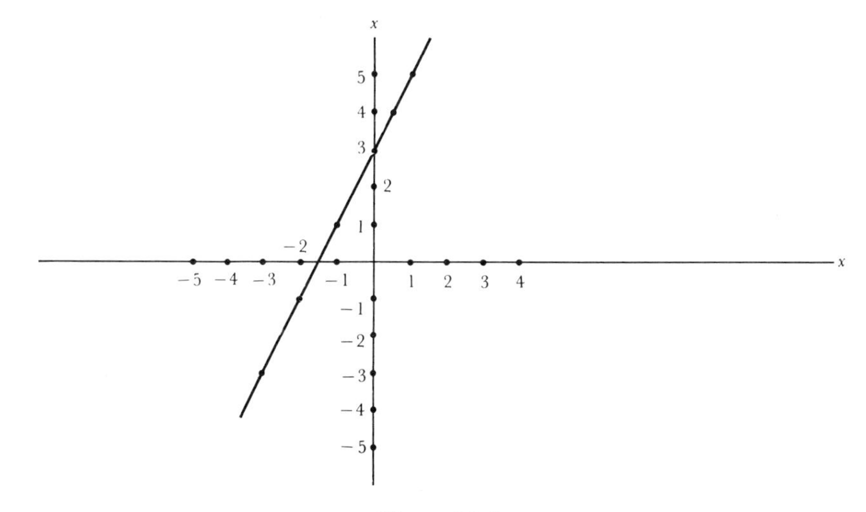

Figure 14-3

plot all possible points, both rational and irrational, that satisfy the given relationship, it would be evident that the graph is a straight line.

An understanding of the graph of an equation is useful in selecting ordered pairs that satisfy a relation. For instance, if we plot several relations of the form $\{(x, y) \mid ax + by + c = 0\}$, where a, b, and c are numbers, we find that we get something that looks like a straight line each time. We state (without proof) that the graphs of equations of the form $ax + by + c = 0$ are straight lines; and for this reason $ax + by + c = 0$ is called a *linear equation.*

Example: Represent or graph the number pairs that satisfy $\{(x, y) \mid y \geq x + 1\}$, where the domain is the set of real numbers.

The solution set consists of all the number pairs that make the statement $y = x + 1$ true, along with all the number pairs that make the statement $y > x + 1$ true. In both cases there is an infinite number of points. The totality of ordered pairs that satisfy $y = x + 1$ is represented by points on a straight line. The ordered pairs that satisfy $y > x + 1$ are represented by points that lie to one side of the straight line. $y = x + 1$ divides the rectangular coordinate plane into two half-planes. Thus, one may locate the half-plane of points that satisfy $y > x + 1$ by finding one point whose coordinates satisfy $y > x + 1$. For example, $(-1, 1)$ is one such point, since $1 > -1 + 1$. All other ordered pairs that satisfy $y > x + 1$ are on the same side of $y = x + 1$ as $(-1, 1)$. The solution set is indicated by the shaded portion of Figure 14-4. Verify this solution by finding additional number pairs in the solution set.

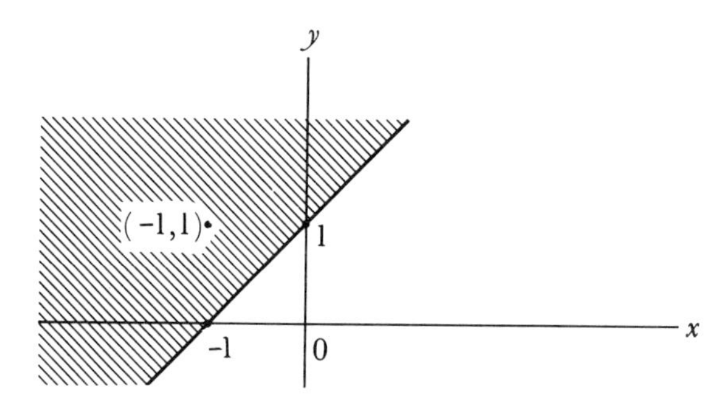

Figure 14-4

Example: Graph the relation $\{(x, y) \mid 3x + y > 6\}$.

First of all, we graph the straight line $3x + y = 6$. This will be represented in Figure 14-5 by a dashed line, since the points on the line do not satisfy the given inequality. However, this line does divide the rectangular coordinate plane into two half-planes, and one of these half-planes contains all of the points in the solution set of $3x + y > 6$. The ordered pair (0, 0) will be tested to see if it satisfies the inequality. $3 \cdot 0 + 0$ is not greater than 6. Therefore, (0, 0) is not in the solution set of $3x + y > 6$. The graph of $3x + y > 6$ (indicated by the shaded region) will be the half-plane that does not contain (0, 0).

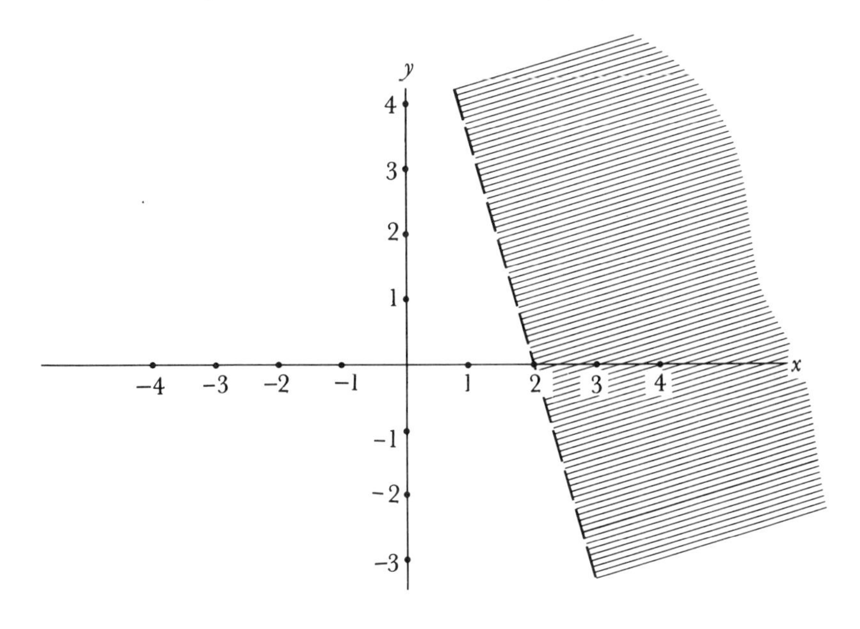

Figure 14-5

Exercise Set 14-2

Construct a graph for each relation, where the domain is the set of real numbers.

1. $\{(x, y) \mid y = 2x - 3\}$
2. $\{(x, y) \mid y > x + 2\}$
3. $\{(x, y) \mid x - y = 2\}$
4. $\{(x, y) \mid y \leq x\}$
5. $\{(x, y) \mid 2x + 3y < 6\}$
6. $\{(x, y) \mid 5x + 3y = 7\}$
7. $\{(x, y) \mid 2x + y = 1\}$
8. $\{(x, y) \mid x + 1 = 0\}$

9. $\{(x, y) \mid y < -2x + 2\}$
10. $\{(x, y) \mid y > 0\}$
11. $\{(x, y) \mid y + 1 = 0\}$
12. $\{(x, y) \mid y < x - 1\}$
13. $\{(x, y) \mid y < x\}$
14. $\{(x, y) \mid y \leq 2x + 1\}$
15. $\{(s, t) \mid 2s + t = 7\}$
16. $\{(x, y) \mid x + y < 9\}$
17. $\{(u, v) \mid u + v = 5\}$
18. $\{(x, y) \mid 5 < y - x\}$

For the sets of points
(a) (0, -10), (b) (4, 6), (c) (2, 3), (d) (-4, -5),
(e) (8, 1), (f) (10, -2), (g) (0, 12), (h) (-1, 15),
which of the points are parts of the graphs of the open sentences in Exercises 19 through 22?

19. $x + (1/2)y > 3/2$
20. $(1/2)x + (1/3)y > 4$
21. $x + (1/2)y \leq 3/2$
22. $x > y$
23. Why is it generally true that $(a, b) \neq (b, a)$? What must be true if $(a, b) = (b, a)$?

3

Function Representation

Functions were introduced earlier in our study. However, because they are so important in mathematics, we consider them again. A function, as defined in Chapter 2, is a special type of relation. Return to Chapter 2 and review the conditions a relation must satisfy to be a function. Note that a function from A to B, usually denoted by f, is a relation having domain A with the property that each element of A is paired with exactly one element of the range, B.

There are several different notations that may be used to denote a function. One commonly used notation involves the symbol $f(x)$. If $(x, y) \in f$, then we write $y = f(x)$. If $f(1) = 3$, then $(1, 3) \in f$; if $f(a) = b$, then $(a, b) \in f$. When the notation $y = f(x)$ is used, x is said to be the *independent variable* and y is said to be the *dependent variable.*

The notation for a function, $y = f(x)$, suggests an alternative definition of a function which is equivalent to the ordered-pair definition given. A function from set A to set B is a rule or procedure that associates with each element of A one and only one element of set B.

Another way of formulating the same definition is to state that a function is a set of ordered pairs, no two of which have the same first element. Graphically, this means that a relation is a function if and only if no vertical line meets the graph of the relation at more than one point.

The following examples illustrate relations, some of which are functions and some of which are not functions.

Example: We say $y = x + 1$ is a function on the set of real numbers because, for every real number we assign to x, there is one and only one value for y. x is the independent variable and y is the dependent variable.

Example: $y^2 = x$, where x is the independent variable and y is the dependent variable, is not a function on the real numbers because both $x = 9, y = 3$ and $x = 9, y = -3$ satisfy this equation.

Example: The relation specified by the table

x	1	0	1	2
y	-2	3	4	5

is not a function because two ordered pairs (1, -2) and (1, 4) have identical first elements and different second elements.

The notation $y = f(x)$ actually states a rule for determining values of y when values are assigned to x. For example, if $f(x) = x^2$, where x is any real number, then $f(2) = 2^2 = 4$, or $y = 4$. If

$$f(x) = x^2 - 3x + 2,$$

then

$$f(3) = 3^2 - 3(3) + 2 = 2.$$

Example: Given $r = 6 + 3t$, find $f(-1)$, $f(2)$, $f(0) - f(-4)$, and $f(1) \cdot f(3)$. Since $r = 6 + 3t$, then $f(t) = 6 + 3t$.

$$f(-1) = 6 + 3(-1) = 3$$
$$f(2) = 6 + 3(2) = 12$$
$$f(0) - f(-4) = [6 + 3(0)] - [6 + 3(-4)] = 6 - (-6) = 12$$
$$f(1) \cdot f(3) = [6 + 3(1)] \cdot [6 + 3(3)] = 135$$

Example: Make a table and draw the graph defined by

$$\{(x, y) \mid y = x^2\}.$$

The domain is the set of real numbers. Is $y = x^2$ a function?

x	0	1	2	–1	–2	1/2
y	0	1	4	1	4	1/4

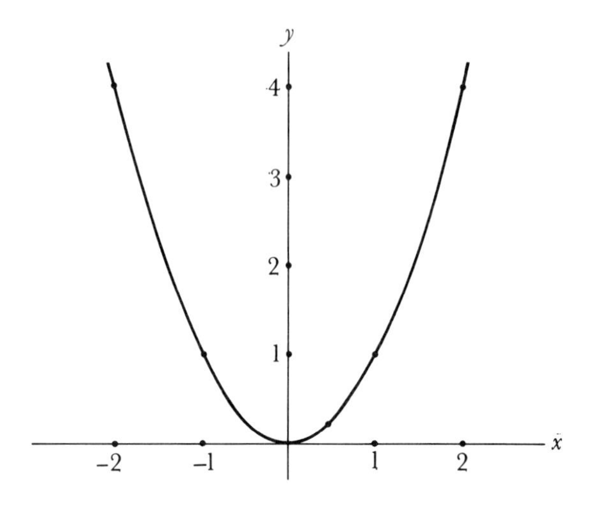

Figure 14-6

Since no vertical line can be drawn which meets the graph in more than one point, $y = x^2$ is a function.

Exercise Set 14-3

1. Which of the following tables define functions?

(a)

x	2	2
y	4	1

(b)

x	1	3
y	-1	-1

(c)

x	1	1
y	1	2

(d)

x	0	0
y	0	1

(e)

x	1	2
y	1	1

(f)

x	0	2	3	5
y	2	4	7	2

(g)

x	-2	1	3	2	1
y	1	2	4	3	4

(h)

x	-1	2	4	2
y	3	4	6	5

2. Given that $y = f(x) = x^2 - x + 2$, make a table showing the values of the function at $x = 0, -1, -2, \frac{1}{2}, 1, 2$. Sketch the graph, assuming that the domain is the set of real numbers.
3. If $y = f(x) = x^3 - 2x$, find $f(-2)$, $f(1)$, $f(0) - f(3)$, $f(-1) \cdot f(1)$, $f(2)/f(4)$, $f(w)$, and $f(2z)$.
4. If $y = f(x) = (x + 2)/(x - 1)$, find $f(3)$, $f(2)$, $[f(-1) - f(0)]/2$, $f(-2) \cdot f(2)$, $f(z)$, and $f(3z)$.
5. The following graphs represent relations on U, the set of all real numbers. Which of these relations are functions?

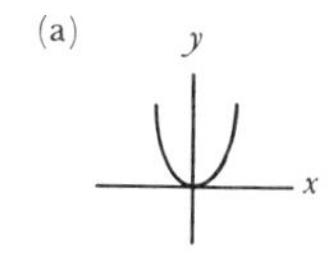

(a)

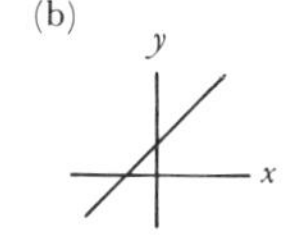

(b)

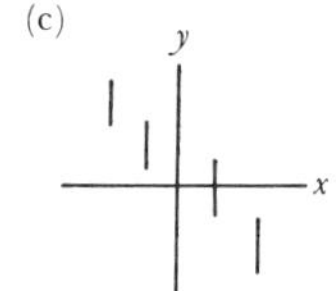

(c)

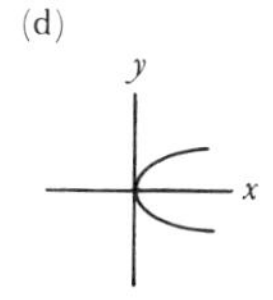

(d)

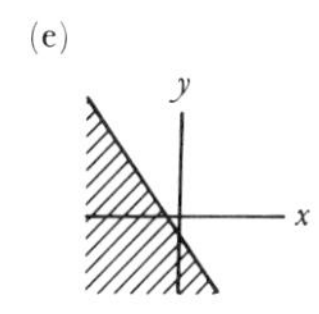

(e)

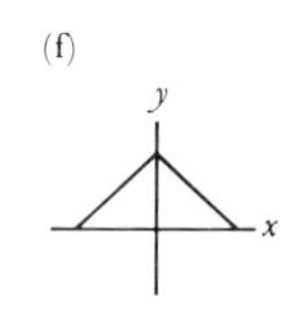

(f)

6. Determine which of the following equations and inequalities represent functions.
   (a) $\{(x, y) \mid x - y = 2\}$ (b) $\{(x, y) \mid y \leq x\}$
   (c) $\{(x, y) \mid x + 6 < 9\}$ (d) $\{(x, y) \mid 5x + 3y = 7\}$
7. Graph each of the following; find the expressions that define functions.
   (a) $x^2 - 3 = y$ (b) $y = 4x^2$
   (c) $x = y^2$ (d) $y = 4x + 3$
   (e) $\{(x, y) \mid x = 2\}$ (f) $\{(x, y) \mid y = -1\}$
   (g) $\{(x, y) \mid x < y + 1\}$ (h) $\{(x, y) \mid x < 2y - 1\}$

*8. The inverse of a relation is itself a relation obtained by interchanging the first and second elements of the original relation. The inverse of a relation with point (x, y) is the relation with point (y, x). Determine if the inverse of each relation below is a function.
   (a) $\{(x, x + 1) \mid x \text{ is a whole number}\}$
   (b) $\{(x^2, x) \mid x \text{ is a whole number}\}$
   (c) $\{(3x + 1, 2x) \mid x \text{ is a whole number}\}$

4

Finding Equations of Linear Functions

Linear functions are important in education, business, sociology, and statistics because they establish what we call trend lines. For this reason, we study linear functions in greater depth than other functions. In Section 2 we learned that every linear equation may be expressed in the form $ax + by + c = 0$, where a, b, and c are real numbers. We now learn to establish such a relationship from certain given information.

Example: Suppose that the enrollment at a college with $300 tuition per student is 3400; enrollment at a college with $400 tuition is 3000; and enrollment at a college with $500 tuition is 2600. Given this trend, what will be the enrollment at a college if the tuition cost is $600? $700?

Make a table and note any relationship existing between the two variables of this function.

x (tuition)	$300	$400	$500
y (enrollment)	3400	3000	2600

Table 14-1

Notice from the table that, as the x increases by $100, the y decreases by 400. Assume that enrollment will continue to decrease in the same rate as tuition increases. Therefore, when the tuition is $600 the enrollment would be 2200; when the tuition is $700, the enrollment would be 1800.

We were able to find the answers in the example above without expressing the relationship between x and y in the form of an equation, but what if we wished to find a projected enrollment for a tuition of $267 or of $523? To do this work, an equation must be found. The processes of finding equations vary from the point-slope form to the two-point form to the slope-intercept form. Each of these methods will be considered in this section.

Geometrically, it is evident that two points determine a straight line

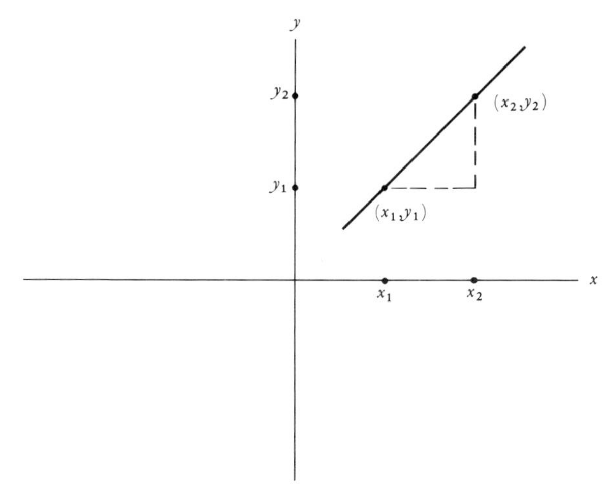

Figure 14-7

(Figure 14-7). We now investigate the relationship between points on a line and the equation of a line. Let (x_1, y_1) and (x_2, y_2) be any two real points on a straight line. The *slope m* of the line is the quotient of the vertical distance divided by the horizontal distance, or of the *rise* divided by the *run*. The slope may also be defined by the fraction

$$m = \frac{y_2 - y_1}{x_2 - x_1} \quad \text{or} \quad m = \frac{y_1 - y_2}{x_1 - x_2}.$$

The slope of a straight line is often called the rate of change of one variable with respect to the other. For example, a slope of 2 when y is a function of x indicates that the rate of change of y with respect to x is 2. When x changes by 1 unit, y must change by 2. When x changes by 3 units, y changes by 6, and so on. The slopes of two parallel lines are equal; and the slopes of perpendicular lines are m and $-1/m$.

Example: The slope of the previous tuition-enrollment example may be found by using any two of the known points.

$$m = \frac{3400 - 3000}{300 - 400} = -4$$

The definition of slope is used in establishing the point-slope formula for finding the equation of a line where at least one point and the slope are given. We know that the graph of a linear function is a straight line and that any number pair (x_1, y_1) that satisfies the equation of the linear function can be made to correspond to a point in the real plane which lies on the line and has coordinates (x_1, y_1). The slope of the line is the ratio of the change in the ordinate along the line to the corresponding change in the abscissa. For example, if the slope is given as 3, one can find a second point, given a fixed point (x_1, y_1), by changing the ordinate by 3 while the abscissa change is held to 1. Thus, the second point is located on the line at the point $(x_1 + 1, y_1 + 3)$.

The equation of a line can be found in a similar manner. For example, let (x_1, y_1) be a fixed point on a given straight line, and let (x, y) represent any other point on the line. The slope m from (x_1, y_1) to (x, y) is given by $(y - y_1)/(x - x_1)$. Then

$$\frac{y - y_1}{x - x_1} = m \quad \text{or} \quad y - y_1 = m(x - x_1).$$

Since the coordinates x and y are variables representing any point along the line, the equality involving x and y represents the functional relationship between x and y—that is, the equation of the line with slope m through the fixed point (x_1, y_1).

Example: Suppose we wish to find the equation of the line in the tuition-enrollment example when given the following facts. When the tuition is \$300, the enrollment is 3400 and the slope is -4. Thus we have a point, (300, 3400), and the slope of the line.

$$\begin{aligned} y - 3400 &= -4(x - 300) \\ y - 3400 &= -4x + 1200 \\ 4x + y - 4600 &= 0 \end{aligned}$$

Example: Find the equation of the line through (2, 1) with a slope of 3.

$$y - 1 = 3(x - 2) \quad \text{or} \quad y = 3x - 5.$$

Sometimes, instead of being given a point and the slope, you are given only two points. The two-point method of finding the equation of a line consists simply of using these two points to determine the slope of the line and then using the point-slope formula to establish the equation.

Example: Suppose that only two points, (300, 3400) and (400, 3000) were given in the tuition-enrollment exercise.

$$m = \frac{3400 - 3000}{300 - 400} = -4$$

Then,

$$y - 3000 = -4(x - 400)$$

and

$$4x + y - 4600 = 0.$$

Example: If (2, 3) and (-1, 4) are on a straight line, find the equation of the line.

$$m = \frac{4 - 3}{-1 - 2} = \frac{-1}{3}$$

$$y - 3 = \left(\frac{-1}{3}\right)(x - 2),$$

or

$$x + 3y - 11 = 0.$$

As a special case, the fixed point on the line may be taken to be the point where the straight line crosses the y axis. The coordinates of such a point, which is called the y-intercept, are usually written as $(0, b)$. In this slope-intercept method, the equation of a line becomes $y = mx + b$. This form of the equation suggests that if a linear function is solved for y in terms of x, one can determine the slope of the line by inspection.

Example: Find the equation of a line that crosses the y axis at -21/8 and has a slope of 1/4.

$$y = mx + b$$

$$y = \frac{1}{4}(x) + \frac{-21}{8}$$

$$2x - 8y - 21 = 0$$

Example: Find the slope of $3x + 2y = 5$.

Write the equation as

$$y = \frac{-3x}{2} + \frac{5}{2}.$$

The slope is -3/2. The point at which the line crosses the y axis is (0, 5/2).

Examples: For each item below, find the equation of the line that satisfies each given condition.

(a) Has a slope of -4 and contains the point (2, -1)
(b) Contains (3, 7) and (2, -1)
(c) Has a slope of -2 and a y-intercept of 4
(d) Has the same slope as $3x + 2y = 7$ and contains the point (1, 1)

The solutions are:

(a) $y + 1 = -4(x - 2)$, or $y + 4x = 7$
(b) The slope of the line is

$$m = \frac{7 - (-1)}{3 - 2} = 8.$$

Thus,

$$y - 7 = 8(x - 3) \quad \text{or} \quad y - 8x = -17.$$

(c) $y = -2x + 4$
(d) Solving for y yields

$$y = \frac{-3x}{2} + \frac{7}{2}.$$

Thus, the slope of the line is -3/2. Now

$$y - 1 = \frac{-3(x - 1)}{2} \quad \text{or} \quad 2y + 3x = 5.$$

Exercise Set 14-4

1. Graph the line and then find the equation if the line has
   (a) a slope of 4 and goes through the point (2, 3).
   (b) a slope of -2 and goes through the point (4, -1).
   (c) a slope of 1/2 and goes through the point (-1, 1).
   (d) a slope of -7/2 and goes through the point (3, 4).
2. For the following linear functions, what is the rate of change of y with respect to x? What are the y-intercepts? Graph the functions, assuming the domains to be the set of real numbers.

(a) $y = 3x + 2$ (b) $y + 2x - 1 = 0$
(c) $x = 3y + 1$ (d) $2y = 1 - x$

(e) $y = \dfrac{x - 4}{2}$ (f) $x = \dfrac{y - 1}{3}$

3. Find the equation of the line characterized by each of the following.
   (a) The line contains the two points (1, -3) and (4, 5).
   (b) The line has a slope of -3 and goes through the point (7, 1).
   (c) The line has a slope of 1 and goes through the point (-7, 1).
   (d) The line contains two points, (0, 1) and (4, 3).
   (e) The line has a y-intercept of 4 and a slope of 5.
   (f) The line has a y-intercept of 6 and a slope of -3.
4. In a certain industrial city, it is believed that the pollution counts increase linearly from 7:00 A.M. to 2:00 P.M. At 8:00 A.M. the pollution count is 140. At 10:00 A.M. the count is 200. Find the linear equation representing this increase and predict the pollution count at 11:00 A.M. At 1:00 P.M.
5. (a) Find the equation of the line parallel to the x-axis which contains the point (1, 7).
   (b) Find the equation of the line parallel to the y-axis which contains the point (1, 7).
6. After traveling two hours on his way home, a college student has covered 125 miles. If he continues at this rate and reaches his destination in $7\frac{1}{2}$ hours, how far is his home from the college?
7. Determine an equation expressing a relationship between centigrade and Fahrenheit measurements. If F = 32, then C = 0, and if F = 68, then C = 20.

5

Systems of Linear Equations and Applications

Now that we have learned to find the equations for lines, we can learn to find the solution set of two linear equations in two variables. The solution set of the given equations is the intersection of the solution sets, where the solution set of each equation is the set of ordered pairs that satisfy the equation.

The representation of a linear function by means of a straight line permits us to solve the system of two linear equations graphically. The graph of the solution set of the first equation is a straight line, and the graph of the solution set of the second equation is also a straight line. Hence, the graph of the solution set of the system of two equations is the intersection of the two lines. That is, the intersection of the lines $y = m_1x + b_1$ and $y = m_2x + b_2$ is represented by the set

$$\{(x, y) \mid y = m_1x + b_1\} \cap \{(x, y) \mid y = m_2x + b_2\}.$$

From a geometric viewpoint, three possibilities may occur: (a) the two lines may intersect in exactly one point, (b) the two lines may coincide, or (c) the two lines may be parallel. As an example of each of these possibilities, consider the following systems of equations.

Example: Solve the following system graphically.

$$\begin{aligned} 3x + y &= 3 \\ x + 2y &= -4 \end{aligned}$$

The first equation is satisfied by infinitely many ordered pairs, three of which are (0, 3), (1, 0), (3, -6). Likewise, some number pairs which satisfy $x + 2y = -4$ are (-4, 0), (0, -2), (2, -3). The graphs of these two equations are given in Figure 14-8. The intersection of the two lines in

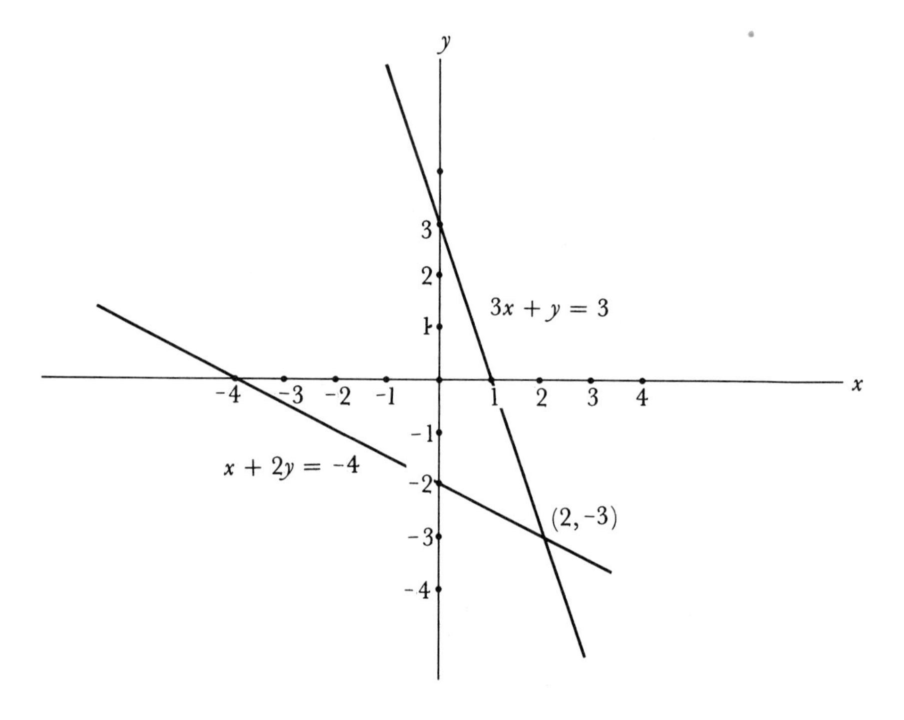

Figure 14-8

Figure 14-8 seems to be the point (2, -3). We can find out if this number pair is a solution of the system of equations by checking it in each equation.

Although the solution of a system of two linear equations with two variables by graphs gives an excellent picture of the relationship between the two variables, the method is time consuming and may not be accurate if the numbers that compose the ordered pairs in the solution set are not integers. Consequently, an algebraic method of solution, usually called the *elimination method*, is more often used. This method seeks to combine the equations in such a way that one of the unknowns appears with a zero coefficient. The derived equation in one unknown is then solved as explained earlier in this chapter. To eliminate one unknown from a system of two equations in two unknowns, multiply one or both of the equations by numbers that will make the coefficients of one of the unknowns the same, or the same except for sign. Then, add or subtract like terms of the equations to obtain an equation with only one unknown. To illustrate the algebraic solution, the solution sets will now be obtained for some of the same systems that were solved graphically in the preceding examples.

Example: Find the solution set of

$$\begin{aligned} 3x + y &= 3 \\ x + 2y &= -4. \end{aligned}$$

In order to make the coefficients of y the same in both equations, multiply each term of the first equation by 2 to obtain

$$6x + 2y = 6.$$

Subtract like terms of the second equation

$$x + 2y = -4.$$

This gives

$$5x = 10.$$

Divide both sides by 5 to obtain

$$x = 2.$$

Substituting $x = 2$ into the first equation gives

$$3(2) + y = 3.$$

Hence, $$y = 3 - 6 = -3.$$

The solution set is $\{(2, \ -3)\}$, and one can prove that this point lies on both lines by substituting its coordinates into both equations and showing that it satisfies both equations.

Substituting $x = 2$ and $y = -3$ into the first equation gives

$$3(2) + (-3) = 6 - 3 = 3.$$

Substituting $x = 2$ and $y = -3$ into the second equation gives

$$(2) + 2(-3) = 2 - 6 = -4.$$

Example: Find the solution set of

$$3x + y = 3,$$
$$6x + 2y = 6.$$

Multiply both sides of the first equation by 2 to obtain

$$6x + 2y = 6.$$

Subtract the second equation term by term,

$$6x + 2y = 6$$

to obtain

$$0 = 0.$$

All number pairs, such as (0, 3), (1, 0), (2, -3), which satisfy $3x + y = 3$ also satisfy $6x + 2y = 6$. Hence, these lines have every point in common; that is, the lines coincide (see Figure 14-9). Every ordered pair that is a solution of one equation is a solution of the other.

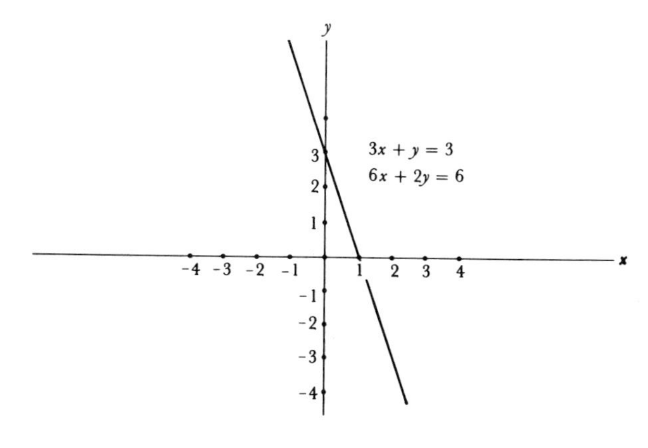

Figure 14-9

Example: Solve (if possible) graphically.

$$2x - 3y = 4$$
$$4x - 6y = -2$$

The graph for each of these equations is shown in Figure 14-10. Figure 14-10 indicates that the two lines are parallel and will never intersect.

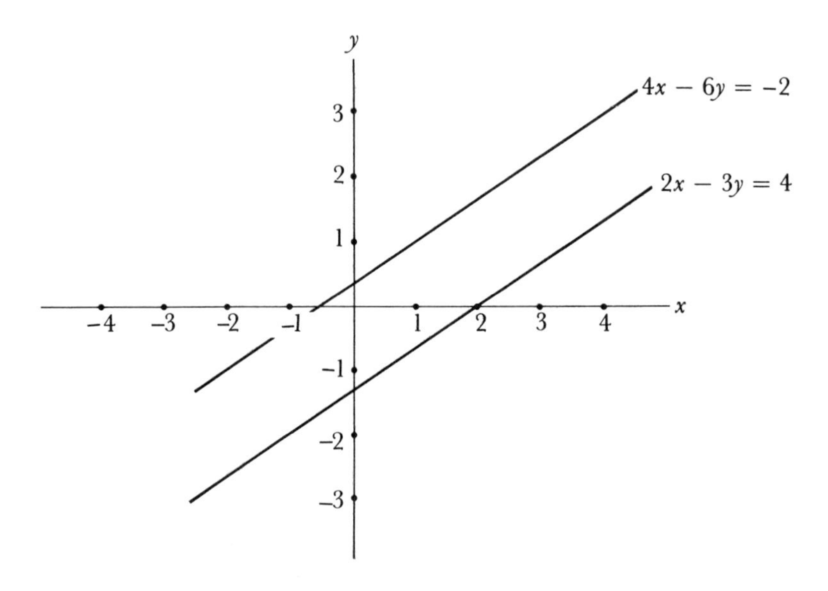

Figure 14-10

Example: Algebraically, find the solution set of

$$3x + y = 3$$
$$6x + 2y = 12.$$

Multiplying the first equation through by 2 gives

$$6x + 2y = 6.$$

Subtracting the second equation from this equation yields zero on the left side of the equation and -6 on the right side of the equation. Since $0 \neq -6$, no numbers x and y can satisfy both equations. This shows that the graphs of these equations must be parallel lines.

We may summarize the relationship between two equations by the following discussion. If two lines intersect, the two equations have a unique simultaneous solution; if the lines are parallel, the system has no solution. If two lines, such as $2x - y = 4$ and $4x - 2y = 8$, coincide, there is an infinite number of solutions, since every ordered pair that satisfies one equation also satisfies the other equation.

Intersections of linear inequalities give very interesting examples when demonstrated by graphs. Since the solution set of each linear inequality is a half-plane, then the solution set of a system of two inequalities is the intersection of two half-planes. A good method to illustrate this idea is by means of a graph.

Example: Graph the solution set determined by

$$2x + y - 1 > 0$$
$$x - 2y + 2 < 0.$$

First draw the two half-planes. We note that $2x + y - 1 > 0$ is a right half-plane and $x - 2y + 2 < 0$ is a left half-plane. The region common to the two is shaded in Figure 14-11.

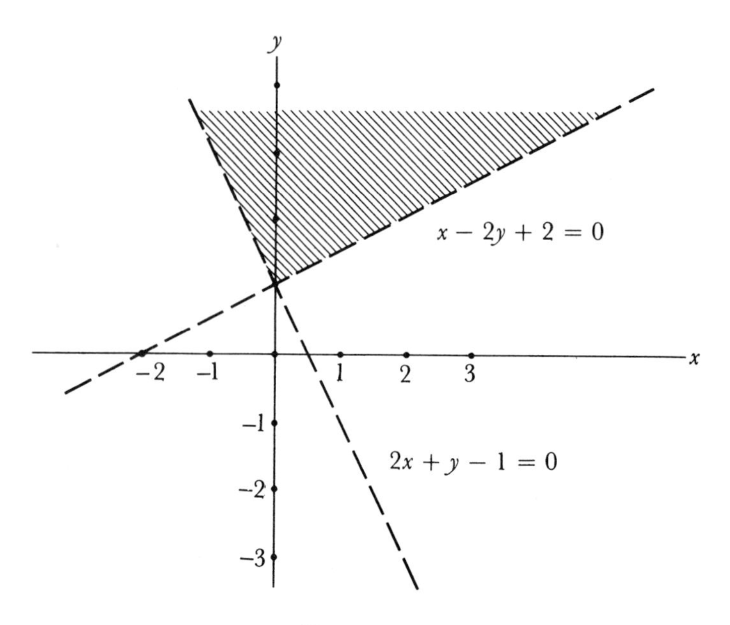

Figure 14-11

If more than two linear inequalities are considered in the system, the solution is often a closed figure.

Example: Graph the set determined by

$$\begin{aligned} x - 2y + 2 &< 0 \\ 2x + y - 1 &> 0 \\ y - 3 &\leq 0 \\ x - 2 &< 0. \end{aligned}$$

The first two linear inequalities are the same as in the preceding example. The third and fourth inequalities are bounded by lines parallel to the coordinate axes. The solution set is shaded in Figure 14-12.

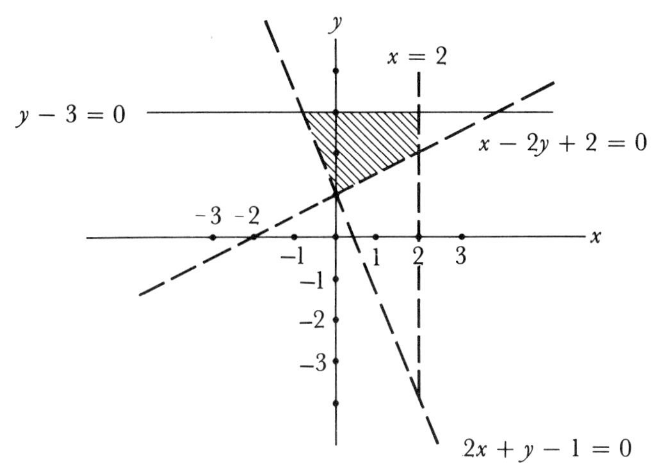

Figure 14-12

A "mixture" problem represents an application of solutions of two equations in two unknowns. The following problem was given earlier as an exercise and solved using one unknown. The solution given now illustrates how two unknowns may be used to solve the same problem.

Example: Suppose a candy store has 40 pounds of candy that sells for \$1.40 a pound. Since this candy has not been selling well lately, the owner believes that the price may be a little too high, but he does not want to lose any of his profit. How much candy that sells for \$1 a pound should he mix with the 40 pounds of \$1.40-a-pound candy to make a mixture which would sell at \$1.25 a pound? This problem can be solved as follows.

Let x = number of pounds of \$1-a-pound candy to be mixed with the \$1.40-a-pound candy,
y = number of pounds of \$1.25-a-pound candy.

The equations representing the conditions of the problem are: Amount of money received from \$1.40-a-pound candy before mixing + amount of money received from \$1-a-pound candy before mixing = amount of money received from mixture at \$1.25 a pound.

$$(\$1.40)(40) + \$1x = \$1.25y \quad \text{or} \quad 56 + x = 1.25y$$

Number of pounds of \$1.40 candy + number of pounds of \$1 candy = number of pounds of mixture, or

$$40 + x = y.$$

Now we need the solution set of the following system of two equations in two unknowns:

$$56 + x = 1.25y,$$
$$40 + x = y.$$

Subtracting gives

$$16 = 0.25y,$$
$$64 = y.$$

Since

$$40 + x = y,$$

substituting $y = 64$ gives

$$40 + x = 64.$$

Therefore,

$$x = 64 - 40 \quad \text{or} \quad x = 24.$$

Thus, if 24 pounds of \$1-a-pound candy are mixed with 40 pounds at \$1.40 a pound, the mixture of 64 pounds could be sold at \$1.25 a pound without any loss of profit. This can be checked by multiplying (24) $\cdot$ \$1 to obtain \$24 from the \$1-a-pound candy; (\$1.40)(40) = \$56 to be received from the \$1.40-a-pound candy. The sum

$$\$24 + \$56 = \$80$$

would be the amount received if the candy were sold separately.

Now, if 64 pounds of the mixture were sold at \$1.25 a pound, (64)(\$1.25) = \$80 would be obtained also, so there would be no loss of profit.

Exercise Set 14-5

1. Solve the following systems of linear equations graphically.

   (a) $x + 2y = 3$, $x = 6$

   (b) $x - y = 4$, $y + 3 = 0$

   (c) $2x + y = 6$, $x - 3y = 6$

   (d) $5x + y = -3$, $x - 3y = -3$

   (e) $y = 3x$, $x - y = 4$

   (f) $x = 2y$, $x + y = 8$

   (g) $3x - 2y = 4$, $9x = 6y$

   (h) $2x - y = -4$, $3y = 6x + 12$

   (i) $2x + 3y = 5$, $x = \dfrac{15 - 9y}{6}$

   (j) $3x - 5y = 3$, $y = \dfrac{6x - 6}{10}$

2. Tabulate the following sets and check your results graphically.

   (a) $\{(x, y) \mid x + y = 5\} \cap \{(x, y) \mid 2x - 3y = -5\}$

   (b) $\{(x, y) \mid x + 2y = 1\} \cap \{(x, y) \mid 3x + 4y = 5\}$

   (c) $\{(x, y) \mid 5x - 2y = -4\} \cap \{(x, y) \mid 3x + 5y = 10\}$

   (d) $\{(x, y) \mid 2x + y = -1\} \cap \{(x, y) \mid 3x - 2y = -12\}$

3. Graph the following systems of inequalities.

   (a) $3x + y - 4 \geq 0$, $x - 2y + 1 < 0$, $y < 4$

   (b) $4x - 5y + 25 > 0$, $5x + 3y - 15 \leq 0$, $y > -3$

   (c) $2x + y - 4 > 0$, $x - 2y + 2 \leq 0$, $y - 3 < 0$, $x + y - 4 \leq 0$

   (d) $3x + y - 3 \geq 0$, $x - 3y + 2 < 0$, $y - 4 < 0$, $x - 2 \leq 0$

4. Mr. Black invests \$4000, part at 3% annual interest and part at 4% annual interest. How much does he invest at each of these rates if he earns \$135 in interest in one year?
5. A coin collector has \$45 in quarters and dimes. How many coins of each kind does he have if the total number of his coins is 240?
6. A chemist has a 90% solution of sulphuric acid and a 70% solution of the same acid. How many gallons of each must be mixed to make 200 gallons of 75% solution?

7. A candy store proprietor wishes to mix candy that sells at \$1.20 a pound with candy that sells for \$1.50 a pound to make a mixture that would sell for \$1.40 a pound. How many pounds of each kind of candy should be used if he wishes to make 80 pounds of the \$1.40 candy?
8. A grocer wishes to mix nuts that sell for \$0.48 a pound with nuts that sell for \$0.76 a pound in order to make a mixture that would sell for \$0.60 a pound. How many pounds of the \$0.48-a-pound nuts should be mixed with 50 pounds of the \$0.76-a-pound nuts?
9. Solve Exercise 7 if the proprietor wishes to make 100 pounds of the \$1.40-a-pound candy.
10. Solve Exercise 8 if the \$0.48-a-pound nuts are to be mixed with 80 pounds of the \$0.76-a-pound nuts.

6

Quadratic Functions

Now consider certain functions termed *quadratic functions*, which are of the form $y = f(x) = ax^2 + bx + c$, where $a \neq 0$. Since the largest exponent of x is 2, the equation is said to be of the *second degree*. Quadratic functions have many interesting properties that could be discussed at length. However, for the purposes of this course, we will consider only briefly the graph of the quadratic function and then determine a procedure for solving quadratic equations.

To draw the graph of $y = 2x^2 - 20x + 46$, we construct the following table and plot the points (Figure 14-13).

A curve is sketched through these points. This curve is called a *parabola*. Notice the broken line drawn in the middle of the graph. This line is called the *line of symmetry* of the graph. Notice also that the graph seems to have its minimum (sometimes maximum) value on this line of symmetry. The graph for this example opens upward. This is because the number multiplied by x^2 is positive. If it were negative, the graph would open downward. This fact will be illustrated in the next example.

x	2	3	4	5	6	7	8
y	14	4	-2	-4	-2	4	14

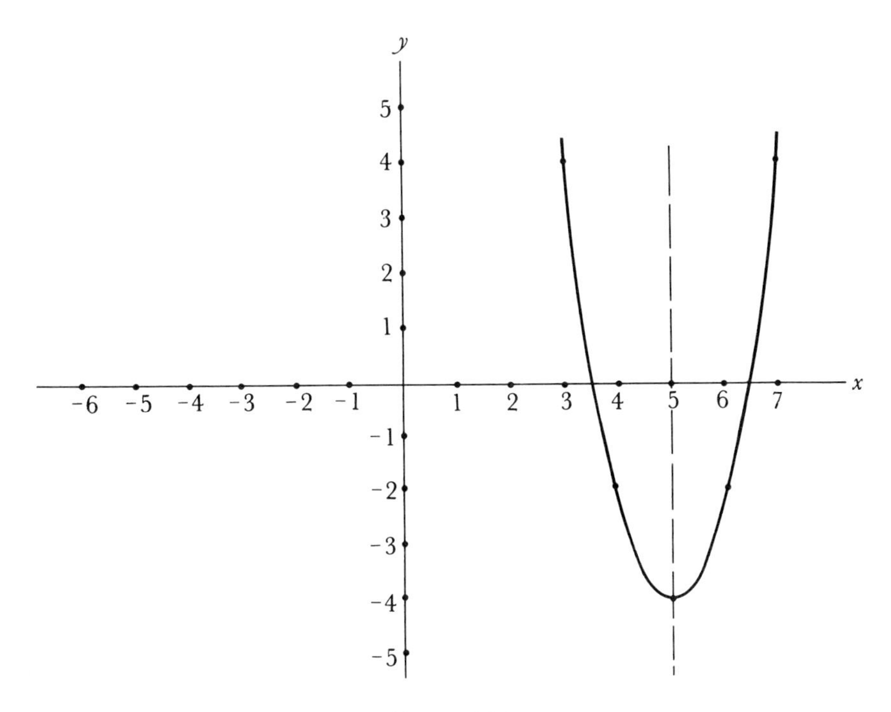

Figure 14-13

Example: Find the graph of $y = -2x^2 + 8x - 5$. A table for this function may be constructed as follows.

x	0	1	2	3	4
y	-5	1	3	1	-5

In Figure 14-14 the graph of this function has been sketched through the points listed in the table. Notice that this curve turns downward and that the line of symmetry is 2 units from the y axis.

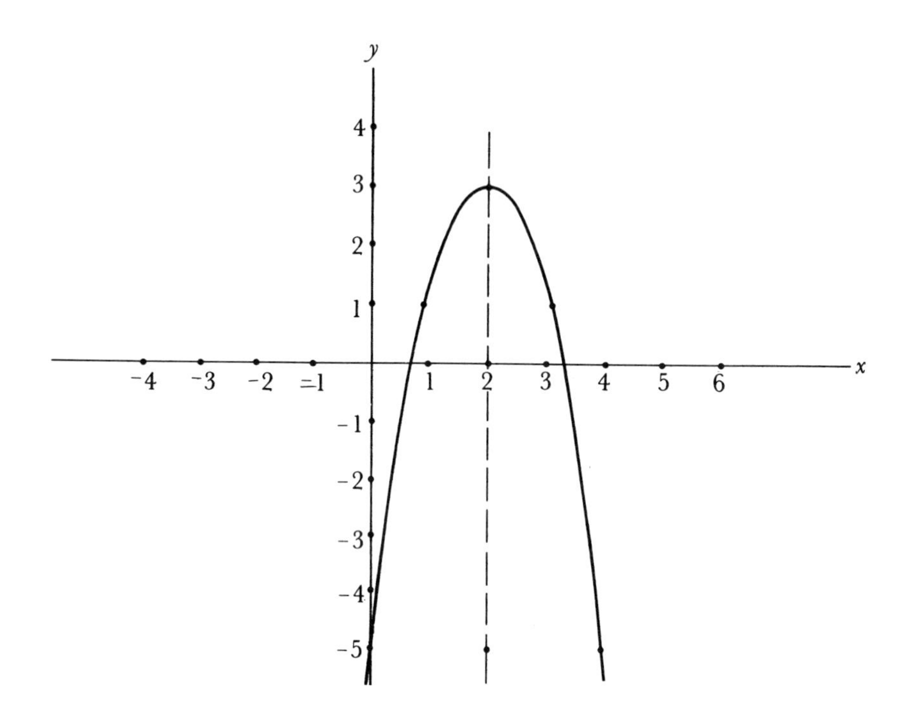

Figure 14-14

It can be shown that the function $f(x) = ax^2 + bx + c$ assumes a maximum or minimum value when $x = -b/2a$. For example, in the preceding illustration, the function assumed a maximum value when

$$x = \frac{-8}{2(-2)} = 2.$$

In the first example, $y = 2x^2 - 20x + 46$, the function assumed a minimum value when

$$x = \frac{-(-20)}{2(2)} = 5.$$

Since the line of symmetry must contain the point at which a function assumes a maximum or minimum value, it is useful to remember that the line of symmetry is the vertical line

$$x = -\frac{b}{2a}.$$

When the quadratic function is set equal to 0, we have a *quadratic equation*. Thus a quadratic equation is of the form $ax^2 + bx + c = 0$, where $a \neq 0$. The solution set of a quadratic equation consists of the values of x that make $ax^2 + bx + c = 0$. These are, of course, the same values that make $y = 0$ in the quadratic function. Therefore, y is 0 on the graph at the points where the curve crosses the x axis. This means that the solution set of a quadratic equation is identical to the values of x at the points where the graph of the corresponding quadratic function crosses the x axis. For example, in Figure 14-13 the solutions for $2x^2 - 20x + 46 = 0$ are between 3 and 4 and between 6 and 7. Similarly, in Figure 14-14, the solutions for $-2x^2 + 8x - 5 = 0$ are between 0 and 1 and between 3 and 4.

In order to find the solution set for $ax^2 + bx + c = 0$, without using graphs, we utilize a method called "completing the square." Since $a \neq 0$, we can divide both sides of the equation by a and write the equation as $x^2 + bx/a + c/a = 0$. We want to write $x^2 + bx/a + c/a$ in the form $(x - s)^2 + t$ or $x^2 - 2sx + s^2 + t$. For the two expressions, $x^2 + bx/a + c/a$ and $x^2 - 2sx + s^2 + t$, to be identical, $-2s = b/a$; then $s = -b/2a$. Also, $s^2 + t = c/a$; then $t = c/a - s^2 = c/a - b^2/4a^2 = (4ac - b^2)/4a^2$. Thus the quadratic equation $x^2 + bx/a + c/a = 0$ can be written in the form $(x + b/2a)^2 + (4ac - b^2)/4a^2 = 0$ or $(x + b/2a)^2 = (b^2 - 4ac)/4a^2$. Now, by extracting the square roots of both sides of the equation, we obtain

$$x + \frac{b}{2a} = \frac{\pm\sqrt{b^2 - 4ac}}{2a},$$

or

$$x = \frac{-b + \sqrt{b^2 - 4ac}}{2a} \quad \text{and} \quad x = \frac{-b - \sqrt{b^2 - 4ac}}{2a}.$$

This proves the following theorem.

Theorem 14-1: The quadratic equation

$$ax^2 + bx + c = 0, \qquad a \neq 0,$$

where a, b, and c are real numbers, has as solutions

$$x = \frac{-b + \sqrt{b^2 - 4ac}}{2a} \quad \text{and} \quad x = \frac{-b - \sqrt{b^2 - 4ac}}{2a}.$$

We should note at this time that if $b^2 - 4ac < 0$, we have the square root of a negative number, which is impossible to compute in the system of real numbers. Why? Thus, it is not possible to find the solution of *all* quadratic equations if we restrict our number system to real numbers. This inadequacy of the real number system will lead to an extension of the real number system in the next section. Note also that, when $b^2 - 4ac = 0$, we have two solutions that are identical.

Example: Solve $3x^2 - 21x = -36$.

Solution: If one finds the two solutions by substituting for a, b, and c in the formulas as given in Theorem 14-1, we say we are finding the solutions by using the quadratic formula. In this case, $a = 3$, $b = -21$, and $c = 36$. Thus,

$$x = \frac{-(-21) + \sqrt{(-21)^2 - 4(3)(36)}}{2(3)}$$

and

$$x = \frac{-(-21) - \sqrt{(-21)^2 - 4(3)(36)}}{2(3)},$$

so

$$x = \frac{21 + \sqrt{441 - 432}}{6} \quad \text{and} \quad x = \frac{21 - \sqrt{441 - 432}}{6},$$

$$x = \frac{21 + 3}{6} = 4 \quad \text{and} \quad x = \frac{21 - 3}{6} = 3.$$

Example: Solve $2x^2 + 2x = 12$. The two solutions are

$$x = \frac{-2 + \sqrt{(2)^2 - 4(2)(-12)}}{2(2)} \quad \text{and} \quad x = \frac{-2 - \sqrt{(2)^2 - 4(2)(-12)}}{2(2)},$$

so

$$x = \frac{-2 + \sqrt{100}}{4} = 2 \quad \text{and} \quad x = \frac{-2 - \sqrt{100}}{4} = -3.$$

Exercise Set 14-6

1. Solve each equation and verify the answers approximately (if they exist) by sketching the graph for $y = ax^2 + bx + c$.
   (a) $3x^2 + 9x = 12$ (b) $2x^2 + 6x - 30 = 0$
   (c) $6x^2 - 7x = 3$ (d) $20x^2 + 11x - 3 = 0$
   (e) $8x^2 - 5x = 3$ (f) $6x^2 - 13x + 6 = 0$
   (g) $x^2 + 3x - 5 = 0$ (h) $2x^2 - 5x - 3 = 0$
2. Find the equation of the line of symmetry and the point at which a maximum or minimum occurs for each part of Exercise 1.
3. Solve each equation by the quadratic formula.
   (a) $x^2 - 2x = 3$ (b) $x^2 - 2x = 15$
   (c) $2x^2 - 6x = 56$ (d) $x^2 + 2x = 24$
   (e) $12x^2 - x - 1 = 0$ (f) $6x^2 - 5x + 1 = 0$
   (g) $3x^2 + 11x = 4$ (h) $2x^2 - 7x + 3 = 0$
   (i) $5x^2 - 7x = 1$ (j) $3x^2 + 8x + 1 = 0$

*4. The formula for determining s, the number of feet a body falls from a resting position, is $s = 16t^2$ where t represents the number of seconds the body falls. Determine the time a body falls if the distance is given below.
   (a) 16 (b) t (c) $1 - 9t$
   (d) $-13t + 2t^2 - 3$ (e) $13t - 2$ (f) $-3t + 4$

7

Complex Numbers

It is evident that there are quadratic equations having no solutions in the system of real numbers. From a geometric point of view, the non-existence of real solutions is indicated when the graph of $y = ax^2 + bx + c$ has no points in common with the x axis. The following example illustrates algebraically that quadratic equations exist which have no real solutions.

Example: Let $x^2 - 2x + 5 = 0$. Using the quadratic formula, where $a = 1$, $b = -2$, and $c = 5$,

$$x = \frac{2 + \sqrt{4 - 4(1)(5)}}{2} \quad \text{and} \quad x = \frac{2 - \sqrt{4 - 4(1)(5)}}{2},$$

so

$$x = \frac{2 + \sqrt{-16}}{2} \qquad \text{and} \quad x = \frac{2 - \sqrt{-16}}{2}.$$

Since $\sqrt{-16}$ does not exist in the system of real numbers, there are no real solutions to this quadratic equation.

To verify geometrically that there are no real solutions of $x^2 - 2x + 5 = 0$, we graph the quadratic function $y = x^2 - 2x + 5$. Five ordered pairs that are solutions of $y = x^2 - 2x + 5$ are (-1, 8), (0, 5), (2, 5), (3, 8), and (1, 4). The graph of this function is shown in Figure 14-15. Notice that the graph does not cross the x axis.

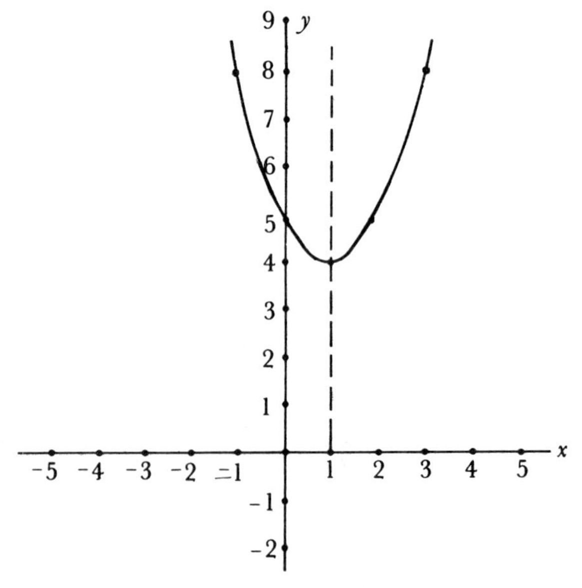

Figure 14-15

A very simple quadratic equation with no real roots is $x^2 + 1 = 0$. Let us compare this equation to $x + 1 = 0$, introduced when we were working with whole numbers. In the system of whole numbers, $x + 1 = 0$ had no solutions; but we introduced new numbers, called negative numbers, so that this equation could have a solution. In the same way, $x^2 + 1 = 0$ has no real solutions; but we will define a new number system in which it will have solutions. Thus, we are concerned with an extension of the real number system in order to provide solutions for equations such as $x^2 + 1 = 0$. As with all such extensions, the enlargement will be made in a way that will preserve the fundamental properties of addition, subtraction, multiplication, and division of real numbers.

Write $x^2 + 1 = 0$ in the form $x^2 = -1$. We now introduce (indeed, we *invent*) two new numbers that satisfy this equation; namely, numbers i and $-i$. Thus i and $-i$ are defined to be numbers such that $i^2 = (-i)^2 = -1$. These new numbers are called *imaginary numbers.* When they are combined with real numbers in the form $a + bi$, where a and b are real numbers, then the number $a + bi$ is called a *complex number.*

Example: Solve $x^2 - 2x + 5 = 0$.

Solution: Since $a = 1$, $b = -2$, and $c = 5$,

$$x = \frac{2 + \sqrt{4 - 4(1)(5)}}{2} \quad \text{and} \quad x = \frac{2 - \sqrt{4 - 4(1)(5)}}{2},$$

$$x = \frac{2 + \sqrt{-16}}{2} \quad \text{and} \quad x = \frac{2 - \sqrt{-16}}{2},$$

$$x = \frac{2 + 4\sqrt{-1}}{2} \quad \text{and} \quad x = \frac{2 - 4\sqrt{-1}}{2},$$

$$x = \frac{2 + 4i}{2} \quad \text{and} \quad x = \frac{2 - 4i}{2},$$

$$x = 1 + 2i \quad \text{and} \quad x = 1 - 2i.$$

We have seen how complex numbers can be generated as solutions of quadratic equations. However, this approach to the definition of a complex number is lacking in mathematical appeal. Let us consider instead the following.

Definition 14-1: *Complex numbers* $a + bi$ and $c + di$ are ordered pairs of real numbers (a, b) and (c, d) with the following properties:

(a) Equality: $(a, b) = (c, d)$ if and only if $a = c$ and $b = d$.
(b) Addition: $(a, b) + (c, d) = (a + c, b + d)$.
(c) Multiplication: $(a, b) \cdot (c, d) = (ac - bd, bc + ad)$.

Examples:

(a) $(3+4i)+(5-7i)=(3, 4)+(5, -7)$
$$=(3+5, 4+-7)=(8, -3)=8-3i$$

(b) $(3+4i)\cdot(5-7i)=(3, 4)\cdot(5, -7)=(15+28, 20-21)$
$$=(43, -1)=43-i$$

The complex number $(a, 0)$ is called the *real part* of the complex number. It is perfectly clear that there is a one-to-one correspondence between the complex numbers $(a, 0)$ and the real numbers a. For example,

$$\begin{array}{cccccc} (1, 0) & (2, 0) & (3, 0) & (4, 0) & \cdots & (a, 0) \\ \updownarrow & \updownarrow & \updownarrow & \updownarrow & & \updownarrow \\ 1 & 2 & 3 & 4 & \cdots & a. \end{array}$$

Under this correspondence, sums correspond to sums and products to products.

$$\begin{array}{ccc} (a, 0)+(b, 0)=(a+b, 0) & \text{and} & (a, 0)\cdot(b, 0)=(ab, 0) \\ \updownarrow \quad\quad \updownarrow \quad\quad \updownarrow & & \updownarrow \quad\quad \updownarrow \quad\quad \updownarrow \\ a \quad + \quad b \quad = a+b & & a \quad \cdot \quad b \quad = \quad ab \end{array}$$

A complex number of the form $(0, c)$ is called the *imaginary part* of a complex number. The addition and the multiplication of imaginary parts are found, from Definition 14-1, to be $(0, b)+(0, d)=(0, b+d)$ and $(0, b)\cdot(0, d)=(-bd, 0)$. Thus we note that the sum of two imaginary numbers is an imaginary number and that the product of two imaginary numbers is a real number. For example, $3i+4i=(0, 3)+(0, 4)=(0, 7)$ and $(3i)\cdot(4i)=(0, 3)\cdot(0, 4)=(-12, 0)$.

Identifying a with $(a, 0)$ and i with $(0, 1)$ is natural to do, since $i^2=i\cdot i=(0, 1)(0, 1)=(-1, 0)=-1$. This, of course, is the result stated at the beginning of this section in an attempt to introduce new numbers to satisfy the equation $x^2+1=0$.

Theorem 14-2: The following properties hold for the system of complex numbers where (a, b), (c, d), and (e, f) represent the three complex numbers $a+bi$, $c+di$, and $e+fi$.

Addition

(a) Closure: The sum of two complex numbers is a complex number.
(b) Commutative property: $(a, b)+(c, d)=(c, d)+(a, b)$.
(c) Associative property:

$$[(a, b) + (c, d)] + (e, f) = (a, b) + [(c, d) + (e, f)].$$

(d) The additive identity for complex numbers is $(0, 0)$.
(e) The additive inverse of (a, b) is $(-a, -b)$.

Multiplication

(f) Closure: The product of two complex numbers is a complex number.
(g) Commutative property: $(a, b) \cdot (c, d) = (c, d) \cdot (a, b)$.
(h) Associative property:

$$[(a, b) \cdot (c, d)] \cdot (e, f) = (a, b) \cdot [(c, d) \cdot (e, f)].$$

(i) The multiplicative identity for complex numbers is $(1, 0)$.
(j) The multiplicative inverse of (a, b) is $[a/(a^2 + b^2), -b/(a^2 + b^2)]$.
(k) Distributive property:

$$(a, b)[(c, d) + (e, f)] = (a, b) \cdot (c, d) + (a, b) \cdot (e, f).$$

Proof: The addition of complex numbers $(a, b) + (c, d) = (a + c, b + d)$ is closed as a result of the definition of addition. Since a, b, c, and d are real numbers, then $a + c$ and $b + d$ are real numbers; and thus, the answer, as a number pair, is in the form of a complex number. Next,

$$(a, b) + (c, d) = (a + c, b + d) = (c + a, d + b) = (c, d) + (a, b),$$

where $c + a = a + c$ and $b + d = d + b$, because a, b, c, and d are real numbers. Thus, addition of complex numbers is a commutative operation. Also,

$$\begin{aligned} [(a, b) + (c, d)] + (e, f) &= [(a + c, b + d)] + (e, f) \\ &= (a + c + e, b + d + f) \\ &= (a, b) + [(c + e, d + f)] \\ &= (a, b) + [(c, d) + (e, f)]. \end{aligned}$$

This demonstrates that addition of complex numbers is associative.

$(0, 0)$ is an additive identity for complex numbers, since $(a, b) + (0, 0) = (a, b)$.

$(-a, -b)$ is the additive inverse of (a, b) because

$$(a, b) + (-a, -b) = (a + -a, b + -b) = (0, 0).$$

The proofs of the other parts of Theorem 14-2 are left as exercises.

Definition 14-2: The following are inverse operations for complex numbers $a + bi$ and $c + di$. If (a, b) and (c, d) are complex numbers, then

(a) Subtraction: $(a, b) - (c, d) = (a - c, b - d)$.

(b) Division: $\dfrac{(a, b)}{(c, d)} = \left(\dfrac{ac + bd}{c^2 + d^2}, \dfrac{bc - ad}{c^2 + d^2}\right)$,

where $(c, d) \neq (0, 0)$.

This definition agrees with the concept that subtraction is the inverse of the operation of addition: $z_1 - z_2 = z_3$, where z_1, z_2, and z_3 are complex numbers, if and only if $z_1 = z_2 + z_3$. Similarly, division is the inverse of multiplication because

$$\frac{(a, b)}{(c, d)} = \left(\frac{ac + bd}{c^2 + d^2}, \frac{bc - ad}{c^2 + d^2}\right),$$

since

$$\begin{aligned}(a, b) &= (c, d) \cdot \left(\frac{ac + bd}{c^2 + d^2}, \frac{bc - ad}{c^2 + d^2}\right)\\ &= \left(\frac{cac + cbd - dbc + dad}{c^2 + d^2}, \frac{dac + bd^2 + cbc - cad}{c^2 + d^2}\right)\\ &= \left(\frac{ac^2 + ad^2}{c^2 + d^2}, \frac{bd^2 + bc^2}{c^2 + d^2}\right) = (a, b).\end{aligned}$$

Examples: (a) Simplify i^8.

$$i^8 = (i^2)^4 = (-1)^4 = 1$$

(b) Subtract.

$$\begin{aligned}(16 - 3i) - (8 + 7i) = (16, -3) - (8, 7) &= (16 - 8, -3 - 7)\\ &= 8 - 10i\end{aligned}$$

(c) Divide.

$$\frac{6 + 4i}{2 + 3i} = (6, 4) \div (2, 3) = \left(\frac{12 + 12}{13}, \frac{8 - 18}{13}\right) = \frac{24}{13} - \frac{10i}{13}$$

Exercise Set 14-7

1. Indicate by a check the sets in which each number is a member.

	Natural Numbers	*Whole Numbers*	*Integers*	*Rationals*	*Reals*	*Complex*
0						
1						
$i + \sqrt{3}$						
$-(5/2)\sqrt{529}$						
$-5/2)\sqrt{-1024}$						
$-(5/2)\sqrt{-(-576)}$						

2. Simplify. (Leave answer in the form $a + bi$.)
   (a) i^{11} (b) $(-i)^6$
   (c) $(-i)^7$ (d) $(4i)^3$
   (e) $(3+2i)(4-i)$ (f) $(2+3i)(3-2i)$
   (g) $(6+4i)+(7-3i)$ (h) $(2+7i)-(3-4i)$
   (i) $\dfrac{6+5i}{2-3i}$ (j) $\dfrac{5+i}{2+5i}$
   (k) $6 - \dfrac{4}{2+i}$ (l) $3i - \dfrac{6}{i-3}$
3. Determine the additive and multiplicative inverses.
   (a) $2+5i$ (b) $5-2i$
   (c) $-3-4i$ (d) $-\sqrt{7}+\sqrt{9}i$
   (e) $-2-i$ (f) $\frac{1}{2}-\frac{1}{3}i$

*4. Is the system of complex numbers a group? A ring? A field? An integral domain?

*5. Prove that multiplication of complex numbers is
   (a) commutative.
   (b) associative.
   (c) distributive over addition.

*6. Prove that the multiplicative identity for complex numbers is (1, 0).

*7. The *conjugate* of a complex number (a, b) is $(a, -b)$. Prove that the product of a complex number and its conjugate is always real.

8

Summary

We began our discussion of number systems in Chapter 3 with the whole numbers. Since an equation of the form $x + a = 0$ did not have a solution, an extension was made to the system of integers. However, in the system of integers, the equation $ax = 1$ did not always have a solution until we created the rational numbers. But in the system of rational numbers, the equation $x^2 = N$, where N was not a square, still did not have a solution. To get a solution for this equation we considered the real numbers. Finally, in the system of real numbers, we could not solve $x^2 = -1$; so we created the system of complex numbers.

Now that we have established the system of complex numbers, we find that no further extension of our system is needed. Karl Friedrich Gauss proved, in what has become known as the Fundamental Theorem of Algebra, that every equation of the form $a_0 + a_1x + a_2x^2 + \cdots + a_nx^n = 0$ has a solution in the system of complex numbers. A schematic diagram is drawn to show the subsets of each set of numbers we have developed (Figure 14-16).

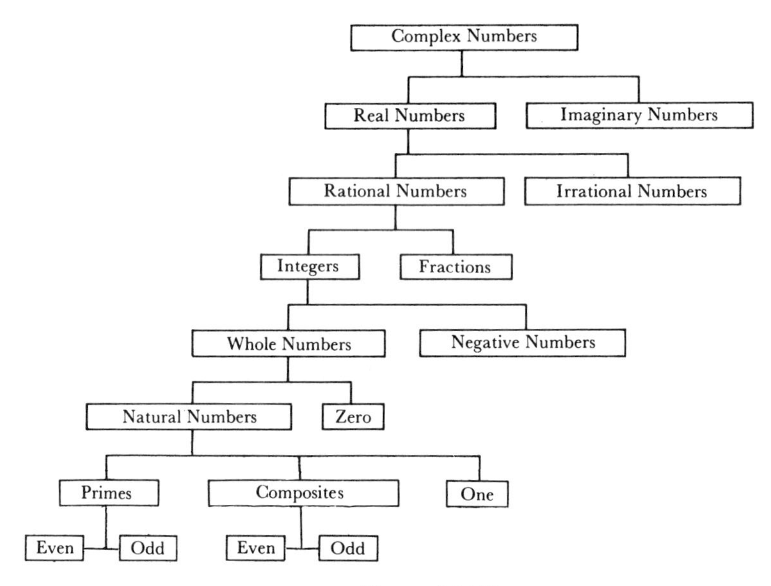

Figure 14-16

Review Exercise Set 14-8

1. Classify the following as true or as false.
   (a) The real numbers are a proper subset of the complex numbers.
   (b) 1 is not a complex number.
   (c) Two linear equations are either coincident or parallel.

(d) There is no point on a coordinate axis such that $(a, b) = (b, a)$.
(e) All quadratic equations graph in the form of a parabola.
(f) The set of real numbers allows the solution of sets impossible to solve with complex numbers.

2. Show that $3i$ is a square root of 9.
3. Solve the systems of equations and check by graphs.
   (a) $3x + 2y = 7$, $x - 4y = -7$ (b) $6x - 4y = -2$, $4x + y = -5$
4. Draw a graph, determine the equations for the points given below, and find the solution set of the lines in each part.

(a)

x	0	-4
y	2	4

and

x	1	-1
y	5	2

(b)

x	-2	3
y	-3	1

and

x	1	3
y	0	-2

(c) Point $(-2, 4)$, $m = 3/2$; and point $(4, -2)$, $m = 2/3$
(d) Point $(0, -5)$, $m = 4/5$; and point $(-5, 0)$, $m = 4/5$

5. If $f(x) = x^2 - 5x + 1$, compute:
   (a) $f(2)$ (b) $f(-1)$ (c) $f(c) - f(2)$ (d) $f(2y)$
6. Graph:
   (a) $x + 7y - 4 < 2$
   (b) $2x + y - 5 < 3$, $x + y - 4 > 1$
   (c) $6x + 4y < 5$, $x + 1 > 7$, $y - 3 < 5$
   (d) $3x + y - 4 < 0$, $6x - 2y > 3$, $x < 4$
7. Solve the following equations, and check your answers by graphing the corresponding function.
   (a) $3x^2 - 25x + 16 = -12$ (b) $5x^2 - 21x + 4 = 0$
8. In the field of complex numbers, find the solution sets of the following.
   (a) $x^2 + 9 = 0$ (b) $3x^2 + 48 = 0$
   (c) $x^2 - 2x + 5 = 0$ (d) $x^2 - 6x + 13 = 0$
   (e) $9x^2 - 6x + 2 = 0$ (f) $4x^2 - 4x + 5 = 0$
   (g) $3x^2 - x + 4 = 0$ (h) $5x^2 - 2x + 9 = 0$

Suggested Reading

Ordered Pairs, Functions, Graphs: Allendoerfer, pp. 139, 646–649, 652. Bouwsma, Corle, Clemson, pp. 35, 205–207. Byrne, pp. 17–18. Campbell, p. 32. Cooley, Wahlert, pp. 231–234, 265–285. Graham, pp. 27–28, 39,

41–42. Hutton, pp. 97–98. Meserve, Sobel, pp. 218–234. Moore, Little, p. 395. Nichols, Swain, pp. 100, 108, 388, 404–406. Peterson, Hashisaki, pp. 37, 61, 295–296. Scandura, pp. 76, 79. Weaver, Wolf, pp. 122–123. Willerding, pp. 150–153. Wheeler, Peeples, pp. 47, 50–51, 55–62. Wren, pp. 311–315. Zwier, Nyhoff, pp. 47–48, 299–301, 319–327.

Linear Equations: Apostle, pp. 122, 130. Boyer, p. 224. Hayden, p. 397. Wren, pp. 327–340.

Systems of Equations: Apostle, p. 131. Wren, pp. 333, 334–340.

Quadratic Equations: Boyer, pp. 235, 240. Hayden, pp. 317–332. Rees, pp. 134–149.

Complex Numbers: Apostle, pp. 16, 74. Boyer, p. 17. Copeland, pp. 223–233.

Bibliography

Adams, Joe Kennedy. *Basic Statistical Concepts*. New York: McGraw-Hill, 1955.

Allendoerfer, Carl B. *Principles of Arithmetic and Geometry for Elementary School Teachers*. New York: Macmillan, 1971.

Apostle, H. G. *A Survey of Basic Mathematics*. Boston: Little, Brown, 1960.

Armstrong, James W. *Mathematics for Elementary School Teachers*. New York: Harper & Row, 1968.

Beck, Anatole, Michael N. Bleicher, and Donald W. Crowe. *Excursions into Mathematics*. New York: Worth Publishers, 1969.

Bouwsma, Ward D., Clyde G. Corle, and Davis F. Clemson, Jr. *Basic Mathematics for Elementary Teachers*. New York: Ronald Press, 1967.

Boyer, Lee Emerson. *An Introduction to Mathematics: A Historical Development*. New York: Holt, Rinehart & Winston, 1955.

Brumfiel, Charles F., and Eugene F. Krause. *Elementary Mathematics for Teachers*. Reading, Mass.: Addison-Wesley, 1969.

Byrne, J. Richard. *Modern Elementary Mathematics*. New York: McGraw-Hill, 1966.

Byrne, J. Richard. *Number Systems: An Elementary Approach*. New York: McGraw-Hill, 1967.

Campbell, Howard E. *The Structure of Arithmetic*. New York: Appleton-Century-Crofts, 1970.

Cooley, Hollis R., and Howard E. Wahlert. *Introduction to Mathematics*. New York: Houghton Mifflin, 1968.

Copeland, Richard W. *Mathematics and the Elementary Teacher*. Philadelphia: W. B. Saunders, 1966.

Freund, John E. *Modern Elementary Statistics*. Englewood Cliffs, N. J.: Prentice-Hall, 1952, 1960.

Fujii, John N. *Numbers and Arithmetic.* Waltham, Mass.: Xerox (Blaisdell), 1965.

Garner, Meridon Vestal. *Mathematics for Elementary School Teachers.* Pacific Palisades, Calif.: Goodyear, 1969.

Garstens, Helen L., and Stanley B. Jackson. *Mathematics for Elementary School Teachers.* New York: Macmillan, 1967.

Graham, Malcolm. *Modern Elementary Mathematics.* New York: Harcourt Brace Jovanovich, 1970.

Hayden, Seymour. *Introductory Mathematics.* New York: Dodd, Mead, 1967.

Huntsberger, David V. *Elements of Statistical Inference.* Boston: Allyn & Bacon, 1961.

Hutton, Rex L. *Number Systems: An Intuitive Approach.* Scranton, Pa.: Intext Educational Publishers, 1971.

Jacobs, Harold R. *Mathematics: A Human Endeavor.* San Francisco: W. H. Freeman, 1970.

Keedy, Mervin L. *A Modern Introduction to Basic Mathematics.* Cambridge, Mass.: Addison-Wesley, 1963.

Lewis, E. Vernon. *Statistical Analysis Ideas and Methods.* Princeton, N. J.: D. Van Nostrand, 1963.

Mack, Sidney F. *Elementary Statistics.* New York: Holt, Rinehart & Winston, 1957.

McFarland, Dora, and Eunice M. Lewis. *Introduction to Modern Mathematics for Elementary Teachers.* Boston: Heath, 1966.

Mendenhall, William. *Introduction to Statistics.* Belmont, Calif.: Wadsworth, 1963.

Meserve, Bruce E., and Max A. Sobel. *Introduction to Mathematics.* Englewood Cliffs, N. J.: Prentice-Hall, 1964.

Mode, Elmer B. *Elements of Probability and Statistics.* Englewood Cliffs, N. J.: Prentice-Hall, 1966.

Moore, Charles G., and Charles E. Little. *Basic Concepts of Mathematics.* New York: McGraw-Hill, 1967.

Nichols, Eugene D., and Robert L. Swain. *Understanding Arithmetic.* New York: Holt, Rinehart & Winston, 1965.

Ohmer, Merlin M., and Clayton V. Aucoin. *Modern Mathematics for Elementary School Teachers.* Waltham, Mass.: Xerox (Blaisdell), 1966.

Ohmer, Merlin M., Clayton V. Aucoin, and Marion J. Cortez. *Elementary Contemporary Mathematics.* Waltham, Mass.: Xerox (Blaisdell), 1964.

Peterson, John M. *Basic Concepts of Elementary Mathematics.* Boston: Prindle, Weber & Schmidt, 1971.

Peterson, John A., and Joseph Hashisaki. *Theory of Arithmetic.* (2nd ed.) New York: John Wiley, 1967.

Podraza, Charles N., Larry L. Blevins, Arlys W. Hanson, and Harry C. Prall. *Mathematics: An Introduction.* Pacific Palisades, Calif.: Goodyear, 1969.

Rees, Paul K., and Fred W. Sparks. *College Algebra.* New York: McGraw-Hill, 1961.

Scandura, Joseph M. *Mathematics: Concrete Behavioral Foundations.* New York: Harper & Row, 1971.

Smart, James R. *Introductory Geometry: An Informal Approach.* (2nd ed.) Monterey, Calif.: Brooks/Cole, 1972.

Smith, Seaton E., Jr. *Explorations in Elementary Mathematics.* (2nd ed.) Englewood Cliffs, N. J.: Prentice-Hall, 1971.

Spector, Lawrence. *Liberal Arts Mathematics.* Reading, Mass.: Addison-Wesley, 1971.

Ward, Morgan, and Clarence Ethel Hardgrove. *Modern Elementary Mathematics.* Reading, Mass.: Addison-Wesley, 1964.

Weaver, Jay D., and Charles T. Wolf. *Modern Mathematics for Elementary Teachers.* (2nd ed.) Scranton, Pa.: International Textbook Co., 1968.

Webber, G. Cuthbert, and John A. Brown. *Basic Concepts of Mathematics.* Reading, Mass.: Addison-Wesley, 1963.

Wheeler, Ruric E., and W. D. Peeples, Jr. *Modern Mathematics for Business Students.* Monterey, Calif.: Brooks/Cole, 1969.

Willerding, Margaret F. *Elementary Mathematics: Its Structure and Concepts.* New York: John Wiley, 1966.

Wren, F. Lynwood. *Basic Mathematical Concepts.* New York: McGraw-Hill, 1965.

Zwier, Paul J., and Larry R. Nyhoff. *Essentials of College Mathematics.* New York: Holt, Rinehart & Winston, 1969.

Answers to Selected Problems

Exercise Set 1-1

4. (a) 22, 27, 32 (c) E, F, E (e) 18, 1, 23
   (g) ⬡ ⬡ ⬡
5. Dick is Tom's grandfather.
7. 2 7 6
   9 5 1
   4 3 8
8. (a) 8 seconds (c) No; dead men do not marry. (e) One coin is a quarter and the other is a nickel.
9. (a) If John had seen red ribbons on both James and Edward, he would have known that he had a blue ribbon. Since John did not know the color of his ribbon, James and Edward reason that one had a red and the other, a blue, or both were blue. If James had seen a red ribbon on Edward, he would have known that his ribbon was blue. Since James could not name the color of his ribbon, Edward reasons that his ribbon must be blue.
   (c) Name the couples (B_1, G_1), (B_2, G_2) and (B_3, G_3).
   G_1G_2 go across the river, G_1 returns.
   G_1G_3 go across the river, G_1 returns.
   B_2B_3 go across the river, B_2G_2 return.
   B_1B_2 go across the river, G_3 returns.
   G_1G_2 go across the river, G_1 returns.
   Finally, G_1G_3 go across the river.
   (e)

Aea	Brooklyn	black poodle	Fifi
Bea	Atlanta	white poodle	Joe
Sea	Whereitsat	mutt	Spot
Dea	Here	bird dog	Joe

Exercise Set 1-2

1. (a) yes, false (c) yes, false (e) no (g) yes, true
   (i) yes, true (k) yes, false
2. (a) It is not true that Abraham Lincoln was born in Texas.
   Abraham Lincoln was not born in Texas.
   (c) It is not true that all athletes over seven feet tall play basketball.
   Some athletes over seven feet tall do not play basketball.
   (e) Not a proposition.
   (g) It is not true that some students work hard at their studies.
   No student works hard at his studies.
   (i) It is not true that some professors are intelligent.
   No professor is intelligent.
   (k) $2 \cdot 3$ is not equal to 7.
   It is not true that $2 \cdot 3 = 7$.
3. (a) $A \wedge D$ (c) $\sim A \rightarrow \sim B$ (e) $C \wedge \sim B$
4. (a) It is snowing, and the roofs are not white.
   (c) It is snowing, and the roofs are white or the streets are not slick.
   (e) It is not true that it is snowing, and the trees are not beautiful.
5. (a) Converse: If one angle of a triangle is 90°, the triangle is a right triangle. Inverse: If a triangle is not a right triangle, then no angle is equal to 90°. Contrapositive: In a triangle, if one angle is not equal to 90°, the triangle is not a right triangle.
   (c) Converse: If the alternate interior angles are equal, then two lines are parallel. Inverse: If two lines are not parallel, then the alternate interior angles are not equal. Contrapositive: If the alternate interior angles are not equal, the two lines are not parallel.
   (e) Converse: If x is divisible by 5, then x is divisible by 10.
   Inverse: If x is not divisible by 10, then x is not divisible by 5.
   Contrapositive: If x is not divisible by 5, then x is not divisible by 10.
6. (a) If a polygon is a triangle, then it is not a square.
   (c) If a politician is honest, then he will not accept bribes.
7. (a) If I pollute the atmosphere, I use leaded gasoline.
   (c) If I do not use leaded gasoline, I will not pollute the atmosphere.
8. (a) false (c) false (e) true
9. (a) $r \rightarrow (q \wedge p)$, true (c) $\sim p \rightarrow (r \vee \sim q)$, true

Exercise Set 1-3

1. (a) yes (c) no (e) no (g) yes

2. (a)

p	$\sim p$	$\sim(\sim p)$
T	F	T
T	F	T
F	T	F
F	T	F

(logically equivalent)

(c)

p	q	$p \vee q$	$\sim(p \vee q)$	$\sim p$	$\sim q$	$\sim p \wedge \sim q$
T	T	T	F	F	F	F
T	F	T	F	F	T	F
F	T	T	F	T	F	F
F	F	F	T	T	T	T

(logically equivalent)

(e)

p	q	$p \rightarrow q$	$\sim p$	$\sim p \vee q$
T	T	T	F	T
T	F	F	F	F
F	T	T	T	T
F	F	T	T	T

(logically equivalent)

3. valid, rule of detachment 5. invalid 7. valid, rule of detachment 9. invalid 11. invalid (The truth of the conclusion of a true implication does not lead to the truth of the hypothesis.) 13. invalid 15. valid, rule of detachment
16. (a) a triangle (c) probably negative 17. when A is false and B is true

Exercise Set 1-4

1. valid 3. invalid 5. invalid 7. invalid
9. Paul may or may not be a college student. 11. Larry was on time.
13. Henry may or may not be a radical, and Henry may or may not be a Republican.
14. (a) Dr. W. is a male professor who is neither dull nor boring.
    (c) Some members of set R are members of set S, and no members of set R are members of set T.
15. c 17. d

Exercise Set 1-5

2. An axiom is assumed to be valid without proof, but a theorem may be proved to be either valid or invalid.
3. Direct proofs involve the logical use of axioms, previously proved theorems, and definitions, along with the property of substitution. The proof is then constructed by arranging a sequence of deductively valid steps that lead from the hypothesis to the desired conclusion. Theorems may be indirectly proved by assuming the negation of the conclusion to be true and then logically developing a statement not consistent with known facts, or by proving the contrapositive of the theorem. Theorems may be disproved by finding a counterexample or by contradiction.
5. (a) false (c) true (e) true (g) false
6. (a) Hypothesis: $x + 5 = 7$,
   Conclusion: $x = 2$; yes
   (c) Hypothesis: $x = y + 1$ and $y = 4$,
   Conclusion: $x = 5$; yes
7. (a) Glenda Tharp lives in the United States; however, she lives in Talladega, Alabama, not Florida. We therefore have a contradiction, and the theorem is false.
   (c) If $x + 1 = 5$, by solving, one finds that $x = 4$. But $4 \neq 6$, so the statement is disproved.
9. (a) If $x = 1$ by substitution in $x + 6 = 9$, we obtain $1 + 6 \neq 9$. Thus, we have found a value of x such that $x + 6 \neq 9$. Hence, $x + 6 = 9$ is not true for all values of x.

(c)

$x = 3$	Given
$x + 1 = 3 + 1$	Assumed axiom that one can add a number to both sides of an equality
$x + 1 = 4$	Addition
also, $x + 1 = 4$	Given
$(x + 1) - 1 = 4 - 1$	Assumed axiom that one can subtract a number from both sides of an equality
$x + (1 - 1) = 4 - 1$	Assumed axiom that one can group the $1 - 1$ together
$x + 0 = 3$	Subtraction
$x = 3$	Addition of 0

(e)

$x + 3 = y + 4$	Given
$(x + 3) - 3 = (y + 4) - 3$	Assumed axiom that one can subtract a number from both sides of an equality
$x + (3 - 3) = y + (4 - 3)$	Assumed axiom that one can group in the manner indicated

$x + 0 = y + 1$ Subtraction
$x = y + 1$ Addition of zero
Thus, the conclusion that $x \neq y + 1$ is false since we have shown that $x = y + 1$.

Review Exercise Set 1-6

1. (a) Converse: If we play tennis, it will not rain.
   Inverse: If it rains, we will not play tennis.
   Contrapositive: If we do not play tennis, it will rain.
2. (a) inductive (c) inductive 3. (a) invalid (c) invalid
4. (c) not equivalent to (a), but is a tautology (e) not equivalent to (a)
5. (a) Glenda is clever. (c) Angles A and B may or may not be right angles.
6. (a) not a tautology

p	q	r	$\sim q$	$\sim q \wedge r$	$\sim(\sim q \wedge r)$	$\sim p$	$\sim(\sim q \wedge r) \rightarrow \sim p$
T	T	T	F	F	T	F	F
T	T	F	F	F	T	F	F
T	F	T	T	T	F	F	T
T	F	F	T	F	T	F	F
F	T	T	F	F	T	T	T
F	T	F	F	F	T	T	T
F	F	T	T	T	F	T	T
F	F	F	T	F	T	T	T

(c) not a tautology

p	r	$r \wedge p$	$\sim r$	$p \rightarrow \sim r$	$(r \wedge p) \rightarrow (p \rightarrow \sim r)$
T	T	T	F	F	F
T	F	F	T	T	T
F	T	F	F	T	T
F	F	F	T	T	T

8. (a) $x = 1$ Given
   $x + 3 = 1 + 3$ Assumed axiom that we can add a number to both members of an equality
   $x + 3 = 4$ Addition
   Therefore, if $x = 1$, then $x + 3 = 4$.

   (c) $x + 5 = 9$ Given
   $(x + 5) - 5 = 9 - 5$ Assumed axiom that we can subtract a number from both members of an equality

$x + (5 - 5) = 9 - 5$	Assumed axiom that we can regroup for subtraction
$x + 0 = 4$	Subtraction
$x = 4$	Addition of 0

Therefore, if $x + 5 = 9$, then $x = 4$.

Exercise Set 2-1

1. (a) true (c) false (e) false (g) false (i) false
2. (a) $\{x \mid x$ is a student at Hi Lo University$\}$
$\{x \mid x$ is a male freshman student at a four-year college in Georgia$\}$
$\{x \mid x$ is a counting number greater than 4 and less than 100$\}$
(c) $\{x \mid x$ is a counting number greater than 4$\}$
$\{x \mid x$ is a multiple of 5$\}$
$\{x \mid x$ is a star in the universe$\}$
3. (a) $\varnothing$ (c) $\varnothing$ (e) $\{2\}$ 4. (a) infinite
(c) infinite (e) finite 5. (a) $\subset$ (c) $\subset$ (e) $\notin$
6. (a) true (c) false (e) true (g) true (i) true
7. (a) a, f (c) a, d, f (e) a, c 8. (a) A (c) none
(e) A
9. (a) Set A is said to be a *proper subset* of set B, denoted by $A \subset B$, if and only if each element of A is an element of B and there is at least one element of B that is not an element of A.
(c) $E = \{\ \}$ or $\varnothing$ is the empty or null set containing no elements.
10. (a) no difference (c) no (e) no 11. (a) $\{2, 4, 6\}$, $\{2, 4\}$, $\{2, 6\}$, $\{4, 6\}$, $\{6\}$, $\{4\}$, $\{2\}$, $\varnothing$ 12. (a) yes
13. (a) $\{2, 4, 8, 16, 32, 64\}$ or $\{2, 4, 6, 8, 10, 12, \ldots, 64\}$
(c) $\{A, B, E, F\}$ or $\{A, B, C, D, E, F\}$
(e) $\{1, 2, 3, 4, 5, 6, \ldots\}$ or $\{1, 2, 3, 11, 12, 13, 21, 22, 23, \ldots\}$

Exercise Set 2-2

1. (a) $R \cap T = \{15\}$ (c) $A \cap B = \{10, 100\}$ (e) $A \cap B = \{x, y\}$
2. (a) $A \cap B = \{4, 5\}$ $B \cap C = \{4, 5, 6\}$
$(A \cap B) \cap C = \{4, 5\}$ $A \cap (B \cap C) = \{4, 5\}$
Therefore, $(A \cap B) \cap C = A \cap (B \cap C)$

3. (a)

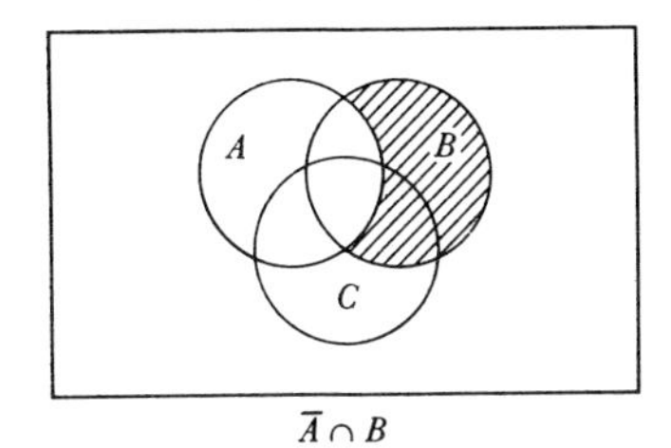

$\overline{A} \cap B$

(c)

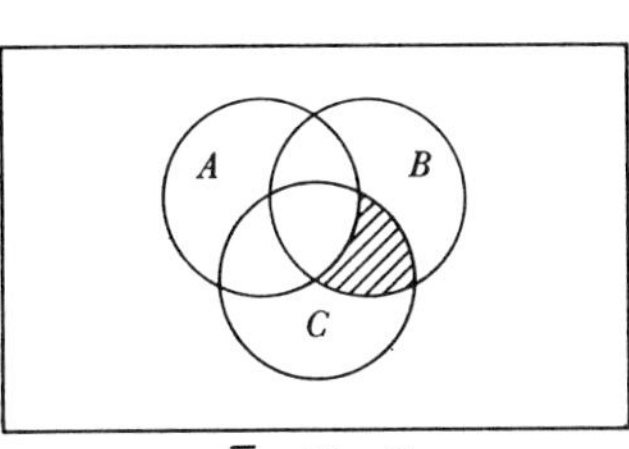

$\overline{A} \cap (B \cap C)$

4. (a) {7} (c) {2, 4, 6, 8, 9, 10, 11, 12} (e) {2, 4, 5, 6, 7, 9, 10, 11, 12} (g) $\varnothing$
5. (a) {all cars not red convertibles}
   (c) {all cars built in 1973 that are not red convertibles}
   (e) {all red convertibles not built in 1973}
6. (a)

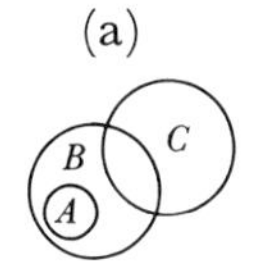

(c)

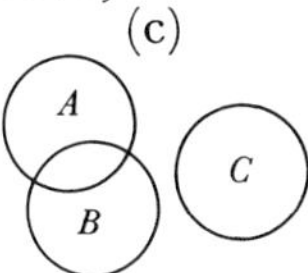

7. (a) No set has any elements in common with the null set. Therefore, $\varnothing \cap (A \cap B)$ is the empty set.
   (c) The intersection of A and B is empty. The intersection of the null set with another set is always $\varnothing$. $(A \cap B) \cap U = \varnothing$.
   (e) The intersection of any set with the universe is the given set. Therefore, $\overline{A} \cap U = \overline{A}$, or the set of all drivers not female drivers under 18.
8. (a) A and B are disjoint or $A = \varnothing$ or $B = \varnothing$ (c) B is a subset of A
   (e) for any set B (g) A is $\varnothing$ 9. (a) $A \cap B = A$
   10. (a) $\overline{B}$ (c) $\overline{B} \cap A$ 11. 35
13.

$x \in A$	$x \in B$	$x \in C$	$x \in [A \cap B]$	$x \in [(A \cap B) \cap C]$	$x \in (B \cap C)$	$x \in [A \cap (B \cap C)]$
T	T	T	T	T	T	T
T	T	F	T	F	F	F
T	F	T	F	F	F	F
T	F	F	F	F	F	F
F	T	T	F	F	T	F
F	T	F	F	F	F	F
F	F	T	F	F	F	F
F	F	F	F	F	F	F

$(A \cap B) \cap C$ and $A \cap (B \cap C)$ are logically equivalent.

Exercise Set 2-3

1. (a) $R \cup T = \{5, 10, 15, 20\}$ (c) $G \cup H = \{1, 2, 3, 4, \ldots, 99\}$
2. (a) $\{1, 3, 4, 5, 6, 7, 8, 9, 11, 12\}$ (c) $\{1, 2, 3, 5, 7, 10\}$
   (e) $\{4, 6, 8, 9, 11, 12\}$
3. (a) (c)

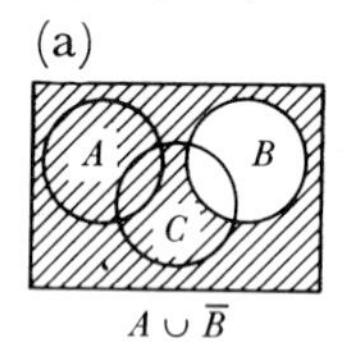

$A \cup \overline{B}$

$\overline{A} \cap (\overline{B} \cup C)$

4. (a) $X \cup Y = \{a, b, c, d, e, f\}$ (c) $\overline{X \cup Y} = \{g\}$ (e) $\overline{Y} = \{a, b, g\}$
5. (a) $M \cup P = \{v, w, x, s, r\}$ $P \cup N = \{v, w, x, y, r\}$
   $(M \cup P) \cup N = \{v, w, x, y, s, r\}$ $M \cup (P \cup N) = \{v, w, x, y, r, s\}$
   Therefore, $(M \cup P) \cup N = M \cup (P \cup N)$
   (c) $P \cap N = \{w, x\}$ $M \cup P = \{v, w, x, s, r\}$
   $M \cup (P \cap N) = \{v, w, x, s\}$ $M \cup N = \{v, w, x, y, s\}$
   $(M \cup P) \cap (M \cup N) = \{v, w, s, x\}$
   Therefore, $M \cup (P \cap N) = (M \cup P) \cap (M \cup N)$
6.

(a)

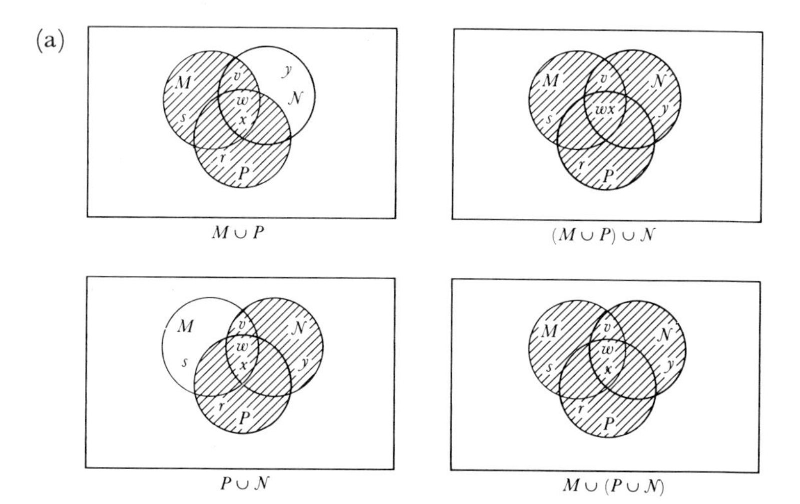

$M \cup P$ $(M \cup P) \cup N$

$P \cup N$ $M \cup (P \cup N)$

Thus, $(M \cup P) \cup N = M \cup (P \cup N)$

(c)

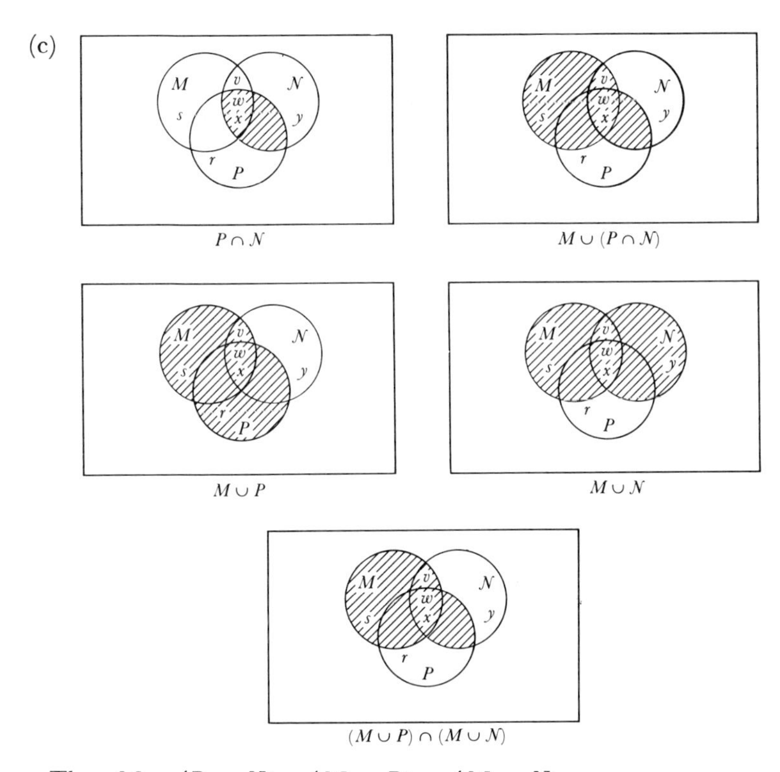

Thus $M \cup (P \cap N) = (M \cup P) \cap (M \cup N)$

7.

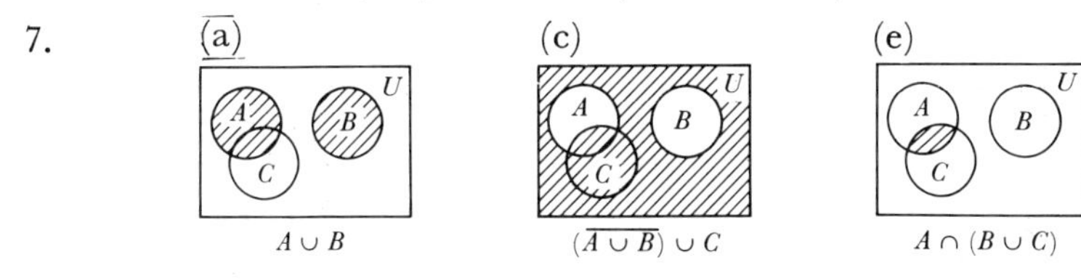

8. (a) The set of all girls who had dates last Saturday night and/or dates this Saturday night.
   (c) The set of all girls who had dates any night except last Saturday night.
   (e) The set of all girls without dates either last Saturday night or this Saturday night.

9. (a) $B \cap C = \{1, 9\}$ | $\overline{B} = \{2, 4, 5, 6, 8, 10\}$

$\overline{B \cap C} = \{2, 3, 4, 5, 6, 7, 8, 10\}$ | $\overline{C} = \{3, 4, 5, 6, 7, 8\}$

$A \cap (\overline{B \cap C}) = \{2, 3, 4, 5\}$ | $A \cap \overline{B} = \{2, 4, 5\}$

$A \cap \overline{C} = \{3, 4, 5\}$

$(A \cap \overline{B}) \cup (A \cap \overline{C}) = \{2, 3, 4, 5\}$

Therefore, $A \cap (\overline{B \cap C}) = (A \cap \overline{B}) \cup (A \cap \overline{C})$

(c) $A \cup B = \{1, 2, 3, 4, 5, 7, 9\}$ $\quad \overline{A} = \{6, 7, 8, 9, 10\}$

$\overline{A \cup B} = \{6, 8, 10\}$ $\quad \overline{B} = \{2, 4, 5, 6, 8, 10\}$

$\overline{A} \cap \overline{B} = \{6, 8, 10\}$

Therefore, $\overline{A \cup B} = \overline{A} \cap \overline{B}$

10. (a)

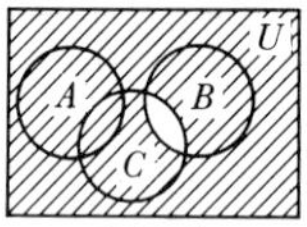

$\overline{B \cap C}$ 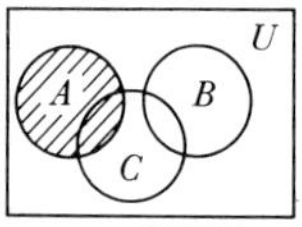 $\quad A \cap (\overline{B \cap C})$

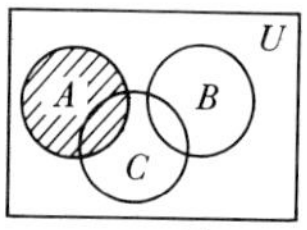

$A \cap \overline{B}$ 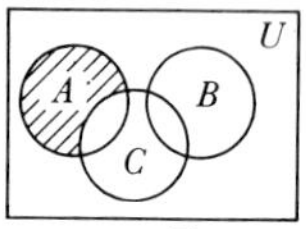 $\quad A \cap \overline{C}$ 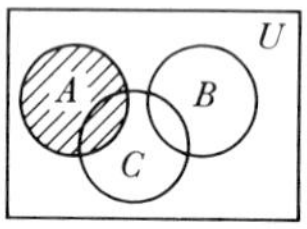 $\quad (A \cap \overline{B}) \cup (A \cap \overline{C})$

Therefore, $A \cap (\overline{B \cap C}) = (A \cap \overline{B}) \cup (A \cap \overline{C})$

(c)

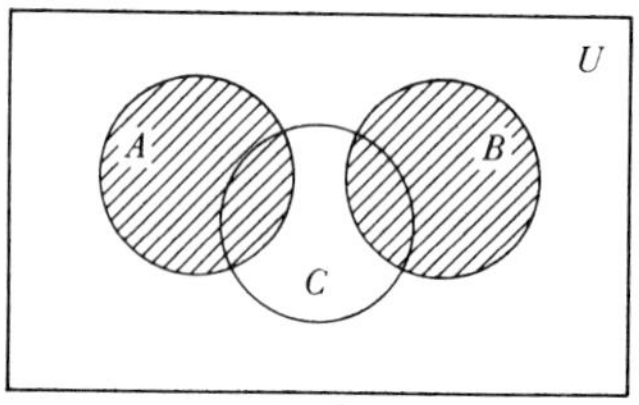

$A \cup B$ 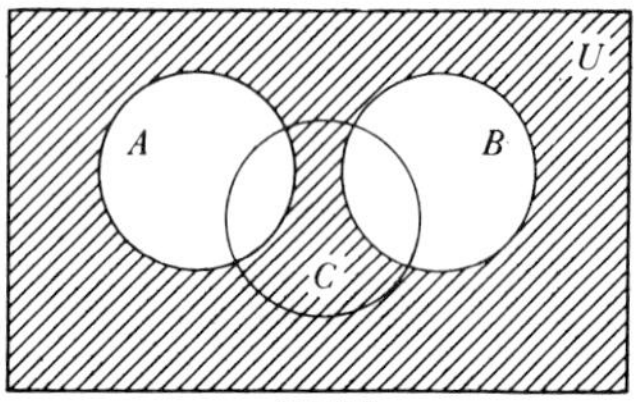 $\quad \overline{A \cup B}$

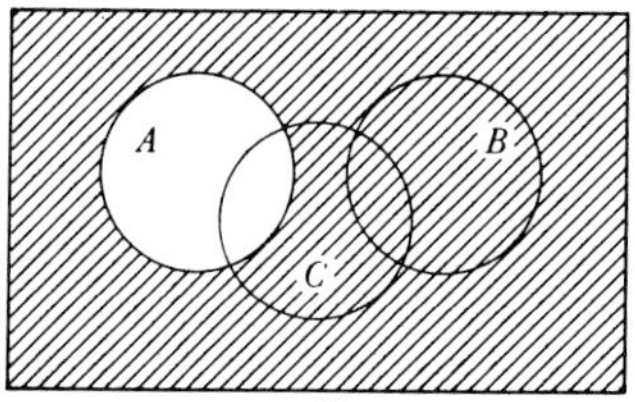

$\overline{A}$ 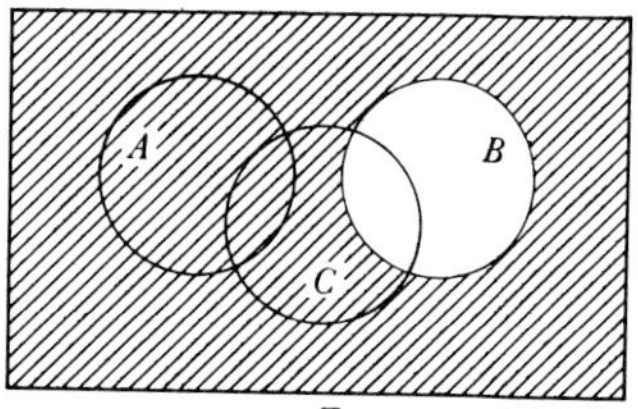 $\quad \overline{B}$

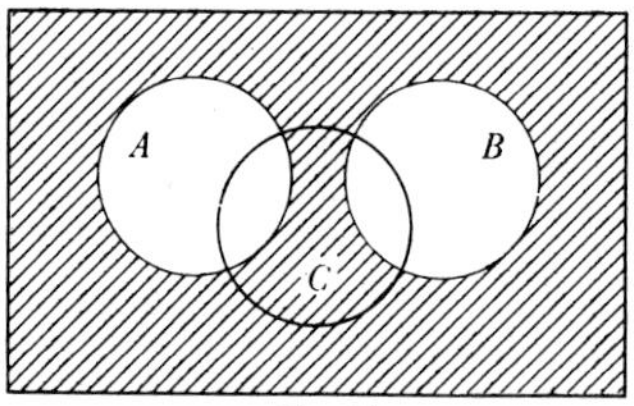

$\overline{A} \cap \overline{B}$

Therefore, $\overline{A \cup B} = \overline{A} \cap \overline{B}$

11. (a) false (c) true (e) false
12. (a) A is a subset of the union of A and B.
    (c) A union B contains the same elements as A; thus, B is a subset of A.
13. (a) $\overline{B}$
15. (a)

$x \in A$	$x \in B$	$x \in (A \cup B)$	$x \in (\overline{A \cup B})$	$x \in \bar{A}$	$x \in \bar{B}$	$x \in (\bar{A} \cap \bar{B})$
T	T	T	F	F	F	F
T	F	T	F	F	T	F
F	T	T	F	T	F	F
F	F	F	T	T	T	T

Columns 4 and 7 are identical. Therefore, $\overline{A \cup B} = \overline{A} \cap \overline{B}$.

(c)

$x \in A$	$x \in B$	$x \in C$	$x \in (B \cap C)$	$x \in [A \cup (B \cap C)]$	$x \in (A \cup B)$	$x \in (A \cup C)$	$x \in [(A \cup B) \cap (A \cup C)]$
T	T	T	T	T	T	T	T
T	T	F	F	T	T	T	T
T	F	T	F	T	T	T	T
T	F	F	F	T	T	T	T
F	T	T	T	T	T	T	T
F	T	F	F	F	T	F	F
F	F	T	F	F	F	T	F
F	F	F	F	F	F	F	F

Columns 5 and 8 are identical. Therefore, $A \cup (B \cap C) = (A \cup B) \cap (A \cup C)$.

Exercise Set 2-4

1. (a) $A \times B = \{(a, r), (a, s), (a, t), (b, r), (b, s), (b, t), (c, r), (c, s), (c, t)\}$
   (c) $B \times A = \{(r, a), (r, b), (r, c), (s, a), (s, b), (s, c), (t, a), (t, b), (t, c)\}$
3. (a) $A \times (B \cup C) = \{(a, c), (b, c), (c, c)\}$
   (c) $(A \cap B) \times C = \{(c, a), (c, c), (c, x)\}$
4. (a) $B \times C = \{(3, 0)\}$
   (c) $B \times C = \{(3, 3), (3, 4)\}$
5. (a) $B = \{1, 4\}$ $C = \{1, 2, 3\}$
   (c) $B = \{6\}$ $C = \{6, 7, 8\}$
7. (a) true (c) true (e) true
8. (a) $B \times (C \times A) = \{[3, (4, 1)], [3, (4, 2)], [3, (5, 1)], [3, (5, 2)], [4, (4, 1)], [4, (4, 2)], [4, (5, 1)], [4, (5, 2)], [5, (4, 1)], [5, (4, 2)], [5, (5, 1)], [5, (5, 2)]\}$

(c) $C \times (B \times A) = \{[4, (3, 1)], [4, (3, 2)], [4, (4, 1)], [4, (4, 2)], [4, (5, 1)], [4, (5, 2)], [5, (3, 1)], [5, (3, 2)], [5, (4, 1)], [5, (4, 2)], [5, (5, 1)], [5, (5, 2)]\}$

(e) $B \times (C \times C) = \{[3, (4, 4)], [3, (4, 5)], [3, (5, 4)], [3, (5, 5)], [4, (4, 4)], [4, (4, 5)], [4, (5, 4)], [4, (5, 5)], [5, (4, 4)], [5, (4, 5)], [5, (5, 4)], [5, (5, 5)]\}$

9. (a) no, if either A or $B = \varnothing$ (c) yes

Exercise Set 2-5

1. (a) yes (c) no
2. (a) $\{(1, 1), (1, 2), (1, 3), (1, 4), (2, 2), (2, 3), (2, 4), (3, 3), (3, 4), (4, 4)\}$
   (c) b is two more than a where a and b are elements in A.
3. (a) yes (c) no (e) no (g) yes 4. (a) yes (c) yes (e) no
5. (a) transitive (c) reflective, symmetric, transitive
   (e) reflexive, symmetric, transitive
6. (a) symmetric (c) transitive (e) transitive (g) equivalence relation (i) symmetric 7. (a) yes (c) yes
9. $A \times (B \times C) = \{[a_1, (b_1, c_1)], [a_1, (b_1, c_2)], [a_1, (b_1, c_3)], [a_1, (b_2, c_1)],$
   $\updownarrow \quad \updownarrow \quad \updownarrow \quad \updownarrow$
   $(A \times B) \times C = \{[(a_1, b_1), c_1], [(a_1, b_1), c_2], [(a_1, b_1), c_3], [(a_1, b_2), c_1],$
   $[a_1, (b_2, c_2)], [a_1, (b_2, c_3)]\}$
   $\updownarrow \quad \updownarrow$
   $[(a_1, b_2), c_2], [(a_1, b_2), c_3]\}$
11. (a) $\{(1, 3), (3, 5), (1, 5)\}$ (c) $\{(1, 3), (1, 1), (3, 3), (3, 1), (5, 7), (5, 5), (7, 7)\}$ 12. (a) not equivalent (c) equivalent
13. 24

Review Exercise Set 2-6

1. (a) $\{1, 2, \ldots, 10\}$ (c) $\{4, 5, 6, 7, 8\}$ 2. (a) false (c) false (e) true (g) false (i) true
3.

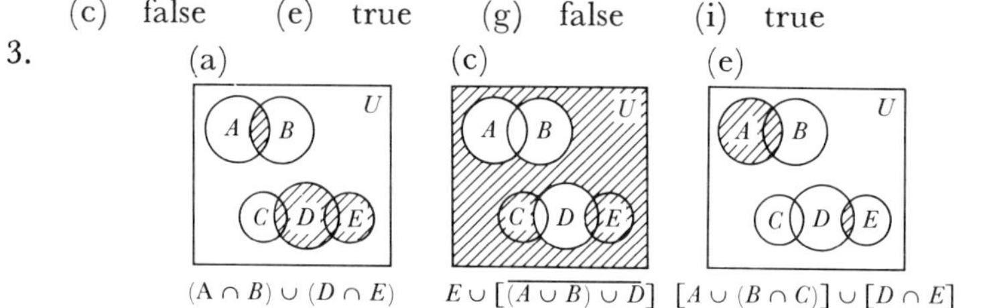

$(A \cap B) \cup (D \cap E)$ $E \cup [\overline{(A \cup B) \cup D}]$ $[A \cup (B \cap C)] \cup [D \cap E]$

4. (a) $\{4, 5\}$ (c) $\{1, 2, \ldots, 15\}$
5. (a) 120; the set of all males preferring either football or baseball
   (c) 110; the set of all females preferring either basketball or hockey
   (e) $A \cap D =$ the set of all males who prefer baseball; 40
   (g) $(D \cup E) \cap A =$ the set of all males who prefer baseball; 40
6. (a) $\{(2, 1), (2, 2), (2, 3), (2, 4)\}$ (c) $\{(1, 1), (2, 2), (3, 3), (4, 4)\}$
7. (a) $B \subset A$ (c) $A \subset B$ (e) $A = B$ 8. (a) $T \cap \overline{M}$
   (c) $\overline{M} \cup (T \cap \overline{B})$ 9. (a) $\overline{(A \cup B)} \cap C$ (c) $\overline{A} \cup B$
   (e) $\overline{(A \cup B)} \cup C \cup (A \cap B)$ 11. (a) B is a subset of A.
   (c) A is a subset of B. (e) Either $B = C$ or $A = \varnothing$.
12. (a) yes (c) yes

Exercise Set 3-1

1. (a) C (c) O (e) C (g) C (i) C (k) C
2. (a) $\{b\} \leftrightarrow \{1\}$ (c) $\{a, b, c, d, e\}$ ↕ $\{1, 2, 3, 4, 5\}$ (e) $\{10, 40, 30, 50, 70\}$ ↕ $\{1, 2, 3, 4, 5\}$
3. (a) 1 (c) 5 (e) 5 4. (a) 4 (c) 7 (e) 2
5. (a) 7 (c) 0 (e) 0 (g) 28 (i) 0
6. (a) false (c) true (e) false 7. (a) false
   (c) false (e) true
8. (a) $n(A \cup B) = 9$, $n(A) = 5$, $n(B) = 5$, $n(A \cap B) = 1$, $9 = 5 + 5 - 1$
   (c) $n(A \times A) = 25$, $n(A) = 5$, $5 \cdot 5 = 25$
9. yes, yes, $n(P \cup Q) = 5$, $n(P \cap Q) = 5$
10. (a) $n(A) = 3$, $n(B) = 2$, $n(A \cup B) = 5$, $n(A \cap B) = 0$, $n(A \times B) = 6$
   (c) $n(A) = 3$, $n(B) = 2$, $n(A \cup B) = 3$, $n(A \cap B) = 2$, $n(A \times B) = 6$
   (e) $n(A) = 3$, $n(B) = 1$, $n(A \cup B) = 4$, $n(A \cap B) = 0$, $n(A \times B) = 3$
12. (a) $\varnothing$ (c) $\{0\}$ (e) N 13. 4 14. (a) 20
   (c) 20

Exercise Set 3-2

1. (a) $n(R) = 4$, $n(S) = 3$, $n(R \cup S) = 7$, $4 + 3 = 7$, yes
   (c) $n(R) = 4$, $n(S) = 4$, $n(R \cup S) = 7$, $4 + 4 \neq 7$, no

2. (a) We may associate with 2 the set $A = \{a, b\}$ and with 4 the set $B = \{c, d, e, f\}$. $A \cup B = \{a, b, c, d, e, f\}$. Thus, $n(A) + n(B) = n(A \cup B)$. Hence, $2 + 4 = 6$.
   (c) We may associate with 5 the set $A = \{a, b, c, d, e\}$ and with 3 the set $B = \{x, y, z\}$. $A \cup B = \{a, b, c, d, e, x, y, z\}$. Thus, $n(A) + n(B) = n(A \cup B)$. Hence, $5 + 3 = 8$.
   (e) We may associate with 0 the null set, $\varnothing$, and with 6 the set $A = \{m, n, o, p, q, r\}$. The union of $\varnothing$ and A is A. Thus, $n(\varnothing) + n(A) = n(A \cup \varnothing) = n(A)$. Hence, $0 + 6 = 6$.
3. (a) commutative (c) additive identity (e) associative
   (g) commutative (i) commutative
4. (a) closed, identity 0 (c) not closed, no identity (e) closed, no identity (g) not closed, no identity
5. (a) $(2 + 4) + 3 = 3 + (2 + 4)$ Commutative property of addition
   $= (3 + 2) + 4$ Associative property of addition
   $(2 + 4) + 3 = (3 + 2) + 4$ Transitive property of equalities
   (c) $8 + (5 + 2) = (8 + 5) + 2$ Associative property of addition
   $= 2 + (8 + 5)$ Commutative property of addition
   $8 + (5 + 2) = 2 + (8 + 5)$ Transitive property of equalities
   (e) $(a + b) + (c + d)$
   $= (a + b) + (d + c)$ Commutative property of addition
   $= [(a + b) + d] + c$ Associative property of addition
   $= [a + (b + d)] + c$ Associative property of addition
   $= [(b + d) + a] + c$ Commutative property of addition
   $= (b + d) + (a + c)$ Associative property of addition
   $(a + b) + (c + d) = (b + d) + (a + c)$ Transitive property of equalities
6. (a) $a + (c + b) = a + (b + c)$ Commutative property of addition
   (c) $(b + c) + a = a + (b + c)$ Commutative property of addition
   (e) $c + (a + b) = (a + b) + c$ Commutative property of addition
   $= a + (b + c)$ Associative property of addition
   $c + (a + b) = a + (b + c)$ Transitive property of equalities
7. We could let set A be the partition $\{l, m, n, o\}$ of the given set and set B be the partition $\{p, q\}$ of that set. $n(A) = 4$ and $n(B) = 2$, while $n(A \cup B) = 6$. Then let $B = \{l, m\}$ and $A = \{n, o, p, q\}$. $n(B) = 2$, $n(A) = 4$, and $n(B \cup A) = 6$. This illustrates the commutative law of addition. $n(A) + n(B) = n(B) + n(A)$.
   $\{l, m, n, o, p, q\} = \{l, m\} \cup [\{n, o, p\} \cup \{q\}]$, and $6 = n\{l, m, n, o, p, q\} = n\{l, m\} + [n\{n, o, p\} + n\{q\}] = 2 + [3 + 1]$. Likewise, $\{l, m, n, o, p, q\} = [\{l, m\} \cup \{n, o, p\}] \cup \{q\}$, and $6 = n\{l, m, n, o, p, q\} = n\{l, m\} + n\{n, o, p\} + n\{q\} = [2 + 3] + 1$. Thus, $2 + [3 + 1] = [2 + 3] + 1$, which illustrates the associative property of addition.

9. no
10. (a) One does not add sets. You add numbers and take the union of sets.
 (c) The union of two cardinal numbers is not defined. You add cardinal numbers and take the union of sets.
11. To maintain the closure property, an operation on a set must hold the properties of uniqueness—that only one answer exists when any elements are operated upon—and of existence—that at least one answer in the set exists when any pair of elements are operated upon.

Exercise Set 3-3

1. (a) commutative property of multiplication
 (c) commutative property of addition
 (e) associative property of addition
 (g) commutative property of addition
5. (a) commutative property of addition
 (c) commutative property of addition and commutative property of multiplication
 (e) additive identity
 (g) multiplicative identity and commutative property of addition
6. (a) not closed, identity 1 (c) closed, no identity (e) closed, no identity (g) closed, identity 1
7. \$7.75 9. (a) yes (c) no [for example, $(3 \otimes 2) \otimes 4 \neq 3 \otimes (2 \otimes 4)$]

Exercise Set 3-4

1. (a) false (c) false (e) false
2. (a) $3(5+4) = (3 \cdot 5) + (3 \cdot 4) = 15 + 12 = 27$
 $3(5+4) = 3(9) = 27$
 (c) $(2+7)9 = (2 \cdot 9) + (7 \cdot 9) = 18 + 63 = 81$
 $(2+7)9 = (9)9 = 81$
3. (a) 20 (c) 10 (e) 14 4. (a) $(2 \cdot 3) + (2 \cdot 4)$
 (c) $4(2+3)$ (e) $2x(2ay+z)$
5. (a) commutative property of addition
 (c) distributive property of multiplication over addition
6. (a)

$3 \cdot 4 = (1+1+1)4$	3 may be rewritten as $1+1+1$ (our numeration system)
$= 4+4+4$	Generalized distributive property of multiplication over addition
$3 \cdot 4 = 4+4+4$	Transitive property of equalities

(c) $5 \cdot 6 = (1+1+1+1+1)6$ — Our numeration system
$= 6+6+6+6+6$ — Generalized distributive property of multiplication over addition
$5 \cdot 6 = 6+6+6+6+6$ — Transitive property of equalities

7. No. For example, $2+(3 \cdot 4) = 2+12 = 14$, $2+(3 \cdot 4) = (2+3) \cdot (2+4) = 30$, and $14 \neq 30$.

9. (a) additive identity
(c) distributive property of multiplication over addition
(e) transitive property of equalities

10. (a) $(2 \cdot 3)4 = 4(2 \cdot 3)$ — Commutative property of multiplication
$= (4 \cdot 2)3$ — Associative property of multiplication
$(2 \cdot 3)4 = (4 \cdot 2)3$ — Transitive property of equalities
(c) $(6+1)5 = (6 \cdot 5) + (1 \cdot 5)$ — Distributive property of multiplication over addition
$= (1 \cdot 5) + (6 \cdot 5)$ — Commutative property of addition
$= (1 \cdot 5) + (5 \cdot 6)$ — Commutative property of multiplication
$(6+1)5 = (1 \cdot 5) + (5 \cdot 6)$ — Transitive property of equalities
(e) $2(3 \cdot 1) = (2 \cdot 3)1$ — Associative property of multiplication
$= 1(2 \cdot 3)$ — Commutative property of multiplication
$= (1 \cdot 2)3$ — Associative property of multiplication
$2(3 \cdot 1) = (1 \cdot 2)3$ — Transitive property of equalities

Exercise Set 3-5

1. (a)

(c)

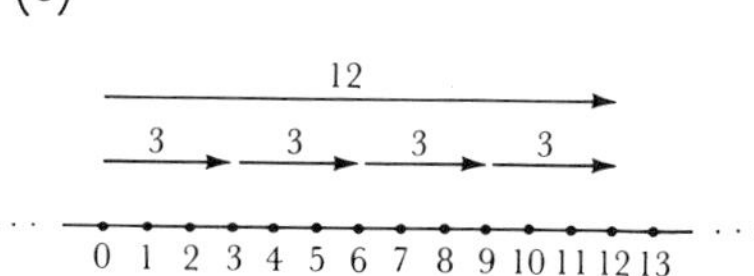

(e)

(g)

(i) (k)

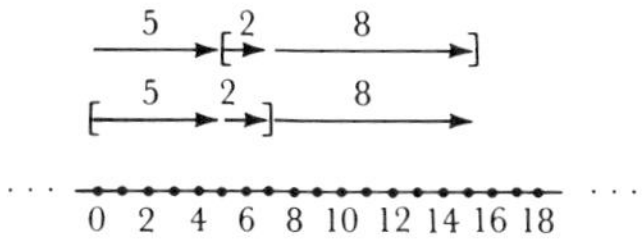

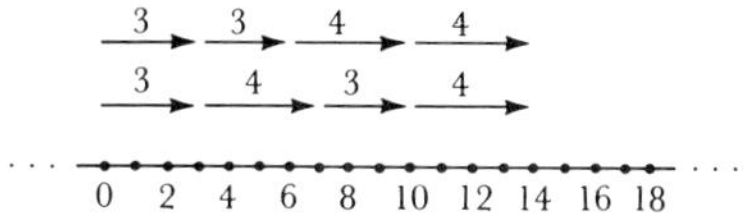

2. (a) $a = 2$, $b = 15$, $k = 8$ (c) $a = 5$, $b = 12$, $k = 3$
3. (a) true (c) false (e) true (g) true
4. (a) $(10 \div 2) = 5$ because $10 = 2 \cdot 5$. (c) $(12 \div 3) = 4$ because $12 = 3 \cdot 4$. 5. (a) $x = 5$ (c) $x = 4$ (e) $x = 7$ (g) $x = 0$ (i) $x = 4$ 6. (a) undefined (c) 0 (e) 0 7. (a) 2 (c) no answer (e) 3
8. (a) $\{0, 1, 2, 3, 4, 5, 6\}$ 9. (a) $4 + 3 < 6 + 3$ (c) $4 + b < 6 + b$ 12. (a) $21 - 16 = 5$ and $21 - 5 = 16$ (c) $a = f + c$ and $c = a - f$
13. (a) associative property of division
(c) distributive property of multiplication over division
(e) commutative property of division

Review Exercise Set 3-6

1. (a) $5 - 5 = 0$ (c) not defined (e) $(5 - 5) \div 2 = 0$
2. (a) distributive property of multiplication over addition
(c) commutative property of addition
(e) commutative property of multiplication
(g) distributive property of multiplication over addition
(i) transitive property of equalities
3. (a)

Statement	Reason
$(110)(9) - (9)(6) = (9)(110) - (9)(6)$	Commutative property of multiplication
$= 9(110 - 6)$	Distributive property of multiplication over subtraction
$(110)(9) - (9)(6) = 9(110 - 6)$	Transitive property of equalities

(c)

Statement	Reason
$4(23 \cdot 25) = (23 \cdot 25)4$	Commutative property of multiplication
$= (25 \cdot 23)4$	Commutative property of multiplication
$= 25(23 \cdot 4)$	Associative property of multiplication
$4(23 \cdot 25) = 25(23 \cdot 4)$	Transitive property of equalities

(e) $c(d \cdot e) = (d \cdot e)c$	Commutative property of multiplication
$= (e \cdot d)c$	Commutative property of multiplication
$= e(d \cdot c)$	Associative property of multiplication
$c(d \cdot e) = e(d \cdot c)$	Transitive property of equalities

5. $(a + b)(a + b)$

$= (a + b)a + (a + b)b$	Distributive property of multiplication over addition
$= [(a \cdot a) + (b \cdot a)] + [(a \cdot b) + (b \cdot b)]$	Distributive property of multiplication over addition
$= [(a \cdot a) + (a \cdot b)] + [(a \cdot b) + (b \cdot b)]$	Commutative property of multiplication
$= (a \cdot a) + [(a \cdot b) + (a \cdot b)] + (b \cdot b)$	Associative property of addition
$= (a \cdot a) + [1(a \cdot b) + 1(a \cdot b)] + (b \cdot b)$	Multiplicative identities
$= (a \cdot a) + [(1 + 1)(a \cdot b)] + (b \cdot b)$	Distributive property of multiplication over addition
$= (a \cdot a) + (2 \cdot a \cdot b) + (b \cdot b)(a + b)(a + b)$	Addition of whole numbers
$= (a \cdot a) + (2 \cdot a \cdot b) + (b \cdot b)$	Transitive property of equalities

7. (a) true (c) true (e) false (g) true
8. (a) 2 pieces (c) 17 cups
9. (a) $x \boxdot y = y \boxdot x$ (c) $(x \boxdot y) \boxdot z = x \boxdot (y \boxdot z)$
   (e) $z \boxdot (x \oplus y) = (z \boxdot x) \oplus (z \boxdot y)$

Exercise Set 4-1

1. (a) 4 (c) 34
2. (a) 2 (c) 11 (e) 4862
3. (a) 21 (c) 49 (e) 156 (g) 3888 (i) 20,666
4. (a) 6 (c) 7 (e) 9 (g) 843 (i) 17,175 (k) 20 (m) 50,400

5. (a) LXXVI (c) CLXXXIX (e) CXLVIII

6.

Hindu-Arabic	1	5	10	50	100
Egyptian	|	|||||	∩	∩∩∩∩∩	𓍢
Roman	I	V	X	L	C
Babylonian	▼	▼▼▼▼▼	◀	◀◀◀◀◀	▼ ◀◀◀◀
Mayan	•	—	=	•• over =	— over 𝋠

The Mayan symbol for 0 is 𝋠.

8. (a) XCIV (c) ◀ ◀▼ (e) — over =

Exercise Set 4-2

1. (a) 2^5 (c) $4^6 = 2^{12}$ (e) 3^7
2. (a) 81 (c) 1 (e) 8 (g) 125 (i) 81 (k) k^5
3. (a) $7(10)^2 + 6(10) + 8(10)^0$
   (c) $2(10)^4 + 3(10)^3 + 1(10)^2 + 0(10) + 5(10)^0$
   (e) $7(10)^3 + 6(10)^2 + 0(10) + 4(10)^0$
4. (a) $n = 3$ (c) $n = 11$ 5. (a) $8c^2$ (c) 6000
   (e) 8970 6. (a) 753
7. (a)

$$\begin{array}{rl} 24 & 2(10) + 4(10)^0 \\ 15 & 1(10) + 5(10)^0 \\ \hline 39 & 3(10) + 9(10)^0 = 39 \end{array}$$

(c) 53 $5(10) + 3(10)^0$
-21 $2(10) + 1(10)^0$
32 $3(10) + 2(10)^0 = 32$

(e) 21 $2(10) + 1(10)^0$
32 $3(10) + 2(10)^0$
42 $4(10) + 2(10)^0$
63 $6(10)^2 + 3(10)$
672 $6(10)^2 + 7(10) + 2(10)^0 = 672$

(g) 768 $7(10)^2 + 6(10) + 8(10)^0$
-254 $2(10)^2 + 5(10) + 4(10)^0$
514 $5(10)^2 + 1(10) + 4(10)^0 = 514$

(i) 724 $7(10)^2 + 2(10) + 4(10)^0$
235 $2(10)^2 + 3(10) + 5(10)^0$
959 $9(10)^2 + 5(10) + 9(10)^0 = 959$

9. 4^0, 10^2, 7^3, 2^{10}, 3^7, 9^4, 4^9
11. (a) No. If a and b are 1, x does not necessarily equal y.

Exercise Set 4-3

1. (a) S (c) Definition of addition (e) A_a (g) Definition of addition
2. (a) steps (a) through (e)
3.

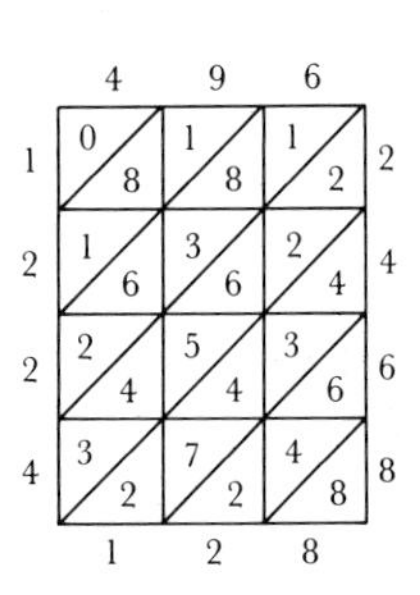

5. (a) S (c) A_m (e) Multiplication by 10
6. (a) S (c) A_m (e) Multiplication by 10

7. (a)

$$\begin{array}{ll}
2(10)+7 & \\
\underline{1(10)+5} & \\
3(10)+12 & \text{By ACD} \\
3(10)+[1(10)+2] & \text{S} \\
[3(10)+1(10)]+2 & \text{A}_{\text{a}} \\
(3+1)10+2 & \text{D}_{\text{a}} \\
4(10)+2 & \text{Addition} \\
42 & \text{S}
\end{array}$$

(c)

$$\begin{array}{ll}
5(10)+3 & \\
\underline{2(10)+4} & \\
7(10)+7 & \text{By ACD} \\
77 & \text{S}
\end{array}$$

(e)

$$\begin{array}{ll}
\qquad\qquad 2(10)+8 & \\
\qquad\qquad \underline{3(10)+2} & \\
\qquad\qquad 4(10)+16 & \\
\underline{6(10)^2+24(10)\qquad\quad} & \\
6(10)^2+28(10)+16 & \text{By ACD} \\
6(10)^2+[2(10)+8]10+[1(10)+6] & \text{S} \\
6(10)^2+[2(10)^2+8(10)]+[1(10)+6] & \text{D}_{\text{a}}\text{, and multiplication} \\
[6(10)^2+2(10)^2]+[8(10)+1(10)]+6 & \text{A}_{\text{a}} \\
(6+2)(10)^2+(8+1)(10)+6 & \text{D}_{\text{a}} \\
8(10)^2+9(10)+6 & \text{Addition} \\
896 & \text{S}
\end{array}$$

8. $a^3+12a^2+47a+60$

Exercise Set 4-4

1. (a) $q=6, r=6$ (c) $q=0, r=11$ (e) $q=3, r=9$
   (g) $q=6, r=0$
2. (a) Method I

$$\begin{array}{r|r}
 & 3(10)+9 \\
\hline
2(10)+7 & 1(10)^3+0(10)^2+7(10)+5 \\
 & \underline{8(10)^2+1(10)\qquad} \\
 & 2(10)^2+6(10)+5 \\
 & \underline{2(10)^2+4(10)+3} \\
 & 2(10)+2
\end{array}$$

Answer: 39 with remainder of 22

Method II

$$\begin{array}{rl}
27\overline{)1075} & \\
\underline{270} & 27\cdot 10\\
805 & \\
\underline{270} & 27\cdot 10\\
535 & \\
\underline{270} & 27\cdot 10\\
265 & \\
\underline{27} & 27\cdot 1\\
238 & \\
\underline{27} & 27\cdot 1\\
211 & \\
\underline{27} & 27\cdot 1\\
184 & \\
\underline{27} & 27\cdot 1\\
157 & \\
\underline{27} & 27\cdot 1\\
130 & \\
\underline{27} & 27\cdot 1\\
103 & \\
\underline{27} & 27\cdot 1\\
76 & \\
\underline{27} & 27\cdot 1\\
49 & \\
\underline{27} & 27\cdot 1\\
22 &
\end{array}$$

$10+10+10+1+1+1+1+1+1+1+1+1=39$
Answer: 39 with remainder of 22
Check: $27(39)+22=1053+22=1075$

Method III

$$\begin{array}{rl}
27\overline{)1075} & \\
\underline{810} & 27\cdot 30\\
265 & \\
\underline{243} & 27\cdot 9\\
22 &
\end{array}$$

$30+9=39$
Answer: 39 with remainder of 22

(c) Method I

$$\begin{array}{r}
5(10)+9\\
2(10)^2+0(10)+3\overline{)1(10)^4+2(10)^3+0(10)^2+6(10)+1}\\
\underline{1(10)^4+0(10)^3+1(10)^2+5(10)}\\
1(10)^3+9(10)^2+1(10)+1\\
\underline{1(10)^3+8(10)^2+2(10)+7}\\
0(10)^2+8(10)+4
\end{array}$$

Answer: 59 with remainder of 84

Method II

```
203|12061
     2030    203 · 10
    10031
     2030    203 · 10
     8001
     2030    203 · 10
     5971
     2030    203 · 10
     3941
     2030    203 · 10
     1911
      203    203 · 1
     1708
      203    203 · 1
     1505
      203    203 · 1
     1302
      203    203 · 1
     1099
      203    203 · 1
      896
      203    203 · 1
      693
      203    203 · 1
      490
      203    203 · 1
      287
      203    203 · 1
       84
```

Method III

```
203|12061
    10150    203 · 50
     1911
     1827    203 · 9
       84
```

$50 + 9 = 59$

Answer: 59 with remainder of 84

$10 + 10 + 10 + 10 + 10 + 1 + 1 + 1 + 1 + 1 + 1 + 1 + 1 + 1 = 59$
Answer: 59 with remainder of 84
Check: $(203 \cdot 59) + 84 = 11{,}977 + 84 = 12{,}061$

3. (a) S (c) D_a (e) A_a (g) theorem that if $a \geq c$ and $b \geq d$, then $(a + b) - (c + d) = (a - c) + (b - d)$ (i) definition of subtraction
4. (a) $q = 5$ $r = 1$ (c) $b = 7$ 8. (a) $q = 0$ $r = 438$ (c) $q = 43$ $r = 8$ 9. (a) $q = 8$ $r = 104$ (c) $q = 810$ $r = 4$ 10. (a) $q = 2, b = 2$ (c) $q = 3$ $b = 237$

Exercise Set 4-5

1. (a) 1, 2, 3, 4, 10, 11, 12, 13, 14, 20, 21, 22, 23, 24, 30 (c) 1, 2, 3, 10, 11, 12, 13, 20, 21, 22, 23, 30, 31, 32, 33
2. (a) 133 (c) 45 (e) 280 (g) 138
3. (a) 20_{seven} (c) 1000_{two} (e) 610_{eight}
4. (a) 1001_{two} (c) 110111_{two} (e) 1111101000_{two}
5. (a) 12_{seven} (c) 106_{seven} (e) 2626_{seven}
6. (a) 9_{twelve} (c) 47_{twelve} (e) $6E4_{\text{twelve}}$
7. (a) 101_{three} (c) ET_{twelve}
8. (a) eleven (c) twelve
9. (a) 39_{twelve}

Exercise Set 4-6

1. (a) 663_{seven} (c) 10100_{two} (e) 1103_{four}
2. (a) 101_{two} (c) 133_{four} (e) 101010_{two}
3. (a)

+	0	1	2
0	0	1	2
1	1	2	10
2	2	10	11

(c)

+	0	1	2	3	4	5	6	7	8	9	T	E
0	0	1	2	3	4	5	6	7	8	9	T	E
1	1	2	3	4	5	6	7	8	9	T	E	10
2	2	3	4	5	6	7	8	9	T	E	10	11
3	3	4	5	6	7	8	9	T	E	10	11	12
4	4	5	6	7	8	9	T	E	10	11	12	13
5	5	6	7	8	9	T	E	10	11	12	13	14
6	6	7	8	9	T	E	10	11	12	13	14	15
7	7	8	9	T	E	10	11	12	13	14	15	16
8	8	9	T	E	10	11	12	13	14	15	16	17
9	9	T	E	10	11	12	13	14	15	16	17	18
T	T	E	10	11	12	13	14	15	16	17	18	19
E	E	10	11	12	13	14	15	16	17	18	19	1T

4. (a) $4T_{\text{twelve}}$ (c) 5_{twelve} (e) $E51_{\text{twelve}}$ (g) 214_{twelve}
5. (a) 222_{five} (c) 101010_{two}
6. (a) $444_{\text{five}} - 100_{\text{five}} = 344_{\text{five}}$ (c) $222_{\text{three}} - 100_{\text{three}} = 122_{\text{three}}$
7. (a) 301_{seven} (c) 10000100_{two}
9. (a) □ (c) $\mathscr{L}\mathscr{L}$ (e) △

Exercise Set 4-7

1. (a) 100111_{two} (c) 11_{two} remainder of 10 (e) 5216_{seven}
   (g) 216_{seven} remainder of 1
2. (a) $39_{ten} = 100111_{two}$ (c) 3_{ten}, remainder of 2 is the same as 11_{two} remainder of 10
3. (a)

·	0	1	2
0	0	0	0
1	0	1	2
2	0	2	11

(c)

·	0	1	2	3	4	5	6	7	8	9	T	E
0	0	0	0	0	0	0	0	0	0	0	0	0
1	0	1	2	3	4	5	6	7	8	9	T	E
2	0	2	4	6	8	T	10	12	14	16	18	1T
3	0	3	6	9	10	13	16	19	20	23	26	29
4	0	4	8	10	14	18	20	24	28	30	34	38
5	0	5	T	13	18	21	26	2E	34	39	42	47
6	0	6	10	16	20	26	30	36	40	46	50	56
7	0	7	12	19	24	2E	36	41	48	53	5T	65
8	0	8	14	20	28	34	40	48	54	60	68	74
9	0	9	16	23	30	39	46	53	60	69	76	83
T	0	T	18	26	34	42	50	5T	68	76	84	92
E	0	E	1T	29	38	47	56	65	74	83	92	T1

4. (a) 341_{five} (c) 31044_{five} (e) 44_{twelve} remainder of 1
   (g) 51_{twelve} remainder of 4
5. (a) 187_{nine}; 242_{seven}
   (c) 15110_{nine}; 20552_{seven}
6. (a) 457_{twelve}, remainder 4
7. (a) / (c) $\mathscr{L}$ (e) □

Review Exercise Set 4-8

1. (a) 810 (c) 718
   (e) 1030_{six}
2. (a) 111_{twelve} (c) 1362_{seven}
3. (a) 729 (c) 4 (e) 128 ·
4. (a) 1487 (c) 191 (e) 43
5. (a) 10442_{five} (c) 1110_{two}
6. (a)

$$\begin{array}{r} 2(10^3) + 3(10^2) + 4(10) + 6 \\ 9(10^2) + 8(10) + 4 \\ \hline 3(10^3) + 3(10^2) + 3(10) + 0 = 3330 \end{array}$$

(c)

$$\begin{array}{r} 5(10) + 4 = 54 \\ 3(10) + 2 \overline{\big) 1(10^3) + 7(10^2) + 2(10) + 8} \\ \underline{1(10^3) + 6(10^2) + 0(10)} \\ 1(10^2) + 2(10) + 8 \\ \underline{1(10^2) + 2(10) + 8} \end{array}$$

7. (a) twelve (c) seven (e) twelve
8. (a) 4974_{twelve} (c) 1011_{two} (e) 21132_{six}
9. (a) $77_{eight} = 63_{ten}$ (c) $314_{eight} = 204_{ten}$
   99_{ten} 314_{ten}
   $EE_{twelve} = 143_{ten}$ $314_{twelve} = 448_{ten}$
   (e) $42160_{eight} = 17520_{ten}$
   39958_{ten}
   $39958_{twelve} = 79124_{ten}$

Exercise Set 5-1

1. (a) $^-5$ (c) 0 (e) ^-a (g) $(a + b)$

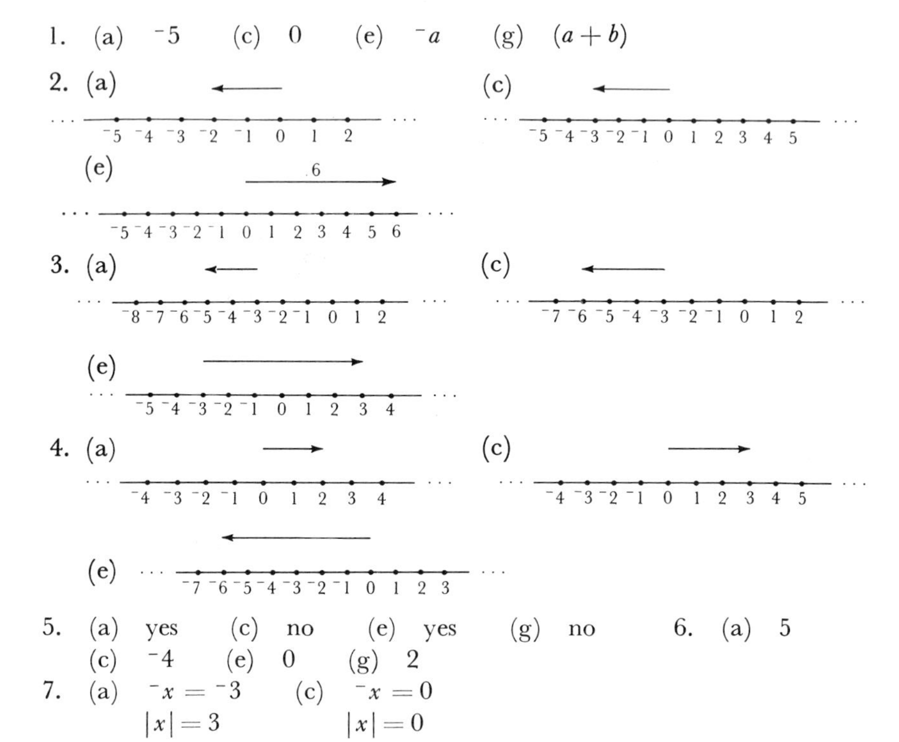

2. (a) (c) (e)
3. (a) (c) (e)
4. (a) (c) (e)
5. (a) yes (c) no (e) yes (g) no 6. (a) 5
   (c) $^-4$ (e) 0 (g) 2
7. (a) $^-x = {}^-3$ (c) $^-x = 0$
   $|x| = 3$ $|x| = 0$

8. (a) no (c) 0
9. (a) true
   (c) false because $C \cap {}^-I = \{1, 2, 3, \ldots\} \cap \{\ldots, {}^-2, {}^-1\} = \varnothing$
   (e) true
   (g) false because $(W \cup {}^+I) \cap C = (\{0, 1, 2, \ldots\} \cup \{1, 2, 3, \ldots\}) \cap C$
   $= \{0, 1, 2, \ldots\} \cap \{1, 2, 3, \ldots\} = \{1, 2, 3, \ldots\} = C$
   (i) true
10. (a) absolute value $= 3$ (c) absolute value $= 3$
    additive inverse $= 3$ additive inverse $= {}^-3$
11. (c) $a + d = b + c$

Exercise Set 5-2

1. (a) $^-3$ (c) $^-8$ (e) $^-13$ (g) $^-11$ (i) $^-16$
   (k) 4 (m) $^-2$ (o) 0 (q) a
2. (a)

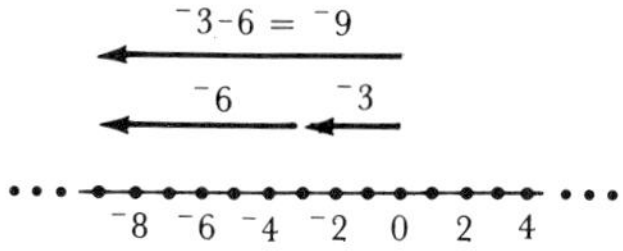

$^-3 - 6 = {}^-9$
$^-6$ $^-3$
$^-8$ $^-6$ $^-4$ $^-2$ 0 2 4

(c)

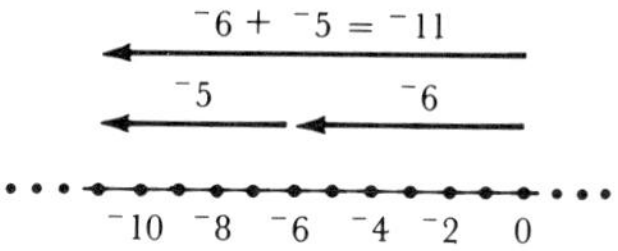

$^-6 + {}^-5 = {}^-11$
$^-5$ $^-6$
$^-10$ $^-8$ $^-6$ $^-4$ $^-2$ 0

(e)

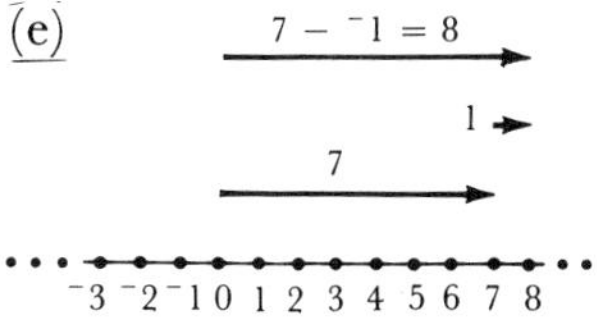

$7 - {}^-1 = 8$
1
7
$^-3$ $^-2$ $^-1$ 0 1 2 3 4 5 6 7 8

3. additive inverse, or closure property of subtraction
4. (a)

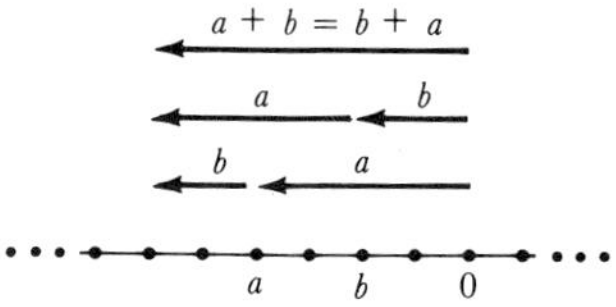

$a + b = b + a$
a b
b a
a b 0

a and b negative integers

(c) $(a + b) + c = a + (b + c)$

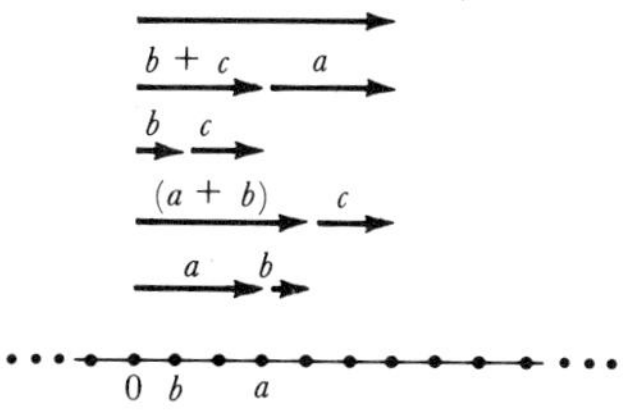

$b + c$ a
b c
$(a + b)$ c
a b
0 b a

5. (a) 6 (c) 2
7. (a) no; $(8 - 2) - 3 = 6 - 3 = 3$, $8 - (2 - 3) = 8 - ({}^-1) = 9$, and $3 \neq 9$ (c) yes; $8 - 0 = 8$

9. $(7+9)+(^-7+^-9)=(7+9)+(^-9+^-7)$ — Commutative property of addition
$=7+(9+^-9)+^-7$ — Associative property of addition
$=7+0+^-7$ — Additive inverses
$=7+^-7$ — Additive identity
$=0$ — Additive inverses
$(7+9)+(^-7+^-9)=0$ — Transitive property of equalities

Thus, $^-7+^-9$ is the additive inverse of $7+9$. Since the unique additive inverse of $7+9$ is $^-(7+9)$, $^-7+^-9=^-(7+9)$.

11. $12+^-17=12+(^-12+^-5)$ — Renaming $^-17$
$=(12+^-12)+^-5$ — Associative property of addition
$=0+^-5$ — Additive inverses
$=^-5$ — Additive identity
$12+^-17=^-5$ — Transitive property of equalities

12. (a) $(a-b)+(c-d)=[(a+c)-(b+d)]$ — Definition of addition of number pairs
$=[(c+a)-(d+b)]$ — Commutative property of addition of natural numbers
$=(c-d)+(a-b)$ — Definition of addition of number pairs
$(a-b)+(c-d)=(c-d)+(a-b)$ — Transitive property of equalities

(c) (a, a)

Exercise Set 5-3

1. (a) $^-45$ (c) 0 (e) $^-322$ (g) xy (i) 2 (k) $^-135$
2. (a) 72 (c) $^-384$ (e) $^-3$ (g) 8 (i) 24 (k) ^-xyz
3. (a) $^-1(3+^-5)=^-1(^-2)=2$
$^-1(3+^-5)=^-1(3)+^-1(^-5)=^-3+5=2$

(c) $6(4\cdot{}^-5)=6(^-20)=^-120$
$6(4\cdot{}^-5)=(6\cdot 4)(^-5)=24(^-5)=^-120$

(e) $(3+^-2)(5)=1(5)=5$
$(3+^-2)(5)=3(5)+^-2(5)=15+^-10=5$

4\. (a)

$(8 + 4) + {}^{-}2 = {}^{-}2 + (8 + 4)$	Commutative property of addition
$= ({}^{-}2 + 8) + 4$	Associative property of addition
$= (8 + {}^{-}2) + 4$	Commutative property of addition
$(8 + 4) + {}^{-}2 = (8 + {}^{-}2) + 4$	Transitive property of equalities

(c)

$({}^{-}4 + 6) + {}^{-}2 = {}^{-}2 + ({}^{-}4 + 6)$	Commutative property of addition
$= {}^{-}2 + (6 + {}^{-}4)$	Commutative property of addition
$= ({}^{-}2 + 6) + {}^{-}4$	Associative property of addition
$({}^{-}4 + 6) + {}^{-}2 = ({}^{-}2 + 6) + {}^{-}4$	Transitive property of equalities

(e)

$(8 \cdot 4) \cdot {}^{-}2 = 8(4 \cdot {}^{-}2)$	Associative property of multiplication
$= 8({}^{-}2 \cdot 4)$	Commutative property of multiplication
$= (8 \cdot {}^{-}2)4$	Associative property of multiplication
$(8 \cdot 4) \cdot {}^{-}2 = (8 \cdot {}^{-}2)4$	Transitive property of equalities

(g)

${}^{-}6(4 + {}^{-}3) = {}^{-}6(4) + {}^{-}6({}^{-}3)$	Distributive property of multiplication over addition

(i)

$2({}^{-}6 \cdot {}^{-}4) = (2 \cdot {}^{-}6) \cdot {}^{-}4$	Associative property of multiplication
$= ({}^{-}6 \cdot 2) \cdot {}^{-}4$	Commutative property of multiplication
$2({}^{-}6 \cdot {}^{-}4) = ({}^{-}6 \cdot 2) \cdot {}^{-}4$	Transitive property of equalities

5\. (a) ${}^{-}5(a) + {}^{-}5(b) + {}^{-}5(c)$ (c) ${}^{-}y(3) + {}^{-}y({}^{-}2x) + {}^{-}y(4z)$

6\. (a) false (c) true 7. (a) 4 (c) ${}^{-}1024$ (e) ${}^{-}1$

8\. (a) $a = b$ or $a = {}^{-}b$, a and $b \neq 0$. (c) Division for integers is not associative. 9. 12.3 feet

11\.

$({}^{-}6)(3) + (6)(3) = ({}^{-}6 + 6)3$	Distributive property of multiplication over addition
$= (0)3$	Definition of additive inverse
$= 0$	Multiplicative property of 0
$({}^{-}6)(3) + 18 = 0$	Renaming $(6)(3)$

Thus, $({}^{-}6)(3)$ is the additive inverse of 18. But ${}^{-}18$ is the unique additive inverse of 18. Therefore $({}^{-}6)(3) = {}^{-}18$.

13. $^{-}1(a) + 1(a) = (^{-}1 + 1) \cdot a$ — Distributive property of multiplication over addition
 $= (0) \cdot a$ — Definition of additive inverse
 $= 0$ — Multiplicative property of 0
 $^{-}1(a) + a = 0$ — Renaming $1(a)$

 Thus, $^{-}1(a)$ is the additive inverse of a. But the unique additive inverse of a is ^{-}a. Therefore, $^{-}1(a) = {^{-}a}$.

15. (a) $^{-}18$, 0, 1 (c) ac, 0, 1

16. (a) $\triangle = 5$, $\square = 8$, $\Theta = 6$, $\sqsubset\!\sqsupset = 4$, $\diamondsuit = 1$

18. (a) $(a, b) \cdot (c, d) = (ac + bd, ad + bc)$ — Definition of multiplication of number pairs
 $= (ca + db, da + cb)$ — Commutative property of multiplication of natural numbers
 $= (ca + db, cb + da)$ — Commutative property of addition of natural numbers
 $= (c, d) \cdot (a, b)$ — Definition of multiplication of number pairs
 $(a, b) \cdot (c, d) = (c, d) \cdot (a, b)$ — Transitive property of equalities

 (c) $(x + 1, x)$, where x is any natural number

Exercise Set 5-4

1. (a) $^{-}7 < {^{-}4} < 0 < 3 < 9$ (c) $^{-}100 < {^{-}5} < {^{-}2} < 0 < 4$
2. (a) $^{-}5 + 3 = {^{-}2}$ (c) $^{-}10 + 2 = {^{-}8}$
3. (a) {0, 1, 2} (c) {0, 5} (e) {0, 1, 2, 3, 4, 5, 6, 7, 8, 9}
 (g) { } (i) {0, 1, 2, 3} (k) {0, 1, 2, 3, 4, 5, 6, 7, 8, 9}
 (m) {2} (o) {5}
4. (a) conditional equation (c) conditional equation
 (e) identity (g) identity (i) identity
5. (a) $\{y \mid y = 14\}$
 (c) $\{z \mid z$ is an integer $\geq 12\}$
 (e) $\{x \mid x = 4\}$
 (g) $\{x \mid x$ is an integer $\geq 2\}$
 (i) $\{x \mid x = {^{-}6}\}$ (k) $\{x \mid x = {^{-}5}\}$
6. (a) {0, 1, 2, 3, 4, 5} (c) {11, 12, 13, 14, 15}
 (e) $\{\ldots, {^{-}3}, {^{-}2}, {^{-}1}, 0, 1, 2, 3, 4\}$
7. (a) definition of "less than"
 (c) distributive property of multiplication over addition
 (e) definition of "less than"

8. (a) definition of "less than"
   (c) distributive property of multiplication over addition
   (e) associative property of addition
   (g) additive identity
   (i) definition of "greater than"

Review Exercise Set 5-5

1. (a) 25 (c) $^-17$ (e) 0 (g) $^-41$ (i) $xz - xy$
   (k) 40 2. (a) 25 (c) $\{^-5, ^-3\}$ (e) $\varnothing$
   (g) $\{^-1, ^-3, ^-9, 1, 3, 9\}$ (i) even, even
4. (a)

$4(^-3) + (^-4)(2) = 4(^-3) + [(^-1)(4)](2)$	Renaming $^-4$
$= 4(^-3) + [(4)(^-1)](2)$	Commutative property of multiplication
$= 4(^-3) + 4[(^-1)(2)]$	Associative property of multiplication
$= 4(^-3) + 4(^-2)$	Multiplication
$= 4(^-3 + ^-2)$	Distributive property of multiplication over addition
$4(^-3) + ^-4(2) = 4(^-3 + ^-2)$	Transitive property of equalities

   (c)

$^-4(^-8 \cdot 2) = (^-8 \cdot 2) \cdot {}^-4$	Commutative property of multiplication
$= [(^-1 \cdot 8) \cdot 2] \cdot {}^-4$	Rewriting $^-8$
$= [(8 \cdot {}^-1) \cdot 2] \cdot {}^-4$	Commutative property of multiplication
$= [8(^-1 \cdot 2)] \cdot {}^-4$	Associative property of multiplication
$= (8 \cdot {}^-2) \cdot {}^-4$	Definition of multiplication
$^-4(^-8 \cdot 2) = (8 \cdot {}^-2) \cdot {}^-4$	Transitive property of equalities

5. (a) additive inverse (c) associative property of addition
   (e) additive inverses (g) Theorem 5-2 (i) additive inverses
   (k) commutative property of addition

Exercise Set 6-1

1. (a) 1, 5, 7 (c) 1, 7, 11 2. (a) divisible by 8 and 7
   (c) divisible by 5 (e) none (g) none (i) divisible by 8

(k) divisible by 11 3. (a) no 4. (a) $24\underline{3}$
(c) $\overline{5}11\underline{2}$ 5. (a) true (c) false (e) true

6. (a) true; $4|24$ and $3|24$, so $12|24$ (c) false; $6|6$ and $2|6$, but $12 \nmid 6$ (e) false; $10 \nmid 6$, but $2|6$.
7. (a) 17,595 is one of many possibilities.
   (c) 175,616 is one of many possibilities.
8. (a) definition of division (c) substitution of xl for z
   (e) integers closed under addition
9. (a) Theorem 6-1(c) (c) commutative property of addition
   (e) additive inverse and additive identity

Exercise Set 6-2

1. (a) prime (c) not a prime (e) not a prime (g) not a prime 2. (a) 2, 3 (c) 2, 3, 5 (e) 3, 47
3. (a) $2 \cdot 2 \cdot 2 \cdot 2 \cdot 3 \cdot 3$ (c) $2 \cdot 2 \cdot 2 \cdot 2 \cdot 2 \cdot 2 \cdot 2 \cdot 2$
   (e) $2 \cdot 2 \cdot 3 \cdot 3 \cdot 3$ 5. (a) 3, 5, 7, 11 (c) 3, 7, 11, 19
6. (a) even (c) even (e) odd (g) even
7. {3, 5, 7, 11, 13, 17, 19, 23}, {1, 2, 3, 5, 7, 9, 11, 13, 15, 17, 19, 21, 23}
8. 15 prime numbers less than 50; 33 composite numbers less than 50
9. 2520 12. 3, 5; 5, 7; 11, 13; 17, 19; 29, 31; 41, 43
13. (a) $61 + 3$ (c) $97 + 3$ 14. (a) $97 + 7$ (c) $97 + 47$

Exercise Set 6-3

1. (a) 15 (c) 6 (e) 2 (g) 6 (i) 4 3. (a) 1
4. (a) 6 (c) 3 (e) 1 5. (a) 3 (c) 22 (e) 5 (g) 24 6. (a) yes (c) yes 7. (a) 1, 2, 3, 4, 5, 6 (c) 1, 2, 3, 4, 5, 6, ..., 22 8. (a) no 9. (a) $D_5 = \{1, 5\}$, and $D_{20} = \{1, 2, 4, 5, 10, 20\}$; therefore, $D_5 \subset D_{20}$

Exercise Set 6-4

1. (a) 40 (c) 168 (e) 462 (g) 990 2. 4, 101, $|a|$
3. (a) 660 (c) 7,341,600 (e) 468 (g) 110,160
4. (a) I. $\left.\begin{array}{l} 44 = 2^2 \cdot 11 \\ 92 = 2^2 \cdot 23 \end{array}\right\} \quad 2^2 \cdot 11 \cdot 23 = 1012 = \text{l.c.m.}$

   II. g.c.d. $= 4$

$$\text{l.c.m.} = \frac{92 \cdot 44}{4} = \frac{4048}{4} = 1012$$

III.
$$\begin{array}{r|rr} 2 & 44 & 92 \\ \hline 2 & 22 & 46 \\ \hline & 11 & 23 \end{array}$$

$2 \cdot 2 \cdot 11 \cdot 23 = 1012 = \text{l.c.m.}$

(c) I. $\left.\begin{array}{l} 146 = 2 \cdot 73 \\ 124 = 2^2 \cdot 31 \end{array}\right\} \quad 2^2 \cdot 73 \cdot 31 = 9052 = \text{l.c.m.}$

II. $\text{g.c.d.} = 2; \ \text{l.c.m.} = \dfrac{146 \cdot 124}{2} = 9052$

III.
$$\begin{array}{r|rr} 2 & 146 & 124 \\ \hline 2 & 73 & 62 \\ \hline & 73 & 31 \end{array}$$

$2 \cdot 2 \cdot 73 \cdot 31 = 9052 = \text{l.c.m.}$

(e) I. $\left.\begin{array}{l} 840 = 2^3 \cdot 3 \cdot 5 \cdot 7 \\ 1800 = 2^3 \cdot 3^2 \cdot 5^2 \end{array}\right\} \quad 2^3 \cdot 3^2 \cdot 5^2 \cdot 7 = 12{,}600 = \text{l.c.m.}$

II. $\text{g.c.d.} = 2^3 \cdot 3 \cdot 5 = 120$

$$\text{l.c.m.} = \frac{840 \cdot 1800}{120} = \frac{1{,}512{,}000}{120} = 12{,}600$$

III.
$$\begin{array}{r|rr} 5 & 1800 & 840 \\ \hline 3 & 360 & 168 \\ \hline 2 & 120 & 56 \\ \hline 2 & 60 & 28 \\ \hline 2 & 30 & 14 \\ \hline 3 & 15 & 7 \\ \hline & 5 & 7 \end{array}$$

$2^3 \cdot 3^2 \cdot 5^2 \cdot 7 = 12{,}600 = \text{l.c.m.}$

5. (a) $3^3 \cdot 4^2 \cdot 2^3$ 7. g.c.d. $(a, a) = a$; l.c.m. $(a, a) = a$
8. (a) $\overline{ab}$ (c) $b \mid a$
9. (a) $M_{15} = \{15, 30, 45, 60, \ldots\}$ and $M_5 = \{5, 10, 15, 20, 25, \ldots\}$; therefore, $M_{15} \subset M_5$.
   (c) If $a \mid b$, then $M_b = \{1, b, 2b, \ldots\}$ and $M_a = \{1, a, b, 2a, 2b, \ldots\}$; therefore, $M_b \subset M_a$.

Exercise Set 6-5

1. (a) true (c) true (e) true (g) true 2. yes
3. (a) g.c.d. $= 2$; $x = 19$ and $y = {}^-7$ (c) g.c.d. $= 2$; $x = 5$ and $y = {}^-2$ (e) g.c.d. $= 1$; $x = {}^-10$ and $y = 7$

Exercise Set 6-6

1. (a) $2^4 \cdot 3$ (c) $2 \cdot 7^4$ (e) $3^3 \cdot 7^4 \cdot 2^3$
2. (a) $3^6 = 729 = 27 \cdot 27 = 3 \cdot 243 = 81 \cdot 9$
   (c) $100 = 10 \cdot 10 = 5 \cdot 20 = 25 \cdot 4$
3. (a) $(2x+1)(2y+1) = 2x(2y+1) + 1(2y+1) = 4xy + 2x + 2y + 1 = 2[2xy + x + y] + 1$
   (c) A prime is a number that is greater than 1 and divisible by only 1 and itself. 3, 5, 7, 11, 13, 17, 19, 29, 31
4. (a) $(4x+1)(4y+1) = 4x(4y+1) + (4y+1) = 16xy + 4x + 4y + 1 = 4[4xy + x + y] + 1$
   (c) A number greater than 1, divisible by only 1 and itself; 5, 9, 13, 17, 21, 29, 33, 37, 41, 49
5. (a) $(5x+1)(5y+1) = 25xy + 5x + 5y + 1 = 5[5xy + x + y] + 1$
   (c) A number greater than 1, divisible by only 1 and itself; 6, 11, 16, 21, 26, 31, 41, 46, 51, 56

Exercise Set 6-7

1. (a) $(6^5 - 6)/5 = (7776 - 6)/5 = 7770/5 = 1554$
   (c) $\dfrac{(3^{11} - 3)}{11} = \dfrac{177{,}147 - 3}{11} = \dfrac{177{,}144}{11} = 16{,}104$
2. (a) $[(1 \cdot 2 \cdot 3 \cdot 4) + 1]/5 = (24 + 1)/5 = 25/5 = 5$
3. $1 + 10 + 22 + 11 + 20 + 2 + 110 + 4 + 55 + 5 + 44 = 284$
   $1 + 2 + 4 + 71 + 142 = 220$; therefore, amicable.
6. (a) $4(2) + 1 = 9$, a composite number
   (c) 2
   (e) $(8)(2) + 1 = 17$
7. (a) $3 + 3 + 3$ (c) $71 + 3 + 3$ 8. (a) deficient: 8 abundant: 12

Review Exercise Set 6-8

1. 2; 36,875,354
3. No, because such an arrangement would result in the seventh number being even; therefore, not prime—41, 42, 43, 44, 45, 46, 47, 48.
4. (a) yes (c) yes
5. It is the product of the two numbers. 6. They would be equal or identical. 7. (a) yes (c) yes 9. 6, 15, 24, 33, 42, 51, 60, 66, 105, 114

Exercise Set 7-1

1. (a) 3 (c) $0, \frac{0}{1}, \frac{0}{2}, \frac{0}{3}, \frac{0}{100}$, etc. (e) $^-6$
2. (a) $\frac{6}{8}, \frac{12}{16}, \frac{18}{24}, \frac{24}{32}, \frac{36}{48}$, etc. (c) $0, \frac{0}{1}, \frac{0}{2}, \frac{0}{3}, \frac{0}{100}$, etc.
3. (a) yes (c) yes (e) no 4. (a) $^-3$ (c) $^-3$
5. (a) 81/44 (c) 14/19 (e) x^2/z^2 (g) x/yz^3
6. (a) 49/21 (c) impossible (e) 30/40 (g) impossible
   (i) $10/^-24$ (k) $^-6/10$
7. (a) (c)

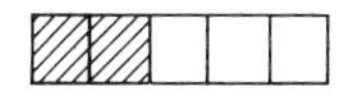

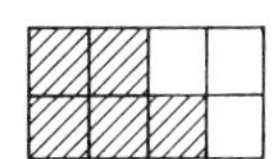

8. (a) $\frac{2}{3}, \frac{3}{2}, \frac{4}{1}, \frac{1}{4}, \frac{0}{5}$ (c) $\frac{2}{5}, \frac{9}{8}, \frac{6}{5}, \frac{2}{1}, \frac{0}{4}$ (e) $\frac{^-7}{^-1}, \frac{^-5}{^-4}, \frac{0}{^-1}, \frac{^-2}{^-2}, \frac{^-8}{^-3}$
9. (a) true (c) true (e) true 11. (a) Yes. Since $a = c$, then $ad = cd$ and $a/d = c/d$. (c) $a = c$
13. {1/8}, {2/8, 1/4}, {3/8}, {4/8, 3/6, 2/4, 1/2}, {5/8}, {6/8, 3/4}, {7/8}, {8/8, 7/7, 6/6, 5/5, 4/4, 3/3, 2/2, 1/1}, {1/7}, {2/7}, {3/7}, {4/7}, {5/7}, {6/7}, {1/6}, {2/6, 1/3}, {4/6, 2/3}, {5/6}, {1/5} {2/5}, {3/5}, {4/5}; 22
15. (a) a/b may be equivalent to e/f. *Example*: $3/4 \not\simeq 2/7$, $2/7 \not\simeq 6/8$, but $3/4 \simeq 6/8$
    (c) $K \neq 1$
16. (a) 2/3 (c) 17/23

Exercise Set 7-2

1. (a) 3/4 (c) 11/36 (e) $3/(2x)$ 2. (a) 8; 19/8
   (c) 54; 10/9 (e) 385; 1187/385 3. (a) $1/8 + 3/16$
   (c) $15/4 + 19/4$
4. (a) $\frac{5}{18} + \frac{7}{24} = \frac{5(24) + 7(18)}{18 \cdot 24} = \frac{246}{432} = \frac{41}{72}$

$$\frac{5 \cdot 4}{18 \cdot 4} + \frac{7 \cdot 3}{24 \cdot 3} = \frac{20}{72} + \frac{21}{72} = \frac{41}{72}$$

(c) $\frac{7}{50}+\frac{41}{210}+\frac{^-6}{20}=\frac{7(210)+41(50)}{(50)(210)}+\frac{^-6}{20}$

$$=\frac{3520}{(50)(210)}+\frac{^-6}{20}$$

$$=\frac{3520(20)+{}^-6(50)(210)}{(50)(210)(20)}=\frac{7400}{210{,}000}=\frac{37}{1050}$$

$$\frac{7}{50}+\frac{41}{210}+\frac{^-6}{20}=\frac{7(42)}{50(42)}+\frac{41(10)}{210(10)}+\frac{^-6(105)}{20(105)}$$

$$=\frac{294}{2100}+\frac{410}{2100}+\frac{^-630}{2100}=\frac{74}{2100}=\frac{37}{1050}$$

(e) $4+\frac{^-2}{3}+7=\frac{4(3)+{}^-2(1)}{(1)(3)}+7=\frac{10}{3}+\frac{7}{1}$

$$=\frac{10(1)+7(3)}{(1)(3)}=\frac{31}{3}$$

$$4+\frac{^-2}{3}+7=\frac{4(3)}{1(3)}+\frac{^-2(1)}{3(1)}+\frac{7(3)}{1(3)}=\frac{12+{}^-2+21}{3}=\frac{31}{3}$$

(g) $\frac{3}{ab}+\frac{5}{a}=\frac{3(a)+5(ab)}{a^2b}=\frac{3a+5ab}{a^2b}=\frac{a(3+5b)}{a(ab)}=\frac{3+5b}{ab}$

$$\frac{3}{ab}+\frac{5}{a}=\frac{3}{ab}+\frac{5(b)}{a(b)}=\frac{3+5b}{ab}$$

5. (a) $10\frac{1}{4}$ (c) $6\frac{2}{9}$ (e) $^-3\frac{59}{60}$ (g) $^-14\frac{59}{63}$
6. (a) Definition 7-4 (c) commutative property of addition of integers
7. (a) Definition 7-4 (b) Definition 7-4 (c) distributive property of multiplication over addition (d) associative property of addition (e) associative property of multiplication
   (f) distributive property of multiplication over addition
   (g) Theorem 7-7 (i) Definition 7-4
12. (a)

$(a, b)+(c, d)=(ad+bc, bd)$	Given definition of addition
$=(bc+ad, bd)$	Commutative property of addition of integers
$=(cb+da, db)$	Commutative property of multiplication of integers
$=(c, d)+(a, b)$	Given definition of addition
$(a, b)+(c, d)=(c, d)+(a, b)$	Transitive property of equalities

(c) $(a, b)+(0, 1)=(a+0, \ )=(a, b)b$

Exercise Set 7-3

1. (a) $\frac{1}{7}; \frac{7}{1} \cdot \frac{1}{7} = \frac{7}{7} = 1$ (c) $\frac{^-5}{^-3}; \frac{^-3}{^-5} \cdot \frac{^-5}{^-3} = \frac{15}{15} = 1$

   (e) $\frac{^-11}{6}; \frac{6}{^-11} \cdot \frac{^-11}{6} = \frac{^-66}{^-66} = 1$ (g) $\frac{y}{x}; \frac{x}{y} \cdot \frac{y}{x} = \frac{xy}{yx} = 1$

   (i) $\frac{w}{^-z}; \frac{^-z}{w} \cdot \frac{w}{^-z} = \frac{^-zw}{^-zw} = 1$

2. (a) 1/4 (c) 5/21 3. (a) a (c) $\frac{ab(ad+be)}{cde}$

   (e) x^2/y^2

4. (a) $7\ 1/8 \cdot 6\ 1/4 = (57/8)(25/4) = 1425/32 = 44\ 17/32$

   $7\ 1/8 \cdot 6\ 1/4 = (7 + 1/8)(6 + 1/4)$
   $= 7(6 + 1/4) + (1/8)(6 + 1/4)$
   $= 7(6) + 7(1/4) + (1/8)(6) + (1/8)(1/4)$
   $= 42 + 7/4 + 6/8 + 1/32$
   $= 42 + 81/32$
   $= 44\ 17/32$

   (c) $2\ 1/10 \cdot {^-8}\ 1/3 = (21/10)(^-25/3) = {^-525/30} = {^-105/6} = {^-17}\ 1/2$

   $2\ 1/10 \cdot {^-8}\ 1/3 = (2 + 1/10)(^-8 + {^-1/3})$
   $= 2(^-8 + {^-1/3}) + (1/10)(^-8 + {^-1/3})$
   $= 2(^-8) + 2(^-1/3) + (1/10)(^-8) + (1/10)(^-1/3)$
   $= {^-16} + {^-2/3} + {^-8/10} + {^-1/30}$
   $= {^-16} + {^-45/30} = {^-17}\ 1/2$

5. (a) 11/36 (c) $^-13$

6. $\frac{3}{4} \cdot \frac{7}{11} = \frac{21}{44}; \frac{6}{8} \cdot \frac{14}{22} = \frac{84}{176} = \frac{21}{44}; \frac{9}{12} \cdot \frac{21}{33} = \frac{189}{396} = \frac{21}{44}$

7. (a) Either r_1 or r_2 is 0. 8. (a) 1/4 lb.

9. y

10. (a) Definition 7-4
    (c) distributive property of multiplication over addition for integers
    (e) Definition 7-5
    (g) commutative property of multiplication of integers
    (i) transitive property of equalities

11. (a) Definition 7-5
    (c) transitive property of equalities

13. (a) no (c) no

15. (a) $(a, b) \cdot (c, d) = (ac, bd)$ Given definition of multiplication
$= (ca, db)$ Commutative property of multiplication of integers
$= (c, d) \cdot (a, b)$ Given definition of multiplication
$(a, b) \cdot (c, d) = (c, d) \cdot (a, b)$ Transitive property of equalities

Exercise Set 7-4

1. (a) $^-25/6$ (c) $3/4$ (e) $25/24$ (g) $^-43/30$
2. (a) $53/6$ (c) $^-91/24$ (e) $^-269/20$ 3. (a) $^-16/21$
(c) $1/6$ (e) $^-36$ (g) $^-2/3$ (i) $^-64/63$
4. (a) $^-59/40$ (c) $^-19/45$ 5. (a) $^-45/26$ (c) $39/14$
6. (a) $\dfrac{rz - sx}{xyz}$ (c) $\dfrac{2b^2 - 3a^2c}{a^3b^3c^2}$
7. (a) $\{x \mid x = {}^-1\}$ (c) $\{x \mid x = {}^-16\}$ (e) $\{x \mid x = 888\ 8/9\}$
(g) $\{x \mid x = 19/6\}$ (i) $\{x \mid x = 2/9\}$ (k) $\{x \mid x = 17/6\}$
8. (a) $3/4$ (c) $(a + 1)/12$ (e) 0 9. (a) definition of additive inverse (c) associative property of addition
(e) associative property of addition (g) unique additive inverse
11. (a) $29/7$ (c) $^-2/7$ (e) $36/91$ 13. (a) no

Exercise Set 7-5

1. (a) false because $^-2(4)$ is not less than $3(^-3)$ (c) true because $1(3)$ is less than $2(7)$ (e) cannot determine because we do not know whether ab is positive or negative
2. (a) $^-3/2 < 71/100 < 23/30$ (c) $16/27 < 11/18 < 2/3 < 67/100$
(e) $156/50 < 22/7 < 10/3 < 7/2$
3. (a) $\{x \mid x < 3\}$ (c) $\{x \mid x < {}^-11/24\}$ (e) $\{x \mid x < 4\}$
(g) $\{x \mid x < {}^-9\}$
4. (a) $361/936$, $181/468$, $363/936$ (c) $^-11/10$, $^-13/15$, $^-19/30$
5. (a) $a/b < c/d$ leads to $ad < bc$ only if the denominators are positive. If $3/{}^-4 < 1/2$, $3 \cdot 2 \not< {}^-4 \cdot 1$.
6. (a) $\{1, 2\}$ (c) 7 7. (a) yes (c) yes (e) no
8. (a) 4 9. (a) 1 (c) $^-1$ 11. $1/x < 1/y$
12. (a) given (c) Theorem 7-1 (e) the product of positive integers is a positive integer (g) associative and commutative properties of multiplication of integers

Review Exercise Set 7-6

1. (a) 3/2 (c) $^-3/4$ 2. (a) $\frac{^-24}{35}, \frac{4}{35}, \frac{4}{35}, \frac{5}{7}$

(c) $\frac{^-3}{4}, 0, \frac{3}{4}, 0$ 3. (a) $x=5$ (c) $x={}^-1$ (e) $x=1$

4. (a) false; $a/0$ is undefined (c) false; division by 0 is undefined (e) false; both 1/2 and 1/3 are less than 1 (g) false; both sets are infinite (i) true (k) false; $^-1/2$ is larger than $^-1$
5. (a) $^-3847/240$ (c) 7/4 6. (a) 3/16 (c) $^-16$

(e) ad/bc 7. (a) $\frac{13}{16}; \frac{^-11}{16}; \frac{3}{64}; \frac{1}{12}$ (c) $\frac{3+2x^2}{6x}; \frac{3-2x^2}{6x};$

$\frac{1}{6}; \frac{3}{2x^2}$ (e) $\frac{26+x}{6x}; \frac{26-x}{6x}; \frac{13}{18x}; \frac{26}{x}$

8. (a) multiplicative inverse
   (c) additive inverse
   (e) commutative property of addition
9. (a) No. It does not have a multiplicative inverse for each element.
   (c) No. It does not have an additive inverse for each element.
   (e) No. It fails to have the properties of closure of addition and multiplication, identities of addition and multiplication, and inverses of addition and multiplication.

Exercise Set 8-1

1. (a) $6(10)^0+7(10)^{-1}+1(10)^{-2}$ (c) $8(10)^1+6(10)^0+6(10)^{-1}$
   (e) $0(10)^{-1}+0(10)^{-2}+1(10)^{-3}$

2. (a) $\frac{85}{1000}$ (c) $\frac{1274}{100}$ (e) $\frac{275}{100}$

3. (a) 6.09 (c) 0.041 (e) 1.2 (g) 5.425 (i) 1.2 (k) 0.375 4. (a) 0.125 (c) 0.225 (e) 0.515625
5. (a) yes (c) no (e) no 6. (a) $0.037 < 0.307 < 0.360 < 0.365 < 0.370$ 7. (a) $b=8$ (c) $b=5$

8. (a) $\frac{7}{12}$ (c) $\frac{140}{27}$ (e) $\frac{1589}{72}$ (g) $\frac{71}{12}$ (i) $\frac{39}{16}$

Exercise Set 8-2

1. (a) 1.142 (c) -0.854 (e) 32.22175 (g) 6.197
2. (a) -19/100 = -0.19 (c) 9/8 = 1.125 (e) 11/4 = 2.75
3. (a) 35.841 (c) 5.30311 (e) 315.87 (g) 683.174
   (i) 0.04230 (k) 6.63
4. (a) 16.3 (c) 17.1
5. (a) 0.375 (c) 0.875
   (e) 3.375
6. (a) 0.055 (c) 0.0031 (e) 0.436
7. (a) 120% (c) 40% (e) 27.5%
8. (a) 25%
   (c) 32 (e) 12.48 (g) 300%
9. 14%
11. 60.8%
13. 63%
15. 95%
17. $9622.64
18. (a) 3.203_{four} (c) 10.0101_{two}
19. (a) 2.45_{six} (c) 1011.001_{two}

Exercise Set 8-3

1. (a) $1.32 \cdot (10)^{-4}$ (c) $1.46 \cdot (10)^{0}$ (e) $8.2 \cdot (10)^{6}$
2. (a) 0.01; 0.0002; accuracy to four significant figures
   (c) 10; 0.002; accuracy to three significant figures
   (e) 0.001; 0.003; accuracy to three significant figures
3. (a) 120 (c) 10
4. (a) 2.832; 2.8
5. (a) 4.24 (c) 17.616

Exercise Set 8-4

1. (a) 0.6875 (c) 0.008 (e) 0.384615384615 ... (g) 16.0
   (i) 12.6
2. (a) 173/500 (c) 2/11 (e) 82/5 (g) 69/55 (i) 62/9
   (k) 595/99 (m) 1/110
3. (a) 0.8 0.84 $0.\overline{8}$ 0.89 $0.\overline{89}$
5. 27

Exercise Set 8-5, page 000

3. (a) no (c) yes
5. (a) irrational (c) irrational
   (e) irrational (g) irrational
7. (a) Geometric proof:

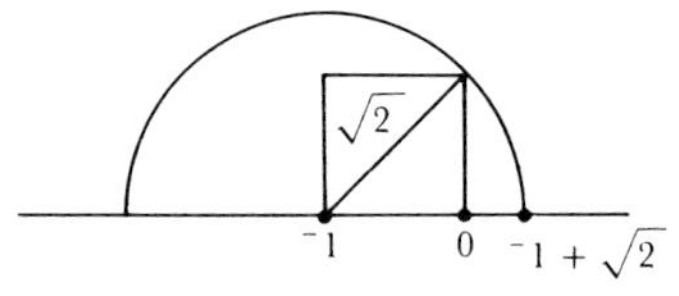

Algebraic proof:

Assume $-1 + \sqrt{2}$ is a rational number, p/q. Then $\sqrt{2} = (p/q) + 1$. The rational number system is closed under addition; therefore, $(p/q) + 1$ is a rational number. This would imply that $\sqrt{2}$ is rational which is a contradiction. Thus, $-1 + \sqrt{2}$ is irrational.

(c) Geometric proof:

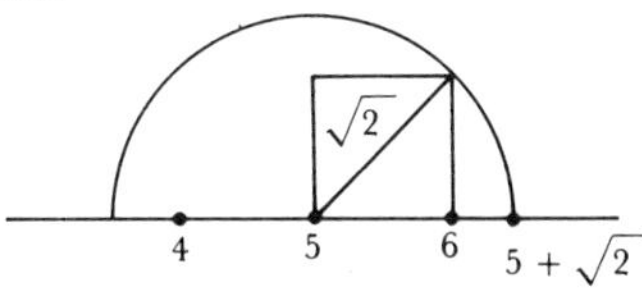

Algebraic proof:

Assume that $5 + \sqrt{2}$ is a rational number, p/q. Then $\sqrt{2} = p/q - 5$. The rational number system is closed under subtraction; therefore, $p/q - 5$ is a rational number. This would imply that $\sqrt{2}$ is rational, which is a contradiction. Thus, $5 + \sqrt{2}$ is irrational.

8. (a) 4 (c) nonexistent
(e) 4.582 (g) 13.221

11. (a) 1.732 (c) 3.000
(i) 21.100 (k) 0.084

Exercise Set 8-6

1. (a) Sum: $1.\overline{70}$
Product: $0.\overline{72}$
Difference: $-0.\overline{07}$
(c) Sum: $3.4\overline{256}$
Product: 0.0451
Difference: $3.3\overline{991}$
(e) Sum: $99.\overline{01}$
Product: $0.\overline{9}$
Difference: $-98.\overline{98}$
(g) Sum: $1.\overline{247}$
Product: $0.\overline{247}$
Difference: $0.\overline{752}$

2. (a) Sum: $\frac{169}{99} = 1.\overline{70}$

Product: $\frac{7128}{9801} = 0.\overline{72}$

Difference: $\frac{-7}{99} = -0.\overline{07}$

(c) Sum: $\frac{34222}{9990} = 3.4\overline{256}$

Product: $\frac{449988}{9980010} = 0.0451$

Difference: $\frac{33958}{9990} = 3.3\overline{991}$

(e) Sum: $99\frac{1}{99} = 99.\overline{01}$

Product: $\frac{99}{99} = 0.\overline{9}$

Difference: $-98\frac{98}{99} = -98.\overline{98}$

3. (a) $2.\overline{3} < 2\ 1/2 < 2.51 < 6000/29$ (c) $2.\overline{71} < 32/11 < 477/154 < 22$
4. (e) 9801.0000... (g) $0.\overline{247}$ 5. yes 8. (a) 28/9
(c) 1046/333 (e) 314,156/99,999

Review Exercise Set 8-7

1. (a) 1409/99 (c) 9904/330 (e) -89/4950 (g) 13/2
2. (a) False, because it is a terminating decimal.
(c) False; $1.76 \cdot (10)^{-4} = 0.000176$.
(e) False, since a real number may be a rational or an irrational number.
(g) False; 231% = 2.31.
(i) False. If a number can be represented by a repeating decimal, it is a rational number.
3. (a) to the left of 3 (c) yes (e) yes
4. (a) $\frac{677}{236} < 3.14 < 3\ 1/2$ (c) $\frac{23}{11} < \sqrt{5} < 2\ 1/2$
5. (a) Sum: $23.\overline{244335}$
Difference: $11.\overline{220311}$
Product: $103.\overline{600934267}$
(c) Sum: 0.517979 ...
Difference: 0.482020 ...
Product: 0.008989 ...

Exercise Set 9-1

1. (a) 1 (c) 6 (e) 6 2. (a) infinite number
(c) none
3. (a) A B C $\overrightarrow{AC} \cap \overline{BC} = \overline{BC}$
(c) A B C $\overline{AB} \cap \overrightarrow{BC} = \{B\}$

4. (a) true (c) true
   (e) False. $\overline{AB}$ is the same set of points as $\overline{BA}$.
   (g) False. $\overline{AB} \subset \overrightarrow{CB}$.
   (i) False. $\overrightarrow{BA} \cap \overline{AC} = \overline{AB}$.
   (k) False. The endpoints of $\overline{BA}$ are B and A.
   (m) False. The two half-planes formed by a line include all the points of the plane except for the points forming the line.
   (o) False. If A, B and C were collinear points, then $\overleftrightarrow{AB} = \overleftrightarrow{AC}$.
   (q) False. A line separates a plane into three disjoint sets.
   (s) False. A ray has one endpoint.
5. (a) $\overline{QR}$ (c) $\overline{RS}$ (e) $\overline{RS}$ (g) $\{R\}$ 6. (a) empty
   (c) not empty (e) empty
7.

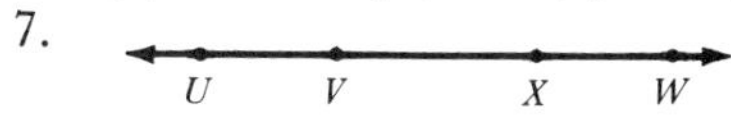

9. (a) (c) (e)

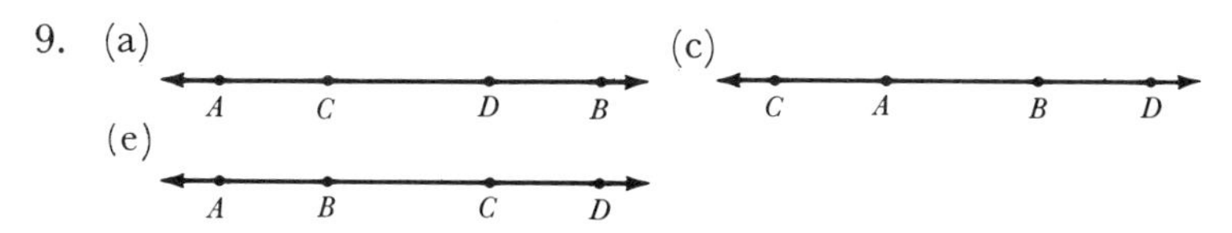

10. (a) $\{A\}$ (c) $\overline{AB}$
11. (a) $\overrightarrow{AB}$, $\overrightarrow{CD}$; $\overrightarrow{AC}$, $\overrightarrow{CD}$; etc. (c) $\overrightarrow{AC}$, $\overrightarrow{DB}$; $\overrightarrow{AB}$, $\overrightarrow{DC}$; etc.
12. (a) $\overline{AC}$, $\overline{BC}$; $\overline{AD}$, $\overline{CD}$; etc. (c) $\overline{AC}$, $\overline{BD}$; $\overline{AD}$, $\overline{BC}$; etc.
13. (a) half line (c) segment 14. (a) yes

Exercise Set 9-2

1. 6, 6 2. one line, one point
3. (a) False. Any three noncollinear points determine a plane and three collinear points are in the plane containing the line. (c) true
   (e) true (g) true
4. (a) $\overleftrightarrow{AB}$ and $\overleftrightarrow{DC}$ (c) $\{B\}$ (e) $\{A\}$ (g) $\varnothing$
5. (a) $\{R\}$ (c) $\overleftrightarrow{RS}$ separates the plane 6. (a) no (c) 10
7. (a) They are the same plane.
9. (a) The intersection will be the one point lying on $\overleftrightarrow{AB}$ and the intersection of β with α.
10. (a) 10 (c) 6

Exercise Set 9-3

1. (a) $\angle z$ and $\angle w$ are adjacent and supplementary.
2. (a) $\angle x$ and $\angle y$ are vertical.
3. (a) yes (c) no
4. (a) (c) (e)

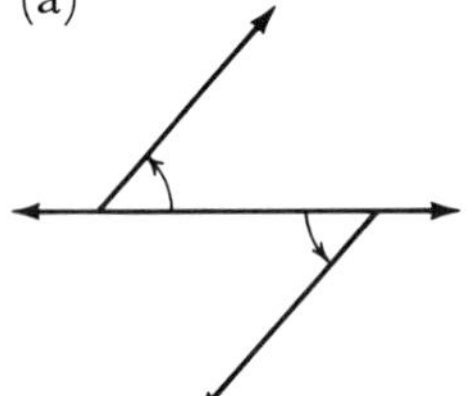

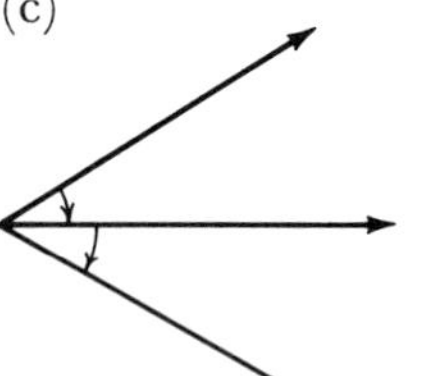

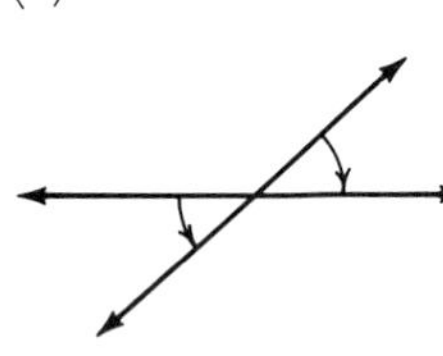

5\.

(a)

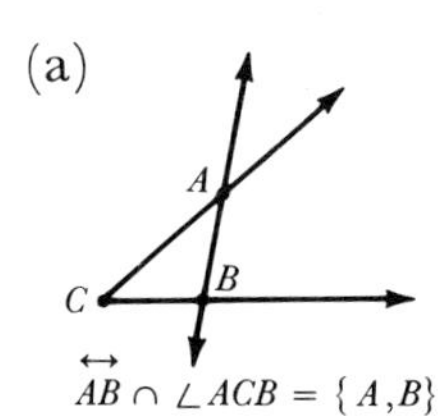

$\overleftrightarrow{AB} \cap \angle ACB = \{A, B\}$

(b)

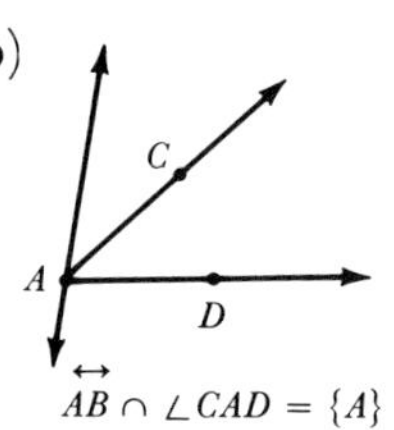

$\overleftrightarrow{AB} \cap \angle CAD = \{A\}$

(c)

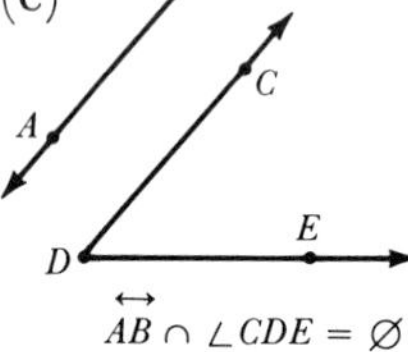

$\overleftrightarrow{AB} \cap \angle CDE = \varnothing$

(d)

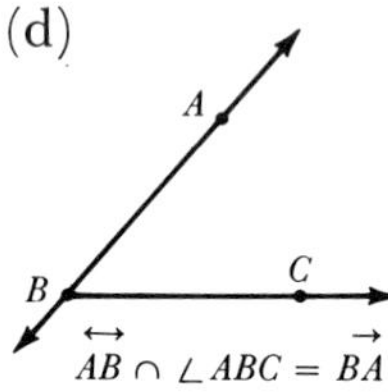

$\overleftrightarrow{AB} \cap \angle ABC = \overrightarrow{BA}$

6. (a) $\{A, C\}$ (c) $\overline{AC}$ and $\{B\}$ (e) $\overline{CB}$
7. (a) 6 or 3 pairs 9. no
10. (a) yes (c) yes

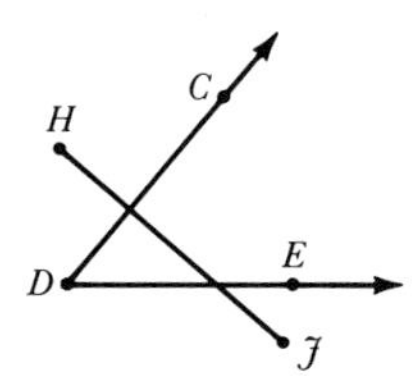

11. (a) 12: $\angle EOC$, $\angle EOB$, $\angle COB$, $\angle COF$, $\angle BOF$, $\angle EOA$, $\angle EOD$, $\angle AOF$, $\angle DOF$, $\angle AOD$, $\angle AOC$, $\angle BOD$ (c) $\angle EOB$ and $\angle AOF$; $\angle EOC$ and $\angle DOF$; $\angle COB$ and $\angle AOD$; $\angle AOC$ and $\angle BOD$; $\angle COF$ and $\angle EOD$; $\angle BOF$ and $\angle EOA$

12. (a) six lines, twelve angles

Exercise Set 9-4

1. (a) simple (c) simple (e) simple
2. (a) not closed (c) closed (e) closed
3. (c) not convex (e) convex
4. (a) $\{X\}$ (c) $\{Z\}$ (e) $\overline{XZ}$ (g) $\overline{XZ}$
6. (a) (c)

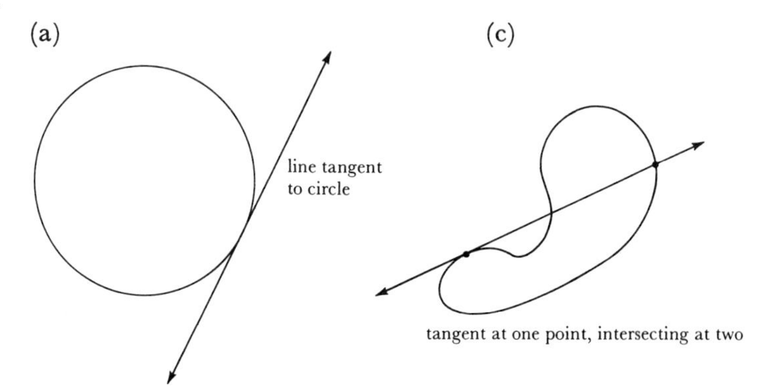

7. (a) (c)

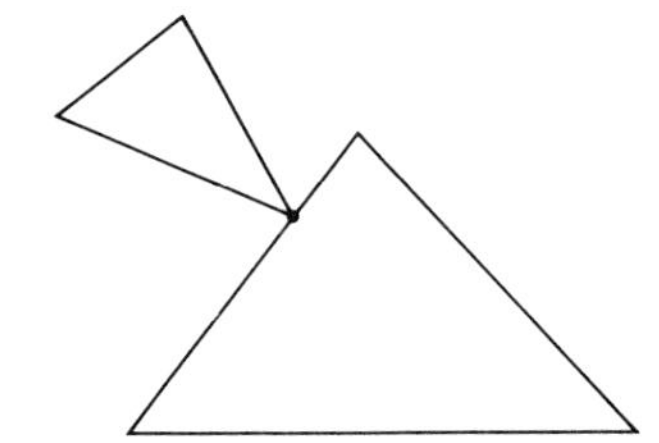

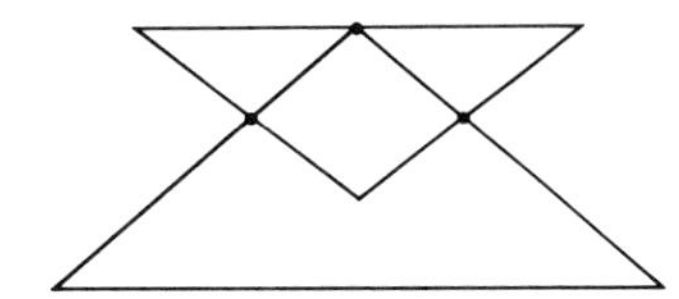

(e)

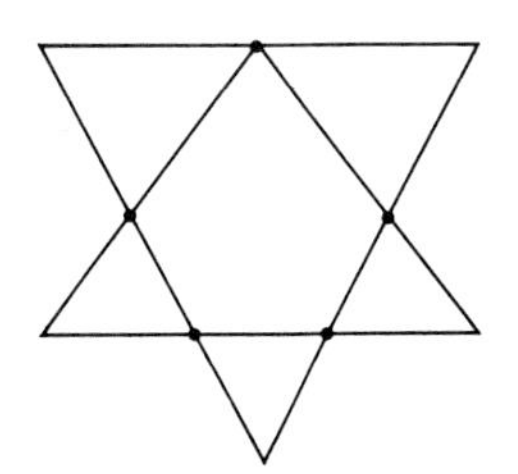

8. yes, interior of $\triangle XYZ$ (c) $\{R, C, S\}$ (e) $\overline{RS}$ 9. 3, 4 10. (a) 11 11. (a) false (c) false (e) false (g) false

12. (a) Let $\overleftrightarrow{AB}$ divide plane α into half-planes α_1 and α_2. α_2 is indicated by horizontal shading in the figure. Let $\overleftrightarrow{BC}$ divide plane β into half-planes β_1 and β_2, with β_2 being indicated in the figure by diagonal lines. Let $\overleftrightarrow{AC}$ divide plane θ into half-planes θ_1 and θ_2, θ_2 being indicated by vertical lines. The interior of $\triangle ABC$ is $\alpha_2 \cap \beta_2 \cap \theta_2$.

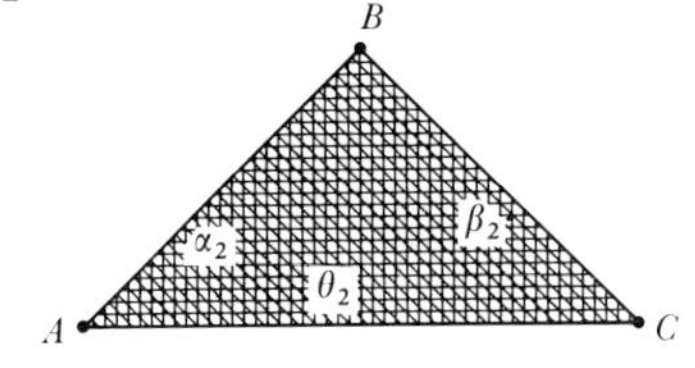

13. (a) 2 (c) 5 14. (a) $\{C\}$ (c) $\overline{CE}$ (e) interior of $\triangle DBF$
16. (a) $\{D, E, G\}$ (c) Figures *DEFGH* and *CAKEH* do not fulfill the requirements because they are not convex.

Exercise Set 9-5

1. (a) true (c) false (e) false (g) true 2. (a) 4 (c) 6 (e) 5
3. (a) quadrilateral pyramid; vertices = $\{A, B, C, D, E\}$; edges, $\overline{AE}$, $\overline{BE}$, $\overline{CE}$, $\overline{DE}$, $\overline{AB}$, $\overline{BC}$, $\overline{CD}$, $\overline{DA}$
8. (a) true; $6 \geq 2 + \frac{1}{2}(6)$ (c) true; $12 \geq 2 + \frac{1}{2}(8)$

Review Exercise Set 9-6

1. (a) False. A plane separates space.
   (c) true (e) true (g) true
   (i) False. 8 is a closed curve but not a simple curve.
   (k) False. A polygon has three or more sides.
   (m) true
   (o) False. The sides of an angle are rays.
2. (a) $\overline{AE}$ (c) $\overline{BE}$ excluding B and E (e) $\overline{BA} \cup \overline{AE}$
   (g) $\{D\}$

3. (a) 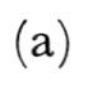 (c)

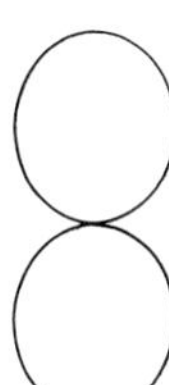

4. (a) 2 distinct rectangles (c) 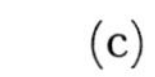 (e)

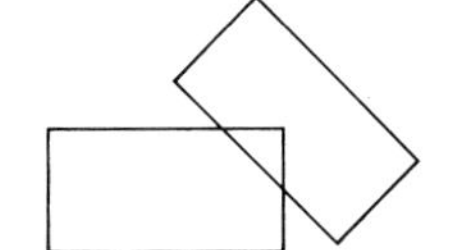

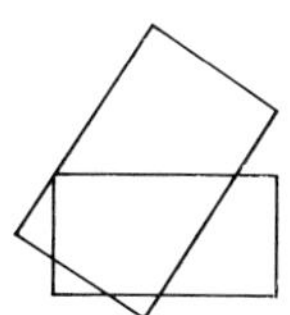

5. (a) (c) 

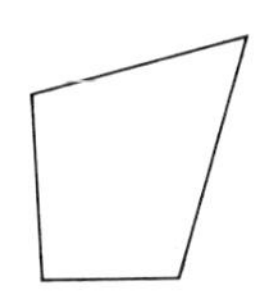

6. (a) (c) 

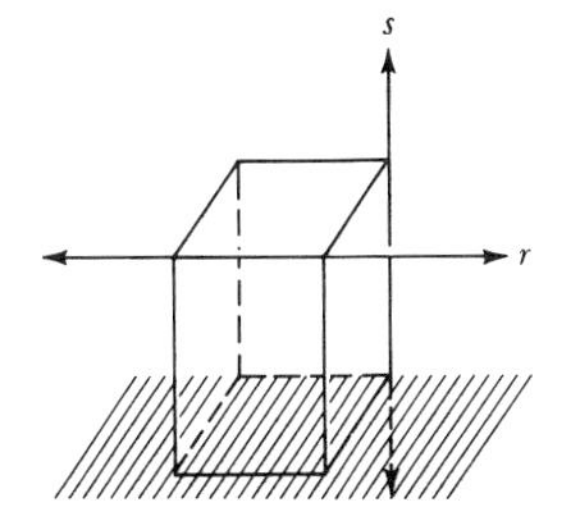

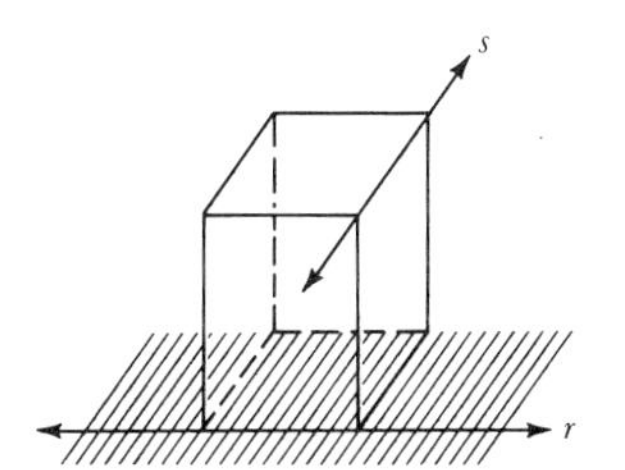

7. point—yes; line—yes; segment—no

Exercise Set 10-1

1. (a) $A \leftrightarrow A'$ $B \leftrightarrow B'$ $C \leftrightarrow C'$ $D \leftrightarrow D'$
   (c) $A \leftrightarrow A'$ $B \leftrightarrow B'$ $C \leftrightarrow C'$
2. (a) true (c) false (e) false (g) true (i) true
   (k) true 3. (a) false (c) false (e) true
4. (a) $m(\overline{AB}) = 1\frac{1}{2}$ (c) $m(\overline{DB}) = 2$ (e) $m(\overline{GA}) = 6$
5. $\overline{DE} \simeq \overline{EF}$, $\overline{CD} \simeq \overline{DF} \simeq \overline{AB}$, $\overline{BD} \simeq \overline{AC}$, $\overline{AD} \simeq \overline{BF}$
6. (a) $x = 3\frac{1}{2}$ or $\frac{1}{2}$ (c) $x = 7$ or -1
7. (a) 12.1 yards (c) 17.6 kilometers (e) 4.265 meters
   (g) 10,560 yards
8. (a) 60 inches, 5/3 yards, 1.524 meters (c) 8.8 yards, 26.4 feet, 316.8 inches
9. (a) $N(\text{feet}) = \dfrac{N(\text{inches})}{12}$ (c) $N(\text{miles}) = \dfrac{N(\text{feet})}{5280}$
10. (a) 16.8 feet, 12.3 feet
11. (a) $B = 1\frac{1}{2}$ or $B = -\frac{1}{2}$ (c) $B = -3$ or $B = -5$ (e) $B = 0$ or $B = -2$
12. (a) 1 inch = 41,900 wavelengths (c) 1 mile = 2,600,000,000 wavelengths

Exercise Set 10-2

1. (a) true (c) false; $\overline{BA}$ may or may not be congruent to $\overline{FE}$.
   (e) false; $\overrightarrow{BC} \simeq \overrightarrow{FG}$ (g) true (i) true
5. (a) yes (c) no (e) yes (g) yes 6. (a) 140
7. (a) 140 (c) 140
8. (a) complement = 20, supplement = 110
   (c) complement = 48, supplement = 138
   (e) supplement = 66
9. (a) $60° \; 49' \; 1''$ (c) $71° \; 42' \; 18''$ 10. (a) no
12. (a) $25° \; 5' \; 8''$ (c) $9° \; 23' \; 26''$
13. 45 15. 45 16. (a) 3 (c) quadrilateral
    (e) 180 in a triangle, 360 in a quadrilateral
19. 120 at four o'clock, 90 at three o'clock, 150 at five o'clock

Exercise Set 10-3

1. (a) yes; Theorem 10-6(b) (c) yes; Axiom 10-1 (e) yes; Theorem 10-1
2. 

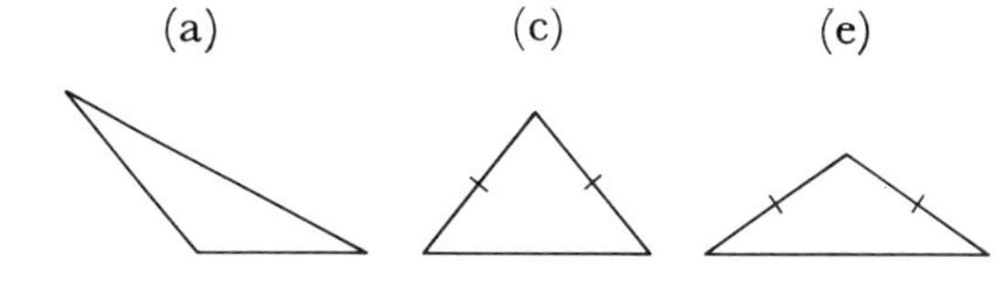

3. (a) $\overline{AB}$ and $\overline{AC}$ (c) 45
5. (a) $(6)^2 + (8)^2 = (10)^2$, $36 + 64 = 100$
   (c) $(7)^2 + (24)^2 = (25)^2$, $49 + 576 = 625$
   (e) $(11)^2 + (60)^2 = (61)^2$, $121 + 3600 = 3721$
   (g) $(28)^2 + (21)^2 = (35)^2$, $784 + 441 = 1225$
6. (a) 11 7. (a) true 8. $\sqrt{2}\ m$

Exercise Set 10-4

1. (a) $m(\angle C) = 70$, $m(\angle F) = 70$ (c) $ST = 8$, $WX = 4$, and $m(\angle WVX) = 60$
2. (a) $x = 16/3$ (c) $z = 27$, $w = 27/2$ 3. (a) 11 1/3
   (c) 76 1/2 4. 13 inches (length); 44 inches (perimeter)
5. 18 feet 7. (a) 8/3 (c) 4
9. 48 feet 11. $\overline{MN}$ is parallel to $\overline{AB}$
12. (a) Yes, because each angle of an equilateral triangle is always equal to a measure of 60, satisfying Theorem 10-9.
    (c) yes
15. (a) Since l and m, n and r are parallel, we know that $ABCD$ is a parallelogram. By Theorem 10-8, $\overline{AB} \simeq \overline{DC}$, $\overline{AD} \simeq \overline{BC}$. Since we are given that $\overline{AB} \simeq \overline{AD}$, by the transitive property, $\overline{AB} \simeq \overline{DC} \simeq \overline{AD} \simeq \overline{BC}$. $ABCD$, then, is a rhombus.

Exercise Set 10-5

1. (a) true (c) false (e) false (g) true (i) true
   (k) true (m) false (o) false (q) true
2. (a) 10,000 square centimeters (c) 1,000,000 square millimeters
3. (a) 0.336 (c) 21/2 (e) $x^2y^2z^2$ 4. (a) 54
5. (a) 16 square inches (c) 32 square feet (e) 28 square inches

(g) 24 square feet (i) 60 square centimeters (k) 75 square feet

6. (a) 8 (c) 2 7. 3/2 8. (a) The area is doubled.
(c) The area is tripled. 9. 28

10. (a) $6\frac{2}{3}$ square yards for the hall and 40 square yards for the living and dining rooms
(c) $15\frac{5}{9}$ square yards; $32\frac{2}{9}$ square yards

Exercise Set 10-6

1. (a) The intersection of two circles may be a circle, two points, one point, or the null set.
2. (a) $A = 9\pi$, $c = 6\pi$ (c) $OB = 4$, $c = 8\pi$
3. (a) tangent (c) radius (e) $\overleftrightarrow{AB}$ perpendicular to $\overleftrightarrow{OD}$
4. (a)

5. (a) 50.2656 square feet (c) 452.3904 square centimeters
6. (a) 7 7. $4\sqrt{3}\pi + 8\sqrt{3}$ 9. $A = 12 + (9/2)\pi$
11. (a) 36π (c) 100π 13. 4 15. $A = c^2/4\pi$
17. 16,807

Exercise Set 10-7

1. 50π
3. It would be four times as much.
It would be twice as much.
It would be eight times as much.
4. (a) $V = 120$, SA $= 148$ (c) $V = 147/64$, SA $= 203/16$
5. (a) 252 6. 282 7. 75π cubic inches
8. (a) The volume doubles. 9. $8\sqrt{2}$ 11. 12π
13. 2 gallons 14. (a) Volume is 360π and surface area is 192π.
15. 59.8 gallons 16. 5.25 square inches 17. $(2/3)\ \pi r^2 h$
19. 6

Exercise Set 10-8

1. (a) 4 (c) 1

3. 

(a)

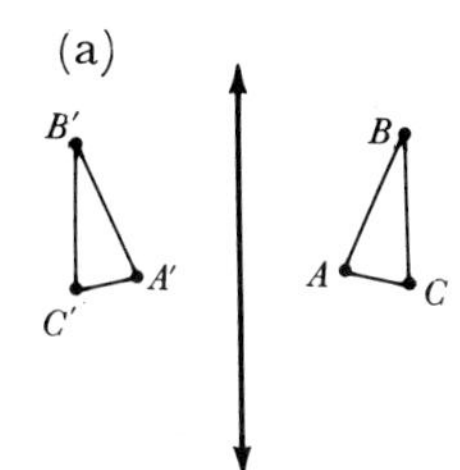

(c)

4. (a)

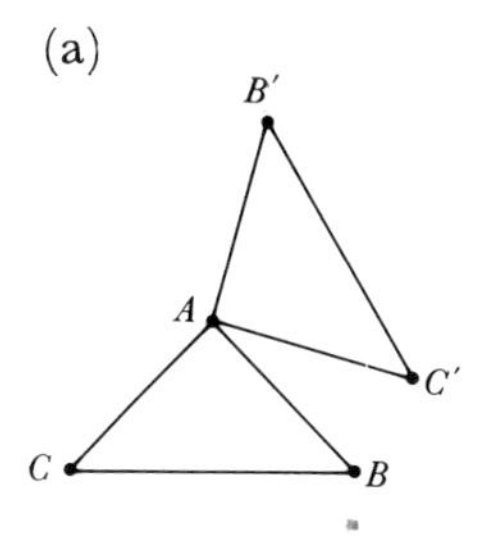

(c)

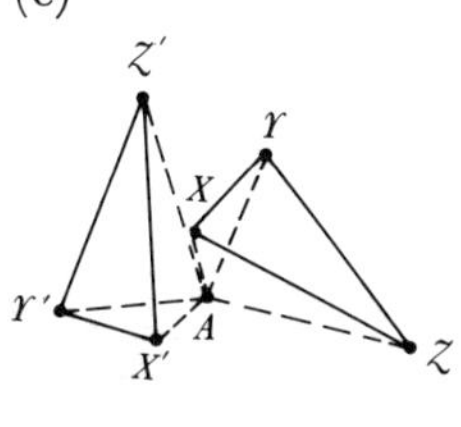

5. (a) (c)

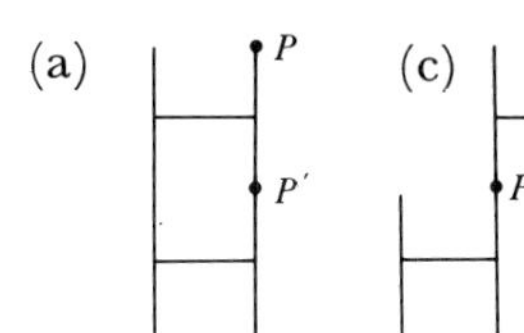

6. (a) reflection (c) rotation
7. (a) (c) (g)

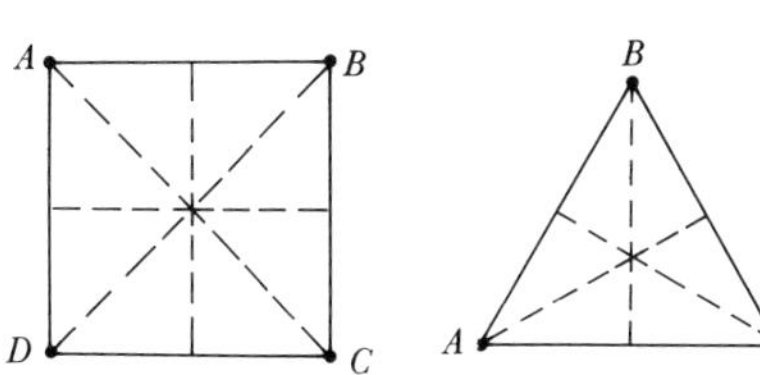

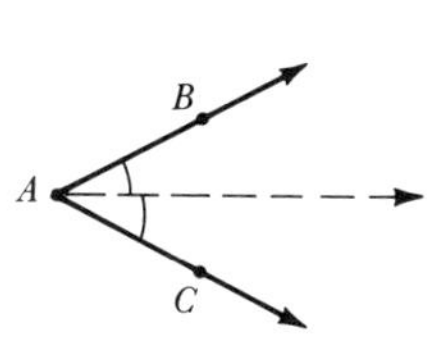

(e) none

8. (a) reflection (c) 180° rotation about B (e) reflection about $\overline{BD}$

9. (a) Translate the figure twice the length of one leg to form $A'B'C'D'$. $A'B'C'D'$ is also formed by reflecting $ABCD$ about $\overline{BC}$ and $BA'D'C$ about $\overline{A'D'}$.

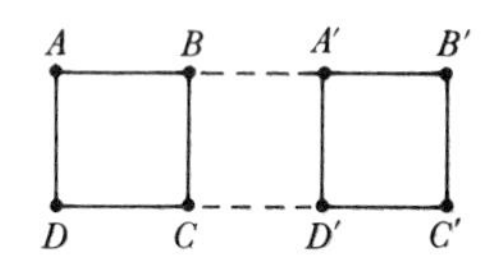

10. (a) Rotate $\triangle ABC$ clockwise 60° about C and translate the new $\triangle BDC$ by the length of leg $\overline{BC}$ to form $\triangle CEF$. $\triangle CEF$ is the same triangle formed by reflecting $\triangle ABC$ about $\overline{BC}$, $\triangle BCD$ about $\overline{DC}$, and $\triangle DCE$ about $\overline{CE}$.

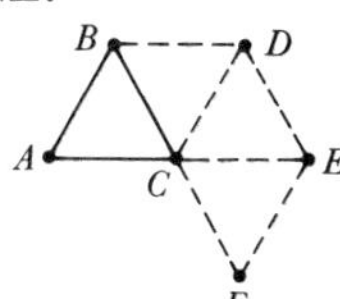

Review Exercise Set 10-9

1. (a) false (c) true (e) true (g) false
2. (a) true (c) false (e) false (g) true (i) true
   (k) false (m) false (o) true (q) true (s) false

3. $12 + \frac{9\sqrt{3}}{4}$ 5. It increases 8 times. 7. $r = 2$ 9. 35

Exercise Set 11-1

1. (a) possible outcomes = {W, R, G}
   (c) possible outcomes = {W, Y, G, B, R, O}
2. (a) {(G, R), (G, B), (W, R), (W, B), (R, R), (R, B)}
   (c) {(G, B), (G, R), (G, G), (G, Y), (R, B), (R, R), (R, G), (R, Y), (W, B), (W, R), (W, G), (W, Y)}
   (e) {(G, R), (G, B), (R, R), (R, B), (B, R), (B, B), (Y, R), (Y, B)}
3. {(H, 1), (H, 2), (H, 3), (H, 4), (H, 5), (H, 6), (T, 1), (T, 2), (T, 3), (T, 4), (T, 5), (T, 6)}.
5. (a) {R, B} (c) {RRR, RRB, RBB, BBB}
6. (a) {Chevrolet, Ford, Pontiac, Oldsmobile, Buick, Plymouth, Dodge}

7. {(1, 1), (1, 2), (1, 3), (1, 4), (1, 5), (1, 6), (2, 1), (2, 2), (2, 3), (2, 4), (2, 5), (2, 6), (3, 1), (3, 2), (3, 3), (3, 4), (3, 5), (3, 6), (4, 1), (4, 2), (4, 3), (4, 4), (4, 5), (4, 6)}
8. (a) no (c) no (e) no

Exercise Set 11-2

1. (a) 1/4 (c) 1/50 2. (a) 5/8 (c) 1/8
3. (a) 1/2 (c) 2/3 (e) 1/2 4. (a) 2/7 (c) 1/2
5. (a) 7/55 (c) 27/55 (e) 7/22 6. (a) 1/6 (c) 1/9
7. (a) 1/4 8. (a) 1/4 (c) 1/52

Exercise Set 11-3

1. (a) 210 (c) 15 (e) $\frac{r(r-1)}{2}$ 3. 120

5. 2,598,960 7. 3/7 8. (a) 2/255,645 (c) 616/2185 9. 969/2530 11. 21/285

Exercise Set 11-4

1. (a) 3/4 (c) 0.35
2. (a) getting 0 or 1 head
(c) getting less than six heads or getting seven heads
3. (a) 1/36 (c) 1/18 4. (a) 4/10 (c) 7/10
5. (a) 1/2 (c) 13/24 6. (a) 1/20 (c) 3/20
8. 15/56 9. 0.85 10. (a) 1/169 11. (a) 4/13

Review Exercise Set 11-5

1. (a) 1/2 (c) 0.55 2. (a) 11/4165 3. (a) 1/2 (c) 1/2 (e) 1/4 (g) 3/4 4. (a) 10/13 (c) 1 (e) 9/13 (g) 1 (i) 1 5. 1/36 6. (a) 1/40425 (c) 912/8085 (e) 0.8847 7. (a) 0.343 (c) 0.189

Exercise Set 12-1

1. (a)

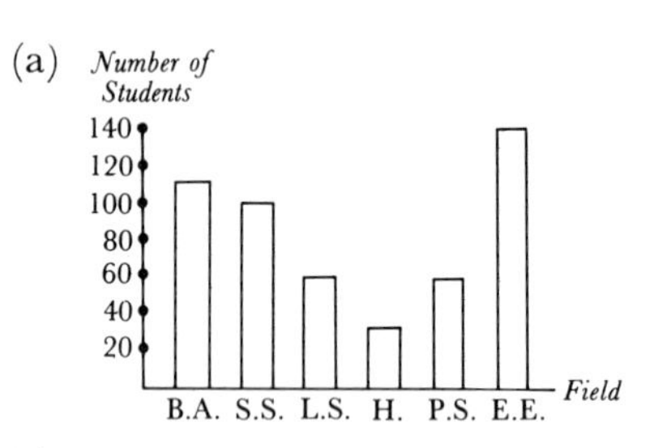

(c)

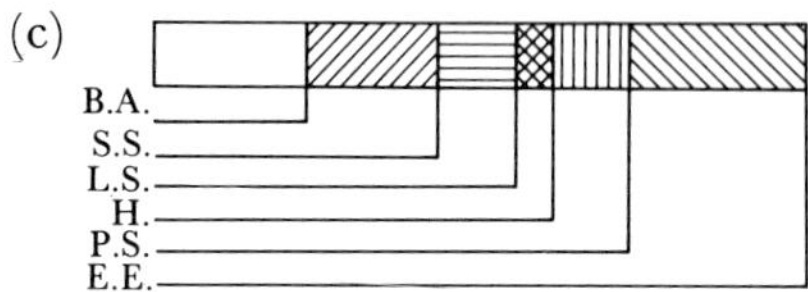

2. (a)

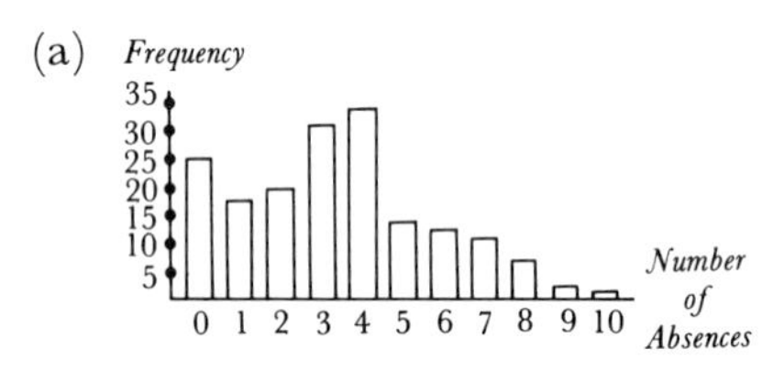

3. (a)

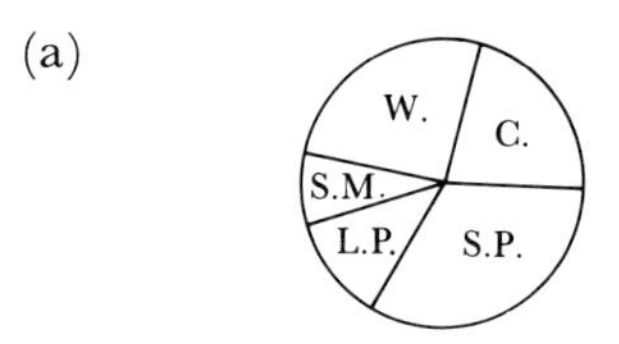

(d)

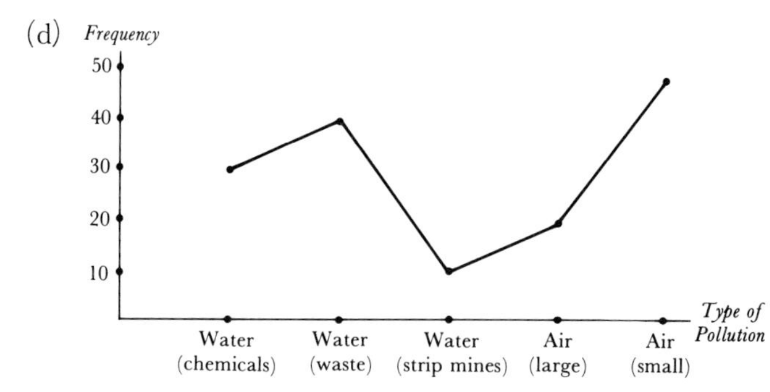

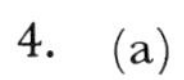

4. (a)

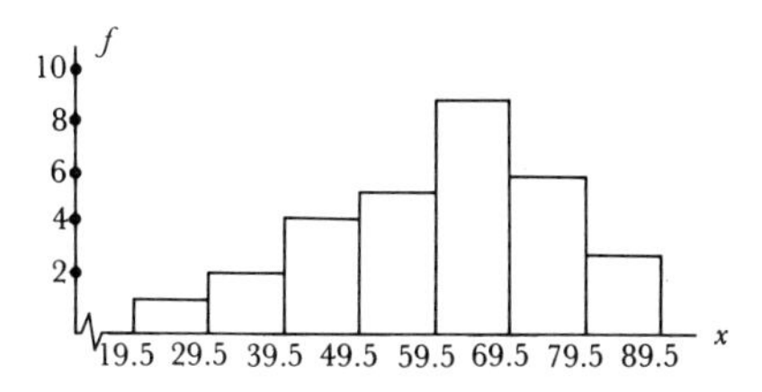

5. (a)

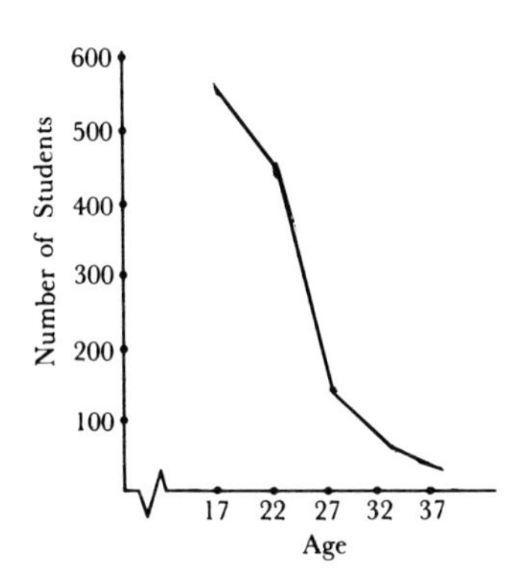

Exercise Set 12-2

1. (a)

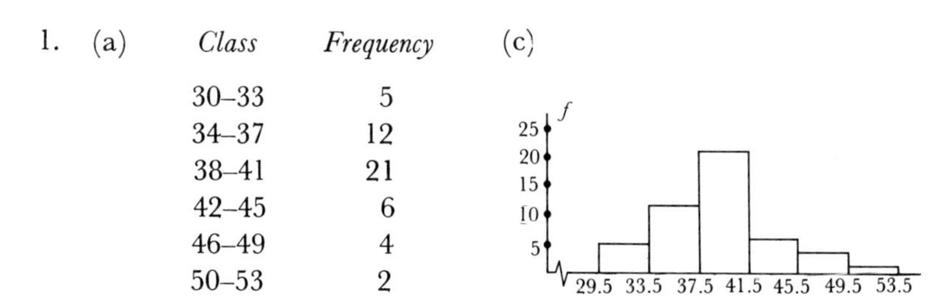

Class	*Frequency*
30–33	5
34–37	12
38–41	21
42–45	6
46–49	4
50–53	2

(c)

2. (a)

Class	*Frequency*
43–50	2
51–58	5
59–66	9
67–74	13
75–82	13
83–90	10
91–98	8

(c)

f
14 12 10 8 6 4 2
42.5 50.5 58.5 66.5 74.5 82.5 90.5 98.5

3. (a)

Class	*Frequency*
75–87	5
88–100	9
101–113	7
114–126	4
127–139	3
140–152	2

(c)

f
10 8 6 4 2
74.5 87.5 100.5 113.5 126.5 139.5 152.5

4. (a)

Class	*Frequency*
60–129	2
130–199	4
200–269	5
270–339	2
340–409	5
410–479	2

(c)

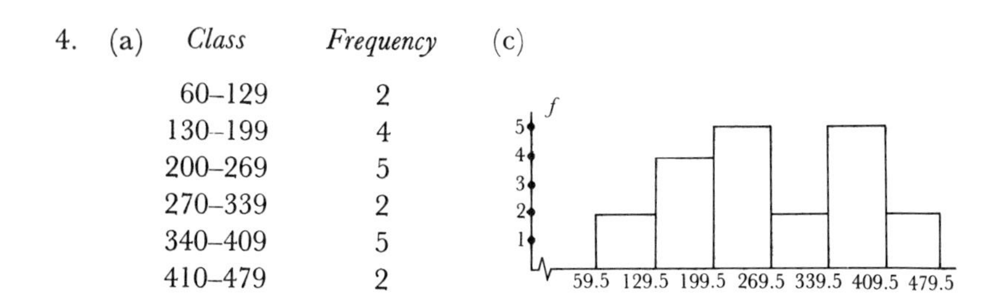

5. (a)

Class	*Frequency*
450–491	4
492–533	7
534–575	6
576–617	5
618–659	3

(c)

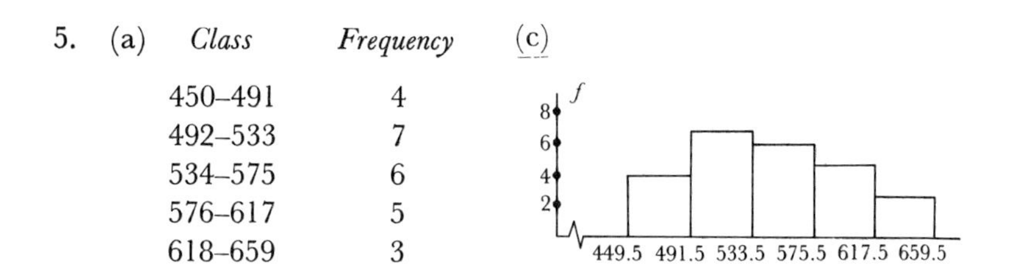

6. (a)

Class	*Frequency*
100–149	5
150–199	21
200–249	14
250–299	5
300–349	3
350–399	4
400–449	2
450–499	2
500–549	4
550–599	2

(c)

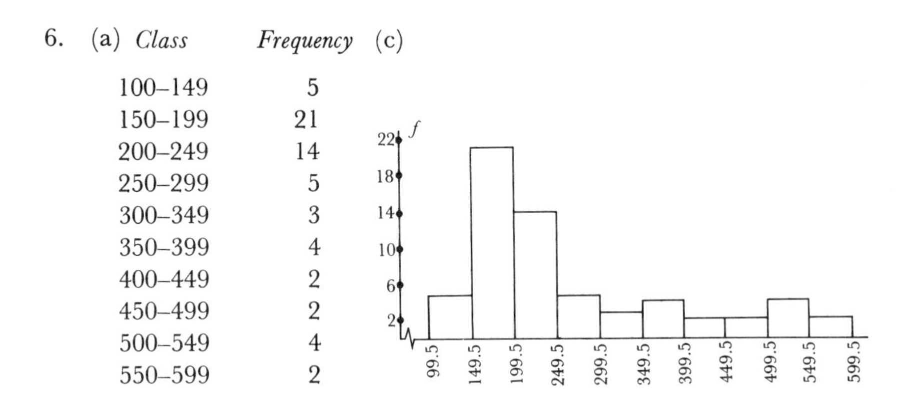

7.

Class	*Class Marks*	*Frequency*
159.5–174.5	167	3
174.5–189.5	182	6
189.5–204.5	197	9
204.5–219.5	212	6
219.5–234.5	227	2

Exercise Set 12-3

1. (a) 5.222 (c) 5.625 2. (a) 118.21 (c) 118.21
3. (a) 223 (c) 223 5. 8.75 7. (a) 39.342
(c) 105.7 (e) 547.78

Exercise Set 12-4

1. (a) median, 4; mode, 6 (c) median, 19.5; mode, 23
2. (a) tenth percentile, 6; ninetieth percentile, 21
3. modal class, \$300,000–\$350,000; median, \$307,258.06
5. modal class, 5,000–15,000; median, 33,571.43
8. (a) 181.73 (c) 180 9. (a) 182.25 (c) 190–199

Exercise Set 12-5

1. (a) 5.24 (c) 5.17 2. (a) 13.44 3. (a) 14.56
(c) 41.81 (e) 1.92 5. 15.17 7. 4.57
10. (a) 4.8 (c) 18.79 (e) 52.7

Exercise Set 12-6

1. (a) 0.9332 (c) 0.3446 (e) 0.0668 (g) 0.9821 (i) 0.0886 (k) 0.5028
2. (a) 0.8413 (c) 0.3830
3. (a) 0.0668
4. (a) 3.4 (c) 6.4 (e) 3
5. .62%
6. (a) 68.53%
7. A's .62% B's 15.25% C's 68.26% D's 15.25% F's .62%

Exercise Set 12-7

1. (a) 11 (c) 11.67 (e) .99
2. (a) 9.75 (c) 8.19 (e) .925
3. (a) 57.5 (c) 970.75 (e) .99
4. (a) 4.71 (c) 7.35 (e) -0.104
5. 0.90
6. (a) 0.696
7. 0.92; they are the same

Review Exercise Set 12-8

5. mean, 107.25; median, 107; standard deviation, 4.71
7. 0.995

Exercise Set 13-1

1.

⊕	0	1	2
0	0	1	2
1	1	2	0
2	2	0	1

⊗	0	1	2
0	0	0	0
1	0	1	2
2	0	2	1

2. (a) yes (c) 0

3.

⊕	0	1	2	3	4
0	0	1	2	3	4
1	1	2	3	4	0
2	2	3	4	0	1
3	3	4	0	1	2
4	4	0	1	2	3

⊗	0	1	2	3	4
0	0	0	0	0	0
1	0	1	2	3	4
2	0	2	4	1	3
3	0	3	1	4	2
4	0	4	3	2	1

(a) yes, yes (c) yes, yes (e) 4, 3, 2 (g) no (i) yes

5.

⊕	1	2	3	4	5	6	7
1	2	3	4	5	6	7	1
2	3	4	5	6	7	1	2
3	4	5	6	7	1	2	3
4	5	6	7	1	2	3	4
5	6	7	1	2	3	4	5
6	7	1	2	3	4	5	6
7	1	2	3	4	5	6	7

⊗	1	2	3	4	5	6	7
1	1	2	3	4	5	6	7
2	2	4	6	1	3	5	7
3	3	6	2	5	1	4	7
4	4	1	5	2	6	3	7
5	5	3	1	6	4	2	7
6	6	5	4	3	2	1	7
7	7	7	7	7	7	7	7

(a) yes, yes (c) yes, yes (e) 6, 5, 4 (g) no (i) yes

6. (a) 3 (c) 10 (e) 3 (g) 8 (i) 8 (k) 8 (m) 9

7. (a) 15 (c) 22 (e) 3 (g) 20 (i) 8 (k) 8 (m) 21

9. (a) Abelian group

Exercise Set 13-2

1. (a) false (c) false (e) true (g) true

2. (a) 1 (c) 2 (e) 4 (g) 4 (i) 0 (k) 8 (m) 4

3. (a) 2 (c) 2 (e) 2 (g) 6

4. (a) yes (c) fourteen; seven; two

5. (a) $2(4+3) \equiv 2 \cdot 4 + 2 \cdot 3 \equiv 8 + 6 \equiv 14 \equiv 0 \pmod 7$
   $2(4+3) \equiv 2 \cdot 7 \equiv 2 \cdot 0 \pmod 7 \equiv 0 \pmod 7$

7. The set is not closed under addition. For example, $1 + 5$ is not a member of the set.

9. (a) no (c) no

Exercise Set 13-3

1. (a)

583	$5+8+3 = 16 \equiv 7$	
427	$4+2+7 = 13 \equiv 4$	
1010	$1+0+1+0 = 2$ $\quad$ 11	$2 \equiv 11 \pmod 9$

(c)

4371	$4+3+7+1 = 15 \equiv 6$	
287	$2+8+7 = 17 \equiv 8$	
30597	48	
34968		$48 \equiv 3 \pmod 9$
8742		
1254477	$1+2+5+4+4+7+7 = 30 \equiv 3$	

$$\begin{array}{lrl}
\text{(e)} & 385 & 3+8+5=16\equiv 7 \\
 & \underline{-156} & 1+5+6=12\equiv \underline{3} \\
 & 229 & 2+2+9=13 \quad 4 \quad 4\equiv 13 \pmod 9
\end{array}$$

$$\begin{array}{lrl}
\text{(g)} & 43590 & 4+3+5+9+0=21\equiv 3 \\
 & \underline{11825} & 1+1+8+2+5=17\equiv \underline{8} \\
 & 55415 & 5+5+4+1+5=20 \quad 11 \quad 20\equiv 11 \pmod 9
\end{array}$$

$$\begin{array}{lrl}
\text{(i)} & 1746 & 1+7+4+6=18\equiv 9 \\
 & \underline{-1428} & 1+4+2+8=15\equiv \underline{6} \\
 & 318 & 3 \pmod 9 \\
 & & 3+1+8=12\equiv 3 \pmod 9
\end{array}$$

$$\begin{array}{lrl}
\text{(k)} & 988 & 9+8+8=25\equiv 7 \\
 & \underline{234} & 2+3+4=9\equiv \underline{9} \\
 & 3952 & 63\equiv 0 \pmod 9 \\
 & 2964 & \\
 & \underline{1976} & \\
 & 231192 & 2+3+1+1+9+2=18\equiv 0 \pmod 9
\end{array}$$

2.

$$\begin{array}{lrl}
\text{(a)} & 583 & 3-8+5 \qquad = 0 \\
 & \underline{427} & 7-2+4 \qquad = \underline{9} \\
 & 1010 & 0-1+0-1=-2 \quad 9 \quad -2\equiv 9 \pmod{11}
\end{array}$$

$$\begin{array}{lrl}
\text{(c)} & 4371 & 1-7+3-4=-7 \\
 & \underline{287} & 7-8+2 \qquad = \underline{1} \\
 & 30597 & -7 \\
 & 34968 & 4\equiv -7 \pmod{11} \\
 & \underline{8742} & \\
 & 1254477 & 7-7+4-4+5-2+1=4
\end{array}$$

$$\begin{array}{lrl}
\text{(e)} & 385 & 5-8+3=0 \\
 & \underline{-156} & 6-5+1=\underline{2} \\
 & 229 & -2 \\
 & & 9-2+2=9 \quad 9\equiv -2 \pmod{11}
\end{array}$$

$$\begin{array}{lrl}
\text{(g)} & 43590 & 0-9+5-3+4=-3 \\
 & \underline{11825} & 5-2+8-1+1=\underline{11} \\
 & 55415 & 8 \\
 & & 5-1+4-5+5=8 \quad 8\equiv 8 \pmod{11}
\end{array}$$

(j)

$$\begin{array}{rll}
264 & 4-6+2=0 & \\
876 & 6-7+8=7 & \\
\hline
1584 & \overline{0} & \\
1848 & & 0\equiv 0 \pmod{11} \\
2112 & & \\
\hline
231264 & 4-6+2-1+3-2\equiv 0 &
\end{array}$$

4. (a)

$$\begin{array}{rll}
4TE_{twelve} & 4+T+E=25_{ten}\equiv 3 \pmod{11} & \\
T0E_{twelve} & T+0+E=21_{ten}\equiv 10 \pmod{11} & \\
\hline
12ET_{twelve} & 13\equiv 2 \pmod{11} & \\
 & 1+2+E+T=24_{ten}\equiv 2 \pmod{11} &
\end{array}$$

(c)

$$\begin{array}{rll}
80T_{twelve} & 8+0+T=18_{ten}\equiv 7 \pmod{11} & \\
-1E4_{twelve} & 1+E+4=16_{ten}\equiv 5 \pmod{11} & \\
\hline
616_{twelve} & \overline{2} & 2\equiv 2 \pmod{11} \\
 & 6+1+6=13_{ten}\equiv 2 \pmod{11} &
\end{array}$$

5. (a)

$$\begin{array}{rll}
465_{eight} & 4+6+5=15_{ten}\equiv 1 \pmod{7} & \\
741_{eight} & 7+4+1=12_{ten}\equiv 5 \pmod{7} & \\
\hline
1426_{eight} & \overline{6} & 6\equiv 6 \pmod{7} \\
 & 1+4+2+6=13_{ten}\equiv 6 \pmod{7} &
\end{array}$$

(c)

$$\begin{array}{rll}
78_{nine} & 7+8=15_{ten}\equiv 7 \pmod{8} & \\
64_{nine} & 6+4=10_{ten}\equiv 2 \pmod{8} & \\
\hline
5575_{nine} & \overline{14}\equiv 6 \pmod{8} & 6\equiv 6 \pmod{8} \\
 & 5+5+7+5\equiv 22_{ten}\equiv 6 \pmod{8} &
\end{array}$$

7. (a) Base eight

$$\begin{array}{rl}
465 & 5-6+4\equiv 3 \pmod{9} \\
741 & 1-4+7\equiv 4 \pmod{9} \\
\hline
1426 & 6-2+4-1\equiv 7 \pmod{9} \\
 & 3+4\equiv 7 \pmod{9}
\end{array}$$

(c) Base nine

$$\begin{array}{rl}
78 & 8-7\equiv 1 \pmod{T} \\
64 & 4-6\equiv -2 \pmod{T} \\
\hline
345 & \\
523 & \\
\hline
5575 & 5-7+5-5\equiv -2 \pmod{T} \\
 & 1(-2)\equiv -2 \pmod{T}
\end{array}$$

Exercise Set 13-4

1. (a) 4 (mod 9) (c) 0 (mod 9) (e) 5 (mod 9)
2. (a) 14, 49, 84 (c) 4, 11, 18, 25, . . . , 102 (e) 10 (g) 4

(i) no solution (k) no solution

3. (a) 1 (c) 2 (e) 7, 0

4.

⊞	1	2	3	4	5
1	3	5	1	3	5
2	5	2	5	2	5
3	1	5	3	1	5
4	3	2	1	0	5
5	5	5	5	5	5

No, not closed.

Review Exercise Set 13-5

1. (a) $X \odot W$ means transformation X followed by transformation W.

$\odot$	I	X	Y	W	V	Z
I	I	X	Y	W	V	Z
X	X	I	W	Y	Z	V
Y	Y	V	Z	X	W	I
W	W	Z	V	I	Y	X
V	V	Y	X	Z	I	W
Z	Z	W	I	V	X	Y

(c) no
(e) yes
(g) yes

2. (a) false (c) false (e) true (g) true

3. (a)

⊗	1	2	3	4	5
1	1	1	1	1	1
2	1	2	2	2	2
3	1	2	3	3	3
4	1	2	3	4	4
5	1	2	3	4	5

(c) yes (e) no

5. (a)

⊗	1	2	3	4	5
1	1	2	3	4	5
2	2	2	6	4	10
3	3	6	3	12	15
4	4	4	12	4	20
5	5	10	15	20	5

(c) yes

6.

+	0	1
0	0	1
1	1	0

×	0	1
0	0	0
1	0	1

(a) yes

7. (a) no (c) yes (e) no (g) yes (i) no

9. Zero has no inverse under multiplication.
11. (a) yes (c) yes (e) no
13. (a) no (c) no (e) yes

Exercise Set 14-1

1. (a) $x = -10$ Check: $-10 + 4 = -6$
   (c) $x = 1/3$ Check: $2(1/3) - (-5) = 6 - (1/3)$
   $2/3 + 5 = 6 - 1/3$
   $5\ 2/3 = 5\ 2/3$
   (e) $x = -4$ Check: $-4/5 - 1/3 = -4/3 + 1/5$
   $-12/15 - 5/15 = -20/15 + 3/15$
   $-17/15 = -17/15$
2. (a) $x < 3$ (c) $x > 3$ (e) $x < -21$
3. (a) $r = \dfrac{A - p}{Pt}$ (c) $b = y - mx$ (e) $r = \dfrac{S - a}{S}$
5. 15 dimes, 45 nickels, 3 quarters
7. Joe has 25, Tom has 50
9. Mrs. Bishop — \$26.20, Mrs. Brown — \$39.30
11. \$.51 a pound
12. (a) $w = -9$; $x = 2$; $y = 1/3$; $z = 5$
13. Dollar; Dime swapped 48 of coin A, 144 of coin B, and 12 of coin C; Dollar swapped 96 of coin A, 16 of coin B, and 192 of coin C.

Exercise Set 14-2

1.
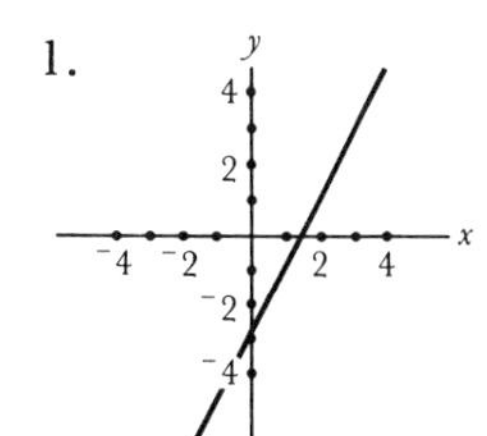

3.
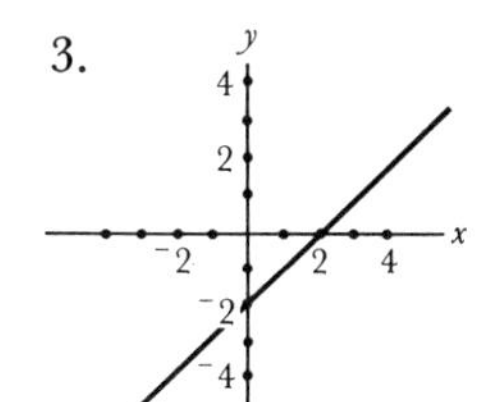

5.
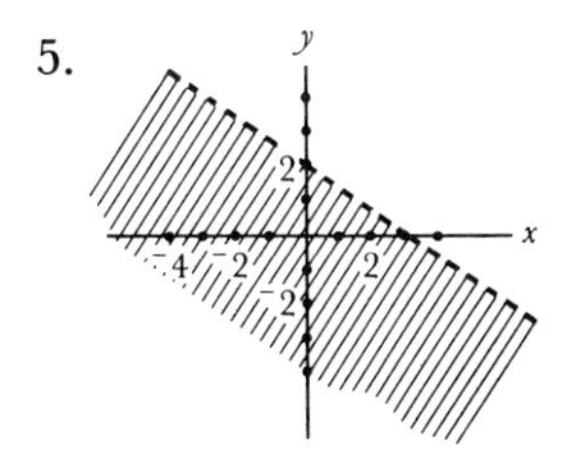

7.
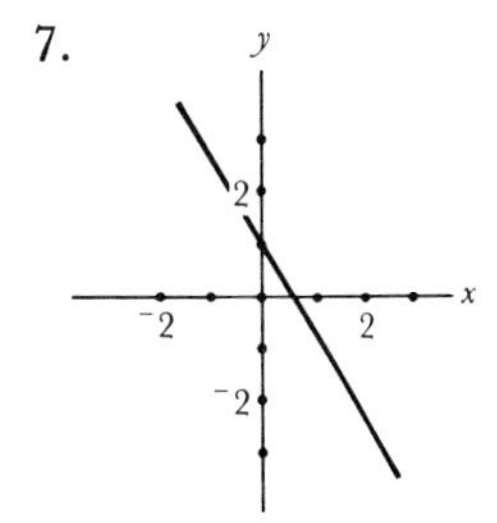

9.
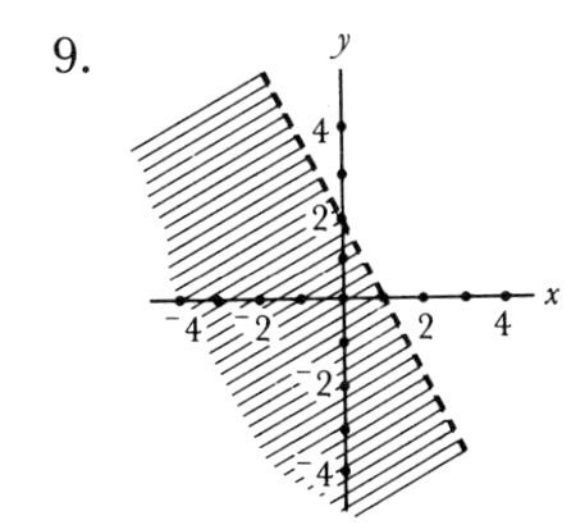

11.

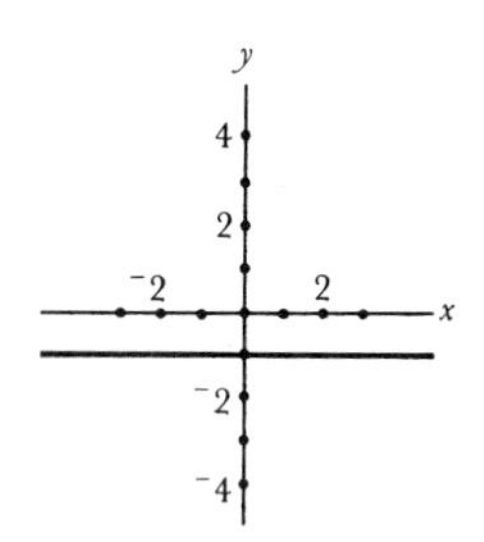

13.

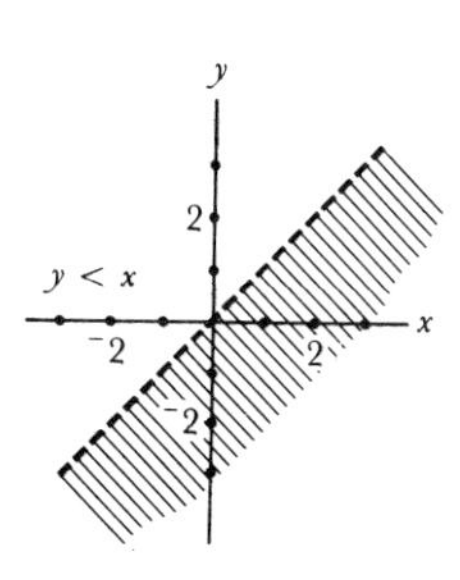

15.

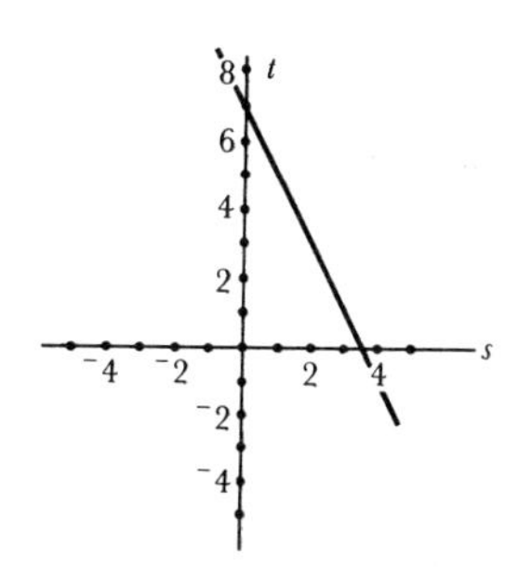

17.

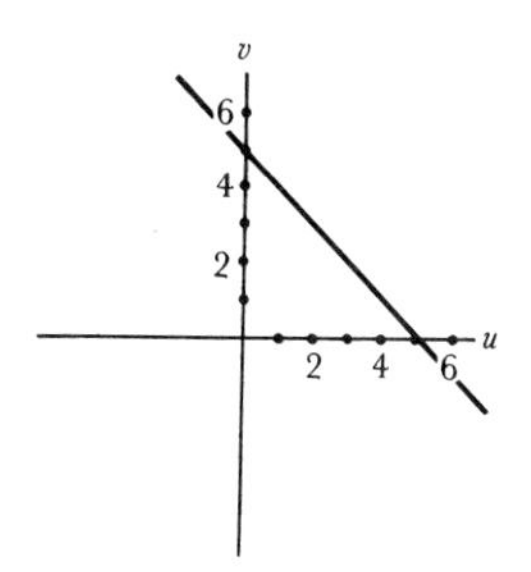

19. *b, c, e, f, g, h* 21. *a, d*

Exercise Set 14-3

1. (a) no (c) no (e) yes (g) no
2.

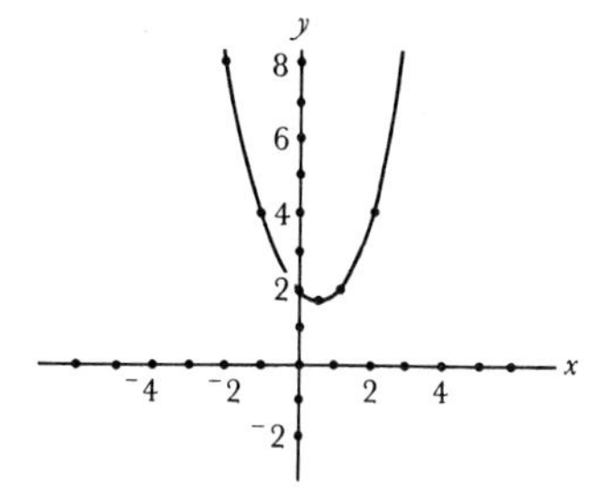

3. (a) -4 (c) -21 (e) 1/14 (g) $8z^3 - 4z$

4. (a) 5/2 (c) 3/4 (e) $\dfrac{z+2}{z-1}$ 5. (a), (b), (f)

6. (a) yes (c) no

7. (a)

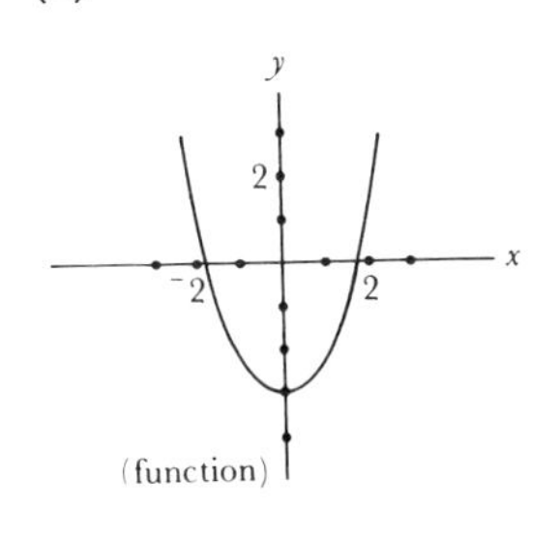

(c)

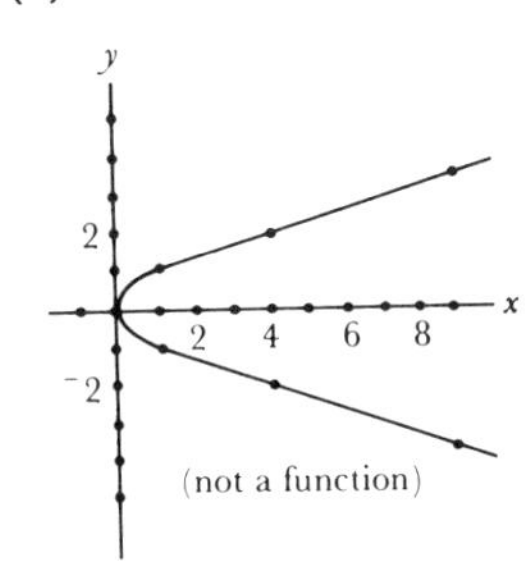

(e)

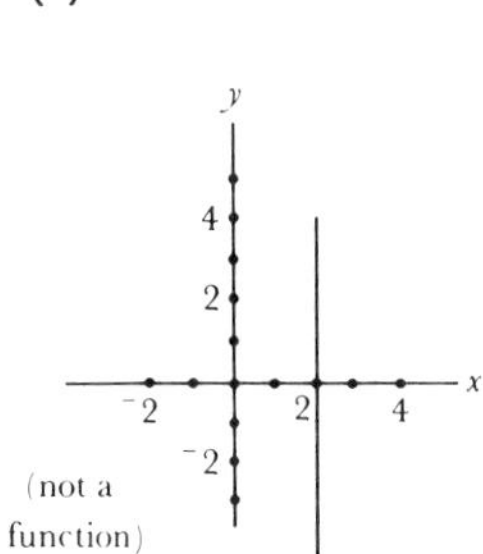

(g)

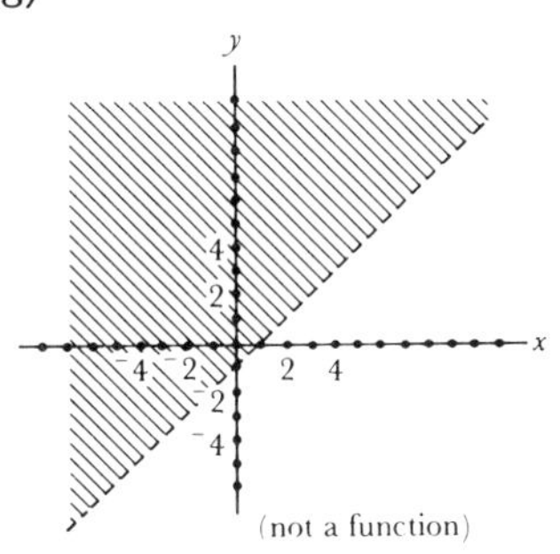

8. (a) yes (c) yes

Exercise Set 14-4

1. (a)

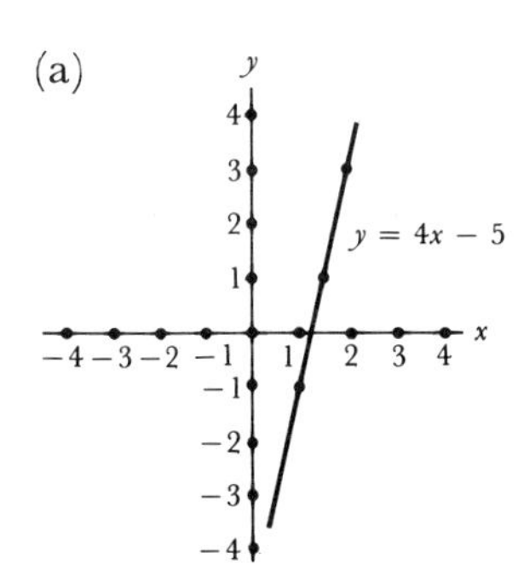

(c)

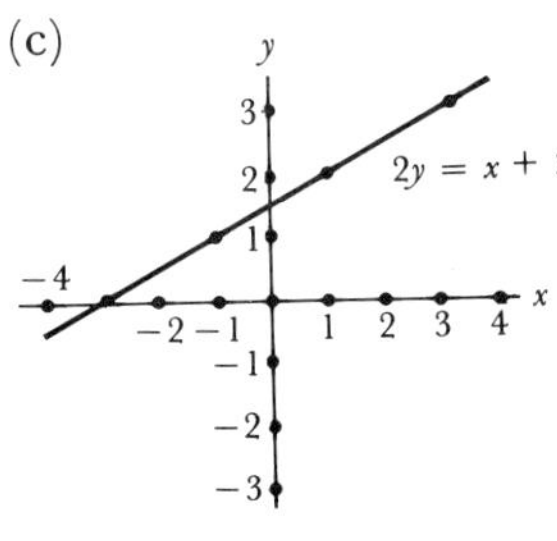

2. (a) $m = 3$, $b = 2$

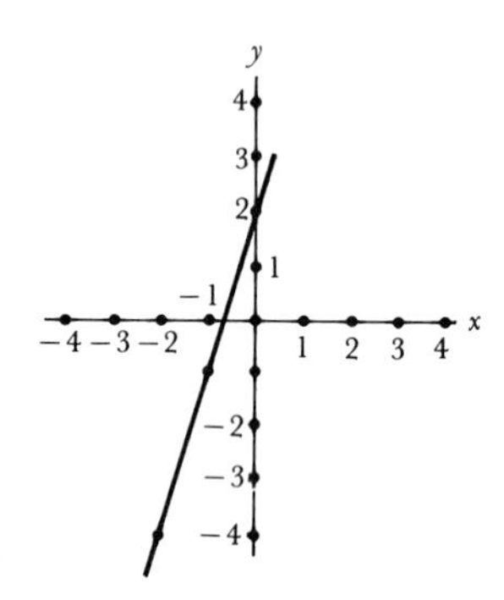

(c) $m = 1/3$, $b = -1/3$

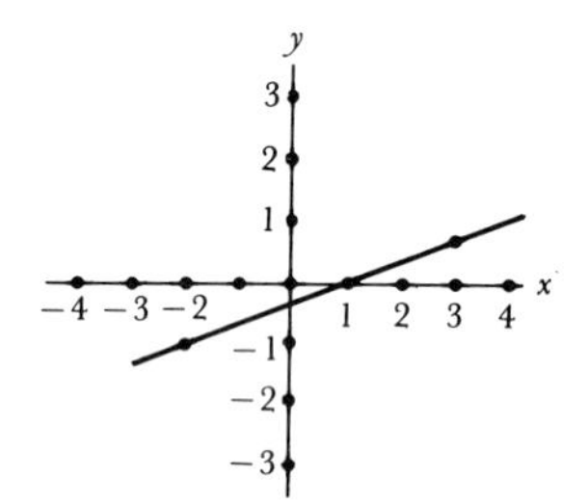

(e) $m = 1/2$, $b = -2$

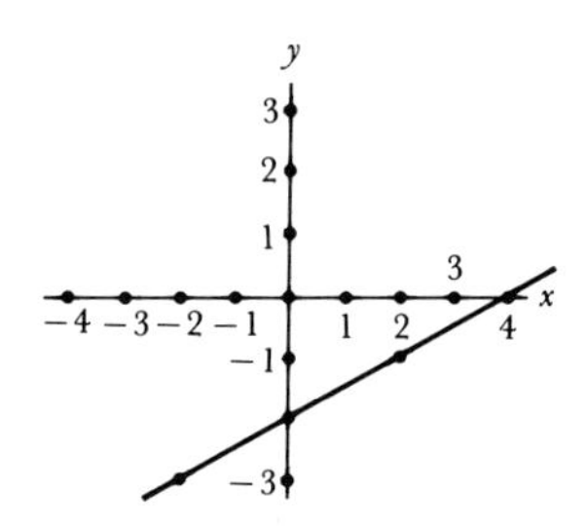

3. (a) $3y = 8x - 17$ (c) $y = x + 8$ (e) $y = 5x + 4$
5. (a) $y = 7$ 7. $F = 1.8C + 32$

Exercise Set 14-5

1. (a)

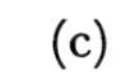
(c)

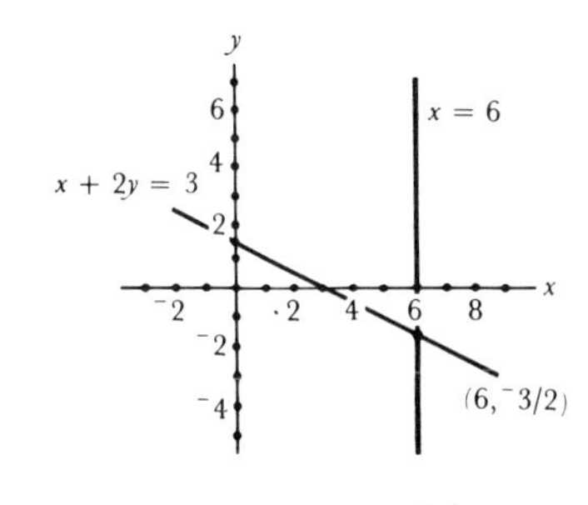

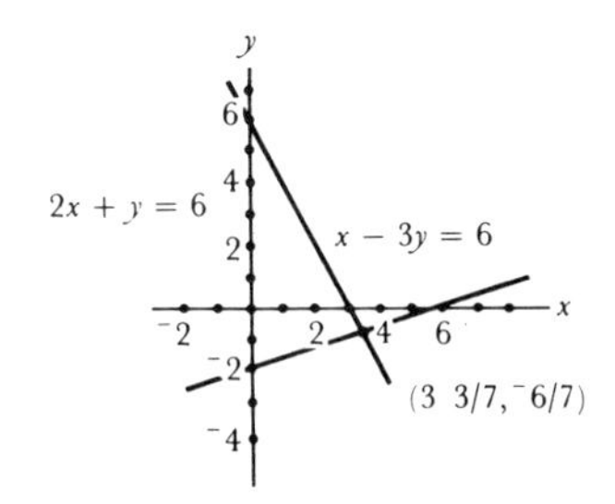

(e)

(g)

(i)

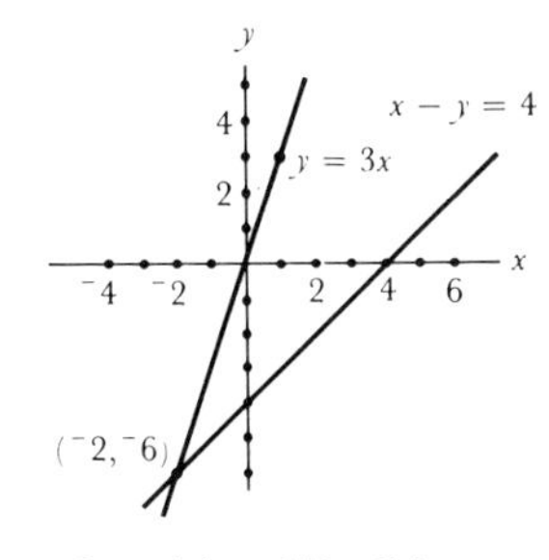

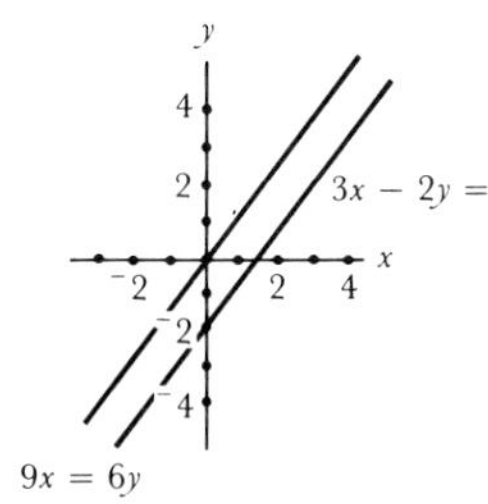

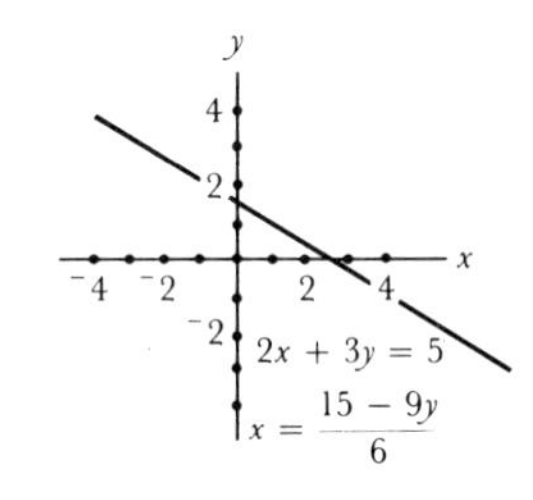

2. (a) $\{(2, 3)\}$

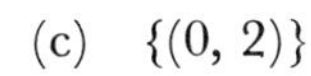
(c) $\{(0, 2)\}$

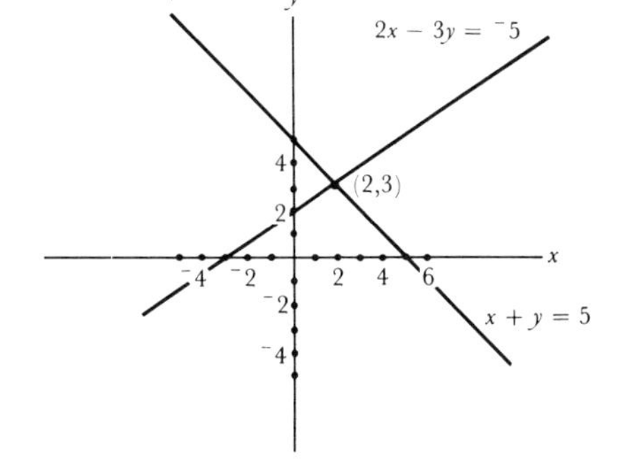

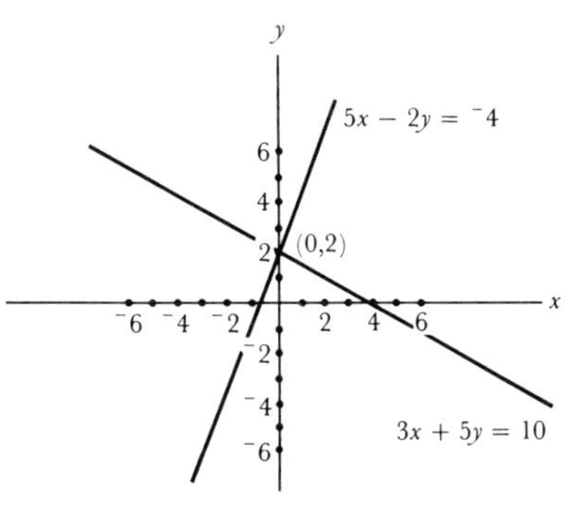

3. (a) (c)

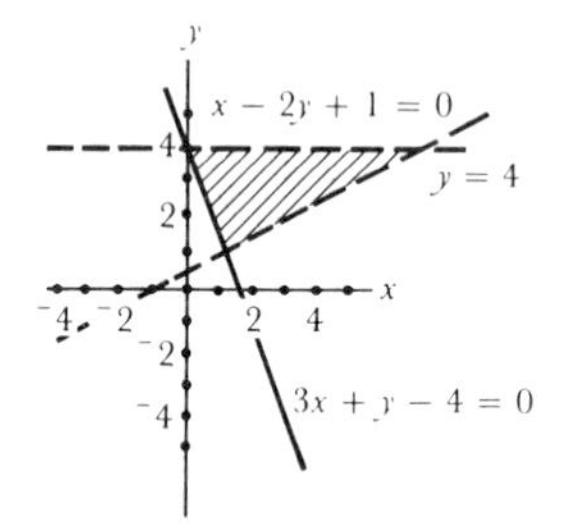

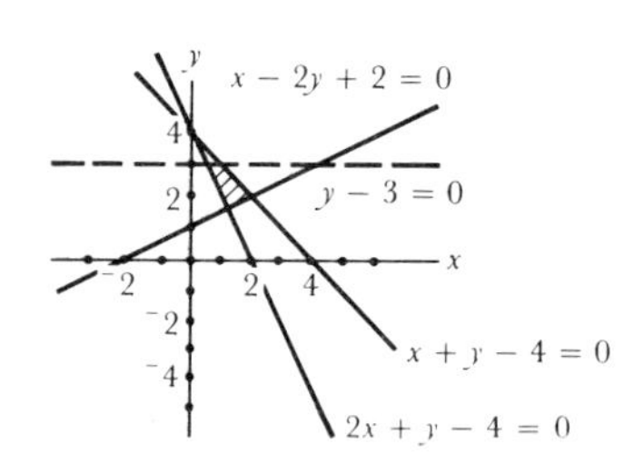

5. 100 dimes, 140 quarters
7. 26 2/3; 53 1/3
9. 33 1/3; 66 2/3

Exercise Set 14-6

1. (a) $x = -4, x = 1$ (c) $x = 3/2, x = -1/3$

   (e) $x = 1, x = -3/8$ (g) $x = \dfrac{-3+\sqrt{29}}{2}, x = \dfrac{-3-\sqrt{29}}{2}$

2. (a) $x = \dfrac{-3}{2}$ $\left(\dfrac{-3}{2}, \dfrac{-75}{4}\right)$

   (c) $x = \dfrac{7}{12}$ $\left(\dfrac{7}{12}, \dfrac{-121}{24}\right)$

   (e) $x = \dfrac{5}{16}$ $\left(\dfrac{5}{16}, \dfrac{-121}{32}\right)$

   (g) $x = \dfrac{-3}{2}$ $\left(\dfrac{-3}{2}, \dfrac{-29}{4}\right)$

3. (a) $x = 3, x = -1$ (c) $x = 7, x = -4$ (e) $x = 1/3, x = -1/4$

   (g) $x = 1/3, x = -4$ (i) $x = \dfrac{7+\sqrt{69}}{10}, x = \dfrac{7-\sqrt{69}}{10}$

4. (a) 1 second (c) $\dfrac{-9+\sqrt{145}}{32}$ seconds

   (e) either $\dfrac{13+\sqrt{41}}{32}$ or $\dfrac{13-\sqrt{41}}{32}$ seconds

Exercise Set 14-7

2. (a) $-i$ (c) i (e) $14+5i$ (g) $13+i$

   (i) $\frac{-3}{13}+\frac{28}{13}i$ (k) $\frac{22}{5}+\frac{4}{5}i$

3. (a) $-2-5i$; $2/29-(5/29)i$ (c) $3+4i$; $-3/25+(4/25)i$
   (e) $2+i$; $-2/5+(1/5)i$
4. It is a group, a ring, a field, and an integral domain.

Review Exercise Set 14-8

1. (a) true (c) false (e) true 3. (a) $x=1, y=2$
4. (a) $2y=4-x$ and $2y=3x+7$ intersect at $(-3/4, 19/8)$.

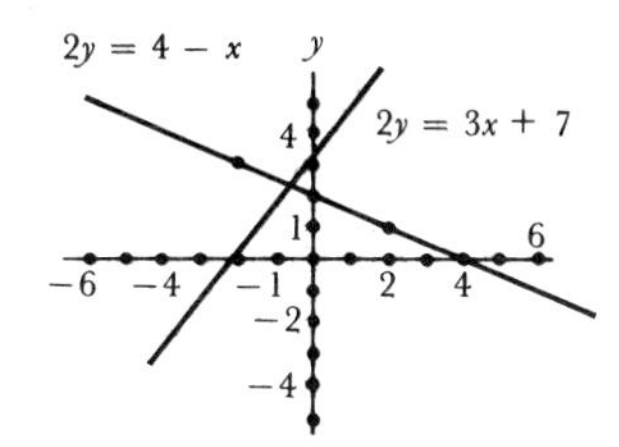

(c) $2y=(3/2)x+7$ and $3y=2x-14$ intersect at $(-14, -14)$.

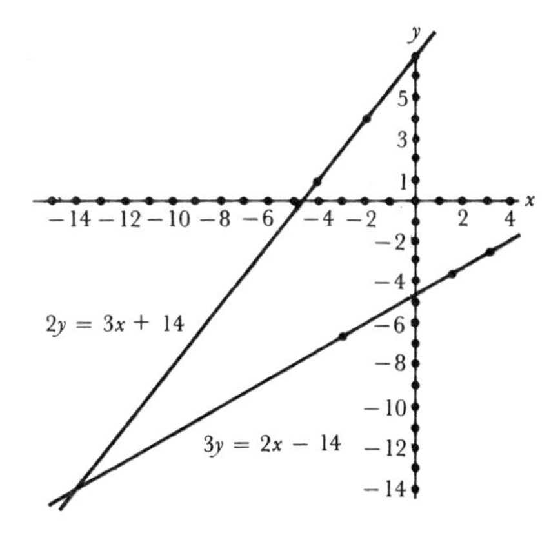

5. (a) -5 (c) $c^2 - 5c + 6$

6. (a) (c)

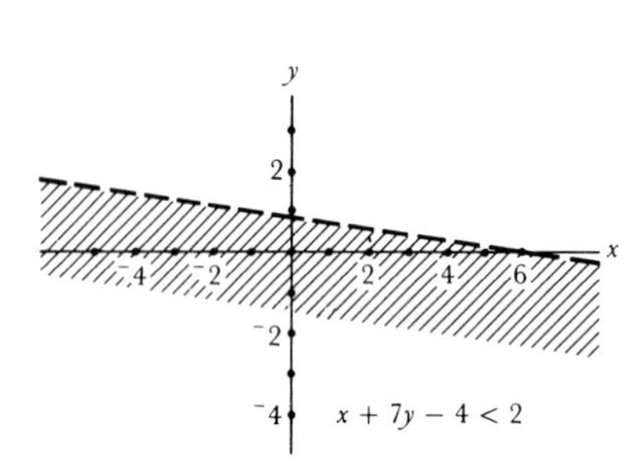

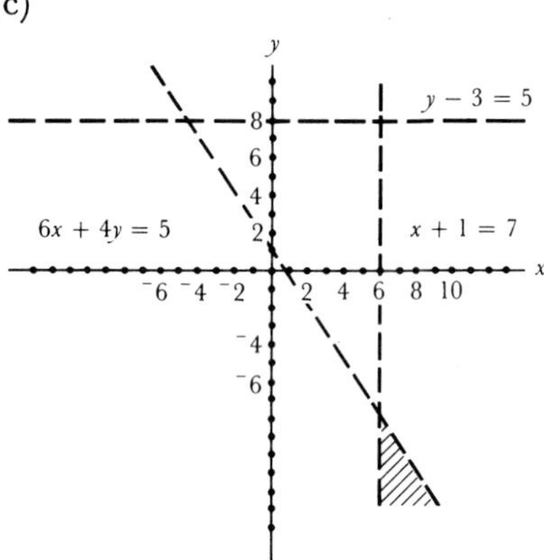

7. (a) $x = 7,\ x = 4/3$

8. (a) $x = \pm 3i$ (c) $x = 1 \pm 2i$ (e) $x = \dfrac{1 \pm i}{3}$

(g) $x = \dfrac{1 \pm i\sqrt{47}}{6}$

Index

List of Symbols

SYMBOL	*MEANING*
$=$	Is equal to
$\neq$	Is not equal to
$\sim p$	It is not true that p
$p \vee q$	Proposition p or proposition q
$p \wedge q$	Proposition p and proposition q
$p \rightarrow q$	If proposition p, then proposition q
$p \leftrightarrow q$	p if and only if q
$\{a, b, c, \ldots\}$	The set whose elements $a, b, c, \ldots$
$\{\ \}$ or $\emptyset$	The empty set
$x \in A$	x is an element of set A
$x \notin A$	x is not an element of set A
$a \leftrightarrow b$	a corresponds to b and b corresponds to a
$\{x \mid x \text{ is a counting number}\}$	x such that x is a counting number (set builder notation)
$A \subseteq B$	A is a subset of B
$A \subset B$	A is a proper subset of B
$\overline{A}$	The complement of set A
$B - A$	Complement of A relative to B
$A \cap B$	The intersection of sets A and B
$A \cup B$	The union of sets A and B
(a, b)	ordered pair
$A \times B$	The Cartesian product of sets A and B
aRb	a is related to b
$n(A)$	The cardinal number of set A
$\{1, 2, 3, \ldots\}$	The set of natural numbers
$\{0, 1, 2, 3, \ldots\}$	The set of whole numbers
$a + b$	The sum of a and b
$a \cdot b,\ ab,\ (a)(b)$	The product of a and b
$<$	Is less than
$\leqslant$	Is less than or equal to
$\nless$	Is not less than
$>$	Is greater than
$\geqslant$	Is greater than or equal to
$\ngtr$	Is not greater than